Progress in
Cell Cycle Research

Volume 2

A Continuation Order Plan is available for this series. A continuation order will bring delivery of each new volume immediately upon publication. Volumes are billed only upon actual shipment. For further information please contact the publisher.

Progress in Cell Cycle Research

Volume 2

Edited by

**Laurent Meijer
Silvana Guidet**

and

Lee Vogel
*Centre National de la Recherche Scientifique
Roscoff, France*

PLENUM PRESS • NEW YORK AND LONDON

Library of Congress Cataloging-in-Publication Data

On file

Front Cover: Double exposure fluorescent photomicrograph of a newt lung cell in mid prophase of mitosis. This cell was fixed with glutaraldehyde after which time the chromatin (blue) was stained by Hoechst 33342 and the microtubules (yellow/green) with a monoclonal antibody against beta tubulin followed by a FITC-congugated secondary. In this example the chromatin in the nucleus is condensing into chromosomes while the two replicated centrosomes, visible as two bright yellow dots adjacent to the nucleus, are just beginning to separate and to generate the radial astral microtubule arrays that will ultimately form the spindle poles. Courtesy of Dr. Conly L. Rieder of the NIH Biological Microscopy and Image Reconstruction Resource and the Wadsworth Center.

ISBN 0-306-45507-2

© 1996 Plenum Press, New York
A Division of Plenum Publishing Corporation
233 Spring Street, New York, N. Y. 10013

10 9 8 7 6 5 4 3 2 1

All rights reserved

No part of this book may be reproduced, stored in a retrieval system, or transmitted in any form or by any means, electronic, mechanical, photocopying, microfilming, recording, or otherwise, without written permission from the Publisher

Printed in the United States of America

PREFACE

Now in its second year, Progress in Cell Cycle Research was conceived to serve as an up to date introduction to various aspects of the cell division cycle. Although an annual review in any field of scientific investigation can never be as current as desired, especially in the cell cycle field, we hope that this volume will be helpful to students, to recent graduates considering a deviation in subject and to investigators at the fringe of the cell cycle field wishing to bridge frontiers.

An instructive approach to many subjects in biology is often to make comparisons between evolutionary distant organisms. If one is willing to accept that yeast represent a model primitive eukaryote, then it is possible to make some interesting comparisons of cell cycle control mechanisms between mammals and our little unicellular cousins. By and large unicellular organisms have no need for intracellular communication. With the exception of the mating phenomenon in *S. cerevisiae* and perhaps some nutritional sensing mechanisms, cellular division of yeast proceeds with complete disregard for neighbourly communication. Multicellular organisms on the other hand, depend entirely on intracellular communication to maintain structural integrity. Consequently, elaborate networks have evolved to either prevent or promote appropriate cell division in multicellular organisms. Yet, as described in chapter two the rudimentary mechanisms for fine tuning the cell division cycle in higher eukaryotes are already apparent in yeast. The work of Moreno and others (see chapter 3) has shown us that B-type cyclins in yeast act both at the G1-S and the G2-M transitions but that among these B cyclins there are those that function primarily at G1 and those that function primarily at G2. This "primitive" mechanism in yeast has evolved to the point that mammalian G1 cyclins are structurally distant from B-type cyclins regulating G2-M. Parallel with the evolution of specific G1 cyclins, higher eukaryotes have developed specific G1 kinases that interact with and are controlled by these cyclins (cyclin dependent kinases: cdk's). The complexities faced by multicellular organisms in relation to cell division are well illustrated by hepatocytes and the regenerating liver model of the cell cycle (chapter 4). This system demonstrates the remarkable capacity of the liver to undergo reversible differentiation in order to proliferate or degenerate according to extracellular signals. While a variety of hormones and growth factors are known to induce the transition from quiescence (G0) to G1, the signal transduction cascade(s) linking extracellular signals to the mechanisms regulating the cell division cycle are complex and have been difficult to elucidate. However, as described in chapter 5, the activation of MAP kinases by the Ras>Raf>MKK cascade has provided some tangible links to the induction of G1 cyclins and cdk's. MAP kinases, as post-translational modifiers, can, of course, only directly regulate those proteins that are already synthesised. Transcriptional regulation, particularly for the cyclins, is also required. The proto-oncogene c-myc is one such transcription factor that has emerged as a principal regulator of mammalian cell proliferation. Chapter 7 describes how this transcription factor acts as an upstream regulator of cdk's as well as an antagonist of one cdk inhibitor.

Control of cell cycle division is obviously not uniquely subject to extracellular influences. It can even be said that the primary condition to be met before successful division can occur, is an intracellular parameter, successful DNA replication. Chapter 8 elaborates some of the more recent observations, including the purification and characterisation of so-called licensing factors that verify complete and accurate replication of the genome. Additional checkpoints and intracellular control factors have also been characterised. For example, observations in *S. cerevisiae* support a DNA licensing function for one of the B-type cyclins, while the exclusively G1 expressed protein, rum1, monitors a critical cell size required to pass a G1 restriction point and reinitiate the cell cycle (Chapter 3). The efficacy of numerous control points and licensing factors relies ultimately on the effector protein interface with the cell cycle engine. Chapters 9 and 10 outline the principal phosphorylation events controlling the activity of cdc2 in *S. pombe* and propose that the regulatory kinase Wee1 may act as one of the principal effectors of numerous cell cycle control points. The newly emerging family of serine/threonine protein kinases termed polo-like kinases, after the *polo* gene

PREFACE

product from *D. melanogaster*, may also fall into the category of checkpoint effector molecules (chapter 11). Diverse function have been described for this family of kinases. However, genetic and biochemical data including a strong cyclic activity peaking in mitosis hints that these proteins may be involved in important regulatory cascades.

Important effectors of cell cycle check points are also to be found in the several endogenously produced inhibitors of cdk's, some of which are discussed in PCCR volume one. The discovery of these endogenous inhibitors has raised considerable interest in targeting cdk's for the discovery of new anti-proliferative cancer therapies. One approach toward this goal is to undertake large scale screening of small molecules that can inhibit cdk activity. Of considerable importance to this direction is the 3-dimensional structure of cdk2 in association with 3 different but related inhibitory molecules (chapter 14). Another aspect of the cell cycle regulatory machinery that is receiving increased attention as a potential cancer therapy target, is the interface between cell cycle checkpoints and apoptosis (chapter 15). Specifically, the interplay between DNA damage checkpoints and apoptosis may be susceptible to exploitation for the development of improved cancer therapies. The p53 tumor suppressor gene, when disrupted, leads to genetic instability and an increased tumorigenesis through loss of the p53 induced pathway to apoptosis. However, as discussed in chapter 16, certain p53 disrupted cell lines demonstrate a G2 checkpoint that can be selectively sensitised to DNA damaging reagents, thus pointing the way to both counteracting the observed resistance of p53 tumors to chemotherapy and also towards a possible means of specifically targeting tumor cells.

The fringe of the cell cycle field can be arbitrarily defined to include molecules which have no apparent involvement in the transition between cell cycle stages but are aligned with cell cycle research either through homology or effects on other parameters of cell division such as membrane dissociation or chromosome separation. For example, it has become apparent that not all cdk-like kinases are involved in cell cycle control. Like pho-85 in *S. cerevisiae* which is involved in controlling phosphate metabolism, cyclin-C/cdk8 is also peripheral to cell cycle control, involved, rather, in regulation of RNA polymerase II (chapter 19). Similarly, chapter 20 describes the neuronal specific cdk5 and how this enzyme has adapted to specifically function in non-proliferating tissue by employing activating subunits unrelated to cyclins. Also on the fringe of, but essential to successful cell division are molecules that regulate aspects of chromosome separation. Chapters 21 and 22 describe topoisomerase and calmodulin-dependent kinases and phosphatases which appear to have cell cycle functions primarily related to spindle formation and chromosome condensation and separation. While said to be on the fringe of cell cycle research, it is often subjects bridging frontiers between two fields that provide the most significant advances. For this reason it is always prudent to read between the lines of articles drawing connections between the cell cycle and seemingly unrelated subjects such as circadian rhythm (chapter 23) or morphology of cellular organelles (chapter 24).

Finally, the editors would like to make a few merited acknowledgements. A work of this nature can never be realised without the effort and support of a wide array of individuals and organisations. We would like to thank firstly all of the contributing authors who have generously given of their time and talent to comprise this second volume. Thanks should also be extended to the cell cycle laboratory in Roscoff for their support and to the librarians of the "Station Biologique", Maryse Collin and Nicole Guyard. It has been a pleasure to continue our collaboration with Plenum Press whose editorial staff was always efficient and professional and offered guidance to us through all the production stages. We wish also to acknowledge the support given by the "Association pour la Recherche sur le Cancer" (ARC 6268) and by the "Centre National de la Recherche Scientifique" (CNRS).

CONTENTS

A quest for cytoplasmic factors that control the cell cycle 1
    Yoshio Masui

G1/S regulatory mechanisms from yeast to man 15
    Steven I. Reed

Regulation of G1 progression in fission yeast by the *rum1*+ gene product 29
    Cristina Martín-Castellanos and Sergio Moreno

Progression through G1 and S phases of adult rat hepatocytes 37
    Pascal Loyer, Guenadi Ilyin, Sandrine Cariou, Denise Glaise, Anne Corlu and
        Christiane Guguen-Guillouzo.

A temporal and biochemical link between growth factor-activated MAP kinases,
    cyclin D1 induction and cell cycle entry 49
    Josée N. Lavoie, Nathalie Rivard, Gilles L'Allemain and Jacques Pouysségur

The plant cell cycle: conserved and unique features in mitotic control 59
    Peter C.L. John

The functions of Myc in cell cycle progression and apoptosis 73
    Philipp Steiner, Bettina Rudolph, Daniel Müller and Martin Eilers

DNA replication licensing factor 83
    James P. J. Chong and J. Julian Blow

Tyrosine kinases wee1 and mik1 as effectors of DNA replication checkpoint control 91
    Jérôme Tourret and Frank McKeon

Regulation of Cdc2 activity by phosphorylation at T14/Y15 99
    Lynne D. Berry and Kathleen L. Gould

The family of polo-like kinases 107
    Roy M. Golsteyn, Heidi A. Lane, Kirsten E. Mundt, Lionel Arnaud and Erich A. Nigg

Ubiquitin-dependent proteolysis and cell cycle control in yeast 115
    Kristin T. Chun, Neal Mathias and Mark G. Goebl

Suc1: cdc2 affinity reagent or essential cdk adaptor protein? 129
    Lee Vogel and Blandine Baratte

CONTENTS

Structural basis for chemical inhibition of CDK2 137
    Sung-Hou Kim, Ursula Schulze-Gahmen, Jeroen Brandsen and Walter Filgueira de Azevedo, Jr.

Apoptosis and the cell cycle 147
    Rati Fotedar, Ludger Diederich and Arun Fotedar

DNA damage checkpoints: Implications for cancer therapy 165
    Patrick M. O'Connor and Saijun Fan

Cellular responses to antimetabolite anticancer agents: cytostasis versus cytotoxicity 175
    Janet A. Houghton and Peter J. Houghton

Telomeres, telomerase, and the cell cycle 187
    Karen J. Buchkovich

The cyclin C/Cdk8 kinase 197
    Vincent Leclerc and Pierre Léopold

Cyclin-dependent kinase 5 (Cdk5) and neuron-specific Cdk5 activators 205
    Damu Tang and Jerry H. Wang

Role of Ca^{++}/Calmodulin binding proteins in *Aspergillus nidulans* cell cycle regulation 217
    Nanda N. Nanthakumar, Jennifer S. Dayton and Anthony R. Means

The roles of DNA topoisomerase II during the cell cycle 229
    Annette K. Larsen, Andrzej Skladanowski and Krzysztof Bojanowski

Circadian rhythm of cell division 241
    Rune Smaaland

The mammalian Golgi apparatus during M-phase 267
    Tom Misteli

Contributors 279

Index 283

chapter 1
A quest for cytoplasmic factors that control the cell cycle

Yoshio Masui

Department of Zoology, University of Toronto, Toronto, Ontario M5S 3G5, Canada

Between 1966 and 1986 the author and his former students carried out an investigation into the cytoplasmic factors that regulate nuclear behaviour during meiotic maturation of oocytes. This anecdotal chronicle traces the development of the problems and the direction in which their solutions were attempted in the course of this investigation. The author examines why he decided to study oocyte maturation, how he discovered progesterone as a maturation-inducing hormone and maturation promoting factor (MPF) and cytostatic factor (CSF) as meiosis-controlling factors, how the idea of the cell cycle without the cell occurred to him, and how it was materialised by invention of a cell-free system.

PROLOGUE

In embryos of multicellular organisms, cells proliferate and differentiate under the control of nucleocytoplasmic interactions. The analysis of relations between the nucleus and cytoplasm in development has long history. It was Hertwig (1) who first asked the question in 1908, "What role does the nucleocytoplasmic relation play in cell proliferation?" (p18), and pointed out that "every [cell] division...depends on [cell] growth (p19)" and that "from one division to another a misrelation between nuclear and cytoplasmic masses develops" (p20). He concluded that "In this nucleocytoplasmic tension I find the cause of cell division" (p20). Two years later, Boveri (2) also wrote that "the case of *Ascaris* offers the simplest paradigm for the way in which reciprocal action of cytoplasm and nucleus in ontogeny is to be conceived. Extremely slight heterogeneities in the egg cytoplasm may act as release mechanisms for the nucleus, followed by feedback (Rückwirkung) from the nucleus to the cytoplasm, finally to produce such prodigious difference among the resulting cells" (p191). Then, Wilson (3) stated that "probably no investigator would today maintain that the nucleus or chromosomes are the sole agents of heredity. On the contrary, both cytological and experimental research have clearly demonstrated that the protoplasm (cytoplasm) plays an important part in development" (p17).

In the past decades, developmental biologists have attempted to characterise cytoplasmic factors that regulate cell proliferation and differentiation in a variety of cells. My investigation into the cytoplasmic factors that regulate nuclear behaviour during meiotic maturation of oocytes was one of these attempts. Looking back on the history, I realise that it was not novel observations or singular ideas, but continuous and systematic synthesis of past observations and discoveries that have led us to the present knowledge of nucleocytoplasmic interactions in development (4).

When I was invited by the editors of this book to write about the discovery of "maturation-promoting factor (MPF)", I thought that it would be necessary to fully discuss the subject in relation to the past research of nucleocytoplasmic interactions in development. However, this is a task difficult to fulfil in the given pages. Nevertheless, I hesitatingly accepted the invitation in the hope that an anecdotal chronicle of the work which I carried out with my former students may be of some interest for readers who are curious about how problems are developed and their solutions are attempted in a field of research.

DIFFERENTIAL GENE ACTIVATION: AN ATTRACTIVE HYPOTHESIS

Cytoplasmic control of nuclear activities was first inferred from observations of synchronous karyokinesis in multinuclear cells, syncytial zygotes, or polyspermic eggs (4 for review). However, its experimental analysis had not been attempted until nuclear transplantation techniques were developed by Briggs and King (5) for the frog egg and by Tartar (6) for the ciliate, and the cell fusion technique by Okada (7). The advent of the nuclear transplantation aroused the hope in my mind that this new technique would make it possible to analyse nucleocytoplasmic interactions during "embryonic induction", which I was studying in those days.

Although this project did not materialise, I understood, at least as a concept, that embryonic induction was nothing but the expression of the genes selectively activated by a signal which was initially given to embryonic cells from the outside and mediated by their cytoplasm. I first learned this idea from Hans Spemann's book (8, p202), but then came across its modern version clearly formulated by Clement L. Markert as a process of "differential gene activation" based on his research on developmental changes of isoenzymes (9). After I had taken my first teaching job at Konan University, in Kobe, George Nace, the late professor of the University of Michigan, visited my laboratory in 1962 during his sabbatical leave. We talked about gene expression in development. One day, Professor Nace recommended that I should go to Markert's lab to study this problem.

Fortunately, I was granted a one year leave from the University to study in Markert's lab at Yale University in 1966. I asked Professor Markert, who was busy as Chairman of the Department of Biology, how to approach the problem of differential gene activation during development through isoenzyme analysis. He gave me packs of deep-frozen penguin embryos with the project of analysing changing patterns of lactic dehydrogenase (LDH) during development of this animal. The penguin project taught me the lesson that even a single enzyme, LDH, must undergo complex changes in the expression of at least 3 genes during development (10). This led me to a rather pessimistic view that the analysis of differential gene activation during embryonic induction was almost impossible by the technologies available in the 1960s.

Several months later, when I finished this project, Professor Markert suggested that I should start something I am interested in. He advised me that I should choose an inexpensive project that I could continue in Japan, because I was supposed to go back to Japan, a poor country. This meant that biochemical analysis of gene expression should be excluded.

WHY OOCYTE MATURATION?

In those days, Professor Markert often talked about many of his ideas in lunch times and seminars. The most interesting idea to me was that if viable and fertile gynogenetic diploids could be produced by suppressing meiotic divisions in farm animal oocytes, it would bring about enormous economic benefit for agriculture. He suggested that I might as well try to suppress meiosis in frog oocytes to produce gynogenetic diploids. However, I thought that in order to suppress female meiosis successfully, it would be necessary to know how meiosis is regulated during oocyte maturation. Oocyte maturation involves two meiotic cell divisions and represents the final stage of differentiation of the female germ cell. It was one of the subjects that I learned from Professor Nace while he was staying in my laboratory. Once Professor Nace showed me Heilbrunn's classical experiment (11), and I saw, for the first time, mature frog oocytes being shed from a piece of ovary suspended in Ringer solution containing a pituitary extract. I was deeply impressed by the fact that the hormone could act directly on its target tissue *in vitro*. I studied the literature of the previous work on oocyte maturation, and wrote a review article on this subject (12). I also found Brachet's comment: "This whole field so interesting both from the point of view of embryology and of cellular physiology, remains to be explored" (13, p146). However, even in 1966, little was known about its regulation.

Studying oocyte maturation using amphibians seemed to offer advantages to the study of nucleocytoplasmic interactions. As Briggs and King (5) showed, amphibian oocytes and eggs are suitable for microsurgery not only because of their large size, but also because of their tolerance to mechanical injuries. It was shown by Subtelny and Bradt (14) that nuclei transplanted into oocytes after the 2nd meiotic metaphase could undergo mitosis following activation of recipient oocytes, whereas those transplanted before 2nd meiosis could not. Also, Iwamatsu (15) using Medaka, the fish, showed that when the oocyte nucleus, called the germinal vesicle (GV), was displaced from the hyaline cytoplasm into yolky cytoplasm by a centrifugal force, it remained intact even after the follicles were stimulated with gonadotropin. Seeing these experiments, I wanted to develop a subcellular "Entwicklungsmechanik" using inexpensive operations, such as nuclear and cytoplasmic transfer and centrifugation. In addition, oocyte maturation, unlike primary embryonic induction, can only be induced by a specific signal, namely "gonadotropin", and has a well defined endpoint "maturation". Thus, oocyte maturation seemed more amenable to experimental analysis. Finally, all changes during oocyte maturation occur in a single cell. This was a very important point for studying induction of developmental cell changes. In those days, many embryologists thought, under the influence of the clonal selection theory in immunology by P. Medawar and M. F. Burnet who won Nobel prize in 1960, that changes induced by an inducer in a group of embryonic cells or a tissue could be a result of clonal selection, rather than of transformation, of the cell by the inducer.

Moreover, oocyte maturation seemed to be the simplest and most well defined cell change in animal development, and its nucleocytoplasmic interactions could be analysed by simple and inexpensive ways. In research I should "conduct my thought, beginning with the simplest objects and the easiest to know, in order to climb gradually" (16, p41). I decided to study maturation of frog oocytes.

During that time, Dennis Smith of the Argonne National Laboratory in Chicago, came to Yale to give a seminar about oocyte maturation, and presented the study by him and his colleague, Robert Ecker, on protein synthesis in *Rana pipiens* oocytes treated with a pituitary extract. I was impressed by his research, but missed the chance to talk with him. I wanted to ask him about the possibility of analysing the control mechanism of oocyte meiosis through biochemical approaches.

HORMONAL CONTROL OF OOCYTE MATURATION

In the spring of 1967, I repeated Heilbrunn's classical experiment using *Rana pipiens*, and observed that oocytes shed from ovarian fragments underwent germinal vesicle breakdown (GVBD) (Fig.1). I confirmed that they could be activated when pricked with a glass needle. Then, I repeated the recent experiment by Dettlaff and her associates (17), but could not reproduce some of their results. According to the authors, pituitary extract should induce maturation in oocytes with or without follicle cells. When I removed the follicular epithelium without Ca-free treatment, oocytes matured in response to the hormone; however, when I removed the follicular epithelium after treatment with EDTA or Ca-Mg-free Ringer's, oocytes did not mature. I looked at these oocytes under a microscope, and found patches of follicle cells on the surface when Ca was present. I thought that the oocytes might need these follicle cells to mature. To test this, I cultured oocytes completely free of follicle cells together with isolated follicle cells in the presence of pituitary extract. These oocytes were induced to mature (Fig.2).

By that time, I had learned that ovulation was enhanced by progesterone in the frog (18, 19, 20) and that a progesterone surge in the rabbit was induced prior to ovulation (21). Therefore, I treated follicle cell-free oocytes as well as those enclosed in the follicular epithelium with progesterone, and found that the hormone induced oocyte maturation in either case. Oocytes induced to mature by progesterone were capable of cleavage when a blastula nucleus was transplanted into them. From these experiments, I concluded that "pituitary gonadotropin acts on the follicle cells to stimulate them to release a hormone that directly acts on the oocyte", (22, p365), and that "the follicle cells secrete a progesterone-like substance that causes maturation of oocytes", (22, p374).

I submitted a paper to Journal of Experimental Zoology (JEZ) to report the above results. To my surprise, when I was browsing the 1967 abstract issue of Journal of Cell Biology, I came across Allen Schuetz's abstract reporting results similar to mine. I informed Professor Markert, the chief editor of JEZ, of this coincidence. When my paper was published, I saw Scheutz's paper (23) and mine in the same issue, and I was pleased that our results corroborated each other.

I began a new experiment to clarify how progesterone induces the oocyte to mature. However, in order to finish the experiment, I had to wait for new frogs coming in the next season, while my official leave was drawing near the end. I asked Konan University to extend my leave for one year, but I was granted only a 6 month extension. So I resigned as assistant professor from the University to continue the research.

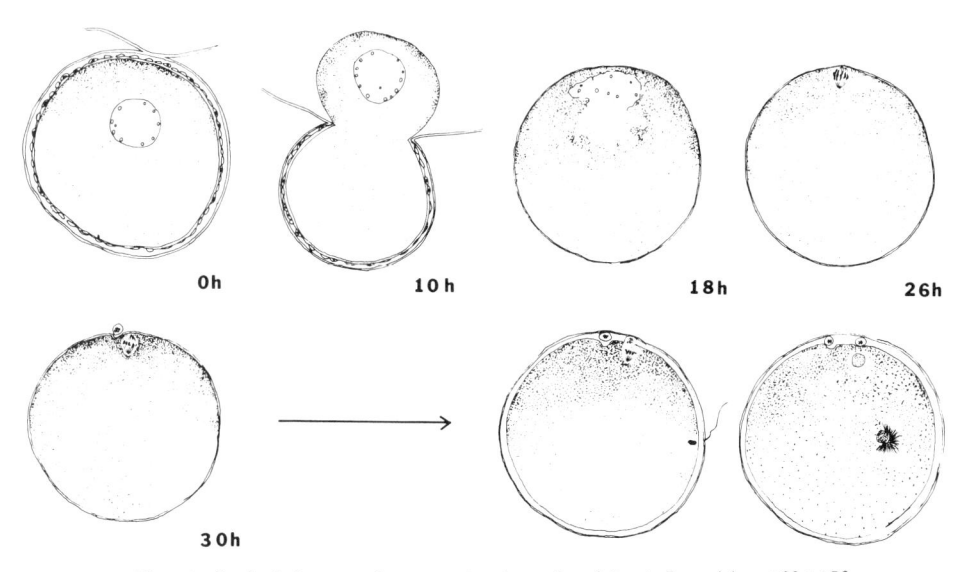

Figure 1. Cytological aspects of oocyte maturation and ovulation in *Rana pipiens* at 20 ± 1C°.

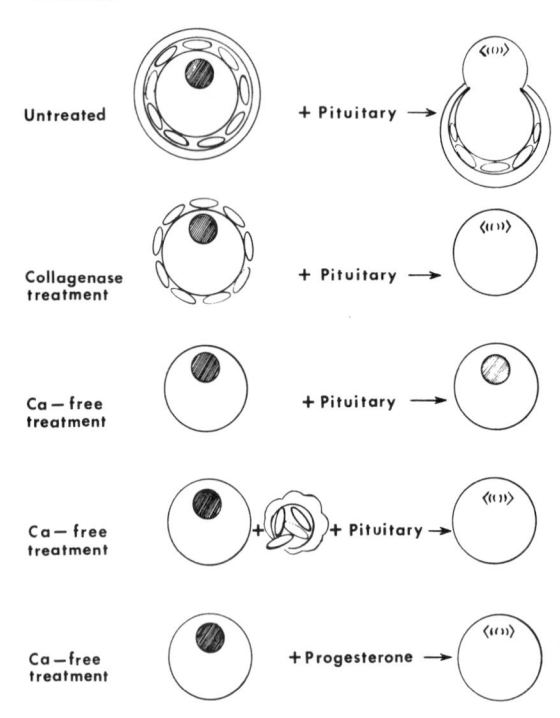

Figure 2. Relative roles of pituitary gonadotropin, follicle cells and progesterone in the induction of oocyte maturation in the frog.

CYTOPLASMIC FACTORS THAT CONTROL OOCYTE MATURATION

Although Dettlaff and her colleagues (17) assumed that gonadotropin could act directly on the nucleus, Gurdon (24) showed that gonadotropin injected into *Xenopus* oocytes had no effect. However, according to the Russian authors, nucleoplasm taken from the GV shortly before GVBD could induce maturation when injected into immature oocytes. I again repeated their experiment, and this time they were right. I thought that progesterone might have direct effect on the nucleus, and to test this, I injected 0.1 µl of a progesterone solution (10 µg/µl) into a *Rana* oocyte (2 µl in volume). However, none of these oocytes matured, while those cultured in a 20 times diluted progesterone solution all matured. The same results were obtained with different doses of progesterone. Progesterone appeared to have an effect only when it acted from the outside of the oocyte. This finding was so interesting that I mentioned it to my old friend, Haruo Kanatani, the late professor from the National Institute of Basic Biology in Okazaki, Japan, when he visited Yale. Several months later he wrote to me, saying that this was also the case in starfish oocyte maturation induced by 1-methyladenine (25).

If this were the case, the signal given by maturation-inducing hormones on the oocyte surface must be transmitted by a cytoplasmic factor to the nucleus. To test this idea, I injected cytoplasm of progesterone-treated oocytes into untreated oocytes, carefully avoiding contamination of the GV content. The donor cytoplasm indeed induced the recipient oocytes to mature. I assumed the presence of a cytoplasmic factor that caused oocyte maturation, and called it "Maturation Promoting Factor (MPF)".

The frequency of maturation induced in recipient oocytes by cytoplasm of the donor oocytes at a same stage of maturation was found to be proportional to the volume of the cytoplasm injected. Therefore, relative activities of MPF could be assayed in oocytes at different maturation stages. From these assays, it became clear that MPF appeared in oocyte cytoplasm by 12 hours after progesterone treatment, which was much earlier than the appearance of maturation-inducing ability in the nucleoplasm, and reached a maximum level at 18 hours just before GVBD.

As well, cytoplasm of oocytes, which were treated with progesterone after the GV had been removed, also became capable of inducing maturation in recipient oocytes (Fig. 3). Clearly, progesterone acted to give rise to MPF in the cytoplasm, independently of the nucleus. Then, I thought that since oocytes injected with only 3 % of the cytoplasm of a maturing oocyte could fully mature, their cytoplasm must have produced the full amount of MPF. If so, MPF must have been amplified in the recipient oocyte. This amplification was also verified by serial transfers of cytoplasm from an oocyte, which had been induced to mature by injected MPF, into the next oocyte (Fig. 4). Thus, MPF was shown to be amplified "autocatalytically" in oocyte cytoplasm.

MPF activity declined sharply after fertilisation. Curiously, however, "even cytoplasm from early embryos retained some capacity to induce oocyte maturation" (26, p129). This finding led me to the idea that MPF might stimulate mitosis in blastomeres. Unexpectedly, when injected into blastomeres of 2-cell embryos, cytoplasm of mature oocytes arrested cleavage (Fig. 3). The arrested blastomeres contained condensed chromosomes, arrested at metaphase and embedded in a well formed spindle lacking asters. I referred to this inhibitory factor as "cytostatic factor (CSF)". CSF was detected in oocytes after 1st meiosis, but not after they were activated, and unlike MPF, it never appeared again (Fig. 5). Based on this result, I hypothesised that CSF is the cytoplasmic factor responsible for the arrest of oocyte meiosis at 2nd metaphase.

In 1968, I met Smith and Ecker at the meeting of American Society of Cell Biology held in Boston.

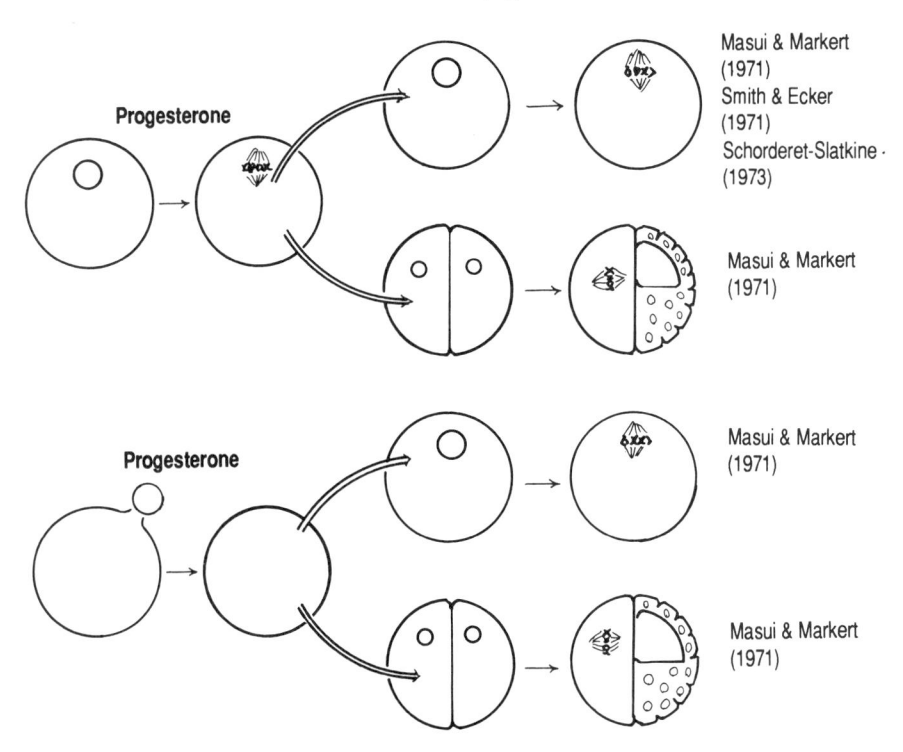

Figure 3. Transfer of cytoplasm from mature oocytes with or without the GV into immature oocytes or into a blastomere of a 2-cell embryo: Demonstration of the nuceus-independent production of MPF and CSF.

We discussed oocyte maturation, and I told them that cytoplasm of maturing oocytes could induce maturation when injected into immature oocytes. In January 1969, I was invited by the Department of Zoology at the University of Toronto to give a seminar for a job interview. I presented all of the results mentioned above in this seminar and was offered a job. In 1969 I moved to Toronto where I completed the manuscript reporting all the results on MPF and CSF. The manuscript was submitted to JEZ and published a year and half later (26).

EXTRACTION AND CHEMICAL CHARACTERISATION OF MPF

MPF activity was found in homogenates prepared from oocytes crushed in a polyethylene tube with a capillary glass pestle (26). So I thought that extraction of MPF from oocytes should be easy. However, I could never detect MPF activity in supernatants of centrifuged homogenates, despite all efforts to prevent protein denaturation in the homogenates. After two years of trials and errors, I gave up my attempts to extract MPF. Rather, I was inclined to think that MPF might be bound to cytoplasmic granules. To determine the subcellular localisation of MPF, oocytes, floated on a Ficoll solution in a tube, were centrifuged and their cytoplasm was stratified without breaking the oocytes. I withdrew the contents of each layer, and injected them into oocytes. Fortunately, the strongest MPF activity was found in the layer devoid of granular components (27). This suggested that it was "homogenisation" that destroyed MPF. To collect the contents of the granule-free layer from a large number of oocytes with stratified cytoplasm, 500 to 1000 unfertilised jellyless eggs were gently packed in a centrifuge tube with a Ca-free saline solution containing sucrose. First, eggs were centrifuged slowly, and excess medium over the packed eggs was withdrawn. Then, they were compressed by a strong centrifugal force to squeeze the liquid component from the eggs. I finally succeeded in extracting MPF in March 1973.

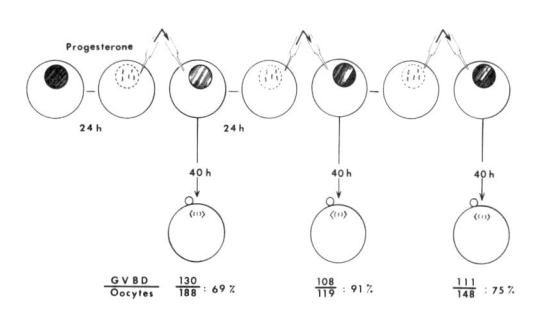

Figure 4. A serial transfer of cytoplasm from a maturing oocyte into immature oocyte demonstrating "autocatalytic" amplification of MPF.

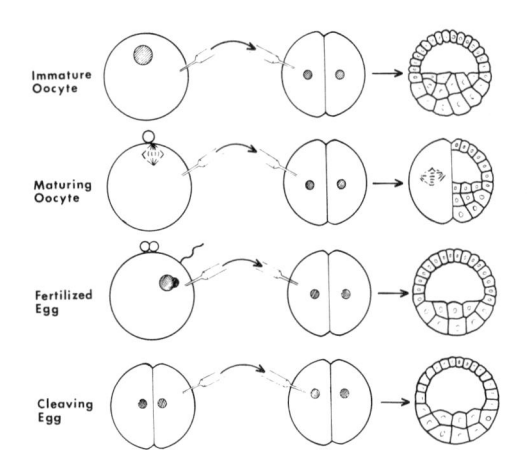

Figure 5. The appearance and disappearance of CSF during oocyte maturation and activation.

William Wasserman, who just started his Ph.D. research, improved this method to characterise MPF. He added a Ca-chelator (EGTA) and $MgCl_2$ to the extraction medium and centrifuged eggs at 150 K x G for 2 h. Bill obtained extracts containing MPF, sufficiently stable, for chemical tests and fractionation through a sucrose density gradient. MPF was inactivated by $CaCl_2$, and Ca- Mg chelator (EDTA) or proteolytic enzymes, but not by RNAases when they were added to extracts. MPF activities were sedimented in 3 different fractions (3S, 13S and 30S). In 1976 we reached the conclusion that MPF is a protein polymerised to different degrees or associated with other proteins and its activity is sensitive to Ca and dependent on Mg ions (28). A few years later, MPF was further stabilised with ATP or protein phosphatase inhibitors by Drury (29), Wu and Gerhart (30). Since then, MPF has been recognised as a phosphoprotein, whose activity is dependent on its phosphorylation.

EXTRACTION AND CHEMICAL CHARACTERISATION OF CSF

CSF was also extracted by the same technique as used for MPF extraction, and its chemical characterisation was attempted in early 1973, but my haste led to mistakes in this study. First, enzyme tests for CSF activity were not well controlled, and I was led to the erroneous conclusion that CSF is sensitive to RNAases, but not to proteases. Secondly, although I correctly inferred that CSF activity was dependent on Mg ions, the test for its Ca sensitivity was confusing. If blastomeres of 2-cell embryos were injected with fresh egg extracts immediately after extraction, they were arrested whether the extraction medium contained Ca or not. However, extracts kept for 24 h arrested blastomeres only when they contained Ca ions. To report these results, I hastily sent a paper to "Rapid Communication" of JEZ without examining changes in CSF activity during the first 24 h after extraction. I drew the erroneous conclusion that "CSF is slowly inactivated in the absence of Ca^{++}. Addition of Ca^{++} to the extraction medium prevents this inactivation" (31, p146).

Fortunately, this mistake was corrected by Peter Meyerhof who started his Ph.D. research in 1975. Peter made Ca-free extracts containing EGTA and found that the extracts retained CSF activity for at least one day at $2^{o}C$, whereas those containing Ca lost the activity completely within one hour of extraction. In these extracts, however, CSF activity reappeared after a few days of storage. At his request I repeated his experiment and confirmed it. We realised that there were two kinds of CSF, the Ca-sensitive one and the Ca-insensitive one, which could develop in the presence of Ca. We called the former "primary CSF ($1^{o}CSF$)" and the latter "secondary CSF ($2^{o}CSF$)".

Effects of the two CSFs were cytologically indistinguishable; that is, both caused the same metaphase arrest when injected into blastomeres. However, we observed that $1^{o}CSF$ could not arrest cleavage when injected shortly after fertilisation, whereas $2^{o}CSF$ arrested cleavage whenever it was injected. We suggested that Ca-sensitive $1^{o}CSF$ was inactivated by Ca ions released at early times of fertilisation, while Ca-insensitive 2° CSF was not. We published these results in 1977 (32).

Ten years later, both CSFs were further characterised in Ellen Shibuya's Ph.D. research. She found that $1^{o}CSF$ could be recovered in 3S fraction of the cytosol (33), and that its activity was protease-sensitive, dependent on ATP, and stabilised by protein phosphatase inhibitors (34). This suggested that $1^{o}CSF$ was a labile phosphoprotein. In contrast, $2^{o}CSF$ activity was found to be associated with a large protein molecule produced by Ca-catalysed polymerisation of smaller proteins in the cytosol (33, 35).

PROTEIN SYNTHESIS AND MPF

In 1966, as mentioned before, Smith and his colleagues found that protein synthesis activity in oocytes significantly increased during maturation (36). Dettlaff (37) also reported that maturation could be inhibited by inhibitors of protein synthesis. However, inhibition of RNA synthesis had no effects on progesterone-induced oocyte maturation (23). Therefore, when we found that MPF was protease-sensitive, I immediately thought that the inhibition of oocyte maturation by protein synthesis inhibitors must be caused by inhibition of translation of MPF mRNA. To test this assumption, Bill Wasserman first repeated the experiments of previous workers using *Xenopus laevis* oocytes. He confirmed their results that both RNA and protein synthesis inhibitors inhibited maturation induced by gonadotropin of follicle-enclosed oocytes. RNA synthesis inhibitors had no effect on oocyte maturation induced by progesterone, while protein synthesis inhibitors inhibited it (38).

However, Bill found that injection of a small amount of MPF induced not only GVBD, but also production of the full amount of MPF in oocytes treated with protein synthesis inhibitor, although the oocytes could not produce MPF when treated with progesterone. Further, he showed that MPF activity did not decrease through transfers of cytoplasm from donor to recipient oocytes even when protein synthesis had been inhibited. When we submitted a paper to Experimental Cell Research to report the above results in 1975, a reviewer requested that we present a hypothesis to explain them. In response, we concocted in a great hurry the following hypothesis: For progesterone-treated oocytes to mature the oocytes must synthesise a new protein "Initiator" in response to the hormone, which in turn activates MPF precursor protein (nowadays called pre-MPF) stored in the oocyte. However, after the initial MPF was activated, its autocatalytic amplification required no protein synthesis (39).

After we published the paper, we saw a paper in Cell which also dealt with the same problem as ours (40). However, the authors' conclusion was different from ours. According to them, protein synthesis is required for MPF amplification as well as for its initial activation, since MPF activity had diminished during serial transfers of cytoplasm through oocytes treated with a protein synthesis inhibitor. Although the exact reason for the discrepancy between the two experiments has not been clarified, we imagined that it could be due to the difference in the interval of cytoplasmic transfers (2h in their case and 7 h in ours). At any rate, our conclusion was corroborated 9 years later by Gerhart and his collaborators (41).

In the mouse, it was known that inhibition of protein synthesis activates eggs arrested at metaphase II, leading to chromosome decondensation (42). In 1978, Hugh Clarke started his M.Sc. research by repeating this experiment and confirmed the results. Further, he found that protein synthesis inhibition not only prevented condensation of chromosomes in maturing oocytes, but also decondensed chromosomes once condensed. Hugh concluded that both the maintenance of condensed chromosomes as well as the induction of chromosome condensation depended on continuous synthesis of a short-lived protein at both metaphase I and II (43).

NUCLEOCYTOPLASMIC INTERACTIONS DURING MEIOSIS AND MITOSIS

Cytoplasmic control of nuclear activities during the cell cycle had been analysed by various ways around the time when I began the study of oocyte maturation. Gurdon using the nuclear transplantation method in *Xenopus* eggs (24, 44), and Harris (45) and Johnson and Rao (46) using the cell fusion technique in mammalian cells showed that nuclei at any phase of cell cycle could be induced to synthesise DNA when exposed to S cell cytoplasm or to undergo chromosome condensation when exposed to M cell cytoplasm.

In 1971 David Ziegler repeated Gurdon's experiments (24, 44) during his M.Sc. research using *Rana* oocytes by transplanting frog brain nuclei and sperm nuclei into progesterone-treated oocytes at various stages of maturation. Dave found that the cytoplasm of oocytes acquired chromosome condensation activity (CCA) to transform nuclei to metaphase chromosomes when GVBD began (47). Chromosome condensation occurred rapidly and each condensed chromosome appeared to have only a single chromatid, suggesting that CCA could transform G_1 nuclei directly to metaphase chromosomes without intervening DNA synthesis (48). Ten years later, Hugh Clarke also investigated behaviour of sperm nuclei introduced into mouse oocytes at various stages of maturation for his Ph.D. thesis. Hugh found that sperm nuclei directly condensed to a metaphase state, instead of forming pronuclei, when oocytes were prematurely inseminated and remained unactivated (49).

In Dave Ziegler's Ph.D. research, he investigated the interactions of chromosomes with nucleoplasm and cytoplasm during oocyte maturation. However, Dave found that although nuclei injected into oocytes which had been induced to mature after enucleation, could also be transformed into condensed chromosomes, these chromosomes disintegrated in less than 3 hours. On the other hand, nuclei injected, together with GV nucleoplasm, into cytoplasm of immature oocytes never condensed chromosomes. The conclusion Dave arrived at in 1976 was that transformation of nuclei into metaphase chromosomes could be induced by maturing oocyte cytoplasm, but the GV nucleoplasm is required for stabilising condensed chromosomes (50). The requirement for the GV nucleoplasm for chromosome condensation was again demonstrated by Hugh Clarke in mouse oocytes during his Ph.D. research ten years later (51).

Dave Ziegler further investigated interactions of chromosomes with *Rana* oocyte proteins. He first labelled donor oocytes with radio-active amino-acids, and then injected their cytoplasm or GV nucleoplasm together with brain nuclei into recipient oocytes treated with a protein synthesis inhibitor. He found that radio-labelled proteins of maturing donor oocytes accumulated preferentially over the brain nuclei or their condensed chromosomes compared with proteins of immature donor oocytes. From these results we concluded that proteins synthesised as well as those accumulated in the GV during oocyte maturation could interact with the nucleus when chromosome condensation was induced (52).

Since CCA in maturing oocytes was capable of inducing chromosome condensation not only for the oocyte's own nucleus, but also for brain and sperm nuclei, we thought that the action of MPF may not be cell type-specific, but may affect both germ and somatic cell nuclei. The previous observation that MPF activity was detected in early embryos (26) suggested its possible role in embryonic cells as well. Therefore, MPF appeared to play a role in mitosis as well as in meiosis. The same idea was also suggested by John Gurdon at Medical Research Council in Cambridge, when I talked with him during his visit to York University in Toronto in 1977. He recommended that I extract MPF from synchronised HeLa cells at metaphase. A year later, however, the appearance of MPF during mitosis was demonstrated by Bill Wasserman, who left Toronto in 1976 for a postdoc. in Dennis Smith's lab at Purdue University, in blastomeres of frog embryos (53) and by Sunkara and his associates in P. N. Rao's laboratory in HeLa cells (54). Around that time, CCA was found in the CSF-arrested blastomeres by Peter Meyerhof who injected brain nuclei (55), and by Ellen Shibuya who injected sperm nuclei (56) into the blastomeres (Fig. 6). I also injected cytoplasm of CSF-arrested blastomeres into immature oocytes and found MPF activity in it (56). We were convinced that MPF could play a role in chromosome condensation during both mitosis as well as meiosis.

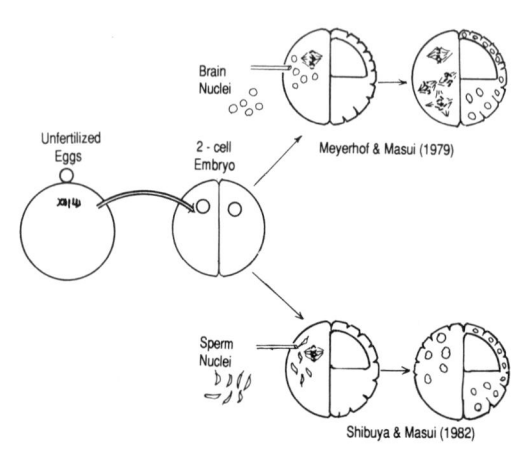

Figure 6. Chromsome condensation activity (CCA) in CSF-arrested blastomeres. Injection of a supernumerary sperm nuclei causes a loss of CCA.

NUCLEOCYTOPLASMIC RATIO AND CHROMOSOME CONDENSATION

The dose-dependent effect of CCA of M phase cytoplasm had been reported by Johnson and Rao (46) based on their cell fusion experiments. However, it remained unclear whether this effect resulted from dilution of M phase cytoplasm by the cytoplasm of fused interphase cells, or from changes in the ratio between the number of nuclei to M phase cytoplasm to which they were exposed. Sometime around 1973 Dave Ziegler told me that the more nuclei that were injected into a maturing oocyte, the less frequently chromosome condensation took place.

Ellen Shibuya investigated chromosome condensation in nuclei injected into CSF-arrested blastomeres for her M.Sc. thesis in 1979 (57 cited in 58). Her findings were as follows: Sperm nuclei injected into CSF-arrested blastomeres all condensed to a metaphase state when their number was less than 100 per blastomere. However, when the number exceeded 200 per blastomere, they formed pronuclei and began the cell cycle, and the recipient blastomere resumed cleavage (Fig. 6).

Hugh Clarke working on his Ph.D. thesis also noticed that if more than 3 sperm nuclei entered into an oocyte during polyspermy of mouse oocytes, they all formed a pronucleus, and this nucleocytoplasmic ratio critical for chromosome condensation of sperm nuclei was well maintained if the oocyte cytoplasmic volume was varied by bisection or fusion (Fig. 7, 51). The response of the sperm nucleus to oocyte cytoplasm was found to be "all or none" in that sperm nuclei, which reside in a single cytoplasm, either all condensed to a metaphase state or all decondensed to form a nucleus. We concluded that "the oocyte cytoplasmic factor which transforms the sperm nucleus into metaphase chromosomes may be stoichiometrically titrated by sperm nuclear material" (51, p 838). These observations suggest that chromosome condensation can occur only when cytoplasm contains the factor above a threshold level needed to saturate the nuclear material.

THE "MPF" CYCLE *IN VITRO*

In 1978, I was trying to stabilise MPF in extracts of *Rana pipiens* eggs by varying the composition of the extraction medium.

In one experiment, I centrifuged eggs at a high speed (150 K X G, 2 h) in a Ca-free sucrose saline solution containing Mg ions (10 mM) and EGTA (5 mM) at a higher concentration than usual, and assayed MPF activity in the extracts kept at 2º C on day 0, 2, 4, 8, and 16 after extraction. I found high MPF activities on all of these days. The result was

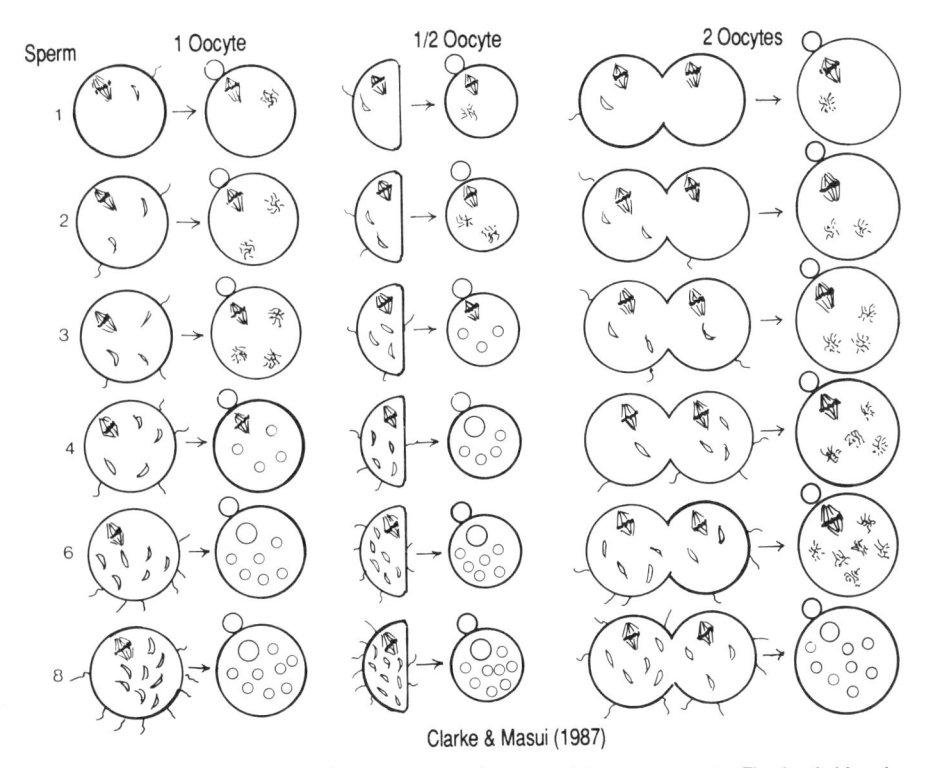

Figure 7. Dose-dependent relation between metaphase cytoplasm and sperm nuclei in mouse oocytes. The threshold nucleocytoplasmic ratio for chromosome condensation is 3 nuclei per oocyte.

so exciting that I talked about this remarkable success in stabilising MPF to Bill Wasserman, who happened to visit his old lab in Toronto. However, he was rather sceptical about my result and said that his extracts always lost the activity completely on day 3, although he never checked them on day 4. I repeated the experiment to assay MPF activity every day. He was right! This time all of the extracts showed no MPF activity on day 3. However, to my surprise, the activity reappeared in the extracts on day 4, but disappeared on day 7, and so on. That is, MPF activity actually oscillated with a fairly regular periodicity of 4 to 6 days (Fig. 8). The days when I happened to assay MPF activities in the preliminary experiment were, in fact, those very days when the activity was rising. Although I reported these results first at an INSERM Conference in Domaine de Seilac, France in 1978, I was still sceptical about my observation of this strange phenomenon. In the following 3 years, I examined 44 extracts and confirmed the result that MPF-like activity in egg extracts oscillated during storage at 2° C.

In those days, the only *in vitro* cell cycle-related oscillation that I knew of was the oscillation of protein synthesis in homogenates of sea urchin embryos (59). I asked Thomas Chen for his help in investigating the correlation between the MPF cycle

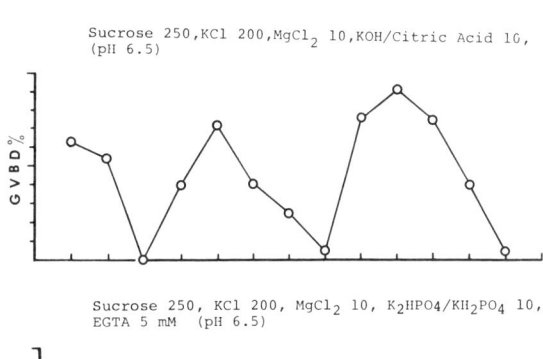

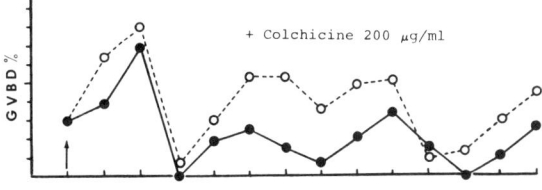

Figure 8. "MPF" cycle *in vitro*.
Top: A *Rana pipiens* egg extract prepared with a saline solution containing sucrose was assayed for MPF activity during storage at 2°C by injection into immature oocytes.
Bottom: An aliquot of *Rana pipiens* egg extract prepared with a saline solution containing sucrose was stored at 2°C after addition of colchicine (open circles) and assayed for MPF as above to compare control aliquote (closed circles).
Ordinates: % GVBD in recipient oocytes; abscissae: days after extraction.

and the oscillation of protein synthesis. I spent 3 months in his molecular biology lab at McMaster University in Hamilton, Ontario during my 1980 research leave. However, little protein synthesis activity was found in extracts prepared by high speed centrifugation at 150 K x G. Although some protein synthesis activities were found in extracts prepared by low speed centrifugation at 10 K X G, they showed no oscillation. Tom, being frustrated, finally suggested to me the daunting task of first isolating MPF mRNA and then translating it *in vitro*.

In 1981 I visited John Gerhart's lab at the University of California, Berkeley to learn how to prepare partially purified MPF from *Xenopus* eggs. Michael Wu and Ken Drury taught me the whole procedure. While I was staying there, I showed them my data of the MPF cycle *in vitro*. Although everybody was puzzled by this enigmatic phenomenon, they seemed to be convinced of the reproducibility of the result. John suggested that I should look into protein phosphorylation in extracts to see whether it cycles or not.

Encouraged by their response to my data, I presented these results at a NATO Advanced Research Conference held in Gargonza, Italy in 1981 and explained them by a home-made limit cycle model (Lotka-Volterra's equation). Richard Noyes, a professor of physical chemistry, University of Oregon, who happened to be in the conference, cautioned that the cycle may not be as simple as I thought. Nevertheless, I wrote a paper to report this strange phenomenon and explained it by my model. When I sent the manuscript to JEZ, a reviewer, who must have been a biophysicist, commented that I did not have to present the model, since I seemed to him to be a man not very comfortable with math any way. My conclusion was that "a homogeneous cell-free system can exhibit cyclic behaviour of a cytoplasmic factor controlling nuclear activities", and I suggested "the possibility that a modification of the system may allow MPF to oscillate at a higher temperature with a periodicity similar to that of - - zygotes" (60, p397).

THE CELL CYCLE WITHOUT THE CELL

After coming back from my leave, I contemplated what I had learned from the visits to the two laboratories. The conclusion I reached was rather pessimistic. The molecular biological approach to the problem of MPF seemed to me logistically difficult in my situation. The reasons were as follows: First, the grant that I could obtain was too small to operate a molecular biology lab. Secondly, students and postdocs. interested in studying zoology in our department were not interested in biochemistry or molecular biology. In those days no colleagues in my department were working in these fields. Thirdly, the time left for me was too short to accomplish anything significant if I were to switch over to molecular biology.

However, I thought that in order to study the mode of MPF action at the molecular level, we must have an *in vitro* system that would allow us to analyse effects of MPF on the nucleus. I decided to go on this line. When Manfred Lohka started his Ph.D. program, I told him of my unsuccessful attempts to induce chromosome condensation *in vitro*. In these attempts, I used brain nuclei and a clear supernatant extracted by high speed centrifugation. I suggested to him the possibility that important "goodies" may be granular components of the egg which had been discarded, and that sperm nuclei may be better than brain nuclei, since in nature, the sperm nucleus is the only nucleus that has the opportunity to interact with egg cytoplasm.

Fred incubated demembranated sperm nuclei with *Rana pipiens* egg extracts prepared by low speed centrifugation (9 K x G) under various conditions, and stained nuclei in various ways. Although he devised a variety of methods to prepare egg extracts and demembranate sperm nuclei, no conspicuous condensation of chromosomes to a metaphase state was observed. However, one day when I had a chance to look at his preparations, I thought I could see some faintly stained bodies, which were difficult to discern from stained cytoplasmic granules, but which looked like chromosomes. I drew his attention to these obscure things. To see clearly these objects, he devised a new staining method and ran a new experiment. Now it became clear that these bodies were indeed chromosomes.

Finally, Fred succeeded in inducing chromosome condensation to a metaphase state *in vitro*. He further went on to demonstrate that sperm nuclei could undergo the cell cycle in activated egg extracts, and presented the results at the meeting of American Society of Cell Biology held in San Francisco in 1981. We sent a paper to Nature to report these discoveries, but it was rejected right away without being reviewed. We next sent it to Science. This time, fortunately it was accepted and published in 1983. We concluded that "this cell-free system may be useful in biochemical analysis of the interactions of nucleus and cytoplasm that control nuclear behaviour" (61, p719). Using this system Fred proved that demembranated sperm required membrane vesicles in egg cytoplasm to form pronuclei (62). He also demonstrated that whereas sperm nuclei could be transformed into metaphase chromosomes in the presence of MPF under a Ca-free condition, they were transformed into pronuclei after inactivation of MPF and CSF by Ca ions (63).

In the summer of 1983, just after our first paper was published in Science, I met Professor Mitsuki Yoneda of Kyoto University, who was visiting Woods Hole Marine Biology Laboratories (MBL). I was shaken when he told me that the results of an experiment similar to ours had been reported by Yasuhiro Iwao and Chiaki Katagiri, at Hokkaido University, at the meeting of the Japanese Society of Developmental Biologists in 1982. Their paper was published a year later, reporting similar results to ours (64).

EPILOGUE

In the summer of 1983, I was teaching in the embryology course at MBL, and I met Tim Hunt of Cambridge University, who had just published the paper reporting the discovery of cyclin (65). He was thinking that MPF and cyclin were perhaps the same molecule. John Gurdon, who was also staying in Woods Hole, suggested that I should inject Tim's cyclin into oocytes of *Xenopus*, which I was rearing in my room at MBL. John thought that Tim had already purified cyclin, which was not the case. However, Tim was rather anxious about the correlation between the level of cyclin and the chromosome cycle, so I showed him how to stain chromosomes with aceto-orcein that I brought from Toronto.

Near the end of the summer, the students in the MBL courses presented the results of their projects. It was Elayne Bornslaeger, one of the students in Tim Hunt's course, who reported a close correlation between the cyclin cycle and the chromosome cycle (66). A few years later, the cyclin gene was cloned by Katherine Swenson of Joan Ruderman's lab at Harvard Medical School, and its mRNA was injected into *Xenopus* oocytes. Thus, it was shown that cyclin could induce oocyte maturation (67). Two years later, Fred Lohka, who had left Toronto in 1984 for a postdoc with James Maller at the University of Colorado, and his colleagues finally purified MPF (68). In the same year, Noriyuki Sagata in George VandeWoude's lab at the National Cancer Institute, in Frederick, Maryland and his colleagues identified the "Initiator" of oocyte maturation as the c-mos proto-oncogene protein (69) and a year later 1°CSF as hyperphosphorylated c-mos protein (70).

This was the beginning of the story of how the cytoplasmic factors that control the cell cycle left embryology and was handed over to molecular biology. Young molecular biologists made my dream finally come true. I saw my old favourite problems suddenly leave my hand, like a lost balloon, moving away into the realm to which my hand cannot reach (Fig. 9).

ACKNOWLEDGEMENT

The author thanks Dr. Richard P. Elinson, Department of Zoology, University of Toronto, for his reading the manuscript and for his encouraging comments.

Figure 9. The dawn of molecular biology of MPF and CSF 1988-1989.

REFERENCES

1. Hertwig, R. (1908) *Arch. f. Zellforsch.* **1**, 1-32.
2. Baltzer, F. (1967) *Theodor Boveri: Life and work of a great biologist 1862-1915* (translated by D. Rudnick), Univ. Calif. Press, LosAngels.
3. Wilson, E. B. (1925) *The Cell in Development and Heredity*, 3rd ed., McMillan, New York.
4. Masui, Y. (1992) *Bioch. Cell Biol.* **70**, 920-945.
5. Briggs, R. and King, T.J. (1952) *Proc. Natl. Acad. Sci. USA* **38**, 455-463.
6. Tartar, V. (1953) *J. Exp. Zool.* **124**, 63-103.
7. Okada, Y. (1958) *Biken's J.* **1**, 103-110.
8. Spemann, H. (1938) *Embryonic development and induction.* Yale Univ. Press, New Haven.
9. Markert, C. L. (1958) in *A symposium on the chemical of development* (McElroy, W. D. and Glass, B. eds.) pp 3-16, The Johns Hopkins Univ. Press.
10. Markert, C.L. and Masui, Y. (1969) *J. exp. Zool.*, **172**, 121-146.
11. Heilbrunn, L.V., Daugherty, K. and Wilbur, K.M. (1939) *Physiol. Zool.* **12**, 97-100.
12. Masui, Y. (1967) *Jap. J. exp. Morphol.* **21**, 245-255 (Japanese text with English abstract).
13. Brachet, J. (1951) *Chemical Embryology* (translated by L. Barth) Interscience Publishers, New York.
14. Subtelney, S. and Bradt, C. (1961) *Dev. Biol.* **3**, 96-114.
15. Iwamatsu, Y. (1966) *Embryologia* **9**, 205-221.
16. Descartes, R. (1637) *Discourse on Method* (translated by F. E. Sutcliffe), Penguin Books, Middlesex, U.K.
17. Dettlaff, T.A., Nikitina, L.A. and Stroeva, O.G. (1964) *J. Embryol. exp. Morphol.* **12**, 851-873.
18. Zwarenstein, H. (1937) *Nature* (London) **139**, 112-113.
19. Burgers, A. C. J. and Li, C. H. (1960) *Endcrin.* **66**, 255-259.
20. Wright, P.A. (1961) *Gen. Comp. Endocrinol.* **1**, 20-23.
21. Hilliard, J., Endroczi, E. and Sawyer, C. H. (1961) *Fed. Proc.* **20**, 187.
22. Masui, Y. (1967) *J. exp. Zool.* **166**, 365-376.
23. Schuetz, A.W. (1967) *J. Exp. Zool.* **166**, 347-354.
24. Gurdon, J.B. (1967) *Proc. Natl. Acad. Sci. USA* **58**, 545-552.
25. Kanatani, H. and Hiramoto, Y. (1970) *Exp. Cell Res.* **61**, 280-284.
26. Masui, Y. and Markert, C.L. (1971) *J. exp. Zool.* **177**, 129-145.
27. Masui, Y. (1972) *J. exp. Zool.* **179**, 365-378.
28. Wasserman, W.J. and Masui, Y. (1976) *Science* **191**, 1266-1268.
29. Drury, K. C. (1978) *Differentiation* **10**, 181-186.
30. Wu, M. and J. C. Gerhart (1980) *Dev. Biol.*, **79**, 465-477.
31. Masui, Y. (1974) *J. exp. Zool.* **187**, 141-147.
32. Meyerhof, P. G. and Y. Masui (1977) *Dev. Biol.*, **61**, 214-229.
33. Shibuya, E. K. and Y. Masui (1989) *Development* **106**, 799-808.
34. Shibuya, E. K. and Y. Masui (1988) *Dev. Biol.*, **129**, 253-264.
35. Shibuya, E. K. and Y. Masui (1989) *Dev. Biol.* **135**, 212-219.
36. Smith, L.D., Ecker, R.E. and Suhtelney, S. (1966) *Proc. Natl. Acad. Sci. USA* **56**, 1724-1728.
37. Dettlaff, T.A. (1966) *J. Embryol. exp. Morphol.* **16**, 183-195.
38. Wasserman, W.J. and Masui, Y. (1974) *Biol. Reprod.* **11**, 133-144.
39. Wasserman, W. J. and Y. Masui (1975) *Exp. Cell Res.*, **91**, 381-388.
40. Drury, K.C. and Schorderet-Slatkine, S. (1975) *Cell* **4**, 268-274.
41. Gerhart, J., Wu, M. and Kirschner, M. (1984) *J. Cell Biol.* **126**, 1247-1255.
42. Siracusa, G., D. G. Whittingham, M. Molinaro and E. Vivarelli (1978) *J. Embryol. Exp. Morphol.*, **43**, 147-166.
43. Clarke, H. J. and Y. Masui (1983) *Dev. Biol.*, **97**, 291-301.
44. Gurdon, J.B. (1968) *J. Embryol. exp. Morphol.* **20**, 401-414.
45. Harris, H. (1965) *Nature* (London) **206**, 583-588.
46. Johnson, R.T. and Rao, P.N. (1970) *Nature* (London) **226**, 717-722.
47. Ziegler, D.H. and Masui, Y. (1973) *Dev. Biol.* **35**, 283-292.
48. Ziegler, D.H. and Masui, Y. (1976) *J. Cell Biol.* **68**, 620-628.
49. Clarke, H.J. and Masui, Y. (1986) *J. Cell. Biol.*, **102**, 1039-1046.
50. Ziegler, D.H. and Masui, Y. (1976) in *Progress in Differentiation Research* (Muller-Berat, N. ed.), pp. 181-188, North Holland, Amsterdam.
51. Clarke, H.J. and Masui, Y. (1987) *J. Cell Biol.* **104**, 831-840.
52. Masui, Y., Meyerhof, P.G. and Ziegler, D.H. (1979) *J. Steroid Biochem.*, **11**, 715-722.
53. Wasserman, W.J. and Smith, L.D. (1978) *J. Cell Biol.* **78**, R12-R22.
54. Sunkara, P.S., Wright, D.A. and Rao, P.N. (1979) *Proc. Natl. Acad. Sci. USA* **76**, 2799-2802.
55. Meyerhof, P. G. and Y. Masui (1979) *Exp. Cell Res.* **123**, 345-353.
56. Shibuya, E. K. and Y. Masui (1982) *J. Exp. Zool.*, **220**, 381-385.
57. Shibuya, E.K. (1981) *Studies on the behavior of injected brain and sperm nuclei in*

58. *blastomeres arrested by extracts of unfertilized Rana pipiens eggs containing cytostatic factors (CSF).* M.Sc. thesis, University of Toronto.
58. Masui, Y., M. J. Lohka and E. K. Shibuya, 1984. *Symp. Soc. Exp. Biol.*, **38**, 45-66.
59. Mano, Y. (1970) *Dev. Biol.* **22**, 433-460.
60. Masui, Y., (1982) *J. Exp. Zool.* **224**, 389-399.
61. Lohka, M.J. and Masui, Y. (1983) *Science* **220**, 719-721.
62. Lohka, M.J. and Masui, Y. (1984) *J. Cell Biol.* **98**, 1222-1230.
63. Lohka, M.J. and Masui, Y. (1984) *Dev. Biol.* **103**, 434-442.
64. Iwao, Y. and Katagiri, C. (1984) *J. exp. Zool.* **230**, 115-124.
65. Evans, R., Rosenthal, E.T., Youngblom, J., Distel, D. and Hunt, T. (1983) *Cell* **33**, 389-396.
66. Cornall, R., Bornslaeger, E. and Hunt, T. (1983) *Biol. Bull.* **165**, 513-514.
67. Swenson, K.I., Farrell, K.M. and Ruderman, J.V. (1986) *Cell* **47**, 861-870.
68. Lohka, M.J. Hayes, M.K. and Maller, J.M. (1988) *Proc. Natl. Acad. Sci. USA.* **85**, 3009-3013.
69. Sagata, N., M. Oskarsson, T. Copeland, J. Brumbaugh and G. F. Vande Woude (1988) *Nature*, **335**, 519-525.
70. Sagata, N., N. Watanabe, G. F. Vande Woude and Y. Ikawa (1989) *Nature*, **342**, 512-518.

chapter 2
G1/S regulatory mechanisms from yeast to man

Steven I. Reed

Department of Molecular Biology, MB-7, The Scripps Research Institute
10666 North Torrey Pines Road, La Jolla, CA 92037 USA

> Cyclin-dependent kinases play a key role in promoting and regulating the transition from G1 to S phase in all eukaryotic organisms. The kinase activities involved are distinguished from those participating in other cell cycle phase transitions in that they are driven by a class of specialised G1-specific cyclins. Although the G1 regulatory components have diverged structurally in the course of evolution, the regulatory mechanisms and principles remain highly conserved from yeast to vertebrates. An important issue that remains is that of identifying the principal targets phosphorylated by G1 cyclin-dependent kinases.

CYCLIN-DEPENDENT KINASES IN THE G1/S PHASE TRANSITION

Introduction

Cyclin dependent kinases (Cdks) are now accepted as the effectors of most, if not all cell cycle transitions. This, however, was not always the case. Largely based on a bias traceable to work in amphibian and invertebrate egg systems, it was inferred that Cdks generally were effectors only of mitotic induction. Mutational analysis of Cdks in yeast that suggested otherwise, was assumed to represent a phylogenetic anomaly. Specifically, most ts mutations in the Cdc28 kinase of budding yeast conferred G1 arrest (1,2) and ts mutations in the homologous cdc2 kinase of fission yeast conferred both G1 and G2 arrest (3), although because of its relationship to the work in amphibian and invertebrate eggs, the latter (G2 arrest) phenomenology received much greater attention. Finally, the mitosis-centric view of Cdks was reinforced by the finding that a mouse cell line possessing a ts mutation in the gene encoding CDC2 (the only known mammalian Cdk at the time) arrested in G2 at the restrictive temperature but not in G1 (4).

Two developments led to a paradigm shift concerning the relationship of Cdks to cell cycle transitions. The first of these was the clear demonstration in budding yeast, *S. cerevisiae*, that a cyclin-driven accumulation of Cdk activity controlled the primary G1 restriction point of the *S. cerevisiae* cell cycle, known as START (5,6). This represented a homologous and parallel mechanism to the mitotic effector defined by CDC2/cyclin B in fission yeast and metazoan organisms. In addition, the characterisation of yeast G1 cyclins as highly unstable proteins (7-10) and the contribution of this instability to regulation of the transition through START was reminiscent of the inferred properties of unstable proteins that regulated the G1 restriction point in mammalian cells, known as the R point (11,12). The hypothetical unstable proteins that regulated the R point had persistently resisted identification, but the work in yeast led to the hypothesis that these might represent mammalian G1 cyclins and to a search by many investigators to find them.

The second critical development was the discovery of additional Cdks in vertebrates. Thus, the constraint imposed by the G2/M-specific phenotype of a ts mouse *CDC2* mutation was relieved. The demonstration of a highly related kinase, CDK2, not apparently involved directly in mitotic functions, forced the issue of other Cdk-regulated transitions (13-15). The final consolidation of the idea of a Cdk-driven G1/S regulatory system came with the discovery of mammalian G1 cyclins and elucidation of their kinase partners (16-20). The question is no longer whether the major cell cycle transitions are controlled by Cdks, but how Cdks are regulated in the context of these transitions and what the principal downstream targets of Cdks are in effecting these transitions.

The molecular basis of START: a paradigm for G1/S regulation

Historically, the Cdc28 protein kinase of *S. cerevisiae* was the archetype of Cdks. It is therefore interesting that virtually all the initial characterisation of the *CDC28* gene and its product was performed in the context of G1 control. The first ts mutant allele of *CDC28*, a derivative of the original cell cycle mutant screen performed by Hartwell and his colleagues, exhibited a G1 arrest at the restrictive temperature (1). Subsequently, all other ts alleles of *CDC28*, with one notable exception, also were shown to confer G1 arrest (2,21). Although it is now clear that the Cdc28 kinase is responsible for other cell cycle transitions, including G2 to M phase (21,22), the reason for the tendency of mutants to arrest in G1 is likely to be that under conditions of limiting kinase function, the G1 forms of the kinase are the first to be impacted. There is some evidence that association of Cdc28 with G1 cyclins is inefficient, requiring assembly factors (23).

The G1-specificity of most *cdc28* mutants led workers in *S. cerevisiae* to focus on the molecular basis of G1 control while those working on *cdc2* in fission yeast, *S. pombe*, due to the genetic idiosyncrasies of that system, focused on G2/M control. The most significant early advances were the demonstration that the *CDC28* gene encoded a protein with protein kinase homology (24), and shortly thereafter, the demonstration that the *CDC28* product was, in fact, a protein kinase (25). The most important discovery, however, was the demonstration that activation of the Cdc28 kinase at the G1/S transition was accomplished by a specialised class of cyclins termed G1 cyclins or Clns. G1 cyclins were discovered both in response to a direct attempt to identify proteins that regulated the Cdc28 kinase, as well is in the context of mutational searches aimed more generally at cellular growth and regulation. The genes *CLN1* and *CLN2* were identified as high-copy suppressors of a ts *cdc28* mutation (26), meaning that elevated expression of the encoded cyclins could partially rescue the G1 defect of the mutant kinase. This is consistent with the idea that ts *cdc28* alleles are defective in their ability to bind and/or be activated by G1 cyclins, accounting for their tendency to arrest in G1. *CLN3*, on the other hand, was identified independently through screens for mutations conferring small cell size (*WHI1* mutations; 27) and resistance to cell cycle arrest by peptide mating pheromones (*DAF1* mutations; 28), respectively. Both regulation of the cell cycle and the Cdc28 kinase by growth and mating pheromones will be discussed below. However, in neither case was the relationship to the Cdc28 kinase initially obvious based on the genetics, and only after "cyclin homology" had become a parameter that could be assessed and appreciated was the potential identity of *CLN3* as a G1 cyclin addressed (29).

Three properties of Clns established the paradigm of Cdk-dependent G1 control. First genetic analysis indicated that Clns were both essential and sufficient for execution of START, the G1 restriction event. That is, mutational elimination of all three Clns conferred G1 arrest (5) whereas premature expression of any Cln advanced the G1/S phase transition (26-29). Second, Clns were shown directly to be activators of the Cdc28 kinase (6). Thus the ability of Clns to drive the G1/S phase transition could be directly attributed to activation of Cdc28 kinase activity. Finally, Cln1 and Cln2 were shown to accumulate

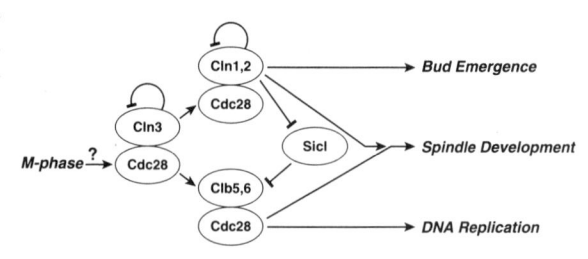

Figure 1. Cyclin dependent kinase activities in the yeast G1/S phase transition. As cells enter G1 from M phase, Cln3/Cdc28 kinase is activated via mechanisms that are not yet understood. This leads to transcriptional induction of Cln1 and Cln2 mRNAs and Clb5 and Clb6 mRNAs. The resultant Cln1,2/Cdc28 kinase activity directly promotes bud emergence and duplication of the microtubule organizing centers known as spindle pole bodies. Cln1,2/Cdc28 kinase also phosphorylates and targets Sic1, an inhibitor of Clb5,6/Cdc28 kinase, for degradation. The presence of Sic1 allows Clb5 and Clb6 to accumulate concomitantly with Cln1 and Cln2 but for the activation of their associated kinases to be delayed relative to Cln1,2/Cdc28 kinase activity. Upon degradation of Sic1 and activation of Clb5,6/Cdc28, DNA replication and spindle pole body separation are initiated. Also depicted is the self-regulation of Cln turnover by the Cln/Cdc28 kinases coupling activation to rapid clearing of activity.

periodically in the cell cycle with peak levels corresponding to the time of execution of START (6). Although, initially, Cln3 was thought not to be periodically expressed, this view has recently been revised, with a peak of expression occurring early in G1 (30). The accumulation kinetics for all appears to based on periodic transcription coupled to a high constitutive level of instability of the Cln proteins (7-10). Taken together, these observations suggest a model where regulated accumulation of labile Cdc28 kinase activators reaches a threshold, allowing a cascade of phosphorylation-driven events to propel cells into the next division cycle (Fig. 1). In budding yeast, this entails commitment both to DNA replication as well to a morphogenetic program required for the budding mode of growth. As stated above, some of the properties of Clns are remarkably similar to those inferred for regulators of the restriction point for mammalian cells, based on earlier physiological experiments. Therefore, once Cdk based G1 control was established for yeast, an intensive search for an analogous system in mammalian cells began. In this context, it is interesting to note that in fission yeast, although it was shown that Cdk activity is required for the transition from G1 to S phase, a system of specialised cyclins designated for regulating a G1 restriction event, analogous to START control in budding yeast, has never been identified. Instead, any B-type cyclin is capable of providing the necessary activation (31-33) and it is clear that most sophisticated cell cycle control occurs at the G2/M boundary in this organism.

G1 control in mammalian cells

Mammalian G1 cyclins were discovered as a result of a determined effort to find them as well as

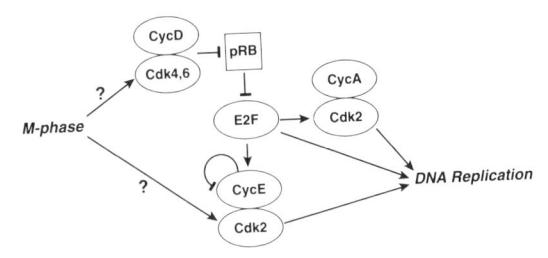

Figure 2. Cyclin dependent kinase activities in the mammalian G1/S phase transition. Accumulation of D-type cyclins after M-phase and association with either Cdk4 or Cdk6 leads to activation of a pRb kinase. Phosphorylation of pRb relieves inhibition of positive effectors of cell cycle progression such as E2F transcription factors. E2F promotes the transcription of cyclin E and A mRNAs, as well as those encoding other proteins required for DNA replication. It is likely, however, that other factors control the transcription of cyclin E and A mRNAs, as well as the accumulation of these cyclins. As is the case with yeast Clns, cyclin E turnover is coupled to activation of the cyclin E/Cdk2 kinase via autophosphorylation. Thus, cyclin E/Cdk2 kinase activity is self-limiting.

by accident. Once G1 cyclins were an established motif in yeast, yeast based screens were developed to isolate mammalian counterparts by genetic means. Specifically, mutants made conditionally deficient for yeast Clns were challenged with libraries of human cDNAs to identify those that could provide Cln function. Cyclins D1 and E were identified in this manner (17,19,20). Cyclin D1, however, was also identified as an oncogene activated as a result of translocations in parathyroid adenomas (16) and certain lymphomas (34) and as corresponding to a transcript induced by treatment of macrophages with CSF1 (colony stimulating factor) (18). The other two known D-type cyclins, D2 and D3, were then cloned by homology using low stringency hybridisation screening (18). The D-type cyclins and cyclin E now represent the two known classes of G1 cyclin, presumably corresponding to Clns in yeast. However, they share minimal primary structure homology with yeast Clns and attempts to find true structural homologs of Clns in metazoans have been unsuccessful. Yet cyclins D and E represent ancient evolutionary motifs, as true structural homologs can be found at least as far back as *Drosophila* (35,36).

Mammalian G1 cyclins have many of the hallmarks of yeast G1 cyclins in spite of their minimal structural homology to Clns (Fig. 2). In many respects, the kinetics of mammalian G1 cyclin accumulation parallel those observed in *S. cerevisiae*. Cyclin E accumulates with strong periodicity near the G1/S phase boundary (37,38) while D-type cyclins appear to exhibit much less cell cycle fluctuation in most cell types (18,39,40), reminiscent of the dichotomy between Cln1 and Cln2 vs. Cln3. Nevertheless, as will be discussed below, the activities associated with D-type cyclins are highly regulated at the post-translational level and at the level of expression in the context of exit and entry from the cell cycle. Second, as is the case with yeast Clns, D-type cyclins and cyclin E have been shown to be both essential and, in a limited sense, sufficient for the transition from G1 to S phase. Antibody microinjection targeting either cyclin D1 (cyclin D2 in some cell types) (39-41) or cyclin E (42) blocks the G1 to S phase transition, whereas ectopic expression of cyclins D1 or E causes advance of the G1/S phase transition (41,43-45). In several critical respects, however, mammalian G1 cyclins differ from their yeast counterparts. Most notably, whereas *S. cerevisiae* G1 cyclins all drive the same catalytic partner, Cdc28, D-type cyclins and cyclin E activate Cdks that are quite divergent. While cyclin E associates with and activates only CDK2 (37,38), a molecule closely related to the Cdks of yeast and to the mitotic kinase of mammalian cells, CDC2 (henceforth referred to as CDK1), D-type cyclins activate CDK4 and CDK6, a pair of closely related kinases that are quite distant structurally from the archetypal Cdks (46,47). The distinct types of catalytic partners of the two classes of mammalian G1 cyclins then raises the issue of redundancy. Whereas any of the three yeast G1 cyclins is sufficient for viability (5), it is now clear that Cln3 normally has a different primary function from Cln1 and Cln2 (to be discussed below). Nevertheless, despite this apparent specialisation, the system contains enough redundancy so that one Cln is sufficient. As stated above, D-type cyclins in themselves are essential, as is cyclin E, suggesting a lack of redundancy between these two classes of G1 cyclin. However, since the two classes of yeast Cln under normal circumstances perform different functions, this distinction between G1 regulation in yeast and mammalian cells may be superficial. In addition, the conclusions concerning the essentiality of mammalian G1 cyclins based on experiments with cells in culture must be interpreted with caution. In the case of cyclin D1, shown to be essential in fibroblasts and other cell types based on microinjection experiments, nullizygous mice developed relatively normally and exhibited minimal phenotypic defects (48). Furthermore, a cyclin D1 and D2 double nullizygous mouse was stunted but viable (49). It is clear, therefore, that a degree of redundancy exists for the various D-type cyclins in a mouse, that is lost under tissue culture conditions. In principle, such redundancy could exist between D- and E-type cyclins, as well.

A final issue that remains to be addressed with mammalian G1 cyclins is their relationship to the G1 restriction point or R point. Both cyclins D1 and E are metabolically unstable proteins (18,50,51). Yet is not clear that either represents the unstable protein(s) required for passage through the R point. Kinetic analysis by videomicroscopy indicates that cyclin E accumulates primarily after passage through the R point, suggesting a downstream function (52). The parallel experiment for D-type cyclins remains to be performed, but the activation of cyclin D-associated kinase activities must be considered as a good candidate for the molecular basis of the R-point.

REGULATION OF G1 CYCLIN-DEPENDENT KINASES

Introduction

Since G1 cyclin accumulation in both yeast and mammalian cells is essential and, to some degree, rate-limiting for the G1/S phase transition, progression from G1 to S phase could, in principle, be controlled at the level of G1 cyclin accumulation. Regulation of cyclin accumulation might be implemented in several different ways, including transcriptional control, post-transcriptional control and control of degradation. In fact, as will be discussed below, each of these modes is utilised. However, regulation of Cdk activity can be regulated in a variety of other ways not involving cyclin accumulation. These include regulatory phosphorylation of Cdk catalytic subunits (reviewed in 53-55) and the modulation of Cdk inhibitory proteins, or CKIs (reviewed in 53-57). The relative importance of each mode of regulation appears to depend on the organism, the cell type and the situation. What can be concluded is that the necessity to integrate G1 control with virtually every other aspect of cellular physiology has led to the evolution of a highly complex and multifaceted regulatory repertoire for G1 Cdks in organisms as disparate as yeast and man.

Regulation of START in yeast

In budding yeast, START, the primary regulatory event of the cell cycle, is controlled in response to both internal and external signals (Fig. 3). Proliferating cells coordinate cell cycle progression with growth by linking START to the attainment of a specific size, or a parameter that correlates closely with size (reviewed in 58). In addition, the passage through START is regulated by the availability of essential nutrients (reviewed in 58) and the action of mating pheromones (reviewed in 59, 60). In the latter situation, secreted peptide hormones trigger internal responses essential for conjugation, including arrest of cells in G1. All of these regulatory systems function by controlling the G1 activity of the single Cdk in this organism, Cdc28 (61). In fact, the G1 Cdk system of *S. cerevisiae* consists of two components. Cln3/Cdc28 is activated first and functions to induce transcription of the two other G1 cyclins, Cln1 and Cln2. These in turn activate Cdc28 rapidly to execute START (62,63). Cdc28/Cln3, while an efficient transcriptional activator for the other G1 cyclins, is a poor inducer of START (62,63). The internal and external regulation of the yeast cell cycle must therefore be integrated into this two-tiered G1 Cdk system.

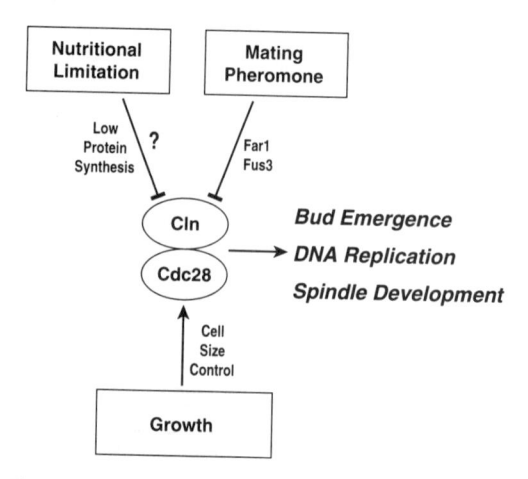

Figure 3. Signals impinging on Cln/Cdc28 kinase activity to control the G1/S phase transition in yeast. Negative regulation is exerted by mating pheromone and nutritional limitation. In the mating pheromone response, the stimulation of a MAP kinase homologous cascade culminates in activation of the Map kinase homolog Fus3 and the mobilization of the Cln/Cdc28 inhibitor Far1. Both effects are required for efficient G1 arrest. When cells are starved for essential nutrients, it has been proposed that a reduced rate of protein synthesis results in a failure to accumulate unstable Clns. However, other direct signaling mechanisms have not been excluded. Cell growth is a positive stimulus for the G1/S transition. When cells reach an appropriate defined size, Cln3/Cdc28 kinase is triggered, leading to initiation of the G1/S program of kinase activation.

Control of cell size in yeast, although described in detail at the phenomenological level, remains mysterious at the mechanistic level. It is not known how cells sense size. However, in light of the hierarchy of G1 cyclin functions, it is likely that detection of size is wired into activation of the Cln3/Cdc28 kinase, which then produces a burst of transcription of the other G1 cyclins and execution of START. Since the pre-START interval of the cell cycle is the only part that can be expanded and contracted significantly (64), it is the timing of START relative to cell growth that ultimately determines the average cell size of a population. The role of G1 cyclins in this program is underscored by the observation that G1 cyclin mutants that interfere with Cln turnover, and therefore lead to premature Cln accumulation, undermine size control, producing abnormally small cells (26-29). In fact, it was through such small size mutants that *CLN3* was initially identified (27). Furthermore, although the timing of START is constant under particular nutrient conditions, it can be adjusted in response to changes in the nutrient environment, leading to a concomitant change in cell size. The most well known example is that cells execute START (and bud at a larger size) when grown under rich nutrient conditions and conversely execute START at a smaller size when grown under poor nutrient conditions (65-67). Although this modulation appears to function via the adenylate cyclase/protein kinase A pathway of yeast, and impinges on the G1 cyclins, it is not yet clear what the molecular mechanisms are.

Deprivation of essential nutrients causes G1 arrest in *S. cerevisiae*. This can be at least partially overridden by expression of hyperstable mutant alleles of *CLN2* (26), suggesting a role for G1 cyclins in the regulatory system. One plausible model is based on the constitutive metabolic instability of Clns. Since nutrient limitation causes a reduction in the rate of protein synthesis, it is likely that the accumulation of unstable proteins such as Clns would be severely impacted. This model accommodates the observation that hyperstable Cln2 can override the G1 arrest normally conferred by nutrient deprivation. However, the actual effects of starvation on protein synthetic rates in general and on those of Clns, in particular, have not been reported, so this hypothesis must be taken as tentative.

The mating pheromone response represents the best-understood cell cycle regulatory system in yeast, although mechanistic uncertainties still remain. Cell-type specific pheromones activate G-protein coupled receptors on responsive cells, triggering a cascade of signalling events that culminate in activation of a pair of MAP kinase homologs known as Fus3 and Kss1, respectively (reviewed in 59,60). Activation of these kinases has at least two consequences. First, pheromone-responsive genes are transcriptionally activated, one of which encodes a Cdk inhibitory protein, Far1, which targets Cdc28/Cln2 and Cdc28/Cln1 (68). Second, Fus3 and Kss1 also activate the inhibitory functions of Far1 by phosphorylating it (69,70). However, even in this detailed scenario of cell cycle control, several critical pieces of information are missing. Most notably, it is not known how Fus3/Kss1 inhibits Cdc28/Cln3, which is apparently not a target of Far1. Yet, inhibition of Cln3 function is clearly critical to proper function of the mating pheromone response pathway, as hyperstable mutants of Cln3 confer resistance to pheromone-induced cell cycle arrest. In fact, it was through a screen for such mutants that *CLN3* was independently identified (28). Since no specific inhibitor of Cdc28/Cln3 kinase activity has been identified, it has been proposed (but not yet demonstrated) that Fus3/Kss1 might exert inhibition by direct phosphorylation of Cln3.

One final mode of regulation that has implications for regulation of START in *S. cerevisiae* is that of G1 cyclin turnover. It has been demonstrated for Cln2 that activation of the Cln2/Cdc28 kinase leads to autophosphorylation of a number sites on Cln2 (10). These phosphorylation events, in turn, lead to Cln2 turnover (7-10). The same regulatory motif is inferred for both Cln1 and Cln3. This coupling of Cln half-life to kinase activation is predicted to have the effect of rendering the lifetime of each active kinase complex short, thus establishing a highly regulable system. It also is likely to contribute to the production of a pulse of Cln/kinase activity followed by rapid clearing once transcription of the *CLN1* and *CLN2* genes is downregulated upon execution of START. It is interesting to note that a similar regulation of turnover has been demonstrated for at least one of the mammalian G1 cyclins (to be discussed below), cyclin E (51). Although this cyclin has little structural homology to yeast G1 cyclins, the conservation of their respective modes of regulating turnover suggests that such a mechanism is advantageous for maintaining a highly dynamic control system. The negative consequences for cell cycle control alluded to above when stabilised *CLN* mutants are expressed underscores this point.

Regulation of the G1/S phase transition in mammalian cells

Like yeast cells, mammalian cells must respond to a variety of external and internal signals in the course of deciding to enter a new cell cycle. Since the normal state of most mammalian cells is to be in a non-proliferative mode, or resting state, much of this decision process involves re-entry into the cell cycle from G0. However, some environmental signals block cycling cells in G1 and maintain them within cycle. As in yeast, all of these regulatory phenomena need to integrated with the cell cycle machinery, the G1 cyclin-dependent kinases (Fig. 4). Because there are two different types of Cdk activity in mammalian cells, part of the question to be resolved for any signal is which kinase is being impacted.

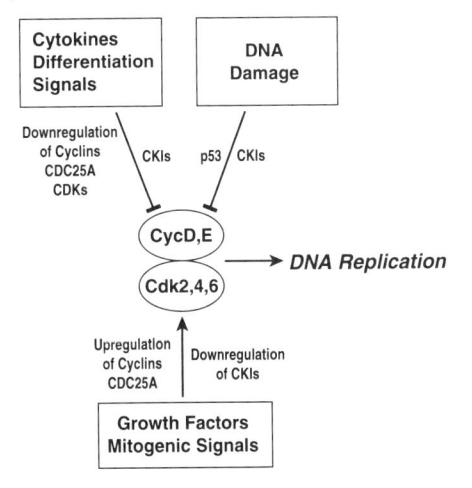

Figure 4. Signals impinging on G1 Cdk activities in mammalian cells. Negative regulation is exerted by damage to the genome, a number of cytokines, and differentiation signals. The DNA damage reponse is mediated via p53, in part through the accumulation of the Cdk inhibitor (CKI) p21^{Cip1}. Cytokines and differentiation signals also exert many of their antiproliferative effects on G1 Cdks by mobilization of CKIs. However, other mechanisms include downregulation of levels of cyclins, Cdks and the activating phosphatase, CDC25A. Conversely, growth factors and mitogens downregulate CKIs and increase levels of cyclins, Cdks and CDC25A.

Extracellular signals in the form of cytokines and growth factors mediate the entry into cycle from the resting state, or G0, for many cell types. Three processes appear to render cells of many types competent to cycle. First, synthesis of G1 cyclins is increased (18,71,72). Second, Cdk inhibitory activities are dissipated (73-75) and third, synthesis of the Cdk activating enzyme, CDC25A, is increased (76-77). CDC25 isoforms reverse inhibitory phosphorylations near the aminotermini of Cdks (78-80). The contribution and relative importance of these diverse regulatory mechanisms probably depends on the cell type. In fibroblasts and epithelial cells, where the majority of the studies to date have been performed, all three mechanisms appear to come into play. Transcription of cyclin D1 and cyclin E mRNAs are reduced in G0 and increase as cells are stimulated to re-enter the cell cycle (71,72,81,82). Likewise, transcription of CDC25A mRNA is strongly stimulated when fibroblasts are induced to return to the cell cycle from G0 (76-77). At the same time, quiescent fibroblasts contain high concentrations of the Cdk inhibitor $p27^{Kip1}$ (74,75,83). Thus, the turnover, and reduction in levels, of this inhibitor is likely to be an important determinant in cell cycle entry. Mechanistically, it is not yet clear how these diverse regulatory events are mediated. Up-regulation of cyclin D1 transcription has not yet been linked to any specific growth factor-controlled transcription factor. In fact, much of the increase in synthesis of cyclin D1 under conditions of growth factor stimulation has been attributed to an increase in nucleocytoplasmic transport of cyclin D1 mRNA (84). On the other hand, transcriptional control of cyclin E has been linked to the transcription factor E2F (85). It has been suggested that because up-regulation of cyclin E transcription occurs later than that of cyclin D1, cyclin E transcription is downstream of activation of cyclin D1 associated kinase activities. This is reasonable in the context of D-type cyclins controlling E2F and is consistent with the yeast paradigm of one class of G1 cyclin controlling transcription of the other. However, while it is clear that overexpression of E2F can stimulate cyclin E transcription and that cyclin D function, via mobilisation of E2F, may contribute to cyclin E expression, other factors, presently unknown, must contribute to cyclin E synthesis in the context of growth factor stimulation. Furthermore, a significant contribution to regulation of cyclin E levels is made at the level of cyclin E turnover and is coupled to the activation of cyclin E/CDK2 complexes (51). When active, cyclin E/CDK2 complexes autophosphorylate at a particular residue, which then targets the protein for ubiquitination and degradation. Inhibition of the cyclin E/CDK2 kinase, therefore, leads to an increase in cyclin E half-life, allowing the protein to accumulate. Thus, the accumulation of cyclin E is governed both by its transcription rate as well as the activation state of cyclin E/CDK2 kinase. In fact, cyclin E persists in quiescent cells due, presumably, to the effects of Cdk inhibitors (75) even though transcription of its message is strongly down-regulated. The mechanism of regulating CDC25A transcription has not yet been elucidated.

In contrast to the transcriptional control of cyclins and CDC25A, control of $p27^{Kip1}$ in the context of exit from and entry into the cell cycle is not transcriptional. In quiescent cells, both the translation rate and half-life of this inhibitor are increased significantly, leading to its accumulation (75,86). Growth factor stimulation reverses both of these regulatory modifications, leading to dissipation of $p27^{Kip1}$, allowing cells to activate G1 Cdks. The molecular basis for these events remains to be elucidated. Similar regulation appears to regulate exit from and entry into the cell cycle in diverse cell types. For example, cyclin transcription and $p27^{Kip1}$ accumulation have been implicated in regulating the transition between cycling and quiescence in T lymphocytes (73,74) and, at least some, epithelial cells (72,87).

Cdk inhibitors also appear to be involved in mediating or contributing to other types of cell cycle control. $p21^{Cip1}$ is induced in numerous lineages as cells differentiate and exit the cell cycle during mouse development (73). In myocytes that can be differentiated in cell culture, $p21^{Cip1}$ is strongly induced as cells cease proliferating and fuse to form myotubes (73,89). Likewise, $p27^{Kip1}$ accumulates as HL60 (promyelocytic leukemia) cells are induced to differentiate into non-proliferating monocytes by treatment with vitamin D3 (75). Temporary exit from the cell cycle, as opposed to quiescence, can be mediated by growth-inhibitory cytokines such as TGFβ (90) and interferons (reviewed in 91). In the best-understood system, mink lung epithelial cells treated with TGFβ arrest in G1 (72,87,90,92). Although $p27^{Kip1}$ levels do not increase in this system, there is a redistribution of this inhibitor from CDK4 and CDK6 to CDK2, which becomes inhibited (92). This mobilisation of $p27^{Kip1}$ appears to be instigated by the rapid induction of the CDK4/CDK6 inhibitor $p15^{INK4}$ (93), which saturates the preferred targets of $p27^{Kip1}$ and thus prevents binding (92). The ultimate outcome is that a single cytokine-mediated signal regulating the accumulation of one CDK inhibitor causes global inhibition of all G1 CDK activities and concomitant G1 arrest. Finally, the first modality of G1 inhibition historically to be elucidated at the molecular level was that involving ionising radiation. Here p53-dependent accumulation of $p21^{Cip1}$ (94-97) is responsible for inhibiting the G1 Cdk activities (98). p53 is a transcription factor that is activated in response to DNA damage, and the gene encoding p21^{Cip1} has been shown to contain p53 response elements, consistent with this model (96). Finally p27^{Kip1} has been shown to accumulate in response to treatment with cytostatic drugs, such as lovastatin and rapamycin (74,75). In the case of lovastatin, the regulation of p27^{Kip1} levels occurs translationally (75).

Although Cdk inhibitors appear, based on cell culture models, to be the dominant global regulators of G1 progression in mammalian cells, a number of observations suggest that this interpretation should be taken with caution. First some systems of G1 arrest do not appear to involve such inhibitors. An example is the effect of interferon α on a Burkitt's lymphoma-derived cell line known as Daudi. These cells exit the cell cycle into a G0-like state in response to interferon-α treatment (100). The mechanism of cell cycle arrest appears to be transcriptional downregulation of mRNAs corresponding to cyclin D3, the only D-type cyclin in the system and CDC25A, the activating phosphatase for G1 Cdk activities (100). As a result, inactive cyclin E/CDK2 complexes accumulate that can be activated *in vitro* by treatment with CDC25. Second, experiments on nullizygous mice support the importance of Cdk inhibitors in some contexts but not in others. For example cells of a p21^{Cip1}-nullizygous mouse are impaired in their response to ionising radiation (101,102). However, development is normal, placing doubt on the inferred critical role of p21^{Cip1} in developmental processes such as myogenesis (89). Even, the G1 response to radiation is not completely lost in cells from the nullizygous mouse, indicating that important controls are likely to be redundant (101,102). The p27^{Kip1} nullizygous mouse, although normal in form, is abnormally large (103-104). Analysis of these animals indicates that many tissues and organs contain an inordinate number of cells, consistent with an inability to properly exit from the cell cycle and with an important role for p27^{Kip1} in this process. Yet division is sufficiently regulated to produce a virtually normal, albeit large, mouse, again suggesting a high level of redundancy of controls. A final question that remains to be resolved is the function of several of the other known inhibitors. p57^{Kip2}, a structural homolog of p21^{Cip1} and p27^{Kip1}, has no known function (105-106). Although it is expressed in a tissue-specific pattern during mouse embryogenesis, a nullizygous mouse has yet to be described. Furthermore, p57^{Kip2} has not yet been associated with cell cycle arrest in a tissue culture model as has p21^{Cip1} and p27^{Kip1}. Of the inhibitors specific for CDK4 and CDK6, members of the INK4 family, a specific regulatory function has been attributed only to p15, as described above (92,93). Nevertheless, a p16-nullizygous mouse is highly cancer prone and mutations in the p16 gene have been implicated in melanoma and other cancers (107-109), suggesting an important regulatory function, perhaps in the maintenance of cell cycle checkpoints.

TARGETS OF G1 CYCLIN-DEPENDENT KINASE ACTIVITY

Introduction

Although much has been learned about regulation of Cdks, few bona fide *in vivo* targets of activated Cdks have been identified. Part of the problem resides in the fact that the consensus for Cdk phosphorylation is not well-defined at the primary structure level and probably involves secondary structure. Therefore, it is difficult to predict likely targets simply based on sequence analysis. Secondly, since it is likely that most G1 Cdks recognise multiple targets essential for transition to S phase, genetic analysis in organisms such as yeast has not been particularly helpful to date. The simplest approach, that of isolating genetic suppressors of Cdk mutants, is not expected to be effective for identifying downstream elements in non-linear pathways (and has not been). Yet, other genetic methods should eventually contribute to the identification of Cdk targets in yeast. In mammalian cells, the genetics of human disease has fortuitously provided the best-known and best-understood target of G1 Cdks, the retinoblastoma protein (pRb) (reviewed in 110,111). Yet, strategies to find other targets have not been, as yet, successful. As a result, a disproportionate effort has been focused on the relationship between G1 Cdks and pRb, although the phosphorylation of pRb is clearly only part of the regulatory story.

G1 Cdk targets in yeast

The manner in which G1 control is organised in budding yeast suggests pathways in which targets must exist, but no particular Cdc28-dependent phosphorylation event has been proven to have a definitive regulatory function, with one notable exception: the Cdk inhibitory protein, Sic1 (to be discussed below). The first pathway that must be targeted involves the transcriptional induction of G1 cyclins Cln1 and Cln2 in response to activation of the Cln3/Cdc28 kinase. The transcription factor shown to mediate transcription of Cln1 and Cln2 mRNAs is SBF, which is composed of two polypeptides, Swi4 and Swi6 (112-114). Swi6 has potential Cdk phosphorylation sites, and it has been proposed, although not proven, that phosphorylation of Swi6 by Cln3/Cdc28 is a key activating event.

Once Cln1 and Cln2 have been produced, four essential processes need to be initiated to commit cells to a new cell cycle (execution of START). These are the initiation of DNA replication, the initiation of budding, the initiation of a mitotic apparatus, and the induction of transcripts needed to support cell cycle progression. The factor responsible for G1/S transcription, MBF, is also composed of two polypeptides, Mbp1 and Swi6 (shared with SBF) (115-117). Again, Swi6 is a probable target of Cdc28/Cln action. Important transcripts regulated by MBF include those encoding enzymes required for DNA precursor synthesis, enzymes required in the replication process (118-120), and two B-type cyclins, Clb5 and Clb6, important for initiation of DNA replication (121-123). Clb5 and, presumably, Clb6 accumulate and bind to the Cdc28 kinase prior to initiation of replication. The actual triggering of S-phase, however, appears to be a Cln-dependent process. Cdc28/Clb complexes capable of initiating DNA replication are stored as inactive complexes with the stoichiometric inhibitor, $p40^{Sic1}$ (124,125). Degradation of $p40^{Sic1}$ is apparently triggered by phosphorylation in trans by Cdc28/Cln kinase complexes, resulting in activation Cdc28/Clb5,6 complexes and, as a consequence, initiation of replication (126). This is the best example of a bona fide G1-specific Cdk substrate of regulatory importance in yeast. The targets of Cdc28/Clb5,6, mobilised by phosphorylation and degradation of Sic1, remain to be identified.

Although many gene products have been shown to be important for initiation of budding, none has been identified as a regulatory target of the Cdc28/Cln kinase. At least two separate pathways appear to be targeted, in fact. The first involves the organisation of cytoskeletal elements for budding (127) and the second modulates growth and secretion for bud morphogenesis, functioning via protein kinase C and a MAP kinase homologous cascade downstream (128). It is likely that at least several proteins are phosphorylated to produce these global changes. Finally, activation of the Cdc28/Cln kinase leads to duplication of the microtubule organising centre, known as a spindle pole body in yeast. The duplication of the spindle pole body, which occurs at the G1/S phase transition, is the first committed step in the organisation of a mitotic spindle. However, no Cdc28/Cln targets have been identified in the context of this essential process. In yeast, however, the prognosis is relatively good that genetic methods will allow the eventual identification of many critical Cdc28/Cln targets.

G1 Cdk targets in mammalian cells

In mammalian cells, only one critical G1 Cdk target has been identified, pRb, the retinoblastoma protein (110,111). The characterisation of pRb as a negative effector of proliferation derived from the discovery that loss-of-function mutations at the human Rb locus were associated with familial retinoblastoma. Subsequently, it was found that pRb was a negative regulator of proliferation only in a hypophosphorylated form and that the protein contained numerous consensus Cdk phosphorylation sites whose occupancy increased as cells progressed through the cell cycle (129-133). Many years of intense investigation by a large number of groups have led to the currently accepted model for pRb function. The central concept is that hypophosphorylated pRb sequesters in an inactive form a number of proteins that are positive regulators of cell proliferation. Upon phosphorylation of pRb by one or several Cdks, these proteins are released to carry out their functions. The issues that remain to be resolved are exactly which Cdks are involved in pRb phosphorylation and what proteins are sequestered.

It is been proposed that both cyclin D/CDK4,6 and cyclin E/CDK2, the two classes of G1 Cdk, are involved in pRb phosphorylation. The arguments in favour of this model are that both types of kinase can phosphorylate pRb *in vitro* and that simultaneous transient transfection of both pRb- and cyclin-expressing plasmids into a tumour cell line demonstrates that D-type cyclins and cyclin E can neutralise the anti-proliferative effects of ectopically-expressed pRb (134). Other experiments, however, suggest that only D-type cyclins have a physiological role in phosphorylating pRb. Conditional expression of either cyclin D1 or E in Rat1 fibroblasts during early G1 advanced the entry into S phase equivalently, but only expression of cyclin D1 produced an immediate phosphorylation of pRb (44,45). Co-expression of cyclin D1 and cyclin E led to a much greater advance of S phase than expression of either alone, suggesting that cyclins D1 and E target independent pathways (45). The association of D-type cyclins with pRb phosphorylation is supported by numerous transfection and microinjection experiments that conclude that cyclin D1/CDK4,6 function is only required in cells that express a functional pRb (135-138). Another link between cyclin D1 and pRb comes from analysis of embryonic development in cyclin D1 nullizygous mice. One of the most obvious pathologies detected in such animals is a severe retinopathy characterised by a drastic reduction in retinal cell number (48). This defect is consistent with the observation that the developing retina expresses much higher levels of cyclin D1 than most other tissues (48), presumably because these cells express unusually high levels of pRb that needs to be regulated. The fact that cyclin D1 is not essential in the whole mouse suggests a level of functional redundancy for D-type cyclins.

On the issue of proteins negatively regulated by pRb, the most established class are the E2F family of transcription factors (reviewed in 139,140). E2F family transcription factors have been associated with transactivation of a number of genes associated with cell cycle progression and DNA replication, supporting the model for pRb control of the cell cycle described above. The issue, however, is complicated by the fact that E2F transcription factors are active only as heterodimers composed of an E2F moiety and a DP1 moiety (reviewed in 139,140). There are at least five isoforms of E2F and two of DP1, leading to many combinatorial possibilities (reviewed in 139,140). Whether these form a highly redundant set or have specific roles has yet to be established. However, a mouse nullizygous for one of the E2F isoforms, E2F1, is viable and develops normally, consistent with a level of redundancy (141). Furthermore, only E2F1, E2F2, and E2F3 are thought to associate with pRB, whereas E2F4 and E2F5 are thought to associate with pRb-related proteins (to be described below) but not pRb (142). Therefore, the details of the relationship of E2F function to pRB remain to be elucidated. Other targets of negative regulation by pRb are the c-Abl tyrosine kinase (143), and another transcription factor Elf-1 (144). The significance of these interactions is less clear than that between pRb and E2F. It is possible that in some instances, the function of pRb is to promote direct interaction between various pRb-binding proteins for specific regulatory functions, rather than mere sequestration (145).

Two pRb-related proteins have been identified, known as p107 and p130, respectively (146,147). In addition to primary structure homology, these proteins share pRb's ability to bind and sequester E2F (146,147). However, as mentioned above, different isoforms are favoured. Also, although both p107 and p130, like pRb are phosphorylated in a cell cycle dependent fashion, the relationship between phosphorylation and E2F sequestration is not as well established and the identity of the kinases responsible needs to be clarified. However, G1 Cdks are the most likely candidates. Whereas mice nullizygous for pRb die before day 16 of gestation with numerous cellular defects (148,149), mice nullizygous for either p107 or p130 are viable with no obvious phenotype (150). Therefore, pRb appears to be the most critical member of this family. However, analysis of doubly nullizygous mice indicates that there is some functional overlap between pRb, p107 and p130, rendering them all likely targets of G1 Cdk action (150).

Whereas pRb and related proteins may be the only important targets of cyclin D-associated kinases, this is not likely to be the case for cyclin E/CDK2. However, no convincing candidates for cyclin E/CDK2 substrates in mammalian cells have been put forth. Work in frog eggs and embryos suggests that cyclin E/CDK2 may be directly involved in replication initiation functions (151). However, these results need to be interpreted with caution since cyclin A/CDK2 does not exist in early amphibian embryos and cyclin E/CDK2 may be performing a function assigned to cyclin A/CDK2 in mature somatic cells. Therefore, finding biologically relevant substrates for cyclin E/CDK2 in the context of the G1/S phase transition is a major remaining challenge facing the cell cycle field.

CONCLUSIONS

The highly conserved nature of cell cycle control has allowed the pooling of resources and data from diverse organisms to develop a general model. The consensus that has emerged has, as its centrepiece, a major cell cycle regulatory event in G1 controlled by cyclin-dependent kinases. In budding yeast, this event has been termed START whereas in mammalian cells it has been referred to as the R-point. While the molecular basis of START in yeast has been established as the activation of the cyclin-dependent kinase, Cdc28, by G1 cyclins, the relationship between the R-point and G1 Cdks in mammalian cells is less clear. Nevertheless, it is difficult to imagine that they are not involved, and the activation of D-type cyclin-associated kinase activities and their phosphorylation of pRb most likely is a component of what constitutes the R-point transition. The parallel between yeast and mammalian cells also extends to the concept of relief of negative control at the G1/S phase boundary. As stated above, a primary target of G1 Cdks in mammalian cells is pRb and its relatives, p107 and p130, all thought to be inhibitors of proliferation because of their ability to sequester positive factors, such as the E2F transcription factors. A primary target of G1 Cdks in yeast is the S-phase Cdk inhibitor Sic1. Although in this case the inhibitor is specific for S phase Cdks rather than transcription factors, the concept of regulating a phase transition by inactivation of an inhibitor appears to be favoured, perhaps because it facilitates a rapid transition. Finally, the principal challenge remaining in both yeast and other organisms is the identification of additional critical targets of cyclin-dependent kinases. In mammalian cells, a clear relationship between cyclin D associated kinases and pRb phosphorylation has emerged. However, the substrates for cyclin E associated kinase are not known. In yeast, similarly, although phosphorylation of Sic1 by G1 Cdks leads to activation of S phase kinases Clb5/Cdc28 and Clb6/Cdc28, it is not yet clear what the targets of these kinases are that result in initiation of DNA replication. It is possible that cyclin E/CDK2 is functionally homologous to Clb5,6/Cdc28 and that the use of yeast genetics to identify the substrates of Clb5,6/Cdc28 will help elucidate the substrates of cyclin E/CDK2. If that is the case, the universality of cell cycle control mechanisms will once again have been demonstrated to be an all-pervasive principle.

REFERENCES

1. Hartwell, L.H., Culotti, J., Pringle, J., and Reid, B. (1974) *Science* **183**, 46-51.
2. Reed, S.I. (1980) *Genetics* **95**, 561-577.
3. Nurse, P., Bissett, Y. (1981) *Nature* **292**, 558-60.
4. Th'ng, J.P.H., Wright, P.S., Hamaguch, J., Lee, M.G., Norbury, C.J., et al., (1990) *Cell* **63**, 313-24.
5. Richardson, H.E., Wittenberg, C., Cross, F. (1989) *Cell* **59**, 11270-133.
6. Wittenberg, C., Sugimoto, K., Reed, S.I. (1990). *Cell* **62**, 225-37.
7. Deshaies, R.J., Chau, V., Kirschner, M., (1995) *EMBO* **14**, 303-312.
8. Yaglom, J., Linskens, M.H.K., Sadis, S., Rubin, D.M., Futcher, B., Finley, D. (1995) *Mol. Cell Biol.* **15**, 731-741.
9. Salama, S.R., Hendricks, K.B., Thorner, J. (1994) *Mol. Cell Biol.* **14**, 7953-7966.
10. Lanker, S., Valdivieso, M.H., Wittenberg, C. (1996) *Science*, **271**, 1597-1601.
11. Pardee, A.B. (1974) *Proc. Natl. Acad. Sci USA* **71**, 1286-1290.
12. Rossow, P.W., Riddle, V.G.H., Pardee, A.B. (1979) *Proc. Natl. Acad. Sci. USA* **76**, 4446-4450.
13. Elledge, S.J., Spottswood, M.R. (1991) *EMBO J.* **10**, 2653-2659.
14. Tsai, L.-H., Harlow, L., Meyerson, M. (1991) *Nature* **353**, 174-177.
15. Ninomiya-Tsuji, J., Nomoto, S., Yasuda, H., Reed, S.I., Matsumoto, K. (1991) *Proc. Natl. Acad. Sci. USA* **88**, 9006-9010.
16. Motokura, T., Bloom, T., Kim, H.J., Juppner, H. Ruderman, J.V., et al. (1991) *Nature* **350**, 512-515.
17. Xiong, Y., Connolly, T., Futcher, B., Beach, D. (1991) *Cell* **65**, 691-699.
18. Matsushime, H., Roussel, M.F., Ashmun, R.A., Sherr, C.J. (1991) *Mol. Cell. Biol.* **11**, 329-337.
19. Lew, D.J., Dulic, V., Reed, S.I. (1991) *Cell* **66**, 1197-1206.
20. Koff, A., Cross, F., Fisher, A., Schumacher, J., Leguellec, K., et al. (1991) *Cell* **66**, 109-137.
21. Piggott, J.R., Rai, R., Carter, B.L.A., (1982) *Nature* **298**, 391-393.
22. Reed, S.I., Wittenberg, C. (1990) *Proc. Natl. Acad. Sci. USA* **87**, 5697-5701.
23. Gerber M.R., Farrell, A., Deshaies, R., Herskowitz, I., Morgan, D.O. (1995) *Proc. Natl. Acad. Sci. USA* **92**, 4651-4655.
24. Lorincz, A.T., Reed, S.I. (1984) *Nature* **307**, 183-185.
25. Reed, S.I., Hadwiger, J.A., Lorincz, A.T. (1985) *Proc. Natl. Acad. Sci. USA* **77**, 2119-2123.
26. Hadwiger, J.A., Wittenberg, C., Richardson, H.E., deBarros Lopes, M., Reed, S.I. (1986) *Proc. Natl. Acad. Sci USA* **86**, 6255-6259.
27. Sudbery, P.E., Goodey, A.R., Carter, B.L., (1980) *Nature* **288**, 401-404.
28. Cross, F.R. (1988) *Mol. Cell, Biol.* **8**, 4675-4684.
29. Nash, R., Tokiwa, G., Anand, S., Erickson, K., Futcher, A.B. (1988) *EMBO J.* **7**, 4335-4346.
30. L. Breeden, personal communication.
31. Fisher, D.L., and Nurse, P., (1996) *EMBO J.* **15**, 850-860.
32. Martin-Castellanos, C., Labib, K. and Moreno, S (1996) *EMBO J.* **15**, 839-849.
33. Mondesert, O., McGowan, C.H., and Russell, P. (1996) *Mol. Cell, Biol.* **16**, 1527-1533.
34. Withers, D.A., Harvey, R.C., Faust, J.B., Melnyk, O., Carey, K., Meeker, T.C. (1991) *Mol. Cell, Biol.* **11**, 4846-4853.
35. Richardson, H.E., O'Keefe, L.V., Reed, S.I., and Saint, R. (1993). *Devel.* **119**, 673-690.
36. Finley, R.L. Jr., Thomas, B.J., Zipursky, S.L., Brent, R., (1996) *Proc. Natl. Acad. Sci. USA*, **93**, 3011-3015.
37. Dulic, V., Lees, E., and Reed, S.I. (1992) *Science* **257**, 1958-1961.
38. Koff, A., Giordano, A., Desai, D., Yamashita, K., Harper, J.W., Elledge, S., Nishimoto, T., Morgan, D.O., Franza, B.R., and Roberts, J.M. (1992) *Science* **257**, 1689-1694.
39. Baldin V., Lukas, J., Marcote, M.J., Pagano, M., Draetta, G. (1993) *Genes Dev.* **7**, 812-8121.
40. Lukas, J., Bartkova, J., Welcker, M., Peterson, O.W., Peters, G. Strauss, M., and Bartek, J. (1995) *Oncogene* **10**, 2125-2134.
41. Quelle, D., Ashmun, R., Shurtleff, S., Kato, J, Bar-Sagi, D. Roussell, M. and Sherr, C. (1993). *Genes Dev.* **7**, 1559-1571.
42. Ohtsubo, M. Theodoras, A.M., Schumacher, J., Roberts, J.M., Pagano, M. (1995) *Mol. Cell Biol.* **15**, 2612-2624.
43. Ohtsubo, M., and Roberts, J. (1993) *Science* **259**, 1908-1912.
44. Resnitzky, D. Gossen, M., Bujard, H., and Reed, S.I. (1994) *Mol. Cell Biol.* **14**, 1669-1679.
45. Resnitzky, D, and Reed, S.I. (1995) *Mol. Cell. Biol.* **15**, 3463-3469.
46. Matsushime H., Ewen, M. Strum, D.K., Kato, J.Y., Hanks, S.K., Roussel, M,. and Sherr, C.J. (1992) *Cell* **71**, 323-334.
47. Meyerson, M. and Harlow, E. (1994) *Mol. Cell. Biol.* **14**, 2077-2086.

48. Sicinski, P., Donaher, J.L., Parker, S.B., LI, T., Fazeli, A., Gardner, H., Haslam, S.Z., Bronson, R.T., Elledge, S.J., Weinberg, R.A. (1995) *Cell* **82**, 621-630.
49. R.A. Weinberg, personal communication.
50. Sherr, C.J. (1993) *Cell* **73**, 1059-1065.
51. Won, K.-A. and Reed, S.I. (1996) *EMBO J.*, in press.
52. A. Zetterberg, personal communication.
53. Morgan, D.O.(1995) *Nature* **374**, 131-134.
54. Nigg, E.A. (1995) *Bioessays* **17**, 471-480.
55. Lees, E. (1995) *Curr. Opin Cell Biol.* **7**, 773-780
56. Hunter, T., Pines, J. (1994) *Cell* **79**, 573-582.
57. Sherr, C.J., Roberts, J.M. (1995) *Genes Dev.* **9**, 1149-1163.
58. Pringle, J.R., Hartwell, L.H., (1981) in *The Saccharomyces cerevisiae Cell cycle* (Strathern, J.N, Jones, E.W, Broach, J.R., eds.) pp97-142, Cold Spring Harbor Laboratory, Cold Spring Harbor, NY.
59. Herskowitz, I. (1995) *Cell* **80**, 187-197.
60. Schultz, J. Ferguson, B. Sprague, G.J. (1995) *Curr. Opin. Genet. Dev.* **5**, 31-37.
61. Wittenberg, C., Reed, S.I. (1988) *Cell* **54**,1061-1072.
62. Stuart, D. and Wittenberg, C. (1995) *Genes Dev.* **9**, 2780-2794.
63. Dirick, L., Bohm, T., Nasmyth, K. (1995) *EMBO J.* **14**, 4803-4813.
64. Johnston, G.C., Pringle, J.R., Hartwell, L.H. (1977) *Exp. Cell Res.* **105**, 79-98.
65. Jagadish, M.N., Carter, B.L.A. (1977) *Nature* **269**,145-47.
66. Carter, B.L.A., Jagadish, M.N. (1978) *Exp. Cell Res.* **112**, 15-24.
67. Carter, B.L.A., Lorincz, A.T., Johnston, G.C. (1978) *J. Gen Microbiol.* **106**:221-225.
68. Chang, F., Herkowitz, I. (1990) *Cell* **63**, 999-1011.
69. Peter, M., Gartner, A., Horecka, J., Ammerer, G., Herskowitz, I. (1993) *Cell* **73**, 747-760.
70. Peter, M., Herskowitz, I. (1994) *Science* **265**,1228-1231.
71. Won, K.A., Xiong, Y., Beach, D. and Gilman, M.Z. (1992) *Proc. Natl. Acad. Sci. USA* **89**, 9910-9914.
72. Slingerland, J.M., Hengst, L., Pan, C.-H, Alexander, D., Stampfer, M.R., and Reed, S.I. *Mol. Cell.Biol.* **14**, 3683-3694.
73. Firpo, E.J., Koff, A., Solomon, M.J., and Roberts, J.M. (1994) *Mol. Cell Biol.* **14**, 4889-4901.
74. Nourse, J., Firpo, E., Flanagan, M., Coats, S. Polyak, C. Lee, M., Massague, Crabtree, G., and Roberts, J. (1994) *Nature* **372**, 570-573.
75. Hengst, L., and Reed, S.I. (1996) *Science* **271**, 1861-1864.
76. Hoffman, I., Draetta, G., and Karsenti, E., (1994) *EMBO J.* **13**, 4302-4310.
77. Jinno, S., Sato, K, Nagata, A, Igarashi, I, Kanaoka, Y, Nojima, H., and Okayama, H. (1994) *EMBO J.* **13**, 1549-1550.
78. Dunphy, W.G., and Kumagai, A. (1991) *Cell* **67**, 189-196.
79. Gautier, J. Solomon, M.J., Booher, R.N., Bazan, J.F., and Kirschner, M.W. (1991) *Cell* **67**, 197-211.
80. Millar, J.B., McGowan, C.H., Lenaers, G., Jones, R., Russell, P. (1991) *EMBO J.* **10**, 4301-4309.
81. Winston, J.T., Pledger, W.J. (1993) *Mol. Bio. Cell* **4**, 1133-1144.
82. Geng, Y., and Weinberg, R.A., (1993) *Proc. Natl. Acad. Sci, USA* **90**, 10315-10319.
83. Resnitzky, D., Hengst, L., and Reed, S.I. (1995) *Mol. and Cell. Biol.* **15**, 4347-4352.
84. Rousseau, D., Kaspar, R., Rosenwald, I., Gehrke, L., and Sonenberg, N. (1996) *Proc. Natl. Acad. Sci. USA* **93**, 1065-1070.
85. DeGregori, J., Kowalik, T., Nevins, J.R. (1995) *Mol. Cell. Biol.* **15**, 4215-4224.
86. Pagano, M., Tam, S.W., Theodoras, A.M., Beer-Romero, P., Del Sal, G., Chau, V., Yew, P.R., Draetta G.F. (1995) *Science* **269**, 682-685.
87. Polyak K., Kato, J.Y., Solomon, M.J., Sherr, C.J., Massague, J. Roberts, J.M., Koff, A. (1994) *Genes Dev.* **8**, 9-22.
88. Parker, S.B., Eichele, G., Zhang, P., Rawls, A., Sands, A.T. Bradley, A., Olson, E.N., Harper, J.W., Elledge, S.J. (1995) *Science* **267**, 1024-1027.
89. Halevy, O., Novitch, B.G., Spicer, D.B., Skapek, S.X., Rhee, J., Hannon, G. J., Beach, D., Lassar, A.B. (1995) *Science* **267**, 1018-1021.
90. Laiho, M, DeCaprio, J., Ludlow, J., Livingston, D., Massague, J. (1990) *Cell* **62**,175-185.
91. Kimchi, A., (1992) *J. Cell Biochem.* **50**,1-9.
92. Reynisdottir, I., Polyak, K., Iavarone, A. Massague, J. (1995) *Genes Dev.* **9**, 1831-1845.
93. Hannon, G.J., Beach, D., (1994) *Nature* **371**, 257-261.
94. Harper, J.W., Adami, G.R., Wei, N., Keyomarsi, K., Elledge, S.J. (1993) *Cell* **75**, 805-816.
95. Noda, A., Ning, Y., Venable, S.F., Pereira, S.O., Smith, J.R. (1994) *Exp. Cell. Res.* **211**, 90-98.
96. El-Deiry, W.S., Tokino, T., Velculescu, V.E., Levy, D.B., Parson, R., Trent, J.M., Lin, D., Mercer, W.E., Kinzler, K.W., Vogelstein, B., (1993) *Cell* **75**, 817-825.
97. Gu, Y., Turck, C.W., and Morgan, D.O. (1993) *Nature* **366**, 707-710.
98. Dulic, V., Kaufmann, W.K., Wilson, S.J., Tlsty, T.D., Lees, E., Harper, J.W., Elledge, S.J., Reed, S.I. (1994) *Cell* **76**,1013-1023.

99. Kato, J-Y., Matsuoka, M., Polyak, K., Massague, J., Sherr, CJ. (1994) *Cell* **79**, 487-486.
100. Tiefenbrun, N., Melamed, D., Levy, N., Resnitzky, D., Hoffmann, I., Reed, S.I., and Kimchi, A. (1996) *Mol. Cell. Biol.*, in press.
101. Deng, C., Zhang, P., Harper, J.W., Elledge, S.I., and Leder, P. (1995) *Cell* **82**, 675-684.
102. Brugaroias, J., Chandrasekaran, C., Gordon, J.I., Beach, D., Jacks, T. andHannon, G.J. (1995) *Nature* **377**, 552-557.
103. Kiyokawa, H., Kineman, R.D., Manova-Todarova, K.O., Soares, V.L., Hoffman, E., Ono, M., Khanam, D., Hayday, A.L., Frohman, L.A. and Koff, A. (1996) *Cell*, in press.
104. Fero, M.I., Rivkin, M., Tasch, M., Porter, P., Carrow, C.E., Firpo, E., Polyak, K., Tsai, L.H., Brondy, V., Perlmutter, R.M., Kanshansky, K. and Roberts, J.M. (1996) *Cell*, in press.
105. Matsuoka, S., Edwards, M.C., Bai, C., Parker, S., Zhang, P., Baldini, A., Harper, J.W., Elledge, S.J. (1995). *Genes Dev.* **9**, 650-662.
106. Lee, M-H., Reynisdottir, I., Massague, J, (1995) *Genes Dev.* **9**, 639-649.
107. Kamb, A., Shattuck, E.D., Eeles, R., Liu, Q, Gruis, N.A., Ding, W., Hussey, C., Tran, T., Miki, Y., Weaver, F.J., et al. (1994) *Nature Genet* **8**, 23-26.
108. Nobori, T., Miura, K., Wu, D.J., Lois, A., Takabayashi, K., Carson, D.A (1994) *Nature* **368**, 753-756.
109. Serrano, M., Lee, H.-W., Chin, L., Cordon-Cardo, C., Beach, D. and DePinho, R.A. (1996) *Cell* **85**, 27-37.
110. Weinberg, R.A. (1995) *Cell*, **81**, 323-330
111. Hinds, P.W. (1995) *Genet. Dev.* **5**, 70-83.
112. Nasmyth K., Dirick, L. (1991) *Cell* **66**, 995-1013.
113. Ogas J., Andrews, B.J., Herskowitz, I. (1991) *Cell* **66**, 1015-1026.
114. Breeden L., Mikesell, G.E. (1991) *Genes Dev.* **5**, 1183-1190.
115. Koch, C., Moll., T., Neuberg M., Ahorn H., Nasmyth, K. (1993) *Science* **261**, 1551-1557.
116. Dirick, L., Moll, T., Auer, H., Nasmyth, K. (1992) *Nature* **357**, 508-513.
117. Lowndes, N.F., Johnson, A.L., Breeden, L., Johnston, L.H. (1992) *Nature* **357**, 505-515.
118. McIntosh, E.M., Atkinson, T., Storms, R.K., Smith, M. (1991) *Mol. Cell. Biol.* **11**, 329-337.
119. Gordon, C.B., Campbell, J.L. (1991) *Proc. Natl. Acad. Sci USA* **88**, 6058-6062.
120. Lowndes, N.F., Johnson, A.L. (1991) *Nature* **350**, 247-250.
121. Epstein, C.B., Cross, F.R. (1992) *Genes Dev.* **6**, 1695-1706.
122. Schwob, E., Nasmyth K., (1993) *Genes Dev.* **7**, 1160-1175.
123. Kuhne, C., and Linder, P. (1993) *EMBO J.* **12**, 3437-3447.
124. Nugroho, T. Mendenhall, M.D. (1994) *Mol. Cell. Biol.* 14, 3320-3328.
125. Schwob E., Bohm, T, Mendenhall, M.D., Nasmyth, K. (1994) *Cell* **79**, 233-244.
126. M. Mendenhall, personal communication.
127. Lew, D.J., Reed, S.I. (1993) *J. Cell. Biol.* **120**, 1305-1320.
128. Marini, N.J., Meldrum, E., Beuhrer, B., Hubberstey, A.V., Stone, D.E., Traynor-Kaplan, A. and Reed, S.I. (1996) *EMBO J.*, in press.
129. Buchkovich K., Duffy, L.A., Harlow, E., (1989) *Cell* **58**,1097-1105.
130. Chen, P.L., Scully, P., Show, J.-Y. Wang, J.Y.J., Lee, W.-H. (1989) *Cell* **58**, 1193-1198.
131. DeCaprio, J.A., Furukawa, Y., Aichenbaum F., Griffin, J.D., Livingston, D. (1992) *Proc. Natl. Acad. Sci. USA* **89**, 1795-1798.
132. Mihara, K., Cao, X., Yen, A., Chandler, S., Driscoll, B. et al. (1989) *Science* **246**, 1300-1303.
133. Ludlow, J.W., Shon, J., Pipas, J.M. Livingson, D.M., DeCaprio, J.A. (1990) *Cell* **60**, 387-396.
134. Hinds, P.W., Mittnacht, S., Dulic, V., Arnold, A., Reed, S.I. and Weinberg, R.A. (1992) *Cell* **70**, 993-1006.
135. Lukas, J., Parry, D., Aagard, L, Mann, D.J., Bankova, J., Strauss, M. Peters, G., Bartek, J. (1995) *Nature* **375**, 503-506.
136. Koh, J., Enders, G.H., Dynlacht, B.D., Harlow, E. (1995) *Nature* **375**, 506-510
137. Lukas, J., Bartkova, J., Rhode, M., Strauss, M., and Bartek, J. (1995) *Mol. Cell. Biol.* **15**, 2600-2611.
138. Lukas, J., Muller, H., Bartkova, J., Spitkovsky, D., Kjerulff, A., Jansen-Durr, P. Strauss, M. and Bartek, J. (1994) *J. Cell Biol.* **125**, 625-638.
139. Muller, R. (1995) Trends Genet 11, 173-178.
140. La Thangue, N.B. (1994) *Curr. Opin. Cell Biol.* **6**, 443-450.
141. E. Harlow, personal communication.
142. Lees, J.A., Saito, M., Vidal, M., Valentine, M., Look, T, Harlow, E., Dyson, N. and Helin, K. (1993) *Mol. Cell. Biol.* **13**, 7813-7825.
143. Welch, P.J., Wang, J.Y.J. (1993) *Cell* **75**, 779-790.
144. Wang, C-Y., Petryniak, B., Thompson, C.B., Kaelin, W.G., Leiden, J.M. (1993) *Science* **260**, 1330-1335.
145. Welch, P.J. and Wang, J.Y.J. (1995) *Genes Dev.* **9**, 31-46.
146 Ewen, M.E., Xing, Y., Lawrence, J.B., and Livingston, D. (1991) *Cell* **66**, 1155-1164.
147. Hannon, G.J., Demetrick, D. and Beach, D. (1993) *Genes & Dev.* **7**, 2378-2391.

148. Jacks, T., Fazeli, A., Schmitt, E.M., Bronson, R.T., Goodell, M.A., and Weinberg, R.A. (1992) *Nature* **359**, 295-300.
149. Lee, E.Y-H.P., Chang, C.-Y., Hu, N., Wang, Y.-C.J., Lai, C.-C., Herrup, K., Lee, W.-H., and Bradley, A. (1992) *Nature* **359**, 288-294.
150. T. Jacks, personal communication.
151. Jackson, P.K., Chevalier, S., Phillipe, M., and Kirschner, M.W. (1995) *J. Cell Biol.* **130**, 755-769.

chapter 3
Regulation of G1 progression in fission yeast by the *rum1+* gene product

Cristina Martín-Castellanos and Sergio Moreno[1]

Instituto de Microbiología Bioquímica, Departamento de Microbiología y Genética,
CSIC/Universidad de Salamanca, Edificio Departamental,
Avda. del Campo Charro s/n, 37007 Salamanca, Spain
[1]To whom correspondence should be addressed

Recently it has been found that B-type cyclins in fission yeast regulate the activation of the cdc2 kinase to promote the onset of both DNA replication and mitosis. cig2 is the major G1 cyclin while cdc13 is the principal mitotic cyclin. cdc13 also has an additional function in G2 phase, preventing more than one round of DNA replication per cell cycle. In opposition to these cyclins the rum1 inhibitor, a protein present exclusively in G1, prevents premature activation of the cdc2/cig2 and the cdc2/cdc13 complexes until cells have reached the critical cell size required to pass Start and initiate a new cell cycle.

INTRODUCTION

During the cell division cycle, DNA replication and mitosis must be coupled in the correct order to ensure that each daughter cell receives a full complement of the hereditary material. Cell cycle progression is principally regulated before the onset of S-phase and before the onset of mitosis, by the activity of one or more cyclin dependent kinases (CDKs). Activation of these protein kinases requires the formation of a complex with a family of regulatory proteins, named cyclins. Specific CDK/cyclin complexes regulate the onset of S-phase from G1 and the initiation of mitosis from G2. In addition, cell cycle checkpoints maintain the temporal order of S phase and mitosis and ensure that S phase takes place only once per cell cycle, that mitosis is not initiated until DNA replication is completed and that the next cell cycle does not begin until the DNA is properly segregated at mitosis.

The cell cycle must also be coordinated with cell growth. In the fission yeast *Schizosaccharomyces pombe*, cells have to reach a critical cell mass before the onset of S-phase and again at the onset of mitosis (1, 2, 3). In exponentially growing wild type cells the mitotic control is limiting, since cell division produces daughter cells with a mass already greater than the minimum required to initiate S-phase. In these conditions G1 is very short and the onset of S-phase is regulated by its dependency upon completion of the previous mitosis (2, 3, 4). In conditions of nutrient limitation, mitosis is initiated at a reduced cell size, producing small daughter cells that must delay the initiation of S-phase until the critical mass is achieved. This

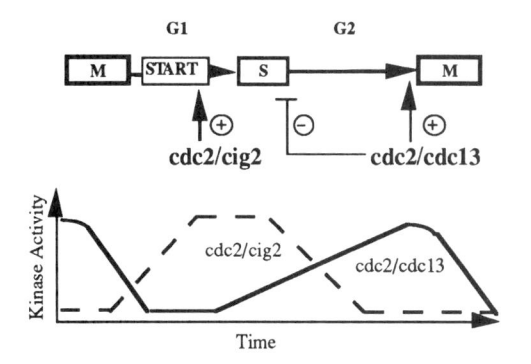

Figure1. CDK/cyclin complexes regulating the fission yeast cell cycle. For the sake of simplicity, only cdc2/cig2 and cdc2/cdc13 are shown. The cdc2/cig2 complex promotes Start and the initiation of DNA replication. In its absence this function could be performed either by cdc2/cig1 or cdc2/cdc13. The cdc2/cdc13 complex has a dual role, triggering mitosis and also preventing more than one round of DNA replication per cell cycle.

sizing control mechanism operates at Start, the point at the end of G1 where cells commit themselves to another round of cell division (5, 6). Before Start, cells can stop the cell cycle and enter stationary phase or undergo sexual differentiation. In fission yeast passage through Start and entry into DNA replication involves the activation of the cdc2 protein kinase in association with B-type cyclins (7, 8, 9). It is possible that this activity could have a role in promoting the activation of transcription factors required in the initiation of DNA replication, such as the cdc10/res1 complex, or perhaps more directly for the activation of initiation factors bound to DNA replication origins. In *S.pombe*, the cdc10/res1 transcriptional complex does not bind DNA in the absence of cdc2 function (10) and cdc2 has recently been shown to interact directly with orp2, a protein similar to the ORC2 replication factor subunit of *Saccharomyces cerevisiae* origin replication complex (ORC) (11). cdc2 activation before S phase has also been shown to depend on cdc10 (8), arguing for an interdependence between cdc2 and cdc10 function in G1. Of the three B-type cyclins described in fission yeast, cdc13 is the principal mitotic cyclin, while cig1 and cig2 seems to have a role earlier in the cell cycle (Figure 1). cdc13 is also required to prevent a second round of DNA replication within the same cell cycle (12), as cells deleted for the $cdc13^+$ gene lose this inhibitory effect over S phase and undergo multiple rounds of DNA replication without intervening mitoses. Deletion of the $cig1^+$ and $cig2^+$ genes abolishes the re-replication phenotype of cells lacking $cdc13^+$, confirming the importance of cig1 and cig2 for the onset of S-phase (8).

In animal cells, a number of specific CDK inhibitors have been identified as important regulators of the cell cycle in G1. These proteins bind either to the CDK or CDK/cyclin complexes blocking protein kinase activity in a yet uncharacterised way. In vertebrate cells these inhibitors block G1 progression in response to DNA damage, contact inhibition, or antimitogenic factors such as TGFß (13, 14, 15). These inhibitors can be divided into two families, the first one composed of molecules that inhibit the CDK/cyclin complexes, such as $p21^{CIP1/WAF1}$, $p27^{KIP1}$ and $p57^{KIP2}$ (17, 18, 19, 20, 21, 22, 23, 24), whilst the second class, such as $p16^{INK4a}$, $p15^{INK4b}$, $p18^{INK4c}$ and $p19^{INK4d}$ (25, 26, 27, 28, 29), compete with D type cyclins for the binding of CDK4 and CDK6 subunits. One of the best characterized cell cycle inhibitors, $p21^{CIP1/WAF1}$, is induced by p53 after DNA damage. p21 molecules bind and inhibit the CDK2/cyclinA and CDK2/cyclin E complexes, blocking cell cycle progression in G1 until DNA has been repaired. Interestingly, CDK/cyclin complexes containing only one molecule of $p21^{CIP1}$ are active suggesting that the stoichiometry of the inhibitor in the complex is important to determine if the complex is active or not (30). These inhibitors are very important to restrain cell proliferation and indeed some of them have been shown to behave as tumour supressor genes (31, 32, 33, 34).

In the budding yeast *Saccharomyces cerevisiae* two CDK inhibitors have been described. $p120^{FAR1}$ protein blocks G1 progression by inhibiting CDC28/CLN kinase activity in response to mating pheromone(35, 36, 37, 38). The other inhibitor $p40^{SIC1/SDB25}$, regulates G1/S progression in a normal cell cycle (39, 40, 41). $p40^{SIC1/SDB25}$ is a specific inhibitor of the CDC28/CLB type complexes preventing DNA replication until other cell cycle events triggered at Start such as bud formation and spindle pole body duplication, have been initiated correctly (42). SIC1/SDB25 deletion is lethal but only in daughter cells that need to grow in G1 before Start. No homologues of FAR1 or SIC1/SDB25 genes have been described so far in other organisms.

$P25^{RUM1}$ IS A CDK INHIBITOR

In fission yeast, the $rum1^+$ gene encodes a CDK inhibitor with a central role in G1 progression, preventing the onset of S-phase until the cell reaches the minimum size needed to pass Start. Cells lacking $rum1^+$ are unable to block the cell cycle in G1 in response to extracellular signals such as nitrogen starvation (43) (Figure 2A). Since G1 arrest after nitrogen starvation in fission yeast is a prerequisite for conjugation in haploids or meiosis in

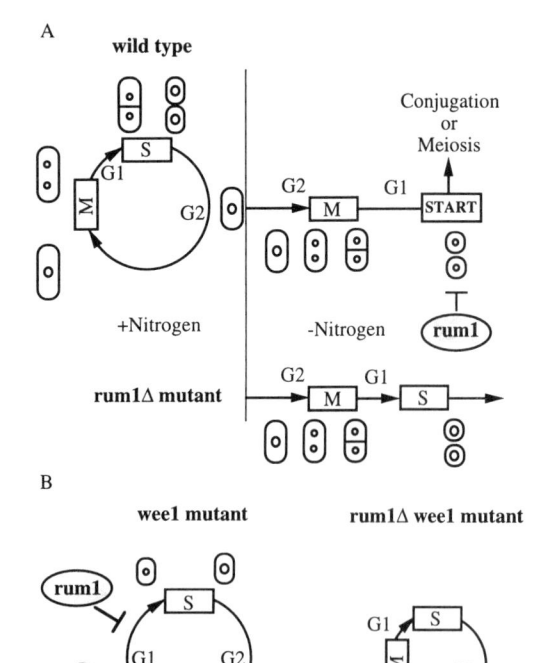

Figure 2. rum1 is a cell cycle inhibitor in G1.
A, exponentially growing wild type cells have a short G1 phase. When wild type cells are shifted to a medium without a nitrogen source, a signal is sent that promotes entry into mitosis at a smaller size and subsequently cell cycle arrest in G1 after one or two divisions. Mutant cells lacking the $rum1^+$ gene also sense this signal but are unable to arrest in G1. Cells lacking the $rum1^+$ gene are sterile and do not undergo meiosis, since mating and meiosis require G1 arrest before Start.
B, *wee1* mutants initiate mitosis at a reduced cell size compared to wild type. This causes a rearrangement of the cell cycle with a shorter G2 phase and an extended G1 phase to allow enough time to couple cell growth and cell cycle progression. *wee1* mutants lacking rum1 cannot delay the cell cycle in G1, and so enter S phase as soon as they complete mitosis. These cells lose about 20% of their mass each cell cycle and die after a few divisions.

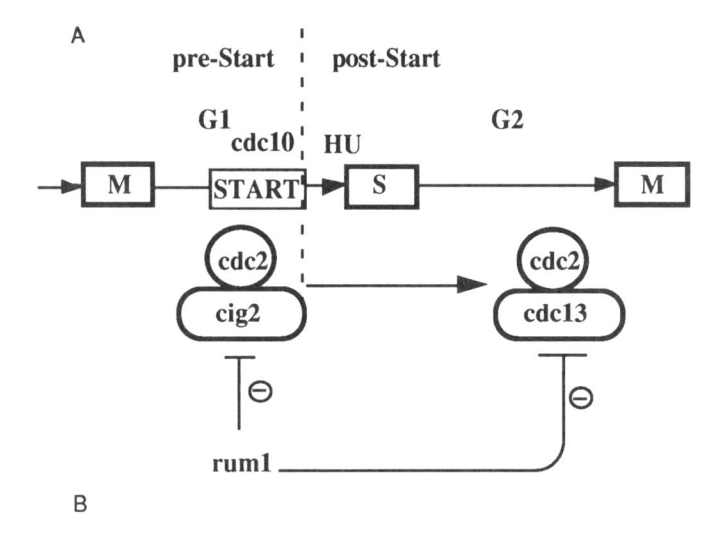

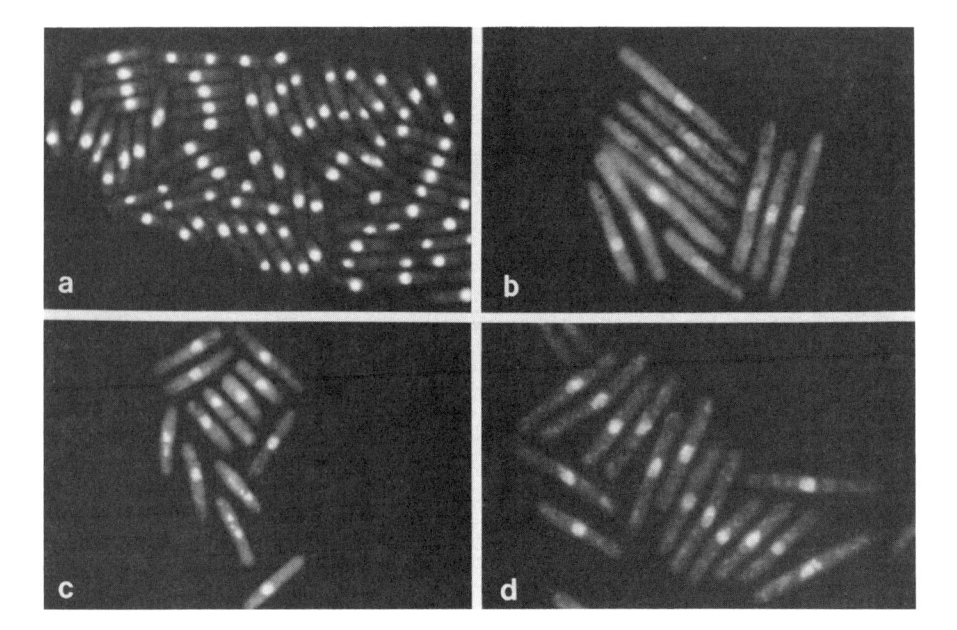

Figure 3. In fission yeast, passage through Start and entry into mitosis is brought about by the cdc2/cig2 and cdc2/cdc13 cyclin complexes, respectively.
A, p25^{rum1} inhibits cdc2/cig2 complex in G1 until the cell mass required for Start is attained. Once cells reach this cell mass the inhibitory effect of rum1 is lost allowing Start to take place and a new complex of cdc2 with cdc13 can be formed, the G2 complex. In a cdc10^{ts} mutant blocked at Start, rum1 activity prevents the formation of a cdc2/cdc13 complex and cells do not initiate mitosis and show a elongated phenotype. In the absence of rum1, a cdc10^{ts} mutant accumulates cdc2/cdc13 complex and mitosis is initiated even in the absence of DNA replication. Once cells pass Start, rum1 protein is degraded and a different sensing mechanism prevents mitosis if DNA replication is blocked with hidroxyurea, for example.
B, DAPI stained cells of cdc10-129 growing at 25°C (a), cdc10-129 after 4 hours at 36°C (b), cdc10-129 rum1Δ after 4 hours at 36°C (c) and rum1Δ after 4 hours in 10 mM hydroxyurea (d).

diploids, these cells are completely sterile and unable to initiate meiosis. The *rum1+* deletion strain undergoes premature S-phase immediately after mitosis and cannot delay G1 progression. In fast growing wild type cells, p25^{rum1} is not important because the cell size at the end of mitosis is larger than the minimal size required to pass Start. Deletion of the *rum1+* gene only has a lethal phenotype in conditions where the G1 phase is expanded because cells are too small after division and have to grow to attain the minimal size (43) (Figure 2B). In this situation, p25^{rum1} is essential to delay the progression through G1 in order to couple the cell cycle with cell growth.

A second role for p25^{rum1} is to prevent inappropriate entry into mitosis from G1. If cells lacking the *rum1*+ gene are blocked in G1 by means of a cdc10^{ts} mutant, they proceed to enter mitosis (43) (Figure 3). This indicates that p25^{rum1} is part of a checkpoint control required to restrain mitosis from G1. This G1/M checkpoint control only works in pre-Start cells since *rum1*+ deleted cells blocked at the initiation of DNA replication with hydroxyurea keep the mitotic inhibition (43) (Figure 3).

Both the cell cycle inhibitory effect in G1 and the G1/M checkpoint control are evident in cells overexpressing the *rum1*+ gene. When p25^{rum1} is overproduced in G2 cells, mitosis is inhibited. p25^{rum1} inhibits the cdc2/cdc13 complex and as a consequence these cells never enter mitosis, but can undergo DNA replication since the feedback mechanism that prevents S-phase from happening until cells have completed mitosis is inactivated. These cells are redirected from G2 to G1 and remain in G1 until rum1 inhibition is overcome by cell growth to pass Start. The result of this permanent inhibition of mitosis and transient inhibition of Start is rounds of S-phase without intervening mitosis, although the coordination between cell growth and cell cycle is mantained (43) (Figure 4). This re-replication phenotype is similar to the phenotype of *cdc13*+ deleted cells (12). It has been proposed that the cdc2/cdc13 complex defines a cell as being in G2 (12), so the effect of *rum1*+ overexpression and *cdc13*+ deletion is probably the same since *rum1*+ overexpression inhibits cdc2/cdc13 kinase activity *in vivo* (7, 43).

RUM1 ACTS AS A CELL CYCLE INHIBITOR IN THE G1/M CHECKPOINT CONTROL

Purified p25^{rum1} protein inhibits cdc2/cdc13 kinase activity *in vitro* in stoichiometric amounts (7, 44). The p25^{rum1} protein seems to be only present in pre-Start G1 cells and absent in S phase and G2 cells (44). Cells blocked in pre-Start G1 using a cdc10^{ts} mutant, accumulate p25^{rum1} protein bound to inactive cdc2/cdc13 complexes (44). This inhibition of the mitotic kinase activity correlates with a decrease in cdc13 protein but not in mRNA levels. It has been proposed that binding of p25^{rum1} to cdc2/cdc13 complexes could lead to p56^{cdc13} proteolysis. Interestingly, when a cdc10^{ts} rum1 deleted cell is shifted to the restrictive temperature, p56^{cdc13} is stable, the cdc2/cdc13 complexes are active and cells undergo mitosis and cell division (44) (Figure 3 B,c).

RUM1 IS A CDK INHIBITOR REGULATING G1 PROGRESSION

When a cdc2^{ts} mutant is placed at the restrictive temperature, cells become arrested in G1 and G2, and

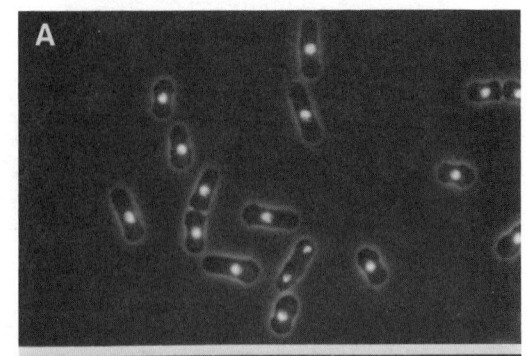

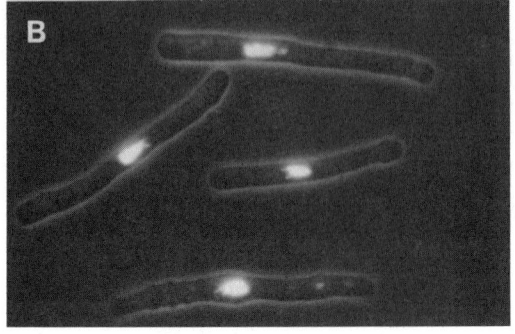

Figure 4. Overexpression of *rum1*+ in G2 cells inhibits the activity of cdc2/cdc13 kinase and causes multiple rounds of S phase in the absence of mitosis. *S.pombe* cells containing an integrated copy of nmt1-*rum1*+ stained with DAPI, promoter off (A) and promoter on (B).

then either remain arrested on continued incubation, or leak past one or both of the block points. This depends on the particular allele chosen, and probably reflects the amount of residual p34^{cdc2} activity at the restrictive temperature. Shifting a cdc2^{ts} deleted for *rum1*+ to the restrictive temperature results in fewer cells becoming arrested in G1, and a more transient arrest for those cells that do arrest before Start. In contrast, the ability of such cdc2^{ts} mutants to arrest in G2 remains unchanged (45). Therefore, in the absence of rum1, cdc2^{ts} mutants are less able to arrest in G1 at the restrictive temperature, implying that p34^{cdc2} has increased Start activity in these cells, consistent with p25^{rum1} being an inhibitor of the pre-Start G1 form of cdc2.

The B-type cyclin, cig1, has been shown to have protein and kinase activity patterns through the cell cycle similar to cdc13 (9, 46). When the *rum1*+ gene is overexpressed in a strain deleted for *cig1*+ we expect no mitotic cdc2 kinase activity. In these conditions the cdc2 immunoprecipitated kinase activity is transiently inhibited and raises as cells go through S-phase causing the re-replication phenotype (7, Martín-Castellanos and Moreno, unpublished results). Since re-replication occurs, there must be other cyclins promoting G1 progression.

Recently it has been shown that the B-type cyclin, cig2, regulates Start in fission yeast, in contrast to budding yeast where G1 progression is regulated by the CDC28/CLN kinase (7, 8, 9). The protein kinase activity of the cdc2/cig2 complex is found to peak at G1/S, and the cig2 protein levels also increases at this point of the cell cycle (7, 9). Using a genetic approach, it has been shown that deletion of $cig2^+$ reduces the cdc2 G1 activity *in vivo* (7). Shifting a $cdc2^{ts}$ strain deleted for $cig2^+$ to the restrictive temperature results in more cells becoming arrested in G1 than in a $cdc2^{ts}$ single mutant, and these cells remain arrested for a longer period. The same effect in G1 progression is seen when $cig2^+$ is deleted in *wee* mutants, where the G1 phase is apparent. This may also explain the enhanced ability of the $cig2^+$ deletion to arrest before Start upon nitrogen starvation (47, 48).

The re-replication phenotype of cells overexpressing $rum1^+$ is blocked by deletion of the $cig1^+$ and $cig2^+$ genes, and instead cells block in G1 and G2 (7). This shows that cig1, cig2 and cdc13 are the main cyclins involved in regulating cell cycle progression in fission yeast. In cells lacking cig1 and cig2, a single oscillation of cdc2/cdc13 kinase activity produces ordered rounds of S and mitosis (8). This suggests that the total level of cyclin/cdc2 kinase activity may be more important than the nature of the cyclin. A quantitative model for cell-cycle control has been proposed (8), in which a low level of cdc2 kinase activity is required for Start, and a high levels is needed for entry into mitosis. In a wild type cell in G1, cdc2/cdc13 kinase activity is very low and the accumulation cdc2/cig2 kinase activity promotes Start and the onset of S-phase. In the absence of cig2, S-phase is delayed until levels of cdc13 begin to accumulate. This shows a paralell with the situation in *S.cerevisiae*, where B-type cyclins normally acting late in the cell cycle can perform the function of early acting cyclins in their absence. In both budding and fission yeasts CDK/B-cyclin activity is needed for the onset of S-phase, and deletion of all B-type cyclin genes blocks cell cycle progression in G1 (8, 42).

The $puc1^+$ gene encodes a fission yeast cyclin with sequence homology to CLN3, that is capable of functionally complementing for lack of the CLN3 gene in budding yeast cells (49). It is unclear what is the role of puc1 in the cell cycle. Studies published so far have failed to show a role for puc1 at Start in the fission yeast cell cycle and, instead suggest a role as a negative regulator of the meiotic cell cycle (50).

RELATIONSHIP BETWEEN RUM1 AND CIG2

$p25^{rum1}$ and cig2 have opposite effects in G1 progression. Whereas rum1 is an inhibitor of cell-cycle progression at Start, cig2 is a G1 cyclin that promotes G1 progression. While $rum1^+$ deleted cells have higher G1 cdc2 activity *in vivo* (45), the opposite is seen when $cig2$ is deleted (7). Cells lacking $rum1^+$ are unable to block in pre-Start G1 in response to nutritional signals (43) while the $cig2^+$ deletion strain blocks better than wild type (47, 48). The inability to arrest in G1 of the $rum1^+$ deletion is suppressed by deleting $cig2^+$ in the same cells, partially rescuing the sterility of haploid cells and the meiosis defect of diploids (7).

All this evidence argues that rum1 and cig2 have an antagonistic effect in the regulation of G1. Indeed, purified $p25^{rum1}$ protein inhibits cdc2/cig2 kinase activity *in vitro* (7, 44) and rum1 overproduction *in vivo* also inhibits the cdc2/cig2 complex (7). A simple model for rum1 and cig2 function in G1 could be that $p25^{rum1}$ restrains cdc2/cig2 activity until cells have reached the critical mass needed to initiate a new cell cycle. At the end of G1, once cells have grown big enough, this inhibition is overcome. This could be achieved in different ways; for example, cig2 protein levels may raise steadily with cell growth and once the level passes a threshold marked by the rum1 inhibitor, the cell cycle is initiated. In this case, growth would correlate with cyclin accumulation and binding of the inhibitor should be stoichiometric. The mammalian CDK inhibitors $p27^{KIP1}$ and $p21^{CIP1/WAF1}$, have been proposed to act in this way (30, 51). It is also possible that the CDK/cyclin complex may directly trigger inactivation of the inhibitor. Once cells reach the required size to enter S-phase, cdc2/cig2 inhibition would be overcome by phosphorylation of the inhibitor which in turn would promote its degradation. Several proteins have been shown to be phosphorylated prior to degradation (IkB, GCN4, etc) and some CDK inhibitors, such as $p27^{KIP1}$ and $p40^{SIC1}$, have been found to be degradated by a ubiquitin dependent proteolytic pathway (42,52).

The cdc2/cig1 complex may have an active role in $p25^{rum1}$ phosphorylation/degradation. When rum1 is overexpressed in a cig1 deleted strain, the re-replication process is slowed down. It could be possible that rum1 is a better inhibitor in this genetic background and that the cdc2/cig1 complex could have an active role in $p25^{rum1}$ phosphorylation/degradation (Figure 5). cig1 kinase activity is the only one of the B-type cyclin kinase activities not inhibited *in vitro* by purified rum1 protein (7, 44). In normal cells, exit from mitosis is triggered by the ubiquitin-dependent cyclin destruction pathway (53, 54, 55). Only the cyclin fraction bound to cdc2 is degradated (56), the remaining cdc13 could bind cdc2 and promote mitosis again, so it must be inhibited by rum1 acting as part of the pre-Start G1/M checkpoint control. cig2/cdc2

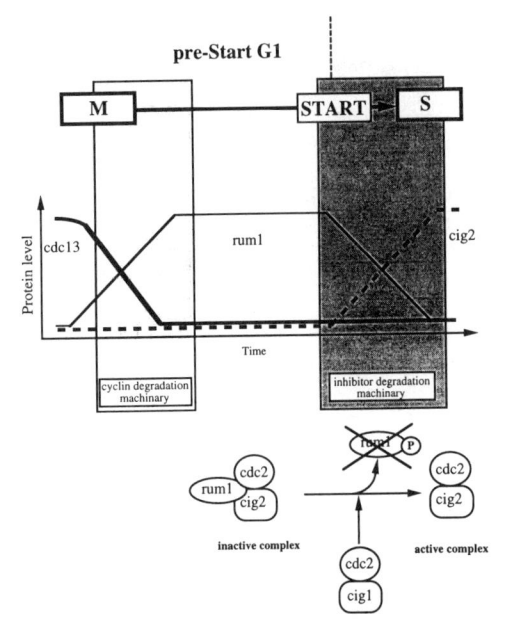

Figure 5. Tentative model for rum1 function as a negative regulator of the G1 phase of the cell cycle. In the late stages of mitosis the activity of cdc2/cdc13 and cdc2/cig1 complexes fall to low level, due to activation of the anaphase degradation machinary. At this time of the cell cycle, the rum1 inhibitor is made (or it becomes stable) preventing the formation of active cdc2/cig2 and cdc2/cdc13 complexes during G1. As cells progress through G1, cig1 cyclin accumulates forming active complexes with cdc2, that is insensitive to rum1 inhibition. At the end of G1, this complex could phosphorylate rum1 and promote its degradation, allowing the activity of the cig2.

complexes could be blocked by rum1 until the cell accumulates enough cig1/cdc2 activity to overcome this inhibition and so pass Start. Active cig1/cdc2 complexes could phosphorylate p25^{rum1} at the end of G1 and thus target it to be degradated by the ubiquitin dependent proteolytic pathway (Figure 5).

ACKNOWLEDGEMENTS

We are very grateful to Karim Labib for comments and corrections to the manuscript. Work in our lab is supported by grants from the DGICYT, the Human Frontier Science Program and the Ramón Areces Foundation.

REFERENCES

1. Nurse, P. (1975). *Nature* **256**, 547-551.
2. Nurse, P. and Thuriaux, P. (1977). *Exp. Cell Res.* **107**, 365-375.
3. Nasmyth, K., Nurse, P. and Fraser, R. S. (1979). *J. Cell Sci.* **39**, 215-233.
4. Nurse, P., Thuriaux, P. and Nasmyth, K. (1976). *Mol. Gen. Genet.* **146**, 167-178.
5. Hartwell, L. H., Culotti, J., Pringle, J. and Reid, B. J. (1974). *Science* **183**, 46-51.
6. Nurse, P. and Bissett, Y.(1981). *Nature* **292**, 558-560.
7. Martín-Castellanos, C., Labib, K. and Moreno, S. (1996). *EMBO J*. **15**, 839-849.
8. Fisher, D. and Nurse, P. (1996). *EMBO J.* **15**, 850-860.
9. Monderset, O., McGowan, C. H. and Russell, D. (1996). *Mol. Cell. Biol.* **16**, 1527-1533.
10. Reymond, A., Marks, J. and Simanis, V. (1993). *EMBO J.* **12**, 4325-4334.
11. Leatherwood, J., López-Girona, A. and Russell, P. (1996) *Nature* **379**, 360-363.
12. Hayles, J., Fisher, D., Woollard, A. and Nurse, P. (1994). *Cell* **78**, 813-822.
13. Hunter, T. and Pines, J. (1994). *Cell* **79**, 573-582.
14. Sherr, C. J. and Roberts, J. M. (1995).*Genes Dev.* **9**, 1149-1163.
15. Massagué, J. and Polyak, K. (1995). *Curr. Opin. Genet. Dev.* **5**, 91-96.
16. El-Deiry, W. S., Tokino, T., Velculescu, V. E., Levy, D. B., Parsons, R., Trent, J. M., Lin, D., Mercer, W. E., Kinzler, K. W. and Volgelstein, B.(1993). *Cell* **75**, 817-825.
17. Xiong, Y., Hannon, G. J., Zhang, H., Casso, D., Kobayashi, R. and Beach, D. (1993). *Nature* **366**, 701-704.
18. Harper, J. W., Adami, G. R., Wei, N., Keyomarsi, K. and Elledge, S. J. (1993). *Cell* **75**, 805-816.
19. Gu, Y., Turk, C. W. and Morgan, D. O. (1993). *Nature* **366**, 707-710.
20. Polyak, K., Kato, J., Solomon, M. J., Sherr, C. J., Massagué, J., Roberts, J. M. and Koff, A. (1994). *Genes Dev.* **8**, 9-22.
21. Polyak, K., Lee, M-H., Erdjument-Bromage, H., Koff, A., Roberts, J. M., Tempst, A. and Massagué, J. (1994). *Cell* **78**, 59-66.
22. Toyoshima, H. and Hunter, T. (1994). *Cell* **78**, 67-74.
23. Lee, M-H., Reynisdóttir, I. and Massagué, J. (1995). *Genes Dev.* **9**, 639-649.
24. Matsuoka, S., Edwards, M. C., Bai, C., Parker, S., Zhang, P., Baldini, A., Harper, J. W. and Elledge, S. J. (1995). *Genes Dev.* **9**, 650-662.
25. Serrano, M., Hannon, G. J. and Beach, D. (1993). *Nature* **366**, 704-707.
26. Hannon, G. J. and Beach, D. (1994). *Nature* **371**, 257-261.
27. Hirai, H., Roussel, M. F., Kato, J-Y., Ashmun, R. A. and Sherr, C. (1995). *Mol. Cell Biol.* **15**, 2672-2681.
28. Chan, F. K. M., Zhang, J., Cheng, L., Shapiro, D. N. and Winoto, A. (1995). *Mol. Cell Biol.* **15**, 2682-2688.
29. Guan, K-L., Jenkins, C. W., Li, Y., Nichols, M. A., Wu, X., O'keefe, C. L., Matera, A. G. and Xiong, Y. (1995). *Genes Dev.* **8**, 2939-2952.
30. Zhang, H., Hannon, G. J. and Beach, D. (1994). *Genes Dev.* **8**, 1750-1758.

31. Nobori, T., Miura, K., Wu, D., Lois, A., Takabayashi, K. and Carson, D. A. (1994). *Nature* **368**, 753-756.
32. Gruis, N. A., van der Velden, P. A., Sandkuijl, L. A., Prins, D. E., Weaver-Feldhaus, J., Kamb, A., Bergman, W. and Frants, R. R. (1995). *Nature Genetics* **10**, 351.
33. Cairns, P., Polascik, T. J., Eby, Y., Tokino, K., Califano, J., Merlo, A., Mao, L., Herath, J., Jenkins, R., Westra, W., Rutter, J. L., Buckler, A., Gabrielson, E., Tockman, M., Cho, K. R., Hedrick, L., Bova, G. S., Isaacs, W., Koch, W., Schwab, D. and Sidransky, D.(1995). *Nature Genetics* **11**, 210.
34. Ranade, K., Hussussina, C. J., Sikorski, R. S., Varmus, H. E., Goldstein, A. M., Tucker, M. A., Serrano, M., Hannon, G. J., Beach, D. and Dracopoli, N. C. (1995). *Nature Genetics* **10**, 114.
35. Chang, F. and Herskowitz, I. (1990). *Cell* **63**, 999-1001.
36. Chang, F., (1993). *Curr. Biol.* **3**, 693-695.
37. Peter, M., Gartner, A., Horecka, J., Ammerer, G. and Herskowitz, I. (1993). *Cell* **73**, 747-760.
38. Peter, M. and Herskowitz, I. (1994) *Cell* **79**, 181-184.
39. Mendenhall, M. D. (1993). *Science* **259**, 216-219.
40. Nugroho, T. T. and Mendenhall, M. D. (1994). *Mol. Cell Biol.* **14**, 3320-3328.
41. Donovan, J. D., Toyn, J. H., Johnson, A. L. and Johnston, L. H. (1994). *Genes Dev.* **8**, 1640-1653.
42. Schwob, E., Bohm, T., Mendenhall, M., and Nasmyth, K. (1994). *Cell* **79**, 233-244.
43. Moreno, S. and Nurse, P. (1994). *Nature* **367**, 236-242.
44. Correa-Bordes, J. and Nurse, P. (1995). *Cell* **83**, 1-20.
45. Labib, K., Moreno, S. and Nurse, P. (1995). *J. Cell Sci.* **108**, 3285-3294.
46. Basi, G. and Draetta, G. (1995). *Mol. Cell Biol.* **15**, 2028-2036.
47. Obara-Ishihara, T. and Okayama, H. (1994). *EMBO J.* **13**, 1863-1872.
48. Connolly, T. and Beach, D. (1994). *Mol. Cell Biol.* **14**, 768-776.
49. Forsburg, S.L. and Nurse, P. (1991). *Nature* **351**, 245-248.
50. Forsburg, S.L. and Nurse, P. (1994). *J. Cell Sci.* **107**, 601-613.
51. Sherr, C. J. (1995). *TIBS* **20**, 187-190.
52. Pagano, M., Tam, S. W., Theodoras, A. M., Beer-Romero, P., Del Sal, G., Chau, V., Yew, P. R., Draetta, G. F. and Rolfe, M. (1995). *Science* **269**, 682-685.
53. Glotzer, M., Murray, A. W. and Kirschner, M. W. (1991). *Nature* **349**, 132-138.
54. Murray, A. W. (1995). *Cell* **81**, 149-152.
55. Glotzer, M. (1995). *Curr. Biol.* **5**, 970-972.
56. Hayles, J. and Nurse, P. (1995). *EMBO J.* **14**, 2760-2771.

chapter 4
Progression through G1 and S phases of adult rat hepatocytes

Pascal Loyer[1], Guenadi Ilyin, Sandrine Cariou, Denise Glaise, Anne Corlu and Christiane Guguen-Guillouzo.

Unite de Recherches Hepatologiques INSERM U49,
Hopital Pontchaillou, Rennes, France
[1]To whom correspondence should be addressed

Regenerating liver, hepatocyte primary cultures and differentiated hepatoma cell lines are widely used to study the proliferation/differentiation/apoptosis equilibrium in liver. In hepatocytes, priming factors (TNFα, IL6) target G0/G1 transition while growth factors (HGF, EGF, TGFα) control a mid-late G1 restriction point. A characteristic pattern of cdk/cyclin expression is observed in hepatocytes, presumably related to their ability to proliferate a limited number of times and to undergo a reversible differentiation. Interestingly, cell-cell interactions between hepatocytes and liver biliary cells in co-cultures, result in a cell cycle arrest in mid G1 of hepatocytes which are insensitive to mitogens. Apoptosis exists in hepatocytes but is still poorly documented. However, hepatoma cell lines stimulated by TGFβ undergo cell death in a p53-independent pathway. In conclusion, the interplay of growth and apoptosis regulators and cell-cell interactions control the proliferation/differentiation/apoptosis balance which is a specific feature of hepatocytes.

INTRODUCTION

In normal adult liver, hepatocytes are highly differentiated and arrested in quiescence or G0 phase. However, hepatocytes have a remarkable capacity to re-enter the cell cycle and to proliferate in response to a reduction of the liver mass in order to restore the global hepatic function. This phenomenon is evidenced in pathologies such as hepatitis, and can be artificially induced in animals by partial hepatectomy (PH) (1) or chemical intoxication (2).

From reports on liver regeneration in animal models as well as clinical observations, it was concluded that there is a close relationship between the global hepatic function (or tissue mass) and body size. This is strikingly illustrated in human liver transplantation. When a patient is transplanted with a smaller liver than the original one, the transplant grows until it reaches the appropriate mass and function for the new host (3). In contrast, when a large liver, compared to the body size of the receiving patient, is transplanted, the transplant undergoes a partial atrophy by apoptosis in order to adjust the global hepatic function to the host size (4).

Therefore, the liver can be considered as an organ which regulates its mass by positive and negative feedback-loops : diminished hepatic function triggers positive factors which lead to proliferation while excess hepatic function triggers inhibitory factors and results in cell cycle arrest and/or apoptosis. Once the ratio adjustment of hepatic activity to the body mass is optimised, then these external factors are turned off. It should be noted that these general considerations represent a working hypothesis rather than a clear and well-defined model. Indeed, it remains unclear, for some of these positive and negative factors if they are produced by the liver itself or other organ(s) sensitive to variations of hepatic activity and how their expression is regulated. However, we know that hepatic mitogens are humoral factors. It has been demonstrated that the sera of animals regenerating their liver after partial hepatectomy, contain factors which can stimulate hepatocyte proliferation in liver of non-hepatectomised animals (5-7).

Rat liver regeneration after a two-third hepatectomy has been the most extensively studied model of hepatocyte proliferation. In this model, it has been clearly demonstrated that differentiated adult hepatocytes are involved in the regenerating process occurring in two major waves of proliferation (8,9). During the first wave, hepatocytes synchronously enter and progress through G1, undergo DNA synthesis and divide with a complete cell cycle lasting 30 to 32 hours. The G0/G1 transition takes place immediately after PH and is followed by a sequential induction of at least 70 genes classified as immediate-early, delayed-early and liver-specific genes (10-15). This is followed by the second wave which involves all the different cell types with asynchronous cell cycles. This compensatory hyperplasia leads to total restoration of liver function within a week after hepatectomy.

The hepatocyte recruitment implies that 1) they are the target of a specific signal(s) leading to the G0/G1 transition, 2) they undergo a partial "retrodifferentiation", a property of only few cell types. This leads to postulate that specific regulations might exist between proliferation and differentiation.

Besides liver regeneration *in vivo* after partial hepatectomy or toxic injury, two ex-vivo cell systems are widely used : primary cultures of hepatocytes and differentiated hepatoma cell lines. In this chapter, using data obtained in these two ex-vivo cell systems, we will mainly focus on the G0/G1 transition, the progression in G1, the G1/S transition and the role of growth factors in these process. We will also discuss some aspects of the balance between proliferation and differentiation as well as cell death.

GROWTH FACTORS AND GROWTH INHIBITORS OF HEPATOCYTES

Hepatocyte growth factors

To date, several proteins have been identified as hepatocyte mitogenic factors. Among them, TGFα (16), EGF (7,17), HGF (18,19), PDGF, and acidic FGF (20) are the most potent for inducing DNA replication of normal hepatocytes in primary culture. Plasma levels of HGF rapidly increase after PH implicating it in stimulation of hepatocyte proliferation (21) and removal of salivary glands in rats, the major organ of EGF production abolishes liver regeneration (7). Therefore, it remains unclear whether only one of these mitogens is important during liver regeneration or if several factors act synergically. The efficiency of these mitogens can be enhanced by a large number of co-mitogens, including Augmenter of Liver Regeneration (ALR) gene product (22), insulin, glucagon, adrenaline, noradrenaline, thyroid hormones, various vitamins and intermediate metabolites (23). Extracellular matrix compounds such as fibronectin, collagens types I and IV, also favour the effect of EGF on hepatocyte DNA replication (24, 25).

It has also been proposed that the G0/G1 transition, progression through G1 and G1/S transition could be controlled by distinct factors (15, 26, 27). Cytokines including TNFα and IL6, would induce G0/G1 transition but with moderate effects, if any, on the DNA synthesis and can be considered as "priming factors". Then, hepatocytes progressing in early G1, also called "primed hepatocytes",

could be stimulated by growth factors in association with comitogens, to induce progression in late G1 and S phases. This hypothesis is based on the observation that in different *in vivo* and *in vitro* model systems, induction of immediate-early genes is not always followed by proliferation of hepatocytes. Indeed, the induction of immediate-early genes observed after hepatectomy and characterising the G0/G1 transition, also occurs, in a lesser extend in sham-operated rats subjected to laparatomy without tissue removal (28,29), during experimentally induced acute phase (29) and after a protein starvation followed by an amino-acid overload (30,31). In these *in vivo* model systems, no significant proliferation of hepatocytes is observed after induction of the immediate-early genes. In addition, the isolation of hepatocytes from the liver by perfusion of enzymatic or chemical solution leads to a spontaneous G0/G1 transition as indicated by overexpression of the immediate early genes (32, 33) but DNA replication does not take place unless growth factors are added into the medium (23).

More recently, TNFα has been proposed to control early steps of G1 phase in hepatocytes. TNFα is promptly produced by liver cells after partial hepatectomy (34) or administration of hepatotoxic reagents (34, 35). On the other hand, TNFα has been shown to activate the PHF/NF-kB (postHP factor/nuclear factor kappaB) complex *in vivo* after hepatectomy (36). PHF/NF-kB is a transcription factor complex inactive in quiescent hepatocytes which is rapidly and transiently activated after hepatectomy (37-40). Its activation is independent of protein synthesis and consists in phosphorylation of the subunit IkB-a resulting in release and nuclear translocation of NF-kB which can regulate transcription of downstream target genes such as early immediate genes (15). These data strongly suggest the involvement of TNFα in the G0/G1 transition of hepatocytes. This hypothesis is reinforced by the observation that injection of anti-TNFα antibodies abolishes hepatocyte proliferation after PH (41) and prevents immediate early gene induction (42).

Another transcription factor stat3 (43), is also rapidly activated and until mid-G1 in remnant liver after hepatectomy (40) and in normal liver of rats injected with EGF or IL6 (44). The predicted target genes of stat3 include the jun family, β actin and c-myc. It has been proposed that NF-kB and stat3 would be activated by a rapid release of cytokines after partial hepatectomy or liver injury. These transcription factors would control the transcriptional activation of immediate and delayed early genes allowing the G0/G1 transition and progression in early G1 (15).

Then, in G1 phase, stimulation by growth factors, i.e. HGF, TGFα and EGF, would target late

G1 genes including cell cycle regulators such as cdks and cyclins. It is interesting to note that liver regeneration, a precisely regulated proliferation *in vivo*, is controlled by a group of growth factors presenting a broad spectrum of targeted cells and biological effects. However, it seems that hepatocytes respond to growth factors only after stimulation by priming factors and G0/G1 transition (45) rising the question of the expression and activation of growth factors receptors and the molecules involved in the signal transduction downstream these receptors, in quiescent hepatocytes and throughout the G1 phase.

Hepatocyte growth inhibitors

Many soluble factors are able to partially or completely inhibit hepatocyte proliferation by antagonising the effects of hepatocyte mitogens. TGFß is a very potent inhibitor of hepatocyte proliferation *in vitro* (46, 47) as well as *in vivo* after partial hepatectomy (48). In addition, TGFß mRNA level is up-regulated during liver regeneration suggesting that it may act as an inhibitory paracrine factor to prevent uncontrolled hepatocyte proliferation, or it could be involved in terminating the regeneration process (49). In liver, TGFβ is produced by nonparenchymal cells suggesting a cooperation between the different cell types in the regulation of hepatocyte proliferation (49).

Interleukin 1ß (50) and prothrombin (51) are also able to significantly reduce hepatocyte growth induced by EGF. Numerous other molecules such as corticoids (23) or dimethyl-sulfoxide (52) both largely used to maintain expression of hepatic specific functions, can inhibit hepatocyte growth activity.

Proliferation of hepatocytes is also regulated by cell density and cell-cell interactions. In primary culture and differentiated hepatoma cell lines, there is an inverse relationship between the cell density and the DNA replication rate (53, 54) involving unidentified plasma membrane protein interactions (54). We have previously shown that loss of cell-cell interactions during liver tissue disruption and hepatocyte isolation leads to the G0/G1 transition of hepatocytes (33) which become sensitive to growth factors after progression in early G1 (45). In addition, when heterotypic cell-cell interactions are re-established in the model of coculture associating hepatocytes and rat liver epithelial cells derived from primitive biliary cells (56), hepatocytes remain differentiated and arrested in G1 (57). These *in vitro* data strongly suggest an important and active role for cell-cell interactions in the control of hepatocyte proliferation. This hypothesis is reinforced by the fact that the expression of connexins, sub-units of gap junctions, and adhesion molecules, is rapidly down-regulated and their cell surface distribution is modified after PH (58-60).

G1 PHASE PROGRESSION OF ADULT RAT HEPATOCYTES AND CELL CYCLE REGULATORS

Mitogen dependent restriction point in G1 phase of rat hepatocytes

In primary culture, in the absence of growth factors, hepatocytes enter G1 phase but are unable to progress through the cell cycle without mitogenic stimulation, suggesting the existence of a mitogenic dependent restriction point in G1 phase. In order to define such a restriction point and to understand the progression of normal hepatocytes through G1 and S phases and its control by growth factors, we use normal rat hepatocytes in primary culture (45).

In this *in vitro* cell system, hepatocytes transit from G0 to G1 during hepatocyte isolation and progress through mid-late G1 in the absence of growth factors in the medium. This is consistent with a sequential overexpression of immediate-early and delayed-early proto-oncogenes such as c-fos and c-jun during cell isolation followed by c-myc, jun B and jun D in the first hours of culture followed by Ki-ras and p53. The progression to the G1/S boundary is strictly dependent on growth factor and consequently, in the absence of a mitogenic signal, cells are blocked in mid-late G1 at a mitogen-dependent restriction point (R-point).

This R-point has been precisely localised in G1 phase by starting EGF stimulations at various times after plating and monitoring DNA synthesis (Figure 1). In our conditions of culture i.e. in absence of any extracellular matrix components as cell attachment support, DNA synthesis occurs at 58 to 60 hours post-plating when stimulation is performed within 42 hours after plating. Later additions of mitogen

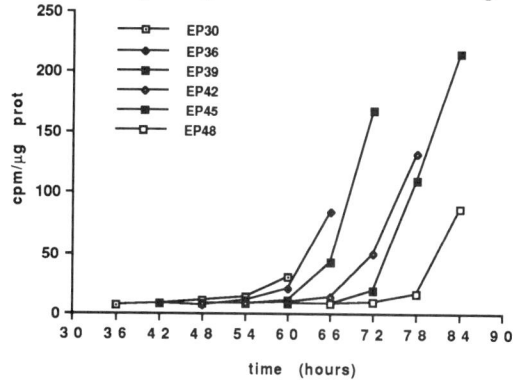

Figure 1 : Efects on hepatocyte DNA synthesis of 12-hr EGF stimulation initiated serially in mid and late G1. EGF (EP) was added to primary cultures at different times : 30, 36, 39, 42, 45 and 48 hours after seeding. For each addition, ^{3}H-thymidine incorporation was performed over 6-hour periods and monitored for 30 hours.

delay the initiation of DNA replication indicating that the R-point in mid-late G1 occurs approximately 42 hours after plating (45). It also appears that the progression in the first-third of G1 is required to make the hepatocytes sensitive to the mitogenic signal. Indeed, EGF stimulation during the first 24 hours of culture induces a very low DNA synthesis while the maximal response is obtained when stimulation is performed between 24 and 48 hours.

It is important to note that the kinetic of hepatocyte progression through G1 depends upon the conditions of culture especially the substrate used for plating the hepatocytes. When hepatocytes are plated on extracellular matrix components that favour cell attachment and spreading such as fibronectin or collagen I or IV, the G1 phase is shortened (32, 61) compared to the G1 phase of cells plated on plastic (45). In these conditions, the sequential activation of proto-oncogenes in G1 is not affected and it can be assumed that the R-point is still located at two-third of the G1 phase.

Hence, our observation that *in vitro*, G0/G1 transition and progression through early G1 occur in the absence of growth factors while the G1/S transition requires stimulation by mitogens, further supports the hypothesis of priming factors and mitogens controlling distinct steps of hepatocyte cell cycle progression (15, 26, 27).

G1-associated cell cycle regulators

Over the past decade, a large number of proteins involved in the cell cycle have been identified. Of the proteins characterised to date, the cyclin dependent serine/threonine kinases (cdks) and their cyclin partners play a crucial role in controlling the progression through a series of checkpoints during the cell cycle (62, 63). Indeed, different cdk/cyclin complexes are sequentially formed during the cell cycle and after activation, their catalytic subunits act by phosphorylating various downstream target proteins which allow progression to the next step of the cycle.

In several mammalian cell types, it has been demonstrated that the D type cyclins in association with either cdk2, cdk4 or cdk6, play a crucial role during G1 progression (64-70) while the cyclin E/cdk2 complex is activated at the end of G1 and is considered to be a limiting step at the G1/S boundary (71, 72). As the cells enter S phase the cyclin A/cdk2 complex is required for DNA replication (73, 74) until the cyclin B/cdk1(cdc2) complex is activated and allows the G2/M transition (75).

In an attempt to understand the mechanisms at play during liver regeneration, we investigated the expression of several cdks and cyclins in both hepatocyte primary cultures and in regenerating liver after PH.

Cdk1 is not expressed in normal or in regenerating liver during G1 phase but it is expressed during S, G2 and M phases with peaks of kinase activity in both S and M phases (76). In contrast, cdk2 is expressed throughout the cell cycle with increased expression at the end of G1 consistent with its activation in S phase. The transient expression of cyclins A and B in S and M phases correlates well with the cdk1 and cdk2 kinase activation. Interestingly, cyclins D1 and E are expressed at relatively constant levels in normal and regenerating liver.

In primary culture of rat hepatocytes, the sequence of cdk1, cdk2 and cyclin A expression is similar to the pattern during liver regeneration (Figure 2). Cyclin E mRNA expression is very low in hepatocytes freshly isolated from liver or during the G1 progression, but is greatly induced in late G1 and S phases and surprisingly cyclin E mRNA level does not correlate with its protein levels. Indeed, cyclin E protein is expressed at similar levels in both unstimulated and stimulated cultures but there is a shift in electrophoretic mobility of the protein in late G1 after mitogenic stimulation, suggesting a modification of its phosphorylation when cells are stimulated. These results are in agreement with findings on cyclin E expression during liver regeneration (45, 77). Recently, it has been shown in BALB/c 3T3 little change in the level of cyclin E during the cell cycle until mitosis at which the expression decreases sharply before to reaccumulate after mitosis is complete (78). It was also observed several migrating forms of cyclin E with shift in electrophoretic mobility through the cell cycle confirming our observations.

Cyclin D3 mRNAs are detectable in quiescent hepatocytes and throughout G1 with an increase in S, G2 and M phases but accumulate in absence of growth factors when the cells are arrested at the R-point. Expression of cyclin D2 mRNA is similar to cyclin D1 expression i.e. low in absence of growth factors and induced by mitogen stimulation while cdk4 messenger is expressed early in G1, regardless of growth factor stimulation and does not vary significantly during the cell cycle in stimulated cells. Interestingly, in unstimulated hepatocytes arrested at the R-point, cdk4 transcripts strongly accumulate in parallel with cyclin D3 expression.

Cyclin D1 mRNA and protein expression is very low in hepatocytes maintained in absence of growth factors. In contrast, its expression is greatly increased in mitogen-stimulated cultures in mid-late G1, showing that its induction is mitogen dependent and suggesting that cyclin D1 expression is associated with the ability of the cells to

EXPRESSION OF CELL CYCLE MARKERS IN HEPATOCYTES IN VITRO

	− MITOGEN							+ MITOGEN						
	T0	18h	24h	36h	48h	60h	66h	72h	30h	36h	42h	48h	54h	60h	66h	72h
cdc2	—	—	—	—	—	—	—	—	—	—	—	—	▬	▬	■	■
cdk4	—	▬	▬	■	■	■	■	■	▬	▬	▬	▬	▬	▬	▬	—
Cyclin A	—	—	—	—	—	—	—	—	—	—	—	—	▬	▬	■	■
Cyclin E	—	—	—	—	—	—	—	—	—	—	▬	▬	▬	▬	▬	▬
Cyclin D1	—	—	—	—	—	—	—	—	—	—	▬	■	▬	▬	▬	▬
Cyclin D3	—	▬	▬	■	■	■	■	■	—	—	—	—	—	—	—	—

G1 — Arrest (Restriction point) | G1 — S-G2 (Restriction point)

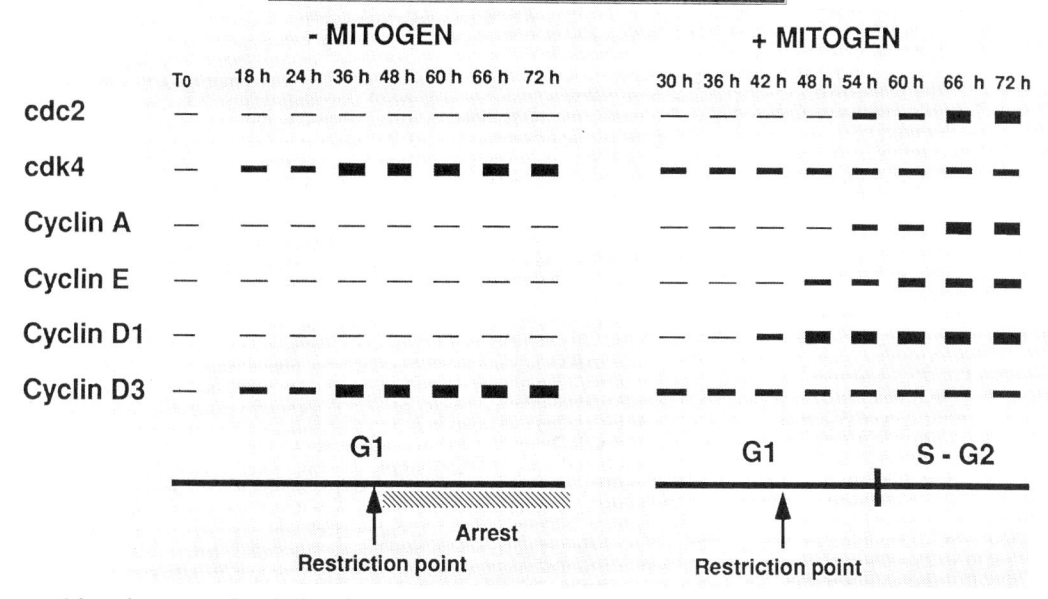

Figure 2: Schematic representation of cdk/cyclin expressions in primary culture of hepatocytes.
The cdks and cyclins represented have been studied in unstimulated and EGF-stimulated cultures during 72 hours. EGF was added into the medium 24 hours after plating and remained present during the entire experiment. At the bottom of the figure, the restriction point is positioned to superimpose cdk/cyclin expression and cell cycle progression.

override the R-point and progress to late G1. This hypothesis is further supported by the fact that when the mitogen stimulation is delayed, the induction of cyclin D1 is also postponed.

In conclusion, the control of hepatocyte cell cycle is characterised by several important features (Figure 3). First, two distinct check-points control G1 progression, i.e. G0/G1 and G1/S transitions respectively targeted by priming factors and growth factors. Secondly, a sequence of cdk and cyclin expressions different to patterns observed in other cell types. Among the major differences, we have to emphasise: 1) the cdk4 induction which occurs in early G1, before the R-point and regardless of growth factor stimulation; 2) the non-coordinated induction of cyclins D1 and D3. Cyclin D3 induction takes place early in G1 in parallel with cdk4 while cyclin D1 is expressed only after mid-late G1 and its expression is dependent of mitogen stimulation. These data suggest that cyclins D1 and D3 could play a role at distinct steps of the G1 progression in hepatocytes; 3) the constant expression of cyclin E throughout the cell cycle.

It can be postulated that these essential features are involved in the precise regulation of hepatocyte proliferation leading to only one or two rounds of division in regenerating liver and the property of mature hepatocytes to undergo a reversible differentiation.

PROLIFERATION AND DIFFERENTIATION EQUILIBRIUM IN HEPATOCYTES

In many cell types, proliferation and differentiation are two exclusive mechanisms. However, during liver regeneration, hepatocytes maintain the expression of the liver-specific genes (79) but undergo changes in the expression of energy metabolism enzymes (80) along with a partial "retro-differentiation" characterised by reexpression of foetal liver markers such as pyruvate kinase M2 (81) and α-foetoprotein (82). Till now, only a few data, if any, have been reported about the mechanisms which govern this balance characterising liver cells.

This reversible differentiation is probably associated with a switch in the expression of transcription factors such as different forms of C/EBP (83-85) and various helix-loop-helix (HLH) proteins (86). It has been shown that after partial hepatectomy at the G0/G1 transition, C/EBPα is down-regulated while C/EBPβ is up-regulated and considered as a immediate early G1 marker. Id (Inhibitor of differentiation) gene products which lack the basic DNA-binding domain, are members of the HLH family and are able to inhibit the binding of other HLH proteins to DNA by forming biologically inactive hetero-dimers (87, 88). Id-1 has been shown to specifically inhibit differentiation of muscle (89), myeloid (90) and

```
                              ┌─────┐
                              │  O  │  G0
                              └─────┘
                                 │
┌────────────────────────────────┼──────────────────────────────┐
│ activation of pre-existing transcriptional factors  ←──── priming factors
│ PHF/NF-kB, Stat3                                         TNFα, IL6, others?
│                                 │
│                                 │        G0/G1 transition
│ induction of immediate-early genes
│ c-jun, c-fos, LRF-1             ▼
│                              ┌─────┐
│                              │  O  │  early G1
│ induction of delayed-early genes
│ c-myc, junD, junB
│
│ constant expression of cyclin E and cdk2
│ induction of cdk4, cyclinD3
│                                 ▼
│                              ┌─────┐
│                              │  O  │
│ priming factor dependent
│ progression                                mid G1
└────────────────────────────────┼──────────────────────────────┘
┌────────────────────────────────┼──────────────────────────────┐
│ induction of cyclins D1 and D2, Id-1 and Id-2  ←──── point de restriction (R-point)
│                                                     growth factor dependence
│                                                     HGF, EGF, TGFα
│                                 ▼
│                              ┌─────┐
│                              │  O  │  late G1
│
│ growth factor dependent
│ progression                                G1/S transition
└────────────────────────────────┼──────────────────────────────┘
                                 ▼
  induction of cdc2, cyclin A  ┌─────┐
                               │  O  │  S
                               └─────┘
```

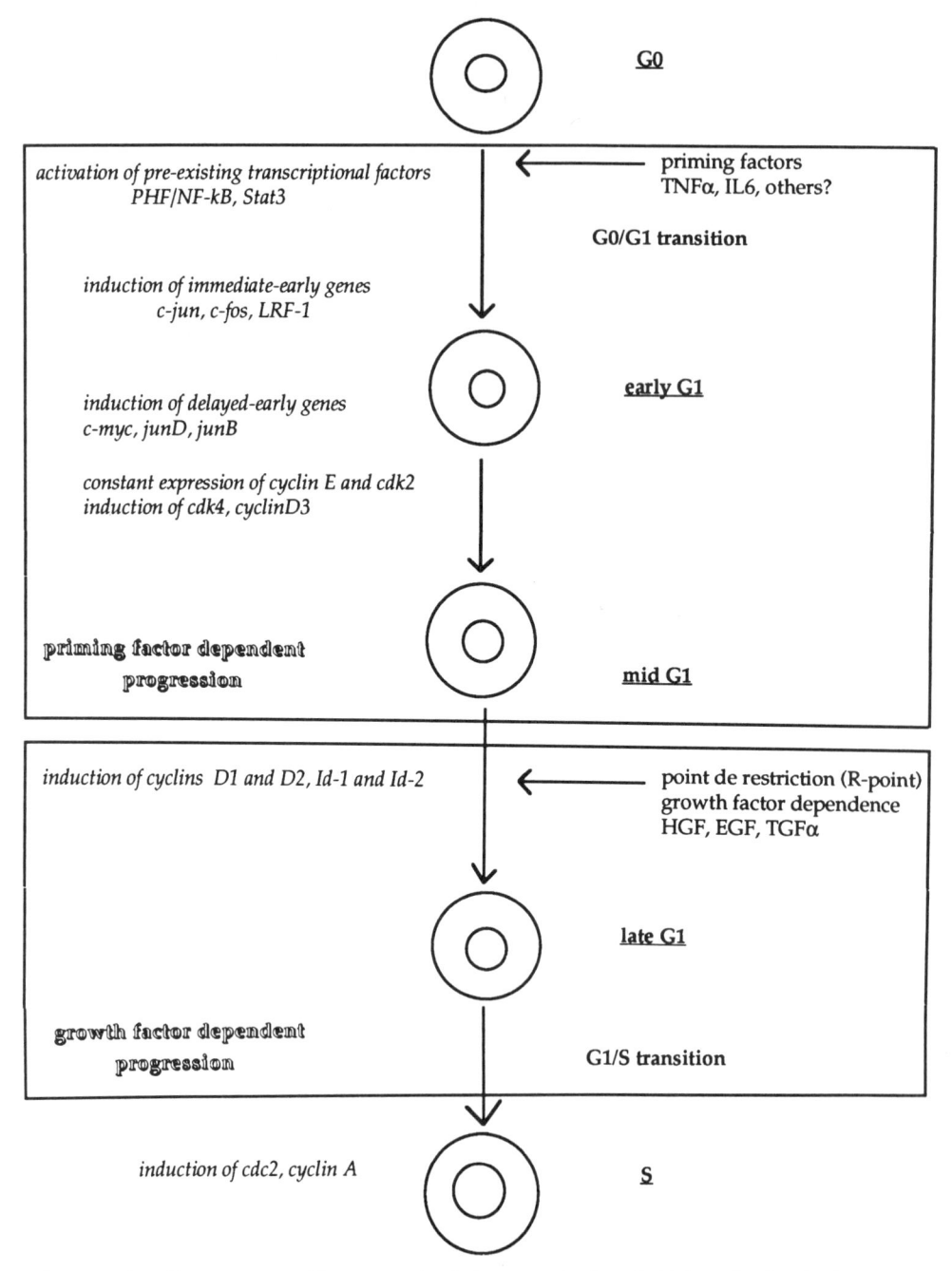

Figure 3 : Representation of important sequential events occuring during progression through G1 phase of rat hepatocytes.
We indicated on this schematic representation, two rectangles indicating the priming factors which would target the G0/G1 transition and the growth factors controlling the G1/S transition. We also represented a sequential activation of genes associated with the progression through G1.

mammary (91) cells. In addition, Id-2 binds to the retinoblastoma protein and its overexpression in the human osteosarcoma cell line U2OS, enhances cell proliferation (92). These data strongly suggest that Id genes have a regulatory effect on the cell cycle, and thus on the decision to proliferate or differentiate.

We therefore investigated the expression of Id-1 and -2 in proliferating and/or differentiated rat hepatocytes (93). Id-1 mRNAs are highly expressed in foetal liver at early stages but gradually decrease during hepatic development and are undetectable in adult liver. Id-1 mRNAs reappear during liver regeneration and in primary cultures of

EGF-stimulated adult hepatocytes but not in unstimulated cells. The induction of Id-1 precisely correlates with the R-point and the progression in late G1. In contrast, Id-2 has a biphasic expression during the cell cycle : Id-2 is induced in freshly isolated hepatocytes during G0/G1 transition, decreases thereafter and is reinduced in parallel with Id-1 in mid-late G1 (Ilyin et al., manuscript in preparation). These data suggest that Id-1 and -2 play different roles during the cell cycle but both may participate in regulating the progression in late G1 and that Id genes are involved in the balance between proliferation and differentiation in hepatocytes.

The model systems used to study the retro-differentiation of hepatocytes are the *in vivo* liver regeneration itself as well as the primary cultures of hepatocytes and hepatoma cell lines. The reexpression of foetal markers in hepatocytes during liver regeneration also occurs in hepatocytes in primary culture (81) concomitantly with a decreased expression of liver specific genes (56) when cells are cultured in conditions stimulating the progression through G1 and S phases. However, when hepatocytes are co-cultured with the rat liver primitive biliary cells, hepatocytes are arrested in G1 and do not respond to growth factors while their differentiated state is maintained along with a much longer survival (56). It appears that the proliferation/differentiation equilibrium in this *in vitro* coculture system is mediated by a cell-cell contact signal between the two cell types (94).

The major advantage of the primary culture and co-culture systems is their normal status compared to the tumourigenic proliferating cell lines. However, their life span is limited to a few days and they do not allow redifferentiation after proliferating. Stable differentiated hepatoma cell lines offer alternative models which can be used in this type of studies.

We have recently established from a human hepatocarcinoma, a set of novel differentiated cell lines named HBG clones (95). At a low density, these cells proliferate until they become confluent and that cell-cell contacts lead to a complete inhibition of proliferation. Interestingly, they undergo a gradual differentiation process in parallel with the decrease of proliferation. These cells can remain confluent, stable and differentiated for several weeks. At this stage, HBG cells are arrested in G1 as argued by reexpression of c-myc and cdk4 along with a loss of cdc2 and Id-1 expression. However, after several weeks of quiescence (cell cycle arrest), cells can be replated ; they immediately reenter the cell cycle and no longer express hepatocyte specific functions.

Therefore, HBG lines provide novel powerful human cell systems for studying the mechanisms that control equilibrium between proliferation and differentiation in human liver cells.

APOPTOSIS IN HEPATOCYTES

Apoptosis does not play a crucial role in liver morphogenesis or in normal adult liver. However, the rate of apoptosis dramatically increases in various pathological situations such as immunodependent chronic and acute hepatitis.

Apoptosis can also be induced in animal model systems. Numerous xenobiotics and environmental pollutants induce liver growth and disruption of the liver and body size ratio (2). After withdrawing the drug, the liver undergoes atrophy by apoptosis of hepatocytes (96), confirming clinical observations (see introduction) indicating that liver maintains tissue homeostasis and functional/mass ratio by reducing the number of cells via apoptosis. Recently, it has been shown in hepatoma cell lines that a proliferation/apoptosis balance also exists in these models and that it can be altered by treating cells with the non-genotoxic hepatocarcinogen nafenopin and EGF (95, 97).

To date, two very potent inducers of apoptosis in hepatocytes have been identified : TGFß and Fas ligand. As mentioned earlier in this chapter, TGFß can inhibit hepatocyte proliferation at very low concentrations (4 to 40 pM) while higher concentrations lead to apoptosis (98) and increase apoptotic rate in regressing hepatic hyperplasia induced by xenobiotics (99).

Fas ligand is a TNF-like molecule (100) which binds to Fas, a cell surface receptor member of the TNF/NGF receptor family (101). Binding of Fas ligand to its receptor induces a cascade of events leading to apoptosis. Fas receptor is expressed on a variety of tissues including liver (102) and injection of anti-Fas monoclonal antibodies into mice activates the Fas receptor followed by a rapid lethal effect due to a massive apoptosis in liver (103).

Little is known about intracellular mechanisms of apoptosis in hepatocytes. However, Oberhammer et al., (104, 105) have showed that apoptosis can occur in epithelial cells, including hepatocytes prior or in absence of internucleosomal DNA degradation but cleavage in very large fragments suggesting that activation of endonuclease(s) is not required in these cells. In contrast, Shinzawa et al., (106) have reported a DNA oligonucleosome fragmentation pattern in high density dependent apoptosis of hepatocytes in primary cultures.

It has previously been showed that p53 controls the transcription of the cyclin-dependent kinase inhibitor p21 (WAF-1/CIP-1) suggesting that p21 would promote p53-dependent cell cycle arrest and apoptosis (107). Recently, Wu et al. (69) have shown that targeted *in vivo* overexpresssion of p21 induces hepatocyte cell cycle arrest but does not affect apoptosis ruling out the concept that hepatocytes undergo apoptosis after a p53-dependent cell cycle arrest and suggesting that cell death, downstream p53 involves effectors distinct from p21.

It has been shown that the Hep3B (108) and HBG hepatoma cell lines (95) undergo apoptosis after stimulation by TGFβ in a p53 inactive context demonstrating that apoptosis occurs through a p53-independent pathway in hepatic cell systems.

CONCLUSION AND FUTURE DIRECTIONS

Besides growth factors, the identification of priming factors, co-mitogens and growth inhibitors associated with a detailed description of a sequential induction of genes throughout the liver regeneration, have given clues on the regulation of hepatocyte proliferation. However, many questions remain unanswered : What are the essential signals targeting the G0/G1 and G1/S transitions? What are the signals regulating the expression of the priming factors, growth factors and growth inhibitors? when precisely during the cell cycle do these factors regulate hepatocyte proliferation? Is there any functional redundancy between them? What are the mechanisms regulating the sequential activation of proto-oncogenes, tumour suppressor genes, cdks and cyclins?.

Apoptosis in hepatocytes is still poorly documented and detailed studies on its cellular and molecular basis are required in order to understand its p53-independent pathway involved and the proliferation/apoptosis balance.

It is clear that liver optimises its activity by regulating its cell number suggesting a crucial role of the proliferation/differentiation/apoptosis equilibrium. This equilibrium involves the hepatocytes and the interplay of many growth factors, mediators of apoptosis and cell-cell interactions resulting in very complex cellular and molecular mechanisms. Thus, a major challenge in liver research will be to define the intracellular actors controlling the commitment of hepatocytes towards differentiation, proliferation or apoptosis as well as the external signals regulating their expression and activity. The identification of these factors will be crucial to understand and perhaps control liver tissue homeostasis and better define molecular basis of hepatocarcinogenesis.

ACKNOWLEDGEMENTS

The authors would like to thank their colleagues of the INSERM unit 49 for helpful comments and suggestions during the course of this research and Dr. J. Daniel for her critical reading of the manuscript. We would also like to acknowledge support from the Institut National de la Sante et de la Recherche Medicale (INSERM).

REFERENCES

1. Higgins, G.M. and Anderson, R.M. (1931) *Arch.Pathol.* **12**, 186-202.
2. Schulte-Hermann, R. (1974) *Crit.Rev.Toxicol.* **3**, 97-158.
3. Van Thiel, D.H., Gavaler, J.S., Kam, I., Francavilla, A., Polimeno, L., Schade, R.R., Smith, J., Diven, W., Penkrot, R.J. and Starzl, T.E. (1987) *Gastroenterol.* **93**, 1414-1419.
4. Kam, I., Lynch, S., Svanas, G., Todo, S., Polimeno, L., Francavilla, A., Penkrot, R.J., Takaya, S., Ericzon, B.G., Starzl, T.E. and Van Thiel, D.H. (1987) *Hepatology* **7**, 362-366.
5. Moolten, F.L. and Bucher, N.L.R. (1967) *Science* **158**, 272-274.
6. Fisher, B., Szuch, P., Levine, M. and Fischer, E. (1970) Science **171**, 575-577.
7. Jones, D.E., Tranpatterson, R., Cui, D.M., Davin, D., Estell, K.P. and Miller, D.M. (1995) *Am.J.Physiol.* **31**, G872-G878.
8. Grisham, J.W. (1962) *Cancer Res.* **22**, 842-849.
9. Fabrikant J.I. (1968) *J.Cell.Biol.* **36**, 551-565.
10. Thompson, N.L., Mead, J.E., Braun, L., Goyette, M., Shank, P.R. and Fausto, N. (1986) *Cancer Res.* **46**, 3111-3117.
11. Sobczack, J., Mechti, N., Tournier, M.F., Blanchard, J.M. and Duguet, M. (1989a) *Oncogene* **4**, 1503-1508.
12. Morello, D., Lavenu, A. and Babinet, C. (1990) *Oncogene* **5**, 1511-1519.
13. Hsu, J.-C., Bravo, R. and Taub, R. (1992) *Mol.Cell.Biol.* **12**, 4654-4665.
14. Harber, B.A., Mohn, K.L., Diamond, R.H. and Taub, R. (1993) *J.Clin.Invest.* **91**, 1319-1326.
15. Scearce, L.M., Lee, J., Naji, L., Greenbaum, L., Cressman, D.E. and Taub, R. (1996) *Cell Death & differentiation* **3**, 47-55.
16. Mead, J.E. and Fausto, N. (1989) *Proc. Natl. Acad.Sci.* USA **86**, 1558-1562.
17. McGowan, J.A., Strain, A.J. and Bucher, N.L.R. (1981) *J.Cell.Physiol.* **180**, 353-363.
18. Nakamura, T., Nishizawa, T., Hagiya, M., Seki, T., Shimonishi, M., Sugimura, A., Tashiro, K. and Shimizu, S. (1989) *Nature* **342**, 440-443.
19. Zarnegar, R. and Michalopoulos, G.K. (1995) *J.Cell.Biol.* **129**, 1177-1180.

20. Kan, M., Huan, J., Mansson, P., Yasumitsu, H., Carr, B. and McKeehan, W. (1989) *Proc.Natl.Acad.Sci.* USA **86**,7432-7436.
21. Lindroos, P.M., Zarnegar, R. and Michalopoulos, G.K. (1991) *Hepatology* **13**, 743-750.
22. Hagiya, M., Francavilla, A., Polimeno. L., Ihara, I., Sakai, H., Seki, T., Shimonishi, M., Porter, K.A. and Starzl, T.E. (1994) *Proc.Natl.Acad.Sci.* USA **91**, 8142-8146.
23. McGowan, J.A. (1986) in *Research in isolated and cultured hepatocytes* (Guillouzo, A. and Guguen-Guillouzo, C. eds) pp 259-283, INSERM Paris and John Libbey Eurotext London.
24. Reid, L.M., Narata, M., Fujita, M., Murray, Z., Liverpool, C. and Rosenberg, L. (1986) in *Research in isolated and cultured hepatocytes.* (Guillouzo, A. and Guguen-Guillouzo, C. eds). pp 225-258.INSERM Paris and John Libbey Eurotext London.
25. Tomomura, A., Sawada, N., Sattler, G.L., Kleinman, H.K. and Pitot, H.C. (1987) *J.Cell.Physiol.* **130**, 221-227.
26. Fausto, N. (1992) in *Liver Regeneration* (Bernuau, D. and Feldmann, G. eds), pp 1-6, John Libbey Eurotext, Paris.
27. Steer, C.J. (1995) FASEB 9, 1396-1400.
28. Kruijer, W., Skelly, H., Botter, F., van der Putten, J.R., Barber, J.R., Verma, I.M. and Leffert, H.L. (1986) *J.Biol.Chem.* **261**, 7929-7933.
29. Sobczack, J., Tournier, M.F., Lotti, A.M. and Duguet, M. (1989b) *Eur.J.Biochem.* **180**, 49-53.
30. Horikawa, S., Sakata, K., Hatenaka, M. and Tsukada, K. (1986) *Biochem. Biophys. Res. Commun.* **140**, 574-580.
31. Mead, J.E., Braun, L., Martin, D.A. and Fausto, N. (1990) *Cancer Res.* **50**, 7023-7030.
32. Ikeda, T., Sawada, N., Fujiniga, K., Minase, T. and Mori, M. (1989) *Exp.Cell.Res.* **185**, 292-296.
33. Etienne, P-L.,Baffet, G., Desvergne, B., Boisnard-Rissel, M., Glaise, D. and Guguen-Guillouzo, C. (1988) *Oncogene Res.* **3**, 225-262.
34. Satoh, M., Adachi, K., Suda, T., Yamazaki, M. and Mizuno, D. (1991) *Mol.Biotherapy* **3**, 136-147.
35. Shinozuka, H., Kubo, Y., Katyal, S.L., Coni, P., Ledda-Columbano, G.M., Columbano, A. and Nakamura, T. (1994) *Lab.Invest.* **71**, 35-41.
36. Fitzgearld, M.J., Webber, E.M., Donovan, J.R. and Fausto, N. (1995) *Cell Growth & Diff.* **6**, 417-427.
37. Tewari, M., Dobrzanski, P., Mohn, K.L., Cressman, D.E., Hsu, J.C., Bravo, R. and Taub, R. (1992) *Mol.Cell.Biol.* **12**, 2898-2908.
38. Cressman, D.E., Greenbaum, L.E., Haber, B.A. and Taub, R. (1994a) *J. Biol. Chem.***269**, 30429-30435.
39. Cressman, D.E. and Taub, R. (1994b) *J. Biol. Chem.***269**, 26594-26597.
40. Cressman, D.E., Diamond, R.H. and Taub,R. (1995) *Hepatology* **21**, 1443-1449.
41. Akerman, P., Cote, P., Yang, S.Q., McClain, C., Nelson, S., Bagby, G.J.and Diehl, A.M. (1992) *Am. J. Physiol.* **263**, G579-G585.
42. Diehl, A.M., Yin, M., Fleckenstein, J., Yang, S.Q., Lin, H.Z., Brenner, D.A., Westwick, J.,Bagby G. and Nelson, S. (1994) *Am. J. Physiol.***267**, G552-G561.
43. Ihle, J.N. and Kerr, I.M. (1995) *TIG* **11**, 69-74.
44. Ruff-Jamison, S., Zhong, Z., Wen, Z., Chen, K., Darnell, J.R. and Cohen, S. (1994) *J.Biol.Chem.* **269**, 21933-21935.
45. Loyer, P., Cariou, S., Glaise, D., Bilodeau, M., Baffet, G. and Guguen-Guillouzo, C. (1996) *J.Biol.Chem.* **271**, 11484-11492.
46. Nakamura, T., Tomita, Y., Hirai, R., Yamaoka, K., Kaji, K. and Ichihara, A. (1985) *Biochem.Biophys.Res.Commun.* **133**, 1042-1050.
47. Carr, B.I., Hayashi, I., Branum, E.L. and Moses, H.L. (1986) *Cancer Res.* **46**, 2330-2334.
48. Russel, W.E., Coffey, R.J., Ouellette, A.J. and Moses, A.L.(1988) *Proc.Natl.Acad.Sci.* USA **85**, 5126-5130.
49. Braun, L., Mead, J.E., Panzica, M., Mikumo, R., Bell, G.I. and Fausto, N. (1988) *Proc.Natl.Acad.Sci.* USA **85**, 1539-1543.
50. Nakamura, T., Arakaki, R. and Ichihara, A. (1988) *Exp.Cell Res.* **179**, 488-497.
51. Carr, B.I., Wang, M., Wang, Z. and Kar. S. (1995) Meeting of *Liver development, gene regulation and disease.* pp 76. Arcachon, France (abstract).
52. Serra, R. and Isom, H.C. (1993) *J.Cell.Physiol.* **154**, 543-553.
53. Nakamura, T., Tomita, Y. and Ichihara, A. (1983) *J.Biochem.* **94**, 1029-1035.
54. Nakamura, T., Nakayama, Y. and Ichihara, A. (1984) *J.Biol.Chem.* **259**, 8056-8058.
56. Guguen-Guilouzo, C., Clement, B., Baffet, G., Beaumont, C., Morel-Chany, E., Glaise, D. and Guillouzo, A. (1983) *Exp.Cell.Res.* **143**, 47-54.
57. Corlu, A., Loyer, P., Cariou, S., Ilyin, G., Lamy, I., Corral-Debrinski, M. and Guguen-Guillouzo, C.(1994) in *Liver carcinogenesis. The molecular pathways* (Skouteris, G.G., eds) vol 88, pp 287-299.
58. Traub, O., Look, J., Dermietzel, R., Brummer, F., Hulser, D. and Willecke, K. (1989) *J.Cell.Biol.* **108**, 1039-1051.
59. Stomatoglou, S.C., Enrich, C., Manson, M.M. and Hughes, R.C. (1992) *J.Cell.Biol.* **116**, 1507-1515.

60. Kren, B.T., Kumar, N.M., Wang, S-Q, Gilula, N. and Steer, C. (1993) *J.Cell.Biol.* **123**, 707-718.
61. Sawada, N. (1989) *Exp.Cell.Res.* **5**, 584-588.
62. Sherr, C.J.(1993) *Cell* **73**, 1059-1065.
63. Morgan, D.O. (1995) *Nature* **374**, 131-133.
64. Xiong,Y., Zhang, H. and Beach, D. (1992) *Cell* **71**, 505-514.
65. Matsushime, H., Ewen, M.E., Strom, D.K., Kato, J-Y., Hanks, S.K., Roussel, M.F. and Sherr, C.J.(1992) *Cell* **71**, 323-334.
66. Baldin, V., Likas, J., Marcotte, M.J., Pagano, M., Bartek, J. and Draetta, G.(1993) *Genes & Dev.* **7**, 812-821.
67. Quelle, D.E., Ashmun, R.A., Shurtleff, S.A., Kato, J-Y., Bar-Sagi, D., Roussel, M.F. and Sherr, C.(1993) *Genes & Dev.* **7**, 1559-1571.
68. Meyerson, M. and Harlow, E.(1994) *Mol.Cell. Biol.* **14**, 2077-2086.
69. Wu, H., Wade, M., Krall, L., Grisham, J., Xiong, Y. and Van Dyke, T. (1996) *Genes & Dev.* **10**, 245-260.
70. Albrecht, J.H., Hu, M.Y. and Cerra, F.B. (1995) *Biochem.Biophys.Res.Commun.* **209**, 648-655.
71. Koff, A., Cross, F., Fischer, A., Schumacher, J., Leguellec, K., Phillipe, M. and Roberts, J.M. (1991) *Cell* **66**, 1217-1228.
72. Dulic, V., Lees, E. and Reed, S.I. (1992) *Science* **257**, 1958-1961.
73. Zindy, F., Lamas, E., Chenivesse, X., Sobczak, J., Wang, J., Fesquet, D., Henglein, B. and Brechot, C.(1992) *Biochem. Biophys. Res. Commun.* **182**, 1144-1154.
74. Pagano, M., Pepperkok, R., Verde, F., Ansorge, W. and Draetta, G. (1992) *EMBO J.* **3**, 961-971.
75. Draetta, G. (1990) *TIBS* **15**, 378-383.
76. Loyer, P., Glaise, D., Cariou, S., Baffet, G., Meijer, L. and Guguen-Guillouzo, C.(1994) *J.Biol.Chem.* **269**, 2491-2500.
77. Fan,G., Xu, R., Wessendorf, M.W., Ma, X., Kren, B.T. and Steer, C. (1995) *Cell Growth & Differentiation* **6**, 1463-1476.
78. Agrawal,D., Dong, F.,Wang, Y-Z., Kayda, D. and Pledger, J. (1995) *Cell Growth & Differentiation* **6**, 1199-1205.
79. Friedman, J.M., Chung, E.Y. and McKnight, S.L. (1984) *J.Mol.Biol.* **179**, 37-53.
80. McKnight, S.L., Lane, M.D. and Gluecksohn-Waelsch, S. (1989) *Genes & Dev* **3**, 2021-2024.
81. Guguen-Guillouzo, C., Szajnert, M-F., Glaise, D., Gregory, C. and Shapira, F. (1981) *In Vitro* **17**, 369-377.
82. Petropoulos, C., Andrews, G., Tamaoki, T., and Fausto, N. (1983) *J.Biol.Chem.* **258**, 4901-4906.
83. Umek, R.M., Friedman, A.D. and McKnight, S.L. (1991) *Science* **251**, 288-292.
84. Mischoulon, D., Rana, B., Bucher, N.L. and Farmer, S.R. (1992) *Mol.Cell.Biol.* **12**, 2553-2560.
85. Rana, B., Mischoulon, D., Xie, Y., Bucher, N.L.R. and Farmer, S.R. (1994) *Mol.Cell.Biol.* **14**, 5858-5869.
86. Kingston, R.E. (1989) *Curr.Opin.Cell Biol.* **1**, 1081-1087.
87. Benezra, R., Davis, R.L., Lockshon, D. Turner, D.L. and Weintraub, H. (1990) *Cell* **58**, 537-544.
88. Sun, X.-H., Copeland, N.G., Jenkins, N.A. and Baltimore, D. (1992) *Mol.Cell.Biol.* **11**, 5603-5611.
89. Jen, Y., Weintraub, H. and Benezra, R. (1992) *Genes & Dev.* **6**, 1466-1479.
90. Kreider, B.L., Benezra, R., Rovera, G. and Kadesch, T. (1992) *Science* **255**, 1700-1702.
91. Desprez, P-Y, Hara, E., Bissell, M. and Campisi, J. (1995) *Mol.Cell.Biol.* **15**, 3398-3404.
92. Iavarone, A., Garg, P., Lasorella, A., Hsu, J. and Israel, M.A. (1994) *Genes & Dev* **8**, 1270-1284.
93. Le Jossic, C., Ilyin, G.P., Loyer, P., Glaise, D., Cariou, S. and Guguen-Guillouzo, C. (1994) *Cancer Res.* **54**, 6065-6068.
94. Corlu, A., Kneip, B., Lhadi, C., Leray, G., Glaise, D., Baffet, G., Bourel, D. and Guguen-Guillouzo (1991) *J.Cell.Biol.* **115**, 505-515.
95. Glaise, D., Ilyin, G.P., Cariou, S., Bilodeau, M., Lucas, J., Osturk, M. and Guguen-Guillouzo, C. submitted
96. Bursh, W., Lauer, B., Timmermann-Trosiener, I., Barthel, G., Schuppler, J. and Schulte-Hermann, R. (1984) *Carcinogenesis* **5**, 453-458.
97. Gill, J.H., Molloy, C.A., Shoesmith, K.J., Bayly, A.C. and Roberts, R.A. (1995) *Cell Death and Differentiation* **2**, 211-217.
98. Oberhammer, F., Bursch, W., Parzefall, W., Breit, P., Erber, E., Stadler, M. and Schulte-Hermann, R.(1991) *Cancer Res.* **51**, 2478-2485.
99. Oberhammer, F.A., Pavelka, M., Sharma, S., Tiefenbacher, R., Purchio, A.F., Bursch, W. and Schulte-Hermann, R.(1992) *Proc.Natl.Acad.Sci.* USA **89**, 5408-5412.
100. Suda, T., Takahasi, T., Golstein, P. and Nagata, S. (1993) *Cell* **75**, 1169-1178.
101. Itoh, N., Yonehara, S., Ishii, A., Yonehara, M., Mizushima, S.-I., Sameshima, M., Hase, A., Seto, Y. and Nagata, S. (1991) *Cell* **66**, 233-243.
102. Watanabe-fukunaga, R., Brannan, C.I., Copeland, N.G., Jenkins, N.A. and Nagata, S. (1992) *Nature* **356**, 314-317.
103. Ogasawara, W.Atanabe-Fukunuga, R., Adachi, M., Matsuzawa, A., Kasugai, T., Kitamura, Y., Itoh, N., Suda, T. and Nagata, S. (1993) *Nature* **364**, 806-809.

104. Oberhammer, F., Fritsch, G., Schmied, M., Pavelka, M., Printz, D., Purchio, T., Lassmann, H. and Schulte-Hermann, R.(1993a) *J.Cell.Sci.* **104**, 317-326.
105. Oberhammer, F., Wilson, J.W., Dive, C., Morris, I.D., Hickman, J.A., Wakeling, A.E., Walker, P.R. and Sikorska, M.(1993b) *EMBO J.* **12**, 3679-3684.
106. Shinzawa, K., Watanabe, Y. and Akaike, T. (1995) *Cell Death & Differentiation* **2**, 133-140.
107. El-Deiry, W.S., Tohino, T., Velculescu, V.E., Levy, D.B., Parsons, R., Trent, J.M., Lin, D., Mercer, W.E., Kinzler, K.W. and Vogelstein, B. (1993) *Cell* **75**, 817-825.
108. Ponchet, F., Puisieux, A., Tabone, E., Midrot, J-P., Frösch, G., Morel, A.P., Frebourg, T., Fontaniere, B., Oberhammer, F. and Ozturk, M. (1994) *Cancer Res.* **54**, 2064-2068.

chapter 5
A temporal and biochemical link between growth factor-activated MAP kinases, cyclin D1 induction and cell cycle entry

Josée N. Lavoie, Nathalie Rivard, Gilles L'Allemain and Jacques Pouysségur

Centre de Biochimie-CNRS, Université de Nice, Parc Valrose, 06108 Nice, France

Cell cycle re-entry requires the growth factor-stimulation of at least two distinct classes of protein kinases: (i) the p42/p44 MAP kinases activated by the Ras>Raf>MKK cascade and (ii) the G1 cyclin-dependent protein kinases (CDKs). Specific inactivation of either class of kinase arrests fibroblasts in G1. Growth factors promote nuclear translocation and persistent activation of p42/p44 MAP kinases during the entire G0/G1 period. Here, we demonstrate that induction of cyclin D1, and therefore cdk4/6 activity associated with, is positively controlled by the p42/p44 MAP kinase cascade whereas the parallel cytokines/stress-activated p38MAP kinase cascade is antagonistic. Finally, using an antisense approach we demonstrate that $p27^{Kip1}$ plays a key role in setting the growth factor-dependency of the G0 state.

INTRODUCTION

Eukaryotic cells rely on both external and internal signals to determine when and where to replicate their chromosomes and divide. Over the last decade, a great deal of attention has been devoted to the understanding of how extracellular signals and particularly growth factors transduce their signal across the plasma membrane. Although many aspects still have to be addressed, it is satisfying to see strong lines of convergence in growth signalling. Two independent signalling networks, not yet fully elucidated, appear to be crucial for the G0 to G1 phase progression: the lipid signalling pathways including PI 3 kinase (1) and subsequent p70S6 kinase activation (2, 3) on the one hand, and the Ras>Raf>MKK>p42/p44MAP kinase cascade on the other (4, 5). Both pathways are commonly activated via a variety of cell surface receptors as diverse as receptor tyrosine kinases (for EGF, FGF, PDGF, CSF1, etc.), G protein-coupled receptors (for α-thrombin, LPA), and T cell receptors.

More recently, a second field of intense research, revealing an extraordinary degree of conservation from yeast to man, established the key role of cyclin-dependent-kinase (CDK) complexes in controlling sequential steps in cell cycle progression (6). One of the most exciting developments of the past two or three years has been the demonstration that growth factors, extracellular matrix, differentiating and anti-proliferative agents appear to operate by promoting the activation or inactivation of the CDK complexes essential for G1 progression and cell cycle entry. However, the molecular mechanisms underlying these effects are still largely unknown.

In this chapter, we will present our first attempt to link the growth factor-activated MAP kinases (p42/p44) to the activation of cyclin D1/cdk4-6, one of the earliest cyclin/CDK complexes involved in setting the G1 phase progression.

PERSISTENT ACTIVATION OF MAP KINASES IS REQUIRED FOR CELL CYCLE ENTRY

Mitogen Activated Protein Kinases (MAP kinases) also described as extracellular signal-regulatory kinases (ERKs), belong to a group of serine/threonine protein kinases that are activated in response to various stimuli (growth factors, neurotransmitters, differentiating agents, and stresses such as UV, osmotic and heat shock) in virtually all cell types (reviewed in 4, 5). In mammalian cells, so far, three MAPK independent modules have been identified: (i) the p42/p44 MAPKs activated by the Ras>Raf>MKK1-2 pathway in response essentially to growth factors, (ii) the p38/HOG MAPK activated by MKK3 in response to cytokines and increased osmolarity (7), and (iii) p54MAPK/JNK1 activated by MKK4 in response to UV and protein synthesis inhibitors (8, 9, 10). Although these MAPKs are commonly activated by dual phosphorylation on tyrosine and threonine residues of the kinase subdomain VII, and are all proline-directed kinases sharing numerous *in vitro* substrates, the activation of each MAPK cascade propagates specific cellular responses. Initial studies revealed that an important aspect in the nature of the biological response resides in the spatial and temporal MAPK activity (11, 12). For example p42/44MAPKs are activated by a wide range of extracellular signals and yet only very few are capable to elicit a mitogenic response when applied to resting fibroblasts. Mitogens are the

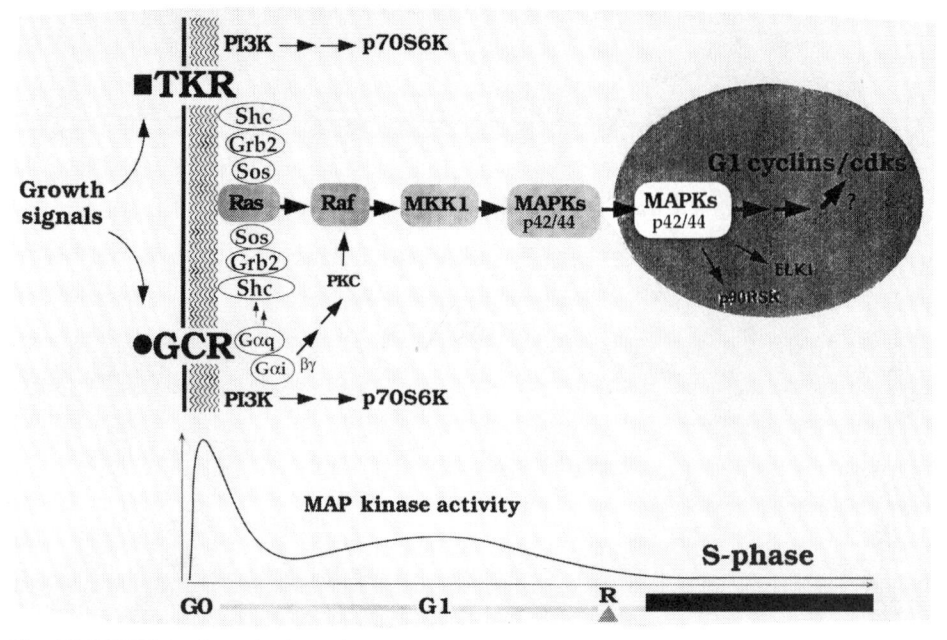

Figure 1. The p42/p44MAPK signal transduction module.
TKR: Tyrosine Kinase Receptors (EGF, FGF...); GCR: G protein-Coupled Receptors (thrombin); Shc and Grb2: adaptors; Sos: Ras GDP/GTP exchanger factor; Raf: Map kinase kinase kinase; MKK1: Map kinase kinase; p42/p44MAPKs: Map kinases (p42, p44 isoforms); p90RSK: Ribosomal S6 kinase. Mitogen activation of G0-arrested fibroblasts leads to the rapid activation of the p42/p44MAPKs which is persistent and biphasic. The long-term activation of p42/p44MAPKs is accompanied by the nuclear translocation of the enzymes and represents 10-30% of the peak activity measured 5 minutes following mitogen addition. This long-term activation usually persists for several hours before declining to barely detectable levels when cells pass the restriction point (R) and progress through S-phase (typical profile of MAPK activation in response to thrombin is shown).

only agents capable of inducing the rapid translocation of p42/44MAPKs into the nucleus, which is accompanied by a prolonged activation of the kinases during the G0/G1 period (fig. 1). Three independent approaches demonstrate that activation of this MAPK module is absolutely required for growth factor-induced G1 progression in fibroblasts. In the growth factor-dependent hamster fibroblast cell line, CCL39, we transiently expressed either the entire p44MAPK antisense RNA or a p44MAPK dominant-negative mutant. Both approaches were found to severely reduce activation of both p42/p44MAPKs (13). As a consequence, we found that both, the antisense and the dominant-negative p44MAPK, suppressed growth factor-stimulated gene transcription and cell proliferation (13). These effects were proportional to the extent of MAP kinase inhibition and reversed by coexpression of the wild type p44MAPK.

MAP kinase activity is down regulated by a set of dual specificity phosphatases that remove phosphate from both threonine and tyrosine residues. These phosphatases are referred to as MAP kinase phosphatase, MKP. MKP family members are induced in response to growth factors and participate in the attenuation of the sustained activation phase of p42/p44MAPKs (14), (JM.

Brondello, F. McKenzie and J. Pouysségur, unpublished results). Forced expression of wild type MKP-1 in serum-starved fibroblasts inhibited serum- or Ras-induced DNA synthesis (15, 16). Hence, we conclude that the persistent activation of the p42/p44MAPK cascade is an obligatory step for growth factor-induced G0 exit.

A LINK BETWEEN MAPK SIGNALLING AND CELL CYCLE MACHINERY

Having established that a prolonged p42/p44MAPK activation is absolutely required for fibroblasts to pass the G1 restriction point, a great challenge is now to define the sequence of events linking this pathway to the G1 cell cycle regulatory components. How does signalling through MAPKs drive specific cell cycle responses (proliferation/differentiation)? In all eukaryotic cells, growth regulatory information from environmental cues such as growth factors, nutrients and cell density, is integrated during a late G1 event termed the restriction point or START (17). Execution of this event and subsequent commitment to cell division is governed by a family of cyclin proteins in association with their catalytic partners, the cyclin-dependent kinases (CDK) (18). During the critical period of the G0/G1 progression, cyclin D1-cdk4, the first cell cycle-regulated complex to be activated in early G1, plays a determinant role. In marked contrast to p42/p44MAPKs, which are rapidly activated following mitogen stimulation, cyclin D1-cdk4 activity in CCL39 fibroblasts emerges rather late in G1-phase and increases progressively as the cell approaches and passes through S-phase (G. L'Allemain, J.N. Lavoie and J. Pouysségur, unpublished results). It has been shown that specific inactivation of either p42/p44MAPKs or of D1-cdk4 activity results in growth arrest in G1. These data strongly suggest that activation of these two protein kinase signalling systems is required for G1 phase progression and cell cycle entry (13, 16, 19, 20, 21).

Based on these results, we postulated that p42/p44MAPKs might control cyclin D1-cdk4 activation following mitogen-stimulation of resting fibroblasts. However, mechanisms that govern cyclin-cdk activation are complex, and involve multiple regulatory steps, including: i) *de novo* synthesis of the regulatory cyclin subunit, which is the most limiting step for G0/G1 progression, ii) phosphorylation and dephosphorylation of appropriate sites on the catalytic subunit by the activating kinase complex, CAK, and the specific cdc25 phosphatase, respectively and iii) inactivation or down regulation of multiple G1-cdk inhibitors (CKIs) (22, 23). Signalling through MAPKs might regulate more than one of these steps. Recent studies have focused on the growth factor-responsive nature of cyclin D1 expression, as a possible target of MAPK regulation (24).

Studies in different cellular systems have revealed that D-type cyclin (cyclins D1, D2, and D3) expression is regulated by cytokines and growth factors (25, 26, 27, 28, 29, 30). In fibroblasts, cyclin D1 expression is strictly dependent on the growth factor supply and consequently, if growth factors are removed, cyclin D1 levels drop rapidly, regardless of the stage of the cell cycle. In contrast, cyclin D2 and D3 do not appear to be so tightly regulated by growth factors (G. L'Allemain, and J. Pouysségur, unpublished results, (31)). It appears that the main role of cyclin D1-associated kinase activity is to phosphorylate the retinoblastoma susceptibility gene product, pRb and to reverse its inhibitory function on cell proliferation, since cyclin D1 becomes dispensable in a cell background deficient in pRb function (32). Phosphorylation of pRb by cyclin D1-cdk4/cdk6 is thought to mediate the release of transcription factor E2F, thus allowing S-phase specific gene expression (23).

Attempts to link the mitogen-regulated p42/p44MAPK activation to the growth factor-sensitive cyclin D1 expression in the Chinese hamster fibroblast CCL39 cell line, revealed a tight correlation between the ability of various agonists to maintain a long-term p42/p44MAPK activation (24), and the extent of cyclin D1 accumulation and pRb hyper-phosphorylation. The D-type cyclin expression profile indicated that cyclin D1 is the only detectable D-type cyclin to be regulated by growth factors in this cell system, being absent in serum-starved cells and detectable 4 to 6 h following mitogen addition. Experiments using various agonists indicated that although cyclin D1 is strongly induced by agents acting either *via* tyrosine kinase receptors (EGF) or through G protein coupled receptors (thrombin), only agents which promote a long-term p42/p44MAPK activation and potently initiate DNA synthesis in arrested CCL39 cells, were able to induce high levels of cyclin D1 and pRb hyper-phosphorylation (J.N. Lavoie and J. Pouysségur, unpublished results), suggesting an interdependent relationship between the long lasting activation phase of p42/p44MAPKs, cyclin D1 expression and pRb inactivation in CCL39 fibroblasts.

THE p42/p44MAPK CASCADE POSITIVELY REGULATES CYCLIN D1 PROMOTER ACTIVITY IN FIBROBLASTS

The role of p42/p44MAPKs in the regulation of the mitogen-sensitive cyclin D1 expression was directly addressed by co-transfection experiments, using a reporter gene driven by the human cyclin D1 promoter (cyclin D1-luciferase) (33), and previously characterised expression constructs which modulate either positively or negatively the endogenous p42/p44MAPK activity in CCL39 fibroblasts (see figure 3) (13, 16, 34, 35). The data obtained clearly indicated that p42/p44MAPKs are major positive regulators of cyclin D1 expression in fibroblasts (24). Inhibition of p42/p44MAPKs by expressing either dominant-negative forms of MAPK kinase (MKK1) or p44MAPK, or the MAPK phosphatase (MKP-1), strongly reduced cyclin D1-driven luciferase expression in exponentially growing cells, indicating that transcription of cyclin D1 depends on p42/p44MAPK activity (fig. 2). Further, in serum-starved cells, mitogen-induced cyclin D1-luciferase expression was dramatically reduced in cells expressing the neutralising constructs, which produced as much as 80% inhibition of the serum-induced cyclin D1-luciferase expression (24). From these experiments and others (36), it is clear that the long-term activation of p42/p44 MAPK isoforms is implicated in growth factor-mediated transcriptional activation of cyclin D1.

ACTIVATION OF THE p42/p44MAPK CASCADE IS NECESSARY AND SUFFICIENT FOR MITOGEN-INDUCED CYCLIN D1 ACCUMULATION IN FIBROBLASTS

An increase in cyclin D1 mRNA does not always lead to a corresponding increase in cyclin D1 protein level in transfected cells. It thus appears that post-

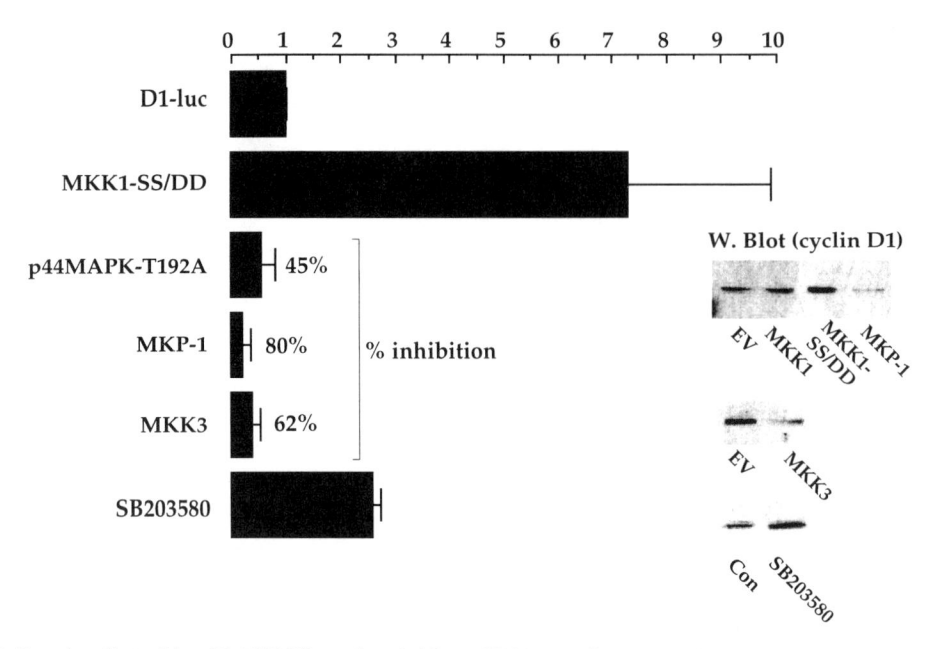

Figure 2. Opposing effects of the p42/p44MAPK cascade and of the p38MAPK on cyclin D1 expression.
Cyclin D1 promoter activity and protein expression were monitored in exponentially growing CCL39 cells following transfection with the indicated constructs. Cyclin D1-dependent luciferase activity was measured 48 h post-transfection and normalized using β-galactosidase as an internal control, by cotransfecting the pCH110 reporter construct. Experiments were performed in duplicate and data are representative of at least three independent experiments. The fold increase in luciferase activity was calculated relative to the basal expression level of the cyclin D1-luciferase, which was set to 1 unit and corrected for empty vector effects. Endogenous cyclin D1 protein expression was analyzed in CCL39 cells transfected with the marker gene pNHE3 encoding an amiloride-resistant Na^+/H^+ exchanger isoform, together with the indicated expression constructs or the appropriate empty vector (EV). Fourthy-eight hours post-transfection, non-transfected cells were selectively eliminated by acid-load selection, as described before (68, 69). Cells were lysed in SDS-sample buffer and equal amounts of each extract were processed for SDS-PAGE and Western blot analysis using a specific antibody against cyclin D1.

transcriptional processes play an important role in the regulation of cyclin D1 expression (37, 38, 39). Although we cannot exclude a possible post-transcriptional control of p42/p44MAPKs on cyclin D1 expression, it seems likely that activation of cyclin D1 promoter by this signalling pathway contributes to the modulation of cyclin D1 protein synthesis in response to growth signals, thus contributing to the regulation of S-phase entry. Noteworthy, inhibition of p42/p44MAPK activation by expression of MKP-1 dramatically reduced endogenous cyclin D1 protein levels in exponentially growing CCL39 cells (fig. 2) and markedly blocked the mitogen-induced accumulation of cyclin D1 in mitogen-stimulated cells (24). The resulting low level of cyclin D1 in MKP-1-expressing cells was not sufficient to maintain appropriate cyclin D1-cdk4 activity since a marked decrease of pRb hyper-phosphorylation was measured in the same cellular extracts (24). This suggests that E2F-regulated gene expression is down-regulated in cells where p42/p44MAPK activity is inhibited and in fact, cdk2 kinase activity was also found to be inhibited in cells expressing constructs which negatively modulate p42/p44MAPK activity (G. L'Allemain, J.N. Lavoie and J. Pouysségur, unpublished results). This is consistent with the fact that cdk2 activation in late G1/early S-phase is dependent upon transcription and synthesis of its regulatory subunits, cyclin E and cyclin A, which in turn require E2F release for their expression (40, 41, 42). It thus appears that the requirement of a long-term p42/p44MAPK activation for S-phase entry relies, at least in part, on its positive regulatory function at the level of cyclin D1 expression.

Analysis of cyclin D1 protein levels upon modulation of p42/p44MAPK activity in CCL39 cells have revealed that not only this signalling cascade is required for appropriate cyclin D1 accumulation in response to growth signals, but also that it is sufficient in itself. Expression of a constitutively active form of MAP kinase kinase (MKK1-SS/DD) produced an increase in endogenous cyclin D1 protein level in exponentially growing cells (fig. 2) as well as in the absence of any growth signal. Constitutively active MKK1, in absence of growth factors, induced cyclin D1 levels equivalent to those measured in control cells stimulated with a strong mitogen (24). More importantly, the use of a CCL39-derived cell line expressing an oestrogen-

CHAPTER 5/ FROM GROWTH FACTORS TO CELL CYCLE

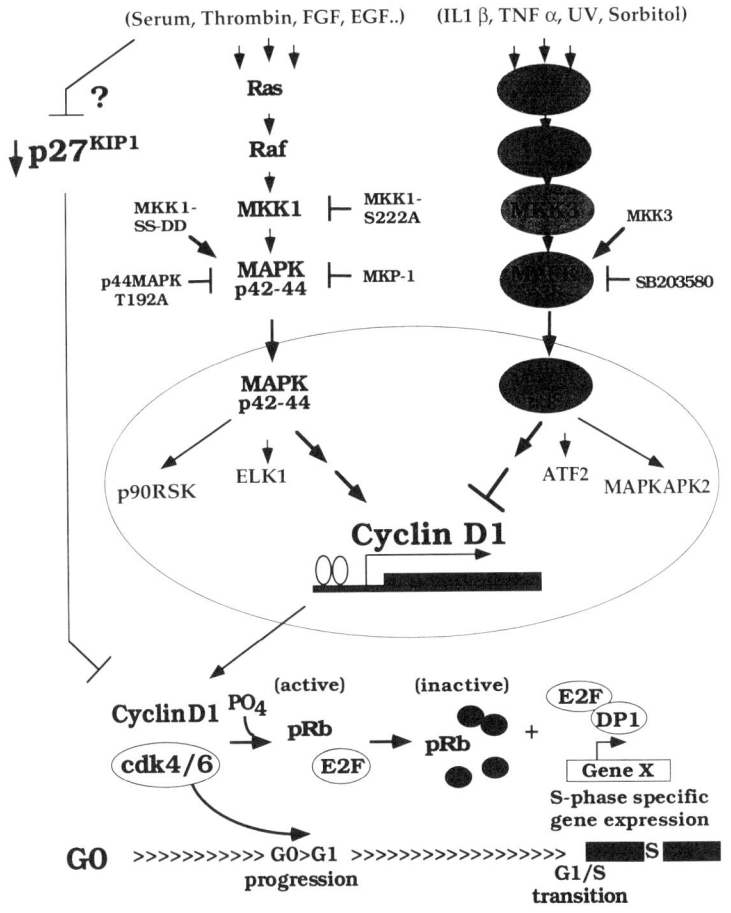

Figure 3. Regulation of cyclin D1 expression and S-phase specific gene expression by MAPK signalling.
Expression constructs which modulate either positively or negatively the MAPK cascades were used to determine the contribution of the p42/p44MAPK module as compared to the p38MAPK cascade to the regulation of cyclin D1 expression. MKK1-SS/DD: constitutively active MKK1; MKK1-S222A: dominant-negative MKK1; p44MAPK-T192A: dominant-negative p44MAPK; MKP-1: MAPK phosphatase 1; MKK3: active form of p38MAPK kinase; SB203580: specific inhibitor of p38MAPK (70, 71). Extensive studies on cyclin D1 promoter activity and protein expression in CCL39 cells led to the conclusion that the mitogenic activation of the p42/p44MAPKs stimulates cyclin D1 transcription and protein synthesis, thus allowing cyclin D1-cdk4 activation, pRb hyper-phosphorylation and E2F-dependent gene expression. In contrast, activation of the p38MAPK cascade following cytokine stimulation negatively interferes with cyclin D1 expression, thus antagonizing the mitogen-dependent pRb hyper-phosphorylation.

dependent human Raf-1 protein kinase demonstrated that the exclusive activation of the Raf>MKK1>p42/p44MAPK cascade was able to induce cyclin D1 protein expression to the same magnitude and with an identical time course as that induced by serum in these cells. However, oestrogen-stimulation of this CCL39-derived cell line (leading to rapid activation of Raf>MKK1>p42/p44MAPKs) could not promote pRb phosphorylation (24), cdk2 kinase activity, nor DNA synthesis (P. Lenormand, G. L'Allemain and J. Pouysségur, unpublished results), suggesting that although activation of the p42/p44MAPK cascade is sufficient for cyclin D1 expression, other signals are required to allow cells to pass the restriction point and to enter S-phase. In these cells serum-induced DNA synthesis is not inhibited by the co-addition of estradiol, excluding the possibility that a constitutive high p42/p44MAPK activity exerts a strong negative effect on G1 progression (P. Lenormand and J. Pouysségur, unpublished results). Although activation of the p42/p44MAPK cascade by itself reduced the growth-factor requirement for DNA synthesis in fibroblasts (35), it appears that multiple signal transduction pathways must collaborate to efficiently drive cells into S-phase.

THE p38/HOG MAPK ANTAGONISES THE MITOGEN-INDUCED EXPRESSION OF CYCLIN D1

In addition to the p42/p44MAPK pathway, at least two other MAPK cascades are implicated in the transduction of external stimuli in mammalian cells, the p38/HOG MAPK (p38MAPK) and the Jun kinases (JNKs), also called 'stress-activated protein kinases' (SAPKs) (10). In contrast to p42/p44MAPKs which are strongly activated by growth factors and growth-promoting hormones, JNKs and p38MAPK are poorly sensitive to growth signals and their activation is preferentially triggered by pro-inflammatory cytokines and environmental stresses (43). Previous studies indicated that the proto-oncogene c-jun can increase expression of a cyclin D1 promoter-controlled luciferase reporter gene, suggesting a positive regulatory effect of the JNK pathway on cyclin D1 transcription (33, 36). However, the functional consequences of such a regulation on cyclin D1-associated kinase activity has not been shown. The p38MAPK is as yet poorly characterised relative to its role in gene expression. The putative contribution of the p38MAPK cascade in the regulation of cyclin D1 transcription and protein expression was investigated, using a co-transfection strategy. In marked contrast to the positive action of p42/p44MAPK activation, the p38MAPK signalling pathway exerts an inhibitory effect on cyclin D1 expression, both at the transcriptional and post-transcriptional levels (24). Activation of the p38MAPK by expression of a p38MAPK kinase (MKK3) led to a decrease in cyclin D1 promoter-controlled gene expression as well as endogenous cyclin D1 protein synthesis in exponentially growing CCL39 cells (fig. 2). Further, activation of the p38MAPK cascade has been found to antagonise the mitogen-induced accumulation of cyclin D1 in CCL39 cells, reducing the serum-induced cyclin D1 promoter activity and protein accumulation by 70% to 50% (24). Experiments using the inhibitor, SB203580 which has been shown to specifically inhibit p38MAPK without inhibitory action on other MAPKs, indicated that the MKK3-mediated inhibition of cyclin D1 promoter activity as well as protein synthesis likely result from activation of p38MAPK. Indeed, inactivation of p38MAPK by incubation of CCL39 cells with SB203580 enhanced cyclin D1-luciferase expression and endogenous protein accumulation in the absence of mitogen (fig. 2). This result further indicates that this signalling pathway exerts an inhibitory action on cyclin D1 expression. More importantly, this negative regulatory effect of the p38MAPK as opposed to the positive p42/p44MAPK control on cyclin D1 expression appears to be physiological. In fact, similar effects could be detected on cyclin D1 expression levels upon simultaneous activation of these two signalling systems in CCL39 cells (24). Using α-thrombin to activate the p42/p44MAPKs and IL-1β to activate p38MAPK with little effect on p42/p44MAPKs, we show that the cyclin D1 protein level was markedly attenuated when these two MAPK pathways were simultaneously activated as compared to the level measured in cells exposed to thrombin alone. This is consistent with a recent study which shows an inhibitory effect of IL-1β on pRb hyper-phosphorylation (44), suggesting that cyclin D1-associated kinase activity, which is mainly dependent upon cyclin D1 protein levels, is negatively regulated by IL-1β, a strong p38MAPK agonist in many cellular systems.

MAPK CONTROL AT THE MOLECULAR LEVEL

Previous studies have shown that activation of the p42/p44MAPK module is sufficient to stimulate early gene transcription in fibroblasts (35, 45). The nuclear translocation of p42/p44MAPKs is thought to result in activation of a range of transcription factors such as Elk1 (46, 47, 48), c-Ets-1 and c-Ets-2 (49, 50), which has been shown to be phosphorylated *in vitro* by these protein kinases on the same residues which increase their transcriptional activation *in vivo*. Elk-1 is also substrate for p38MAPK (J. Raingeaud, J.S. Rogers, A.J. Whitmarsh, B. Dérijard and R.J. Davis, unpublished results), in addition to ATF2 (51), which binds to CRE elements on various promoters. Indeed, increased p38MAPK activation following MKK3 expression in fibroblasts was found to activate reporter gene expression from a promoter containing multiple CRE elements (J.N. Lavoie and J. Pouysségur, unpublished results), indicating that this signalling pathway can positively modulate gene expression.

Activation of cyclin D1 transcription by the p42/p44MAPK module was recently proposed to be mediated by a putative Ets-like binding domain on the proximal region of the human cyclin D1 promoter (36). Albanese *et al.* reported that overexpression of either p42MAPK or c-Ets-2 stimulated cyclin D1 promoter through this same proximal 22 base pairs region. Transactivation of cyclin D1 promoter *via* c-Ets-2 is therefore predicted to be responsible for the positive regulatory effect of p42/p44MAPKs since activation of these protein kinases is associated with activation of the Ets family transcription factors (46, 47, 49). However, the molecular mechanisms underlying the negative regulation of cyclin D1 expression by the p38MAPK is less clear. The cyclin D1 promoter contains multiple regulatory elements (TRE, E2F, Oct, SP1, CRE) and some uncharacterised elements that may also play a role in transcription of the gene (33). Therefore, cyclin D1 expression may be responsive to a large set of transcription factors, whose activities may in turn be modulated by more than one MAPK signal transduction pathway, adding to the complexity of such a transcriptional regulation (51, 52, 53, 54).

In addition, similar to previous observations following transfection of cyclin D1 in some cellular systems, it appears that transcriptional activation of cyclin D1 is not necessarily synonymous with cyclin D1 protein accumulation. Preliminary experiments indicate that although activation of the JNK signal transduction pathway leads to induction of a cyclin D1 promoter-controlled reporter gene expression, no increase in endogenous cyclin D1 protein levels can be detected in CCL39 cells and rather, a decrease in protein expression is measured. At least in cell expressing a constitutively-active form of the JNK kinase kinase (MEKK), although cyclin D1 transcription was largely enhanced as measured by the increased expression of the cyclin D1-luciferase reporter, a marked inhibition of the basal endogenous cyclin D1 protein expression was detected, as well as of the mitogen-induced cyclin D1 protein levels in CCL39 cells (J.N. Lavoie and Jacques Pouysségur, unpublished results). Similar observations were made using an expression construct encoding the JNK kinase (MKK4). From these results, it is clear that any functional conclusion regarding the regulation of cyclin D1 expression by signalling pathways requires an extensive analysis of the effects on cyclin D1 protein synthesis and accumulation.

Finally, there are several mechanisms for the regulation of the respective MAPK signal transduction pathways which can be simultaneously activated by a single external stimulus (10). Negative feedback is a widely used mechanism for signal attenuation and desensitisation, and several reports have clearly shown that such a negative feedback operates in MAPK signalling pathways. In addition, both positive and negative cross talks between these MAPK cascades probably exist. Two groups reported that p42/p44MAPKs phosphorylate SEK (JNK kinase) (55, 56), but the significance of this cross talk among MAPK pathways is still not understood. Thus it cannot be excluded that some cross talk between the p42/p44MAPK and the p38MAPK cascades may be responsible for the antagonising effect of the p38MAPK on the mitogen-induced cyclin D1 expression. Further analysis will be required to understand the regulation of cyclin D1 expression by MAPK cascades at the molecular level.

$p27^{KIP1}$, A PIVOTAL CKI THAT SETS THE QUIESCENT STATE

Growth factor deprivation leads to a G0/G1-growth arrest in fibroblasts and/or to programmed cell death in more differentiated cellular systems. Which are the molecular 'sensors' of the cellular environment that mediate this growth arrest state? In CCL39 cells, growth factor removal leads to rapid inactivation of the p42/p44MAPKs (11, 57, 58), whereas the p38MAPK activity seems to increase with the duration of starvation, as if serum starvation was inducing its own stress (A. Brunet and J. Pouysségur, unpublished results). Cyclin D1 expression is rapidly abrogated in serum-starved fibroblasts (59) and immune complexes prepared with antisera either to cyclin D1, cyclin E, cdk4 or cdk2 lacked kinase activity able to phosphorylate pRb (60), (G. L'Allemain, J.N. Lavoie, and J. Pouysségur, unpublished results). It has been proposed that this cdk inactivation results from an excess of cyclin-dependent kinase inhibitors (CKI) p21 and p27, responsible for titrating out cyclin-cdk complexes (23, 61, 62, 63, 64). In general, p27 level increases in cells arrested by serum-starvation or high cell density, and progressively decreases upon mitogen stimulation, suggesting a role for p27 in the control of the quiescent state. This is also the situation encountered in CCL39 cells (N. Rivard, G. L'Allemain, J. Bartek and J. Pouysségur, unpublished results). To analyze the contribution of p27 in G0-induced state, we specifically abrogated its expression by transfecting a full length cDNA p27 antisense in CCL39 cells that was shown not to affect the expression of the related CKI, p21. Reduction of up to 90% of p27 protein expression increased both basal and serum-stimulated gene transcription of cyclin D1, cyclin A, and DNA synthesis reinitiation. Moreover, overexpression of this antisense allows cells to grow for several generations in a serum-free medium supplemented with insulin and transferrin only, thus suggesting that p27-depleted cells cannot exit the cell cycle (table 1). These effects were fully reversed by coexpression of a plasmid encoding p27 sense (N. Rivard, G. L'Allemain, J. Bartek and J. Pouysségur, unpublished results). We conclude that p27, by setting the level of growth factor requirement, plays a pivotal role in controlling cell cycle exit, a fundamental step in growth control.

It has been shown that p27 undergoes rapid degradation in exponentially growing cells by the ubiquitin-proteasome pathway and that this proteolysis is reduced by several fold in resting cells, a phenomenon which may account for increasing p27 expression in mitogen-starved cells (65). Which critical growth factor-regulated pathway in G1 does signal p27 degradation? Preliminary results indicated that the p42/p44MAPK cascade is not implicated. Indeed, a conditionally sustained p42/p44MAPK activation in the CCL39-derived cell line expressing an oestrogen-dependent human Raf-1 protein kinase (CCL39-ΔRaf-1:ER) under estradiol-stimulation, does not lead to down-regulation of p27 expression in estradiol-stimulated cells (N. Rivard and J. Pouysségur, unpublished results). The maintenance of a high p27 expression level in those cells following oestrogen-stimulation may account for the lack of pRb phosphorylation (24), suggesting the

Table 1. Effect of transient expression of p27 and antisense p27 on cell proliferation (colony numbers after a week).

cDNA constructs	7.5% FCS	ITS
empty vector	61 ± 7	9 ± 3
p27	9 ± 2*	2 ± 1†
ASp27	151 ± 5*	31 ± 5†

PS200 cells were cotransfected with 1 µg of the selection vector pEAP (Na^+/H^+ antiporter cDNA) and 18 µg of either empty vector (pECE), sense (p27) or antisense construct (ASp27). Forty-eight hours after transfection, cells were subjected to an acid-load selection. Cultures were subsequently changed to complete growth medium and allowed to proliferate for 36 hours before repeating one cycle of acid-load selection. Cells were then incubated either into complete growth medium (DMEM with 7.5% FCS) or to DMEM serum-free medium supplemented with ITS (Insulin 5 µg/ml; Transferrin 5 µg/ml; Selenium 5 µg/ml). Visible colonies that developed after one week were stained with Giemsa and counted. Results are the means ± S.E. of three experiments. *, significantly different from control (empty vector 7.5% FCS) at $p<0.05$ (Student's t test). †, significantly different from control (empty vector 7.5% FCS) at $p<0.05$ (Student's t test)

absence of cyclin D1-cdk4 activity, even in the presence of a high cyclin D1 expression level. Interestingly, another study recently reported a similar situation in Balb/c-3T3 fibroblasts which conditionally expressed an oncogenic ras mutant (66). In these cells, forced expression of an oncogenic ras led to accumulation of cyclin D1 mRNA and protein, but no active cyclin D1-cdk4 complexes were found and p27 expression levels persisted. Therefore, other signal transduction pathways are expected to be implicated in the down-regulation of p27 since addition of plasma-derived growth factors in medium of cells expressing oncogenic ras led to a decreased p27 expression, together with activation of cyclin-cdks complexes and progression of cells through S-phase (66).

CONCLUDING REMARKS

MAPK signal transduction pathways are pivotal in the transduction of extracellular signals into biological responses. Here we demonstrated that MAPK cascades play essential and differential roles in the regulation of cyclin D1 expression, thus establishing a link between cell surface receptors and the cell cycle machinery. The positive regulatory role of the p42/p44MAPK cascade further emphasises its essential function as a positive regulator of cell proliferation in response to growth signals. In contrast, the negative effect of the p38MAPK cascade on cyclin D1 expression suggests that this signalling pathway may be detrimental to cell growth. A recent study has reported that activation of p38MAPK together with the JNK cascade is critical for the induction of apoptosis in PC12 cells (67). This is thus in agreement with a possible negative function of the p38MAPK in cell division mechanisms. Interestingly, the p38MAPK activity was found to increase following incubation of cells in serum-free medium (A. Brunet and J. Pouysségur, unpublished results). It can thus be hypothesised that in contrast to the p42/p44MAPK cascade which controls the G0 to G1-phase transition in fibroblasts and enhances cell differentiation in PC12 cells, the p38MAPK cascade could be implicated in the maintenance of cell quiescence in fibroblasts and in induction of programmed cell death in differentiated cellular systems. The signal transduction pathways responsible for the mitogen-induced proteolysis of p27 remain to be identified. Considering the pivotal role of p27 in the maintenance of the quiescent state, the regulation of this degradative process is certainly a key issue in understanding growth control. It remains to be seen whether the p38MAPK cascade contributes to the down-regulation of p27 following mitogen starvation in fibroblasts. Another potential candidate signal for initiating p27 degradation, is a cdk associated with a D-type cyclin, as activation of these complexes represent one of the earliest 'sensor' of G1 progression.

In recent years, a large number of cell cycle regulatory molecules have been identified. However, their precise functions and their modulation by environmental signals remains poorly understood. Further identification and understanding of the complex signalling networks that control the G0/G1 phase of mammalian cells, *the heart of growth control*, represents a great challenge for the future.

ACKNOWLEDGEMENTS

The authors wish to thank their colleagues, J-M. Brondello, A. Brunet, J-C. Chambard, B. Derijard, V. Dulic, F. McKenzie, P. Lenormand, G. Pages and E. Van Obberghen-Schilling for their kind gift of various reagents, helpful discussions and technical help at various stages of this work. We would like to thank V. Baldin, Jiri Bartek for providing us with specific antibodies, R. Müller for cyclin D1 promoter constructs and A. Lundberg for critical reading of the manuscript. This work was supported by grants from CNRS (Centre National de la Recherche Scientifique), the University of Nice, la Ligue Nationale Contre le Cancer et l'Association pour la Recherche contre le Cancer (ARC). Drs J. N. Lavoie and N. Rivard were supported by the Canadian postdoctoral fellowship programs, respectively, MRC and NSRC.

REFERENCES

1. Kapeller, A., and Cantley, L. (1994) *Bioassays* **16**, 565-576
2. Chou, M., and Blenis, J. (1995) *Curr. Opin. Cell Biol.* **7**, 806-814
3. Ferrari, S., and Thomas, G. (1994) *Crit. Rev. Bioch. Mol. Biol.* **29**, 385-413
4. Marshall, C. (1994) *Curr. Opin. Genet. Dev.* **4**, 82-89
5. Waskiewicz, A., and Cooper, J. (1995) *Curr. Opin. Cell Biol.* **7**, 798-805
6. Massagué, J., and Roberts, J. (1995) *Curr. Opin. Cell Biol.* **7**, 769-772
7. Han, J., Lee, J.-D., Bibbs, L., and Ulevitch, R. J. (1994) *Science* **265**, 808-811
8. Dérijard, B., Hibi, M., Wu, I.-H., Barrett, T., Su, B., Deng, T., Karin, M., and Davis, R. J. (1994) *Cell* **76**, 1025-1037
9. Kyriakis, J. M., Banerjee, P., Nikolakaki, E., Dai, T., Rubie, E. A., Ahmad, M. F., Avruch, J., and Woodgett, J. R. (1994) *Nature* **369**, 156-160
10. Cano, E., and Mahadevan, L. C. (1995) *Trends Biochem Sci* **20**, 117-122
11. Meloche, S., Seuwen, K., Pagès, G., and Pouysségur, J. (1992) *Mol. Endocrinol.* **6**, 845-854
12. Lenormand, P., Sardet, C., Pagès, G., L'Allemain, G., Brunet, A., and Pouysségur, J. (1993) *J. Cell. Biol.* **122**, 1079-1088
13. Pagès, G., Lenormand, P., L'Allemain, G., Chambard, J. C., Meloche, S., and Pouysségur, J. (1993) *Proc. Natl. Acad. Sci. USA* **90**, 8319-8323
14. Sun, H., Charles, C. H., Lau, L. F., and Tonks, N. K. (1993) *Cell* **75**, 487-493
15. Sun, H., Tonks, N. K., and Bar-Sagi, D. (1994) *Science* **266**, 285-288
16. Brondello, J.-M., Mckenzie, F.R., Sun, H., Tonks, N.K., and Pouysségur, J. (1995) *Oncogene* **10**, 1895-1904
17. Pardee, A. (1989) *Science* **246**, 603-608
18. Sherr, C. J. (1994) *Cell* **79**, 551-555
19. Quelle, D. E., Ashmun, R. A., Shurtleff, S. A., Kato, J. Y., Bar, S. D., Roussel, M. F., and Sherr, C. J. (1993) *Genes Dev* **7**, 1559-1571
20. Baldin, V., Lukas, J., Marcote, M. J., Pagano, M., and Draetta, G. (1993) *Genes Dev* **7**, 812-821
21. Roussel, M. F., Theodoras, A. M., Pagano, M., and Sherr, C. J. (1995) *Proc Natl Acad Sci U S A* **92**, 6837-6841
22. Lees, E. (1995) *Current Opinion in Cell Biology* **7**, 773-780
23. Sherr, C. J., and Roberts, J. M. (1995) *Genes Dev.* **9**, 1149-1163
24. Lavoie, J. N., L'Allemain, G., Brunet, A., Müller, R., and Pouysségur, J. (1996) *J. Biol. Chem.* (in press)
25. Matsushime, H., Roussel, M. F., Ashmun, R. A., and Sherr, C. J. (1991) *Cell* **65**, 701-13
26. Cocks, B. G., Vairo, G., Bodrug, S. E., and Hamilton, J. A. (1992) *J Biol Chem* **267**, 12307-12310
27. Motokura, T., Keyomarsi, K., Kronenberg, H. M., and Arnold, A. (1992) *J. Biol. Chem.* **267**, 20412-20415
28. Won, K. A., Xiong, Y., Beach, D., and Gilman, M. Z. (1992) *Proc Natl Acad Sci U S A* **89**, 9910-9914
29. Ajchenbaum, F., Ando, K., DeCaprio, J. A., and Griffin, J. D. (1993) *J Biol Chem* **268**, 4113-4119
30. Sewing, A., Burger, C., Brusselbach, S., Schalk, C., Lucibello, F. C., and Müller, R. (1993) *J Cell Sci* **104**, 545-55
31. Sherr, C. J., Matsushime, H., and Roussel, M. F. (1992) *Ciba Found Symp* **170**, 209-219; discussion 219-226
32. Lukas, J., Bartkova, J., Rohde, M., Strauss, M., and Bartek, J. (1995) *Mol Cell Biol* **15**, 2600-2611
33. Herber, B., Truss, M., Beato, M., and Müller, R. (1994) *Oncogene* **9**, 1295-304
34. Pagès, G., Brunet, A., L'Allemain, G., and Pouysségur, J. (1994) *EMBO J.* **13**, 3003-3010
35. Brunet, A., Pages, G., and Pouyssegur, J. (1994) *Oncogene* **9**, 3379-3387
36. Albanese, C., Johnson, J., Watanabe, G., Eklund, N., Vu, D., Arnold, A., and Pestell, R.G.(1995) *J. Biol. Chem.* **270**, 23589-23597
37. Matsushime, H., Roussel, M. F., and Sherr, C. J. (1991) *Cold Spring Harb Symp Quant Biol* **LVI**, 69-74
38. Rosenwald, I. B., Lazaris, K. A., Sonenberg, N., and Schmidt, E. V. (1993) *Mol Cell Biol* **13**, 7358-63
39. Rosenwald, I. B., Kaspar, R., Rousseau, D., Gehrke, L., Leboulch, P., Chen, J-J, Schmidt, E.V., Sonenberg, N., and Lendon, I.M. (1995) *J. Biol. Chem.* **270**, 21176-21180
40. Devoto, S. H., Mudryj, M., Pines, J., Hunter, T., and Nevins, J. R. (1992) *Cell* **68**, 167-76
41. Ohtani, K., DeGregory, J., and Nevins, J.R. (1995) *Proc. Natl. Acad. Sci. USA* **92**, 12146-12150

42. DeGregori, J., Kowalik, T., and Nevins, J. R. (1995) *Mol Cell Biol* **15**, 4215-4224
43. Davis, R. J. (1994) *Trends Biochem. Sci.* **19**, 470-473
44. Muthukkumar, S., Sells, S. F., Crist, S. A., and Rangnekar, V. M. (1996) *J. Biochem. Chem.* **271**, 5733-5740
45. Cowley, S., Paterson, H., Kemp, P., and Marshall, C. J. (1994) *Cell* **77**, 841-852
46. Marais, R., Wynne, J., and Treisman, R. (1993) *Cell* **73**, 381-393
47. Janknecht, R., Ernst, W.H., Pingoud, V., and Nordheim, A. (1993) *EMBO J.* **12**, 5097-
48. Kortenjann, M., Thomae, O., and Shaw, P. E. (1994) *Mol Cell Biol* **14**, 4815-4824
49. Wasylyk, B., Hahn, S.J., and Giovane, A. (1993) *Proc. Natl. Acad. Sci. U.S.A.* **90**, 7739-7743
50. Coffer, P., de, J. M., Mettouchi, A., Binetruy, B., Ghysdael, J., and Kruijer, W. (1994) *Oncogene* **9**, 911-21
51. Raingeaud, J., Gupta, S., Rogers, J.M.D., Han, J., Ulevitch, R.J. and Davis, R.J. (1995) *J. Biol. Chem.* **270**, 7420-7426
52. Whitmarsh, A. J., Shore, P., Sharrocks, A. D., and and Davis, R. (1995) *Science* **269**, 403-407
53. Gupta, S., Campbell, D., Derijard, B., and Davis, R. J. (1995) *Science* **267**, 389-393
54. Livingstone, C., Patel, G., and Jones, N. (1995) *EMBO J.* **14**, 1785-1797
55. Dérijard, B., Raingeaud, J., Barrett, T., Wu, I. H., Han, J., Ulevitch, R. J., and Davis, R. J. (1995) *Science* **267**, 682-685
56. Yan, M., Dai, T., Deak, J. C., Kyriakis, J. M., Zon, L. I., Woodgett, J. R., and Templeton, D. J. (1994) *Nature* **372**, 798-800
57. L'Allemain, G., Pouysségur, J., and Weber, M. J. (1991) *Cell Regulation* **2**, 675-684
58. Kahan, C., Seuwen, K., Meloche, S., and Pouysségur, J. (1992) *J. Biol. Chem* **267**, 13369-13375
59. Sherr, C. J. (1994) *Stem Cells (Dayt)* **12**, 47-55; discussion 55-57
60. Matsushime, H., Quelle, D. E., Shurtleff, S. A., Shibuya, M., Sherr, C. J., and Kato, J. Y. (1994) *Mol Cell Biol* **14**, 2066-76
61. Harper, J. W., Adami, G. R., Wei, N., Keyomarsi, K., and Elledge, S. J. (1993) *Cell* **75**, 805-816
62. Toyoshima, H., and Hunter, T. (1994) *Cell* **78**, 67-74
63. Polyak, K., Kato, J. Y., Solomon, M. J., Sherr, C. J., Massague, J., Roberts, J. M., and Koff, A. (1994) *Genes Dev* **8**, 9-22
64. Kato, J. Y., Matsuoka, M., Polyak, K., Massague, J., and Sherr, C. J. (1994) *Cell* **79**, 487-496
65. Pagano, M., Tam, S., Theodoras, A., Beer-Romero, P., DelSal, G., Chau, V., Yew, R., Draetta, G., and and Rolfe, M. (1995) *Science* **269**, 682-685
66. Winston, J. T., Coats, S. R., Wang, Y.-Z., and and Pledger, W. J. (1996) *Oncogene* **12**, 127-134
67. Xia, Z., Dickens, M. Raingeaud, J., Davis, R.J., and Greenberg, M.E. (1995) *Science* **270**, 1326-1331
68. Pouysségur, J., Sardet, C., Franchi, A., L'Allemain, G., and Paris, S. (1984) *Proc. Natl. Acad. Sci. USA* **81**, 4833-4837
69. Counillon, L., Scholz, W., Lang, H. J., and Pouysségur, J. (1993) *Mol. Pharmacol.* **44**, 1041-1045
70. Lee, J.C., Laydon, J.T., McDonnel, P.C., Gallaguer, T.F., Kumar, S., Green, D., MsNulty, D.et al. (1994) *Nature* **372**, 739-746
71. Cuenda, A., Rouse, J., Doza, Y.N., Meir, R., Cohen, P., Gallaguer, T.F., Young, P.R. and Lee, J.C. (1995) *FEBS letters* **364**, 229-233

chapter 6
The plant cell cycle: conserved and unique features in mitotic control

Peter C.L. John

Plant Cell Biology Group, Research School of Biological Sciences, Australian National University, Canberra, ACT 2600, Australia, and Collaborative Research Centre for Plant Science, GPO Box 475 ACT 2601, Australia.

Somatic plant cells can use a hormone checkpoint in late G2 phase. Here cytokinin stimulates removal of phosphotyrosine from p34^{cdc2} kinase and concurrently capacity for activation of the kinase by Cdc25 phosphatase declines while activity of the kinase increases and cells enter mitosis. Processes unique to plant mitosis are driven by the mitotically active kinase since the enzyme taken from plant cells in metaphase, when injected, can disassemble the preprophase band microtubules that form in G2 phase at the site of the future cross wall. This action is specific, since microtubules are not depolymerised when in interphase cytoplasmic array, or spindle, or phragmoplast. Plant metaphase kinase acts as MPF by accelerating chromosome condensation and nuclear envelope breakdown.

VARIATIONS IN CYCLE CONTROL IN PLANTS AND OTHER CELL TYPES

The diversity of timing and events seen in eukaryote cell cycles does not immediately suggest universal control genes. The major rate limiting division control is at mitosis in fission yeast growing in rich media (1) but is in late G1 in budding yeast (2). Furthermore in budding yeast the late G1 control point termed START is a time when, in parallel with initiation of progress to DNA replication seen in all eukaryotes, there is also an unusual early initiation of cytokinesis in the form of bud emergence (3). Higher plants exhibit further diversity in forming spindles without spindle pole bodies, and having three unique arrays of microtubules detached from the nuclear apparatus; the cortical array, preprophase band and phragmoplast (as described and illustrated later, Figure 3). Recognition that nonetheless some division control elements might be common to all eukaryotes began with detection in fission yeast that a late G1 control operated like the START control of budding yeast (4). In both yeasts adequate cell size is required for START, leading to DNA replication, but in fission yeast the size requirement is hidden in rich media because daughter cells are then formed at a large size that exceeds the requirement for START (5). This understanding prompted the bold experiment of testing the capacity of genes that acted at START in budding yeast (6, 7) to complement the cdc2 gene that is required at START and mitosis in fission yeast. Remarkably the CDC28 gene complemented cdc2 (8) although the yeasts had been separate for 1000 million years, indicating that some cell cycle proteins have been universally conserved (9). The cdc2/CDC28 gene encodes a protein kinase subunit that can be activated by binding to proteins termed cyclins, which have been similarly conserved in evolution (10). Cyclins are generally characterised by brief periods of synthesis in the cell cycle followed by proteolysis (11). When complexed to protein kinases, like the cdc2 gene product p34^{cdc2}, cyclins direct the enzyme to specific substrates (12). An explosion of research, illustrated in the first volume in this series, has now established that the association of cyclin dependent protein kinase (CDK) with cyclins is a universal element in cell cycle control (9), allowing a progression of CDK activities that phosphorylate changing populations of proteins that initiate transcription, DNA biosynthesis and structural changes in nucleus and cytoplasm (9, 10, 13).

Plants conform to this model in showing a cell size dependent START control (14) and containing p34^{cdc2}-like protein that shows cell cycle dependent changes in phosphorylation (15). More significantly plants contain genes that can complement cdc2/cdc28 mutants (16, 17, 18, 19, 20, 21, 22) and are inhibited by a dominant negative mutant form of cdc2 (23). Additionally plants contain genes encoding cyclin proteins that are likely equivalents of G1 and mitotic cyclins (described more fully below). However much is still to be learned about the catalysts of plant cell division; for example only recently has it been detected that plants use inhibitory tyrosine phosphorylation of p34^{cdc2} to regulate onset of mitosis (illustrated here in Figures 1 and 2) therefore confirming that this is a universal mechanism (9). Conformity with animals is less certain in the functions of the multiple cdc2-like genes that have several times been observed in a single plant (18, 20, 21, 22). Multiple cdc2-like

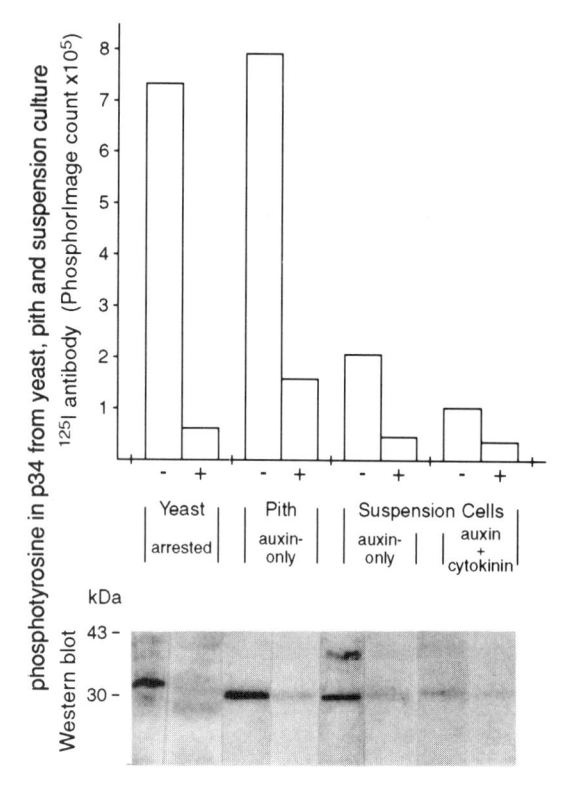

Figure 1. Phosphotyrosine in p34 protein detected with antiphosphotyrosine antibody in Western blot (lower panel). Antibody binding was strongly competed by 1 mM phosphotyrosine (+); non competed lanes (-). The p34 protein was isolated with p13 beads (68, 69) from;
Y-2, *cdc25-22* mutant fission yeast, arrested at 36 °C with tyrosine-phosphorylated p34;
Pith NAA, tobacco pith cultured without cytokinin in the auxin analogue napthylene-1-acetic acid;
suspension cells 2,4-D, cultured cells arrested without cytokinin in the auxin analogue 2,4-dichlorophenoxyacetic acid;
2,4-D + kinetin, cultured cells after cytokinin arrest was relieved by addition of the cytokinin analogue kinetin.
p34 protein was purified from the same amount of total soluble protein in each case. Monoclonal antibody (PY20) was used to probe the blot and detected with iodinated second antibody and PhosphorImager. Quantification of phosphotyrosine in p34 protein is shown in the upper panel and shows a decline when cytokinin allows progress through the control point. Observations of Zhang et al (24).

genes could indicate that each is specialised to perform a particular set of functions within the cell cycle, as diverse CDKs do in mammalian cells, however we cannot yet eliminate the possibility that presence of multiple genes simply allows hormone -induced expression of alternative cdc2 genes in different tissues with different hormone contents. This possibility is discussed below.

Control of cdc2 in plant development

The hormone inducibility of cdc2 expression plays a part in the control of cell division in plants that can now be better understood. Plants grow by the continuing formation of new cells in specialised regions termed meristems and cells displaced from these regions typically cease division and differentiate by expanding and taking on specialised structural and physiological functions. Differentiation involves a decline in cdc2 protein relative to other proteins in leaf (25, 26) and stem pith (27) tissue. When such differentiated cells are stimulated to resume division, as in response to the release of hormones in wound response, it has been observed that cdc2 protein is accumulated. This restores its level relative to other proteins to that seen in active meristems, which is twenty to forty times the basal level in differentiated cells (24, 26, 27). Several laboratories have confirmed that hormones that induce division induce cdc2 gene expression (17, 21, 28). We have argued that low levels of the key cell cycle catalyst represent an economical and secure way of curtailing division (24, 25, 26, 27, 29).

Restoration of cdc2 levels occurs in root cortical cells when root nodules are induced by nitrogen fixing bacteria, and increased levels of cdc2 mRNA have then been detected (21), but interestingly the formation of lateral root primordia may require a smaller shift in cdc2 accumulation than generally occurs in the restoration of division since the pericycle retains levels of cdc2 mRNA significantly above that of surrounding tissues although lower than the active apical meristem (28). This correlates with continuing cdc2 expression in the pericycle, detected with GUS reporter (30). Although levels are above basal in the pericycle, further localised accumulation of cdc2 mRNA has been detected at the initial stages of lateral root development in *Arabidopsis* (28) and in root segments of pea (27). The property of retaining cdc2 expression indicates that the pericycle is an extension of the root apical meristem in the sense that the cells retain significant levels of cell cycle catalysts and are programmed to soon resume division in foci that form lateral root primordia.

Experiments with lateral root formation underline two aspects of the developmental control of cell division in plants. First, adequate cdc2 protein level is necessary for division and is restored prior to division in tissues that have fully ceased dividing and attained low cdc2 level, as in cotyledon (26), pith (24) and root cortex (21). Second, presence of cdc2 enzyme is not in itself sufficient to lead to cell proliferation, indicating that the activity of the enzyme is controlled. One form of control is the regulation of whether cyclin partners are present and it has been noted that initiation of division in lateral root meristems is associated with an increase in cyclin *cyc1At* gene expression detected by GUS reporter (31) or hybridisation (32). However there is reason to believe that elevation of cyclin levels does not provide a full explanation for induction of division

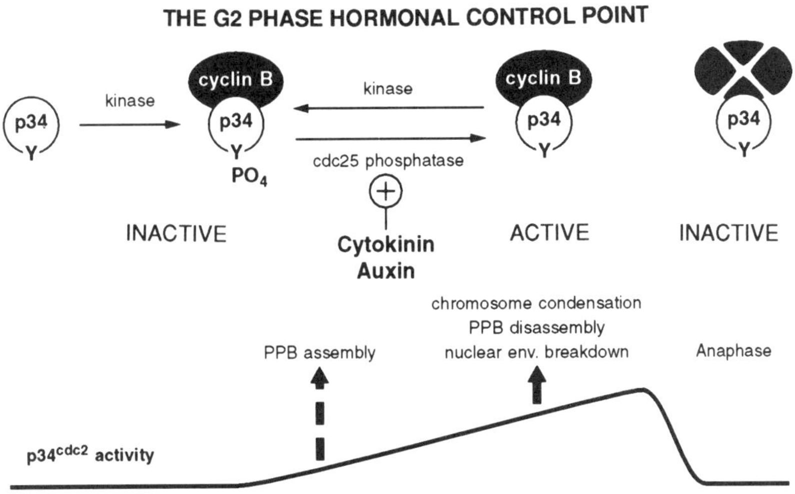

THE G2 PHASE HORMONAL CONTROL POINT

Figure 2. The late G2 control point at which cytokinin is stringently required for activation of cdc2/cyclin B H1 histone kinase in plant cells. The proteins were identified as cdc2/cyclin B because their activity appears in prophase, is recovered in the p13^{suc1} purified fraction and the same activation can be performed *in vitro* by cdc25, the specific cdc2-phosphatase of fission yeast (24, 67). See text for evidence of cdc2 and cyclin B genes in plants. The hormone dependent transition in activity can be referred to as the cytokinin control point because stringent requirement for cytokinin is not detected elsewhere in the cell cycle. However there is evidence for coordination of signal transduction at this point because a contribution from auxin can also be deduced from the behaviour of a cohort of cells arrested in G2 phase by removal of auxin. Indicated below are events influenced by rising activity of the kinase; broken arrow hypothetical effect on PPB assembly; solid arrow direct effects indicated by microinjection experiments illustrated in Figure 3.

since there is evidence that, at the level at which they are normally expressed, cyclins are not rate limiting for entry into mitosis. In fission yeast cells the presence of an additional mitotic cyclin gene (cdc13) with its natural promoter is not sufficient to advance mitotic initiation (33), although mitosis can be catastrophically advanced even into G1 phase if the cyclin is greatly overexpressed (34). Rather the timing of mitosis is regulated by the balance between phosphorylation and dephosphorylation at tyrosine in cdc2 enzyme (9). It is therefore exciting and a little unexpected that the rate of root growth in *Arabidopsis* can be accelerated by the overexpression of the mitotic cyclin *cyc1At* driven by the cdc2 promoter. It will be extremely interesting to learn whether growth enhancement derived from an increase in meristem size or an acceleration of cell cycle progress, and if the latter which phase(s) were affected. Additional restraint to cell division activity still remained since the introduced construct was observed not to induce additional lateral root primordia (32). One possibility is that the restraint is imposed by a further level of control that operates after association of cdc2 with mitotic cyclin and acts through inhibitory phosphorylation in the cdc2, as described more fully below.

This review will give particular attention to mitotic control in plant cells since two significant differences from other eukaryotes have recently emerged. One is the operation of a hormone dependent control at mitotic initiation, which is in contrast with somatic cells of all other eukaryote taxa where hormone influence is restricted to G1/S phase progression. Another difference is the interaction of the cdc2 cell cycle oscillator with a set of G2 cytoplasmic microtubules other than the mitotic spindle. It will be suggested that these differences should be seen as variations to a core of conserved cell cycle events no more extreme than can be recognised in other eukaryote taxa.

Core cell cycle

A set of probably universal events, directly concerned with driving alternate DNA replication and separation, has been identified in the fission yeast (35), where cdc2 is the sole cell cycle CDK and there are relatively few cyclins. In late G1 phase, cdc2 binds the cyclin cig2 and attains a low protein kinase enzyme activity (36) that is optimum for promoting S phase (37). In G2 phase mitotic cyclins, particularly cdc13, bind to cdc2 and allow a higher protein kinase activity, which is expressed when an inhibitory tyrosine phosphate (Y_{15}-PO_4) is removed by the specific phosphatase cdc25, so driving progress to metaphase (reviewed, 9). The activators and inhibitors of cdc2 interact in a way that normally ensures that (i) DNA duplication is not repeated, (ii) that chromatid separation waits for completion of DNA duplication and is also not re-attempted (35). Repeat replication of DNA is inhibited by the raised cdc2 enzyme activity that develops in S and G2 phases and prevents reinitiation (38, 39), probably (from evidence in budding yeast) by stimulating proteolysis of replication initiator protein (40). Completion of genome duplication prior to mitosis is promoted by requirement for completion of S phase before cdc2 is fully activated by dephosphorylation (35, 41). Repetition of mitosis when cells have entered G1 phase is inhibited by high sensitivity of Cdc2/cyclin B activity to the inhibitor protein Rum1 that is abundant in G1 phase(39, 42). Repetition of mitosis can also be prevented (from elegant evidence in budding yeast) by continuing proteolysis throughout G1 phase directed to any mitotic cyclins that might be present (42, 43).

These core events are probably common to all eukaryote cells since the cdc2 inhibitor Rum1 is functionally similar to Sic1 of budding yeast (34, 44) and to a suite of cyclin dependent kinase inhibitors (CKI) in animal cells (13, 45), and furthermore the S phase cyclin cig2 is similar in function to Clb5,6 of budding yeast and cyclin E of mammals (44). The mitotic cyclin cdc13 is equivalent to Clb1,2 of budding yeast and B cyclins of animals. In addition supplementary cyclins, which are present from S to early M phase and are only poorly or not at all able to support mitosis if they are engineered to be the sole cyclin present, may also be universal since they are represented by cig1 in fission yeast, Clb3,4 in budding yeast and cyclin A in animals. Because of the capacity of cyclin A to contribute at S and M phases it has been suggested it might represent an evolutionary persistence from cells with a single CDK and perhaps a single cyclin (34). In plants the available, much less complete, evidence suggests that similar core molecules and events occur since the plant $p34^{cdc2}$ is phosphorylated on tyrosine prior to mitosis (shown here, 24) and B-like cyclins are expressed specifically at mitosis (46, 47) and furthermore ectopically expressed B-like cyclins can accelerate division (32).

However significant differences between taxa in the operation of the core events will be described below. Additionally variation is seen in whether the cell cycle involves CDK activity early in G1 phase, as seen in animal cells and in budding yeast, or lacks this early activity as in fission yeast. These variations raise the question of whether plants have also evolved specific variations in cycle events and the control of cycle progress.

G1 cyclins

Budding yeast and animal cells accumulate in early G1 phase a special class of G1 cyclins that form active complexes with CDKs (13, 45,48). Some equivalence between these yeast and animal G1 cyclins is indicated by the ability of mammalian early G1 cyclins, such as cyclin D, to complement lack of all three of the budding yeast G1 cyclins (Cln1, Cln2, Cln3). However, equivalence of these cyclins may be limited since yeast and animal G1 cyclins differ in sequence more than do members of the mitotic cyclin family and some complementation of cln can be detected with all known classes of cyclin (49). This complementation may simply reflect the requirement for low CDK activity for initiation of S phase (37, 38) and show that this low activity can be supplied by cdc28 in complex with a heterologous cyclin. Clear differences between the functions of mammalian cyclin D and the budding yeast G1 cyclins are indicated by (i) the requirement of growth hormone for induction of cyclin D, (ii) the ability of the animal cyclin to remain at elevated levels through all cycle phases indicative of not directly driving particular core events in DNA replication and separation, and (iii) capacity to form active complexes with more remote variants of the cdc2 kinase, most universally with CDK4 and CDK6 (44, 45, 49). Whereas budding yeast G1 cyclins are not induced by hormone, cln1 and cln2 peak in abundance in G1 phase, are eliminated in other phases and all bind the single cdc28 kinase (42). These differences indicate that the mammalian G1 cyclins participate in signal transduction leading to cell growth after hormone stimulation. Whereas in budding yeast the three Cln cyclins are required for the unusual complexity of sustaining three parallel activities concerned with the core of the cell cycle; early mitosis represented by spindle pole duplication, and early cytokinesis represented by bud emergence, together with DNA synthesis (50). In most cell types events of mitosis and cytokinesis occur after completion of DNA replication.

Plants have cyclins with some structural similarity to the diverse group of G1 cyclins. Genes cloned from *Arabidopsis* by complementation of CLN deficient yeast contain a motif implicated in the binding of G1 cyclins to the Rb protein in animal cells (51). Rb protein restrains DNA replication, by sequestering transcription factors for DNA replication genes, until animal cells are committed to division (45, 49). The possibility that plant G1 cyclins also participate in signal transduction and may act through directing CDK activity to an Rb-like protein is an exciting area of current study.

Mechanisms of CDK activation at S phase

There are also important differences between taxa in the mechanisms by which CDK activity is raised in initiating S phase. In both fission and budding yeast in late G1 phase the single cdc2/cdc28 kinase is in complex respectively with cyclins Cig2 or Clb5,6 and not detectably inhibited by phosphorylation nor dependent upon cdc25 phosphatase for activation (37, 38). The kinase can therefore be activated by reduction in level of CDK kinase inhibitor (CKI); Rum1 in fission yeast (39, 44) or Sic1 in budding yeast (52). In contrast, animal cells use removal of tyrosine phosphorylation from

CDK at S phase as well as using it at mitosis. In animal cells at late G1 phase the cdk2 variant of cdc2 complexes with cyclin E and is held inactive by Y_{15}-PO_4 then activated at initiation of S phase by the specific tyrosine phosphatase cdc25A (53). In animal cells decline of CKI inhibition does not seem to play a role in the normal progression through G1 phase and S phase. Although many CKI proteins are present in mammalian cells their function is to halt the cell cycle in G1 or S phase in response to external signals or DNA damage (13, 45, 49). In plants we cannot predict from the Y-PO_4 control of cdc2 at mitosis whether cdk2 is present and subject to the same mechanism at S phase. This possibility requires investigation since the multiplicity of plant cyclin genes is more consistent with the multiplicity of CDKs in animal cells than with the minimal core exemplified by fission yeast, however the implication of the diversity between eukaryotes being presented here is that plants need not correspond closely with either.

Plant cyclins and CDKs

Sequence comparisons have begun to identify groups of plant cyclins and to align them with cyclins of other eukaryotes. Likely G1 cyclins (51) and mitotic cyclins active in maturation of *Xenopus* oocytes have been cloned from *Arabidopsis* (54) and alfalfa (55) and mitotic cyclins from snapdragon (46). There has been a welcome convergence of classifications since a set of four maize cyclins, tentatively identified from sequence comparison as like animal cyclins A and B (57), has been matched with five from soybean (47), and Kouchi et al (47) have concluded that it is valid to retain the classification of Soni et al (51) and to recognise in plants; D-type G1 cyclins; two groups of likely A-type cyclins α1 and α2; and two of likely B-type cyclins β1 and β2. This classification is supported by detection of differential expression of mRNA in cell cycle phases (56) with examples of putative cyclin B in metaphase cells (46, 47) and cyclin A (*cyc3Gm*) in S phase cells recognised by presence of histone H4 mRNA (47). Interestingly *cyc1Gm* mRNA detection was consistent with presence during only a short proportion of S phase, suggesting that it belongs in a novel class of cyclins (47) and might provide a further instance of variation in plants.

The plant kingdom currently lags behind other taxa in identification of the CDK subunits and associated proteins with which the cyclins associate. A significant observation is the detection of a protein kinase activity that peaks at S phase, in immunoprecipitates obtained with anti-human cyclin A from synchronous cells of alfalfa (58). This is consistent with findings in animal cells but the identity of the CDK(s) in the plant cyclin A complex is not yet clear. Analogy with animal cells indicates it might be CDK2, CDK4 or CDK6.

Further evidence that raises this possibility is the cloning from *Arabidopsis*, by complementation of cln mutants in yeast, of G1 cyclins with a sequence motif characteristic of the Rb binding site in cyclin D (51). Cyclin D can complex with CDK 4 and CDK 6 in animal cells. CDK variants could be present in plants because additional bands in Southern blots of plant DNA probed for cdc2 and also partial sequences observed in PCR products from cdc2-based primers are indicative of multiple cdc2-like genes. However proof that they encode proteins with functions of the animal CDKs is lacking.

The plant genes that might be considered as functional variants of cdc2 for which complete sequences are available, are the cdc2-like genes that have been found not to complement cdc28/cdc2 in yeast when taken from alfalfa (18), rice (22) and wheat. In these plants two cdc2-like genes have been found with 82-92% sequence identity but differing ability to complement. Non-complementation could be due to inadequate expression, or more significantly could indicate that the gene has only a subset of the cdc2 functions, as might be expected of a CDK that is active only in certain cycle phases. Particularly intriguing is the pair of cdc2-like genes in alfalfa, one of which complements a cdc28 mutant defective at the G1/S function while the other complements loss of G2/M function (18). This has lead to consideration of the possibility that non complementing genes could be CDK2, which is close in sequence to cdc2 but activated only at G1/S transition. However deduction from noncomplementation is not clear-cut since some laboratories find cdk2 to complement cdc28 significantly(59). We have used sequence analysis to evaluate two cdc2-like genes that we have isolated from wheat, only one of which could complement cdc28 mutants although both gave detectable protein products when expressed in yeast. Sequence analysis showed that the non-complementing gene has no correspondence of amino acids with animal cdk2 at points where cdk2 differs from cdc2. Neither is there correspondence with the non-complementing alfalfa (18) and rice (22) cdc2-like genes in the location or the identity of amino acids where the complementing and non-complementing members of each pairs differ (Chongmei Dong and P.C.L. John unpublished observations). We conclude that plants could have cdc2-like genes with specialised functions at G1 and S phase but in terms of protein structure there is not yet any indication that they form a distinct class, nor that they resemble animal cdk2.

Variation in preconditions for mitotic activation of cdc2

Plants contribute to the evidence that at mitosis eukaryotes differ in the prior events that must be completed before activation of cdc2/cyclin B.

Extensive evidence from fission yeast and animal cells indicates that a DNA checkpoint, monitoring completion of DNA replication and repair, holds the cdc2 enzyme inactive by preventing dephosphorylation of Y_{15}-PO_4 (9, 11, 41). However, in budding yeast DNA-checkpoint arrest does not require Y-PO_4 but operates by an unknown mechanism. Instead Y-PO_4 in budding yeast cdc28 is coupled to the successful deployment of the actin cytoskeleton in formation of a daughter bud (60). Disruption of the actin cytoskeleton prevents normal bud development and prolongs the inhibitory phosphorylation of cdc28/cdc2. In principle other cells, which use the cdc2 Y_{15}-PO_4 mechanism to prevent mitosis until genome duplication is complete, could additionally use Y_{15}-PO_4 to restrain mitosis until other preconditions are also met. Plants now provide an example of this additional use.

Plants indicate that a hormonal checkpoint in late G2 phase can control cdc2 activity through Y-PO_4, since presence of adequate cytokinin is required for progress through a mitotic initiation checkpoint at which cells can arrest with cdc2 containing Y-PO_4 and requiring only dephosphorylation by cdc25 for activation (illustrated below). Thus plants resemble the lower eukaryote budding yeast in being able to link tyrosine dephosphorylation of cdc2 at mitosis to an event other than DNA replication.

HORMONAL CONTROL OF CELL CYCLE PROGRESSION

Plant and animal cell proliferation hormones

More than ten peptide hormones are known to influence animal cell proliferation and some of these hormones are targeted to particular cell types. Whereas in higher plants only two groups of hormones, the auxins and cytokinins are generally stimulatory to the proliferation of most cells types (61). The importance of auxins and cytokinins is underlined by the fact that in culture they readily support cell proliferation and in the intact plant raising their level by *Agrobacterium tumefaciens* is sufficient to induce tumours. One of the unresolved questions of plant development is how division control can be exerted by the limited set of plant hormones with sufficient subtlety to allow the development of complex structures while the presence of cell walls prevents adjustment of cell position by migration. A partial answer may be that subtlety of hormonal response is increased by two factors; one is that different ratios of hormone concentration, rather than absolute concentrations, can induce different responses. The significance of ratios is underlined by the capacity of lowering auxin level by genetic means to produce a similar developmental effect as raising cytokinin level (62). Another factor is that the developmental identity of the responding cell can greatly influence its sensitivity to a hormone.

Requirement of different concentration ratios of auxin and cytokinin for cell proliferation is clearly illustrated by roots and shoots which, in conformity with different hormone responses, have different tissue distributions and patterns of initiating new lateral outgrowths. Root cell division is inhibited by the high ratios of cytokinin to auxin that stimulate shoot cell division (63). Nevertheless both root and shoot tissues require some cytokinin for proliferation and it should also be recognised that even in shoots cytokinin concentration is at least one order of magnitude lower than auxin. It may be because the cytokinin hormones are present at lower concentrations and not usually directly limiting for growth that plants have evolved a sensitive cytokinin checkpoint and that it is located among the later events of the DNA-division sequence at mitotic initiation (Figure 2).

In contrast, no hormonal influence at mitotic initiation has been detected in mammalian cells or budding yeast, although they are influenced respectively by peptide growth hormones and peptide mating hormones. Both these effectors influence progression through G1 phase, when growth hormones stimulate and mating factors inhibit. In cultured animal cells growth hormone stimulation is required prior to a transition, or restriction point (64), in mid G1 phase after which progress to S phase and mitosis can continue without stimulation (65). In budding yeast a transition has been found in late G1 phase, prior to which presence of mating factor can block commitment to division by inhibiting cdc28/Cln1,2 complex, with the result that cells can fuse with a mating partner (2, 3, 50).

There is a similarity in the G1 arrest of plant cells deprived of auxin, with the G1 arrest of hormone depleted mammalian cells. In neither case is it known if the main function of the growth hormone is to activate division events and perhaps secondarily to stimulate growth by cross talk between activated division kinases and biosynthetic enzymes. Alternatively hormones could act primarily on growth and only trigger division indirectly by stimulating growth to a size that starts division. The latter possibility conforms with control in unicellular organisms, which grow until cell size triggers division (2, 3, 4), but in multicellular tissues there is not a competition between autonomous individual cells for resources. Rather in multicellular tissues growth and division are both restrained in forming organs of controlled size. The possibility of a continuing interaction, with activated division proteins stimulating biosynthesis in higher eukaryotes remains unresolved in both plants and animals. However the hormone dependent mitotic control point appears to be unique to the plant kingdom.

THE G2 PHASE HORMONAL CHECKPOINT

Hormonal requirements for plant cell proliferation are often detectable in culture. The commonly observed requirement for auxin could arise because auxin synthesis is largely restricted to young leaves and developing seeds, therefore the proliferating cells at root and shoot tips and in seedlings receive auxin transported to them from young leaves or seed reserves (66) and this dependence is continued by cells in culture. A contribution from cytokinin is sometimes less obvious in culture since cells often synthesise enough cytokinin to support proliferation. This could arise because cytokinin has consistently been observed to be present in tissues containing proliferating cells (61) and it will be argued below that this is because cytokinin is required and synthesised at mitotic initiation. However, prolonged supplementation with cytokinin in suspension culture allows cytokinin synthesis to be lost, presumably by random mutation. It then becomes experimentally possible, by removal of cytokinin, to study where cytokinin is required in the cell cycle. Experiments of this sort have not detected cytokinin dependent events in G1 phase or S phase but a stringent requirement for cytokinin was detected at the transition from late G2 phase to mitosis (27).

Cells arrested at the cytokinin control point are close to mitotic initiation since addition of cytokinin induced the rapid activation of p34^{cdc2} H1 histone kinase activity recoverable on p13^{suc1} beads, the appearance of mitotic nuclear configurations and then doubling of cell number (24). The increase in enzyme activity occurred without increase in p34^{cdc2} enzyme protein detectable by antibody against the peptide sequence EGVPSTAIREISLLKE (PSTAIR) universally conserved in cdc2 (9). The mechanism of activation was a decline in tyrosine phosphorylation (Y-PO$_4$), which has been clearly established by extensive genetic and biochemical analysis as the means of cdc2 activity control at mitotic initiation in fission yeast and animal cells where dephosphorylation of cdc2 is catalysed by highly specific cdc25 phosphatase (9, 11, 41, 67). Change in plant cdc2 Y-PO$_4$ was detected in two ways. One was by direct probing with antiphosphotyrosine antibody in p34 protein that was purified with p13^{suc1}. This detected Y-PO$_4$ in p34^{cdc2} from *cdc25ts* mutant cells of *S. pombe* at restrictive temperature and detected equal abundance of Y-PO$_4$ in the plant p34 purified with p13^{suc1} from G2 arrested cells and also detected a decline in Y-PO$_4$ when the cytokinin block was released (Figure 1). A second form of evidence was the extent of activation that could be obtained *in vitro* using cdc25-GST enzyme expressed in *E. coli*, which has been shown to specifically remove the Y$_{15}$-PO$_4$ of cdc2 (67). Our preparation of cdc25 enzyme was effective in activating yeast p34^{cdc2} containing Y$_{15}$-PO$_4$ that was obtained from *cdc25ts* mutant *S. pombe* and it was equally effective in activating the plant enzyme from cells arrested at the cytokinin checkpoint. Significantly, as plant cells progressed into mitosis after cytokinin stimulation, preparations of the plant enzyme became progressively less responsive to cdc25 phosphatase, indicating that progressive dephosphorylation of Y-PO$_4$ in p34^{cdc2} is part of plant prophase (24).

The G2 phase hormonal checkpoint may be used generally in plant cells. It is not restricted to cells transferred to suspension culture, since it has also been detected in freshly excised tobacco pith tissue. This tissue, when cultured on auxin without cytokinin (but not in controls containing both hormones), also arrested with fully-induced p34^{cdc2} containing Y-PO$_4$ and with low catalytic activity that could be released by treatment with cdc25. Cytokinin alone did not induce the enzyme, nor increase the amount induced by auxin, although it was essential for enzyme activation (24, 27).

There have been earlier indications of a role for cytokinin in division. The first observations of cytokinin effects included appearance of some binucleate cells when excised tobacco pith tissue was cultured with auxin but without cytokinin (70). This lead to naming of the hormone for a suspected role in cytokinesis, but this has not been borne out and may reflect difficulties in working with freshly excised tissue, where hormones are carried over from the intact plant or released by wounding response. By allowing wound response to subside in excised tobacco pith Simard (71) deduced that auxin without cytokinin could stimulate only DNA synthesis but not completion of the cell cycle, which is consistent with our data. Cultured cells have proved easier to bring to at least partial dependence upon supplied hormone. Some varieties of soybean have yielded callus in which mitosis is delayed by absence of supplied cytokinin, leading to the suggestion of a role for cytokinin in G2 phase (72). This has not been universally accepted (73), perhaps because the soybean callus cells did not arrest entirely in G2 phase. However we note that the results from soybean callus could be consistent with ours if the callus synthesised enough endogenous cytokinin to allow very slow progress through the G2 checkpoint.

Plant cells may synthesise at least part of the cytokinin they require for mitosis at the time it is required, since accumulation of cytokinin in a cytokinin autonomous culture of *N. tabacum* L. cv.

Xanthi shows a three fold peak in cytokinin concentration in late G2 phase but not at other times (74). This is exactly the time when our cells that are not autonomous for cytokinin showed a stringent requirement. This consensus suggests that cytokinin normally increases at the critical G2/M transition. This conclusion is consistent with the general finding of raised concentration of cytokinin in all the meristematic tissues of the plant, such as roots, shoot tips and developing fruits (61).

The endogenous synthesis of cytokinin in proliferating tissues is in conformity with general differences between animal and plant hormone action. Animal hormones are synthesised in specialised tissues and exert their effects after being circulated to distant target tissues. Plant hormones may similarly be transported and exert effects at a distance. Indeed one of the actions of cytokinins provides an example of this since the transport of cytokinins from roots to leaves delays leaf senescence (75). But plant hormones can also exert some of their effects in the tissues in which they are synthesised. Ethylene is an example of a hormone that acts entirely in the tissue in which it is synthesised (76). However ethylene mediates response to changes in the environment and it is not one of the major hormones that regulate vegetative growth (76). It is therefore novel to suggest that cytokinins can not only exert long distance effects but also contribute to the division of the cell in which they are synthesised. Since dividing cells in plant tissue are not usually synchronous it has to be postulated that the mitotic accumulation of cytokinin by an individual cell is significant for its own division. It is not suggested however that cells in the intact plant are fully autonomous for cytokinin, except perhaps in established apical meristems. It is postulated that; (1) the group of cells in a meristem, particularly of the root, raise the local level of cytokinin thus contributing to the autocatalytic nature of meristems; (2) exogenous cytokinin transported from elsewhere can stimulate proliferation by accelerating the G2/M phase transition and that tissues that have become division quiescent are initially particularly dependent on exogenous cytokinin, as seen in tobacco stem pith; (3) the G2 control point can also be acted on by auxins and perhaps other hormones, or by excess hormone concentrations. A contribution from auxin has been detected at this point.

Auxin requirement at the G2/M phase transition is indicated in suspension culture by cells transferred to medium lacking auxin (but with cytokinin) which arrested both in G1 phase and G2 phase (24, 27). Arrest in G1 phase can most simply be explained as resulting from inadequate growth of cells that were in G1 phase when deprived of auxin. The similarity of this with the G1 phase restriction points in animal and yeast cells has been discussed above. However the G2 phase arrest point due to inadequate auxin is indistinguishable from that caused by lack of cytokinin since, on restoration of auxin, the cohort of cells previously arrested in G2 activated $p34^{cdc2}$ and entered mitosis as rapidly as cells did when responding to re-supply of cytokinin.

We conclude that plants integrate hormonal information at the G2 control point (Figure 2). There is evidence for this in the intact plant where arrest at G2 phase can occur both in response to starvation and as part of normal development. For example cells of the root pericycle arrest in G2 phase (77) and normally resume division at an appropriate distance from the apical meristem, forming new lateral root meristems. This process can be induced closer to the root apex by exogenous auxin that has been proposed to act by establishing a favourable ratio of auxin to cytokinin. This ratio is proposed to form the basis of distancing lateral root initiation from the root tip, where the ratio of cytokinin to auxin is suggested to be too high for initiation (63). Arrest of pericycle cells at the hormonal check point is indicated by the rapid increase in $p34^{cdc2}$ activity in root tissue during the induction of lateral root meristems by auxin (27) which is consistent with the presence of $p34^{cdc2}$ awaiting activation by tyrosine dephosphorylation (24).

It is possible that in different tissues plant hormones other than auxin and cytokinin may influence $p34^{cdc2}$ activation in the same way. It is also possible that in the intact plant cytokinin contributes to cell proliferation not only at mitosis but earlier in the cell cycle by promoting growth. A stimulation of long term growth by cytokinin must occur in dependent tissue since absence of cytokinin will arrest proliferation and therefore in turn limit growth. A more direct involvement in early cell cycle events would be indicated if cytokinin acted in the initiation of growth. In this connection it is intriguing that both cytokinin and auxin affect the expression of one of the three cyclin D-like genes in Arabidopsis. Cyclin δ3 is induced by cytokinin or sucrose within 4h in Arabidopsis suspension culture, however simultaneous presence of auxin completely overrides the induction (51). The induction is reminiscent of the induction of cyclin D in animal cells and the parallel is underlined by the presence of Rb binding motif in both these animal and plant cyclins, however the repressive effect of auxin prevents any simple hypothesis concerning signal transduction to proliferation because auxin stimulates rather than inhibits plant cell proliferation.

HORMONES AND THE SWITCH TO APOPTOSIS

Although plant development has given rise to the term apoptosis, to describe the process of leaf death in autumn, the phenomenon of apoptosis seen in animals is very different in its induction and significance for the survival of the organism. Animal apoptosis results in the rapid and complete removal of a cell that is incompatible with further development, or is producing antibodies that are hostile to the animal's own proteins, or has damaged control of cell proliferation due to presence of a virus or mutation and therefore could cause a metastasising cancer (reviewed 78). The same risk is not present in plants where development is largely by tip growth and cell walls prevent metastasis. Absence of apoptotic induction of the type seen in animal cells is indicated by the cytokinin and auxin deprivation experiments described above, in which cells remained viable for many days in a state of hormone imbalance. This is a common finding in plant cell culture but is in sharp contrast with normal animal cells, in which hormonal interruption of proliferation signals abnormal cycle progress and triggers apoptosis that is rapidly executed by proteins that are already present but latent in proliferating animal cells (reviewed 79).

Analysis of processes in plants that could be likened to animal apoptosis reveals crucial differences. In animal embryology programmed cell death is used to completely remove cells that are inappropriate in later stages of development (78). Conversely, in plant development the most common incidence of programmed cell death is the terminal differentiation of water conducting vessels and other lignified (woody) cells, which depends on persistence of the cell wall. Similarly in plants the response to pathogens that is referred to as hypersensitive cell death depends upon leaving the dead cells containing toxic chemicals, as a barrier to the pathogen (reviewed 80). Furthermore, in seed germination the dissolution of the endosperm releases cell contents as nutrients for neighbouring cells in the embryo (81), whereas release of contents is avoided by animal apoptosis because inflammatory to adjacent cells. One similarity that has been noted between plant and animal apoptosis is the presence of DNA fragments in regular sizes (seen as a ladder on electrophoresis) indicative of nuclease attack while still organised in nucleosomes. This is not a universal finding in animal apoptosis (82) and in view of the other differences between the processes in plants and animals should perhaps not be taken as evidence that many animal apoptosis mechanisms will be found in plants.

It is however likely that plants will require proteins that share some of the functions of the proteins that initiate the early stages of animal apoptosis. A common trigger for apoptosis in animal cells is the detection of DNA damage by the DNA-binding protein p53 which then induces the CKI p21 that can arrest DNA synthesis by inhibiting CDK enzymes (79). The molecular mechanisms that subsequently lead from such arrest to the initiation of apoptosis are not yet identified so they cannot easily be sought in plants. However a role for proteins like p53 and p21 can be anticipated in plants since the sessile photosynthetic life style carries a high risk of U.V.-induced DNA damage.

MITOTIC EVENTS DRIVEN BY PLANT p34^{cdc2}

A similar relationship between p34^{cdc2} activity and mitotic events in plants as in other eukaryotes (9) is indicated by the increase in p34^{cdc2}-like H1 histone kinase activity that is coincident with prophase (29) and mediated by decline in Y-PO$_4$ (24). However regulation of the cytoskeleton in division is more complex in plants, at least in terms of the number of elements that must be controlled. Animal cells can be thought of as having a single microtubular cytoskeleton permanently associated with the centriole and changing at mitosis only by increase in number and decrease in length of microtubules (MTs). Whereas plant cells have a succession of four different MT arrays (Figure 3), only the spindle has any direct physical contact with any nuclear structure and forms directly on the nuclear surface not from a centriole. The plant interphase MTs in G1 phase and S phase are not oriented with respect to the nucleus but are arranged in the cortical cytoplasm just below the outer membrane where they form an extensive array responding to physical stress and participating in alignment of cellulose microfibrils that form parallel to them during wall synthesis (83). As mitosis approaches in G2 phase there continues to be no direct physical association of MTs with a nuclear structure, rather MTs form a preprophase band (PPB) that encircles the cell, still at its periphery but usually in the region of the nucleus (84). At prophase the mitotic spindle is then a third new array that forms *de novo*. Spindle formation begins with the assembly of separate MTs on the outer surface of the nuclear envelope, which has been hypothesised to act as an organising role (85) until a coherent spindle has formed. Finally, at telophase the phragmoplast forms between the daughter nuclei and transports materials to the assembling cross wall which extends centrifugally towards the existing wall (Figure 3). The phragmoplast again accepts positional cues not derived from the nucleus since it extends towards the position of the previous PPB, and if the nucleus has been moved during mitosis the phragmoplast will curve and junction at the marked position (86, 87). The importance of this positional information is underlined by the distorted cells and tissues that are formed in mutant *Arabidopsis* lacking PPBs (88).

PLANT CELL DIVISION AND p34^{cdc2} KINASE ACTIVITY

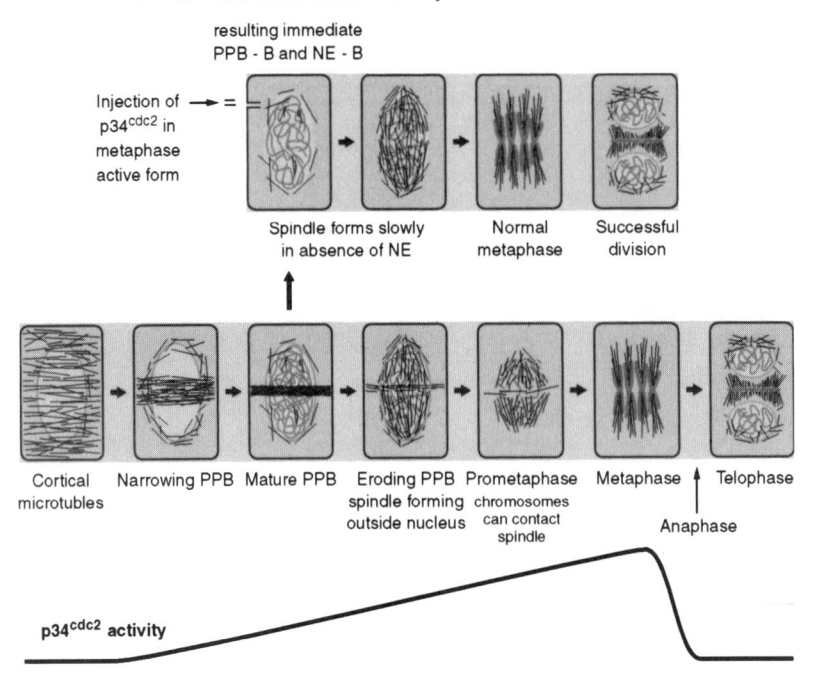

Figure 3. Impact of rising cdc2 kinase activity during prophase on the structural events of mitosis, shown in the lower horizontal sequence. A causal effect of cdc2 activity in driving PPB disassembly is indicated by the effects of injecting (vertical arrow) p13^{suc1}-purified metaphase-active mitotic kinase, containing cdc2-like and cyclin B-like proteins, from conditionally-arrested metaphase cells of the unicellular plant *Chlamydomonas*. The illustration shows effects seen (Figure 4) in stamen hair cells of *Tradescantia* flowers (96). An acceleration of events (shown in upper horizontal sequence) was seen in early prophase cells injected with active metaphase kinase, which within 2-3 min disassembled the PPB, also accelerated chromatin condensation and halved time to nuclear envelope breakdown; all of these occurring before assembly of the mitotic spindle in a dislocation of normal timing. Injections made in other cycle phases showed other microtubule arrays to be insensitive to plant metaphase kinase (96).

There are therefore unique cytoskeletal elements in plants and understanding their control in the cell cycle presents problems that must be solved by investigating directly in plant cells. Some information is now available concerning the PPB. It has been hypothesised that p34^{cdc2} is involved in control of the PPB since immunostaining has detected an association of PSTAIR-containing protein with the PPB in onion root (89) and antibody specific for the maize p34^{cdc2} showed staining in the PPB region in about 10% of cells with PPBs (90). As with all localisations of soluble proteins in fixed cells there must be concerns that the final pattern may have been distorted by localised precipitation during fixation and by uneven losses during immunoprobing (91), however it seems that p34^{cdc2} is at least in the vicinity of the PPB.

MPF and PPB

It has been suggested that p34^{cdc2} plays a part in the assembly of the PPB (90), but conversely the effect of okadaic acid (a protein phosphatase inhibitor) on cells of *Nicotiana plumbaginifolia* in synchronous cultures had earlier indicated that the enzyme drives disassembly of the PPB. Okadaic acid blocked events that normally accompany PPB breakdown but allowed normal timing of p34^{cdc2} activation and PPB breakdown, which were therefore deduced to be causally related (92). Consistent with this hypothesis the protein kinase inhibitor staurosporine blocks PPB breakdown (93). To test the hypothesis directly the technique of injecting fluorescently tagged tubulin into live cells (94,95) was used to measure the effect on microtubules of co-injecting plant p34^{cdc2} (96).

To derive a CDK preparation specifically active at mitosis we isolated a conditional mutant of the unicellular green plant *Chlamydomonas* that arrests in metaphase. The plant metaphase kinase was purified by affinity for the essential mitotic protein p13^{suc1} (69) and it contained proteins recognised by cdc2 and cyclin B antibody, thus in composition it resembled partially purified animal MPF (96). We cannot eliminate the possibility that CDK2-like protein could be present in *Chlamydomonas* but since this form of enzyme is not active at metaphase it is not responsible for effects that we observed. Control enzyme was purified in the same way from cells of the same age that, at permissive temperature, had completed mitosis and inactivation of p34^{cdc2}. Live stamen hair cells of

CHAPTER 6/ MOLECULAR MECHANISMS OF PLANT MITOTIC CONTROL

min. (D), transmission image; (E), (F), fluorescence image of the same cell in mid plane and cortex, showing that no PPB is visible. Observations of Hush et al (96). The effect on the PPB is summarised in Figure 3.

Tradescantia were used because the duration of division stages have been determined and can be recognised by Nomarski DIC microscopy (94, 95). In these cells the technique of injecting carboxyfluorescein-tubulin (CF-tubulin) to study rates of MT assembly/disassembly has been established. CF-tubulin when injected with control enzyme was incorporated normally (within 2-3 min) into the MTs of interphase arrays, spindle and phragmoplast and when injected into early prophase cells was incorporated into the PPB (Figure 4 A,B,C). In contrast the active metaphase kinase injected into early prophase caused such rapid disassembly of the PPB that no incorporation of CF-tubulin occurred (Figure 4 D,E,F). The injected enzyme also caused an acceleration of chromatin condensation spreading from the side of the nucleus closest to the point of injection followed by halving of the time taken to reach nuclear envelope breakdown. The active enzyme therefore accelerated some mitotic events with no evidence that any abnormal process occurred since nuclear division was carried through to normal completion (96). Therefore plants can yield metaphase kinase that contains $p34^{cdc2}$/cyclin B-like protein and in function resembles MPF (maturation or mitosis promoting factor) of animal cells.

Positive marking of microtubules for disassembly by MPF

The plant MPF enzyme was highly selective in its action on MTs since it did not depolymerise microtubules of the interphase cortical array, spindle or phragmoplast. There is a particularly interesting discrimination between the G1 phase cortical microtubules and the PPB, both of which are located just below the outer membrane. The possibility that the PPB microtubules are inherently less stable can be eliminated, since measurement of the rates at which MTs are reassembled with CF-tubulin in photobleached regions of live cells has shown very similar half lives close to 1 minute in PPB and in cortical MTs (95). It is therefore suggested that MTs assembled in the PPB are specifically destabilised at prophase and the insensitivity of other MTs to MPF suggests that those in the PPB must be marked in some way.

Stability of the G1 phase cortical MTs and the phragmoplast MTs could not have been predicted since they do not normally encounter active $p34^{cdc2}$. The stability of these structures suggests that tubulin itself is not acted upon by $p34^{cdc2}$/cyclin B, rather additional proteins must be present in the tubules governing sensitivity to MPF. In principle

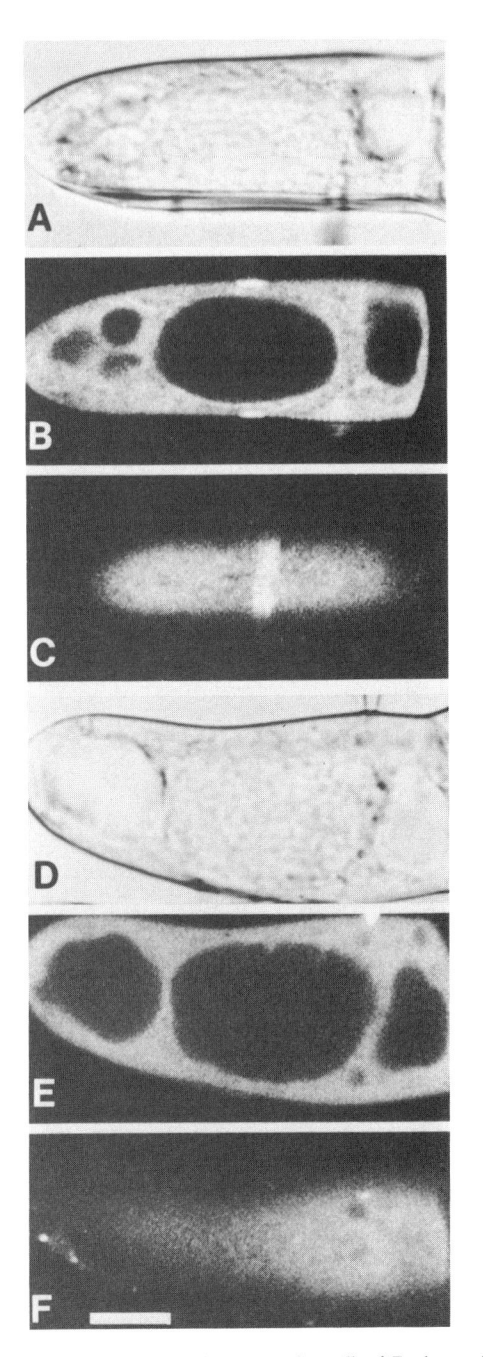

Figure 4. Effects of microinjection into live cells of *Tradescantia* plant metaphase kinase purified by affinity for the yeast mitotic protein $p13^{suc1}$ (68, 69). Illustrated are injections into terminal cells of growing stamen hairs.
(A) (B) (C) control cell seen 11 min after injection with CF-tubulin monomer (94) together with inactive kinase obtained from control *Chlamydomonas* cells that had completed mitosis at permissive temperature. (A) transmission from image, (B), (C) fluorescence image of the same cell in the mid plane and cortex, showing the bright PPB encircling the cortex, made visible because of the incorporation of CF-tubulin.
(D) (E) (F) equivalent cell injected with CF-tubulin and active metaphase kinase from blocked *Chlamydomonas*, seen after 10

the modifying proteins could be stabilising or destabilising. In the spindle, which normally resists active MPF, we cannot discriminate between either presence of stabilising proteins or an absence of proteins targeted by MPF. However we can make a tentative deduction concerning the MTs of the cortical array and of the phragmoplast that they do not contain proteins stabilising against MPF since they do not normally encounter the active enzyme. Therefore their stability in the presence of active MPF is simply due to the lack of positive marker proteins that can be targeted by MPF. In conformity with this deduction it is proposed that PPB MTs do associate with proteins that positively mark them for MPF sensitivity. Two sorts of protein that could confer this MPF targeting are MAPs (microtubule associated proteins) that alter the rates of MT assembly/disassembly, such as the ubiquitous MAP4 of animal cells that is phosphorylated by $p34^{cdc2}$/cyclin B (97), or oligomeric MT severing proteins whose activity is increased by MPF, as detected in *Xenopus* (98).

It should be noted that the converse proposal, that $p34^{cdc2}$ plays a part in assembly rather than breakdown of the PPB (90), cannot be completely discounted from present evidence, since we cannot eliminate the possibility that low $p34^{cdc2}$/cyclin B kinase activity early in G2 phase phosphorylates an unknown MAP that catalyses PPB assembly. There is no direct evidence for this, but low and high levels of $p34^{cdc2}$, albeit normally in complex with different cyclins, have profoundly different effects in driving S phase or M phase in *S. pombe* (34, 44, discussed earlier).

The precedent of metaphase kinase control of the PPB suggests that other MT arrays could be actively disassembled in this way by different CDK activities. There is evidence against the alternative possibility, that MT arrays are passively eroded by the leaching of tubulin to later-forming arrays. This leaching might occur in view of the transfer of tubulin monomer from PPB to spindle during prophase (97) but is not causal from the evidence of direct disassembly of the PPB by MPF, which occurs without waiting for spindle formation (Figures 3, 4). An intriguing possibility is therefore raised that CDK/cyclin complexes that become active at G1/S phase and at telophase/interphase respectively could drive disassembly of the cortical array and the phragmoplast. This predicts that unique cyclins, perhaps unique CDKs, could be found at these arrays. Clearly many fascinating questions remain.

ACKNOWLEDGEMENTS

I thank J. Hayles, B.E.S. Gunning and D.S. Letham for advice and encouragement.

REFERENCES

1. Nurse, P. and Thuriaux, P. (1977) *Exp. Cell Res.* **107**, 365-375.
2. Johnston, G.C., Pringle, J.R. and Hartwell, L.H. (1974) *Exp. Cell Res.* **105**, 79-98.
3. Hartwell, L.H., Culotti, J., Pringle, J.R. and Reid, B. (1974) *Science* **183**, 46-51
4. Nurse, P. and Bissett Y. (1981) *Nature* **292**, 558-560.
5. Nasmyth, K., Nurse, P. and Fraser, R. (1979) *J. Cell Sci.* **39**, 215-233.
6. Reed, S.I. (1980) *Genetics* **95**, 561-577.
7. Reed, S.I. (1992) *Annu. Rev. Cell Biol.* **8**, 529-561.
8. Beach, D., Durkacz, B. and Nurse, P. (1982) *Nature* **300**, 706-709.
9. Nurse, P. (1990) *Nature* **344**, 503-508.
10. Norbury, C. and Nurse, P. (1992) *Annu. Rev. Biochem.* **61**, 441-470.
11. Evans, T., Rosenthal, E.T., Youngblom, J., Distel, D. and Hunt, T. (1983) *Cell* **33**, 389-396.
12. Peeper, D.S., Parker, L., Ewen, M.E., Toebes, M., Hall, F.L., Xu, M., Zantema, A., van der Eb, A.J., Piwnica-Worms, H. (1993) *EMBO J.* **12**, 1947-1993.
13. Hunter, T. and Pines, J. (1994) *Cell* **79**, 573-582.
14. Donnan, L. and John, P.C.L. (1983) *Nature* **304**, 630-633.
15. John, P.C.L., Sek, F.J. and Lee, M.G. (1989) *Plant Cell* **1**, 1185- 1193.
16. Colsanti, J., Tyers, M. and Sundaresan, V. (1991) *Proc. Natl. Acad. Sci. USA* **88**, 3377-3381.
17. Hirt, H., Pay, A., Györgyey, J., Bako, L., Németh, K., Bögre, L., Schweyen, R.J., Heberle-Bors, E., Dudits, D. (1991) *Proc. Natl. Acad. Sci. USA* **88**, 1636-1640.
18. Hirt, H., Pay,A., Bögre, L., Meskiene, I. and Heberle-Bors, E. (1993) *Plant J.* **4**, 61-69.
19. Ferreira, P.C.G., Hemerly, A.S., Vilarroel, R., Van Montagu, M. and Inzé, D. (1991) *Plant Cell* **3**, 531-540.
20. Hirayama, T., Imajuku, Y., Anai, T., Matsui, M. and Oka, A. (1991) *Gene* **109**, 159-165.
21. Miao, G-H., Hong, Z., Verma, D.P.S. (1993) *Proc. Natl. Acad. Sci. USA* **90**, 943-947.
22. Hashimoto, J., Hirabayashi, T., Hayano, Y., Hata, S., Ohashi, Y., Suzuka, I., Utsugi, T., Toh-E, A. and Kikuchi Y. (1992) *Mol. Gen. Genet.* **233**, 10-16.
23. Hemerly A., Engler J.A., Bergounioux C., Van Montagu M., Engler G., Inzé D. and Ferreira, P. (1995) *EMBO J.* **14**, 3925-3936.
24. Zhang, K., Letham, D.S. and John, P.C.L. (1996) *Planta,* **200**, 2-12.
25. John, P.C.L., Sek, F.J., Carmichael, J.P. and McCurdy, D.W. (1990) *J. Cell Sci.* **97**, 627-630.

26. Gorst, J., Sek, F.J. and John, P.C.L. (1991) *Planta* **185**, 304-310.
27. John, P.C.L., Zhang, K., Dong, C., Diederich, L., Wightman, F. (1993) *Aust. J. Plant Physiol.* **20**, 503-526.
28. Martinez, M.C., Jorgensen, J-E., Lawton, M.A., Lamb, C.J. and Doerner, P.W. (1992) *Proc. Natl. Acad. Sci. USA* **89** 7360-7364.
29. John, P.C.L., Zhang, K. and Dong, C. (1993) in *Molecular and Cell Biology of the Plant Cell Cycle*. (Ormrod, J.C. and Francis, D. eds.) pp. 9-34 Kluwer Academic Publishers, Dordrecht.
30. Hemerley, A.S., Ferriera, P.C.G., Engler, J.A., Van Montagu, M., Engler, G., Inzé, D. (1993) *Plant Cell* **5**, 1711-1723.
31. Ferriera, P.C.G., Hemerley, A.S., Engler, J. de A., Van Montagu, M., Engler, G., Inzé, D. (1993) *Plant Cell* **6**, 1763-1774.
32. Doerner, P., Jørgensen, J.-E., You, R., Stepphun, J. and Lamb, C. (1996) *Nature* **380**, 520-523.
33. Hagan, I.M., Hayles, J. and Nurse, P. (1988) *J. Cell Sci.* **91**, 587-595.
34. Stern, B. and Nurse, P. (1996) *Trends in Genet. Sci.* (in press).
35. Nurse, P. (1994) *Cell* **79**, 547-550.
36. Labib, K., Moreno, S. and Nurse, P. (1995) *J. Cell Sci.* **108**, 3285-3294.
37. Hayles, J. and Nurse, P. (1994) *EMBO J.* **14**, 2760-2771.
38. Hayles, J., Fisher, D., Woolard, A. and Nurse, P. (1994) *Cell* **78**, 813-822.
39. Correa-Bordes, J. and Nurse, P. (1995) *Cell* **83**, 1001-1009.
40. Heichman, K.A. and Roberts, J.M. (1996) *Cell* **85**, 39-48.
41. Gould, K.L., Nurse, P. (1989) *Nature* **342**, 39-45.
42. Nasmyth, K. (1993) *Curr. Opin. Cell Biol.* **5**, 166-179.
43. Amon, A., Irniger, S. and Nasmyth, K. (1994) *Cell* **77**, 1037-1050.
44. Labib, K. and Moreno, S. (1996) *Trends in Cell Biol.* **6**, 62-66.
45. Sherr, C.J. and Roberts, J.M. (1995) *Genes Dev.* **9**, 1149-1163.
46. Fobert, P.R., Coen, E.S., Murphy, G.J.P. and Doonan, J.H. (1994) *EMBO J* **13**, 616-624.
47. Kouchi, H., Sekine, M. and Hata, S. (1995) *Plant Cell* **7**, 1143-1155.
48. Richardson, H.E., Wittenberg, C., Cross, F. and Reed, S.I (1989) *Cell* **59**, 1127-1133.
49. Sherr. C.J. (1994) *Cell*, **79**, 551-555.
50. Cross, F.R. (1995) *Curr. Opin. Cell Biol.* **7**, 790-797.
51. Soni, R., Carmichael, J.P., Shah, Z.H., Murray, J.A.H. (1995) *Plant Cell* **7**, 85-103.
52. Schwob, E., Bohm, T., Mendenhall, M.D. and Nasmyth, K. (1994) *Cell* **79**, 233-244.
53. Hoffmann, I., Draetta, G. and Karsenti, E. (1994) *EMBO J.* **13**, 4302-44310.
54. Hemerly, A., Bergounioux, C., Van Montagu, M., Inzé, D. and Ferriera, P. (1992) *Proc. Natl. Acad. Sci. USA* **89**, 3295-3299.
55. Meskiene, I., Bögre, L., Dahl, M., Pirck, M., Ha, D.T.C., Swoboda,I., Heberle-Bors, E., Ammerer, G. and Hirt, H. (1995) *Plant Cell* **7**, 759-771.
56. Ferriera, P., Hemerly, A., Engler, J.D., Bergounioux, C., Burssens, S., Van Montague, M., Engler, G. and Inzé, D. (1994) *Proc. Natl. Acad. Sci. USA* **91**, 11313-11317.
57. Renaudin, J.P., Colasanti, J., Rime, H., Yuan, Z.A., and Sundaresan, V. (1994) *Proc. Natl. Acad. Sci. USA* **91**, 7375-7379.
58. Magyar, Z., Bako, L., Bögre, L., Dedeoglu, D. and Dudits, D. (1993) *Plant J.* **4**, 151-161.
59. Elledge, S.J. and Spottswood, M.R. (1991) *EMBO J.* **10**, 2653-2659.
60. Lew, D.J. and Reed, S.I. (1995) *J. Cell Biol.* **129**, 739-7849.
61. Davies, P.J. (1995) in *Plant Hormones* (Davies, P.J. ed.) pp. 1-12, Kluwer Academic Publishers, Dordrecht.
62. Romano, C.P., Hein, M.B., Klee, H.J. (1991) *Genes Dev.* **5**, 438-446.
63. Torrey, J.G. (1956) *Physiologia Plantarum* **9**, 370-388
64. Pardee, A.B., Dubrow, R., Hamlin, J.L., Kletzein, R.F. (1978) *Annu. Rev. Biochem.* **47**, 715-750.
65. Zetterberg, A., Larsson, O. (1985) *Proc. Natl. Acad. Sci. USA* **82**, 5365-5369.
66. Bandurski, R.S., Schulze, A., Desrosiers, M., Jensen, Y., Epel, B., Reinecke, D. (1990) in *Plant Growth Substances* (Pharis, R.P. and Rood, S.B. eds.) pp. 341-352, Springer-Verlag, Heidelberg.
67. Millar, J.B.A., McGowan, C.H., Lenaers, G., Jones, R. and Russell, P. (1991) *EMBO J.* **10**, 4301-4309.
68. Brizuela, L., Draetta, G., Beach, D. (1987) *EMBO J.* **6**, 3507-3514.
69. John, P. C. L., Sek, F. J. and Hayles, J. (1991) *Protoplasma* **161**, 70-74.
70. Das, N.K., Patau, K. and Skoog, F. (1956) *Physiologia Plantarum* **9** 640-651.
71. Simard, A. (1971) *Can. J. Bot.* **49**, 1541-1548.
72. Fosket, D.E. (1977) in *Mechanisms and control of cell division* (Rost, T.L. and Gifford, E.M. eds.) pp 62-91, Dowden Hutchinson and Ross Inc, Strousberg, Pennsylvania.
73. Bayliss, M.W. (1985) in *The cell division cycle in plants* (Bryant, J.A. and Francis, D. eds.) pp 157-177, Cambridge University Press, Cambridge, UK.
74. Nishinari, N., Syono, K. (1986) *Plant Cell Physiol.* **27**, 147-153.

75. Badenoch-Jones, J., Parker, C.W. Letham, D.S. and Singh, S. (1996) *Plant Cell Envir.* **19**, 504-516.
76. Reid, M.S. (1995) in *Plant Hormones* (Davies, P.J. ed.) pp. 486-508, Kluwer Academic Publishers, Dordrecht.
77. Blakely, L.M. and Evans, T.A. (1979) *Plant Sci. Lett.* **14**, 79-83.
78. Barr, P.J. and Toei, L.D. (1994) *BioTechnology* **12**, 487-493.
79. Evan, G.I., Brown, L., Whyte and M. Harrington, E. (1995) *Curr. Opin. Cell Biol.* **7**, 825-834.
80. Lambe, C.J. (1994) *Cell* **76**, 419-422.
81. Fincher, G.B. (1989) *Annu. Rev. Plant Physiol. Molec. Biol.* **40**, 305-346.
82. Collins, R.J., Harmon, B.V., Gobe, G.C. and Kerr, J.F.R. (1992) *Int. J. Radiat. Biol.* **61**, 451-453.
83. Wymer, C. and Lloyd, C. (1996) *Trends in Plant Sci.* **1** 222-228.
84. Gunning, B.E.S. and Sammut, M. (1990) *Plant Cell* **2**, 1273-1282
85. Stoppin, V., Vantard, M., Schmit, A.-C. and Lambert, A.-M. (1994) *Plant Cell* **6**, 1099-1106.
86. Gunning, B.E.S. (1982) in *The Cytoskeleton in Plant Growth and Development* (Lloyd, C.W. ed.) pp. 230-288, Academic Press, New York
87. Murata, T. and Wada, M. (1992) *J. Cell Sci.* **101**, 93-98.
88. Traas, J., Bellinin, C., Nacry, P., Kronenberger, J., Bouchez, D. and Caboche, M. (1995) *Nature* **375**, 676-677.
89. Mineyuki, Y., Yamashita, M. and Nagahama, Y. (1991) *Protoplasma* **162**, 182-186.
90. Colasanti, J., Cho, S.-O., Wick, S. and Sundaresan, V. (1993) *Plant Cell* **5**, 1101-1111.
91. Melan, M.A. and Sluder, G. (1992) *J. Cell Sci.* **101**, 731-743.
92. Zhang, K., Tsukitani, Y. and John P.C.L. (1992) *Plant Cell Physiology* **33**, 677-688
93. Katsuta, J. and Shibaoka, H. (1992) *J. Cell Sci.* **103**, 397-405.
94. Zhang, D.H., Wadsworth, P. and Hepler, P.K. (1990) *Proc. Natl. Acad. Sci. USA* **87**, 8820-8824
95. Hush, J.M., Wadsworth, P., Callaham, D.A. and Hepler, P.K. (1994) *J. Cell Sci.* **107**, 775-784.
96. Hush, J., Wu, L., John, P.C.L., Hepler, L.H. and Hepler, P.K. (1996) *Cell Biol. Internat. Rep.* **20**, 275-287.
97. Ookata, K., Hisanaga, S.-i., Bulinski, J.C., Murofushi, H., Aizawa, H., Itoh, T.J., Hotani, H., Okamura, E., Tachibana, K. and Kishimoto, T. (1995) *J. Cell Biol.* **128**, 849-862.
98. Shina, N., Gotoh, Y. and Nishida, E (1992) *EMBO J.* **11**, 4723-4731.
99. Cleary, A.L., Gunning, B.E.S., Wasteneys, G.O. and Hepler, P.K. (1992) *J. Cell Sci.* **103**, 977-988.

chapter 7
The functions of Myc in cell cycle progression and apoptosis

Philipp Steiner*, Bettina Rudolph, Daniel Müller and Martin Eilers[1]

Zentrum für Molekulare Biologie Heidelberg (ZMBH), Im Neuenheimer Feld 282, 69120 Heidelberg
*Whitehead Center for Medical Research, 9 Cambridge Center, Cambridge, USA
[1] To whom correspondence should be addressed

c-myc has emerged as one of the central regulators of mammalian cell proliferation. The gene encodes a transcription factor of the HLH/leucine zipper family of proteins that activates transcription as part of a heteromeric complex with a protein termed Max. In mammalian fibroblasts, Myc acts as an upstream regulator of cyclin-dependent kinases and functionally antagonises the action of at least one cdk inhibitor, p27. Myc also induces cells to undergo apoptosis, and the relationship between Myc-induced cell cycle entry and apoptosis is discussed.

INTRODUCTION: TUMORIGENESIS

Viral myc (*v-myc*) genes were discovered as the transforming gene of an avian retrovirus termed MC29. Later, no less than four additional viruses were found to contain a closely related oncogene; closely related and not identical, as viral myc genes differ in whether coding sequences are fused to the viral gag gene or not and they differ by the presence of strain-specific point mutations which somewhat affect the transforming potential of the individual virus. Viruses that express myc genes are potent tumour inducers in chicken and they induce tumours in a wide array of tissues; thus, the transforming function of *v-myc* appears not to be restricted to only a few cell types (for a general review, see 1).

Cellular homologues, *c-myc* genes, were subsequently identified in a wide array of vertebrates and, more recently, in echinoderms. They appear not to be conserved in Drosophila, Caenorhabditis and yeast. It soon became apparent that *c-myc* is part of a small multigene family, which encloses the closely related N-Myc, L-Myc, and s-myc genes; another member, B-myc, encodes a truncated protein (at the carboxy-terminus) of unknown biological functions. Myc genes differ in their pattern of expression: where as *c-myc* is expressed in proliferating tissues throughout the lifetime of an organism, both N-Myc and L-Myc are usually more restricted in expression to embryonic tissues (e.g. 2, 3, 4). The encoded proteins differ functionally, both in their in transforming potential in tissue culture and in their role *in vivo* (5). The remainder of the review will focus on *c-myc*, for which a role in cell proliferation is documented most clearly.

It soon became apparent that mutations at myc gene loci are involved in a wide array of mammalian neoplasias. Most prominent among these are the Ig/myc translocations found in Burkitt's lymphomas. In these translocations, expression of *c-myc* is released from its normal controls and driven by one of the potent Ig-enhancers (different translocations occur to the different coding genes for light and heavy chains). Thus, B cells carrying this translocation produce high amounts of c-Myc protein and this contributes to the genesis of a B cell neoplasia. Experiments in which chimeric Ig-myc genes were expressed in transgenic mice unequivocally established a causal role for myc genes in this process (for review, see 6). A large set of chromosomal aberrations involving myc genes are now known from a variety of tumours; other examples include N-Myc amplification in neuroblastoma and L-Myc translocations and amplification in small cell lung carcinoma (7, 8). However, many tumours with no apparent chromosomal alteration at a myc gene show elevated levels of Myc proteins and this often correlates with poor prognosis; presumably, in at least some of these tumours upstream regulators of Myc expression are deregulated or mutated (for review, see 9).

In culture, expression of *v-myc* genes and deregulated expression of *c-myc* transforms chicken fibroblasts and macrophages and some rodent fibroblast lines. One hallmark of virtually all transformation experiments involving myc genes has been the requirement for additional genetic events in the induction of tumour formation or full transformation. One classic demonstration of this behaviour is the oncogene co-operation assay by Land and Weinberg, who demonstrated that neither activated ras nor deregulated myc by itself was sufficient to transform primary rat embryo fibroblasts; both together, however, led to focus formation and cell lines established from the foci were tumorigenic in nude mice (10, 11). Similarly, tumours that develop in Ig-myc or MMTV-myc transgenic mice are clonal and develop after long latency periods, strongly suggesting that additional genetic changes have to occur before a tumour develops (12, 13). Retroviral tagging subsequently identified a number of co-operating oncogenes in Ig-myc mice, termed (among others) pim-1, bmi-1, and gfi-1 (see for example: 14, 15) (T. Möröy, personal communication). These loci encode either a protein kinase (pim-1), a Zn-finger protein (gfi-1) or a member of the polycomb group of chromatin proteins (bmi-1).

Several not mutually exclusive models have been proposed as to why different oncogenes synergize with c-myc. For example, synergizing oncogenes have been suggested to provide cumulative mitogenic stimuli and might "push" cells together over a restrictive barrier (for example negative growth stimuli exerted by neighbouring cells) (for review, see 16). Alternatively, oncogenes might synergize with Myc because ectopic expression of Myc alone induces cells to undergo active cell death, apoptosis (17). Third, somewhat surprisingly, activated alleles of ras have been shown to arrest proliferation of some cells and nuclear oncogenes may be needed to break this arrest (18). Fourth, in intact animals, oncogenes may also co-operate with myc simply because expression of myc leads to an expansion of a pool of cells susceptible to transformation by other oncogenes (see below). Thus, multiple mechanisms of oncogene co-operation may exist in tissue culture and *in vivo*.

THE NETWORK

c-myc encodes a protein of the helix-loop-helix/leucine zipper (HLH/LZ) family of transcription factors; both motifs are known to mediate protein/protein interactions *in vivo* (19, 20). Like all members of this family, it binds to DNA in a sequence specific manner and recognises so-called E-box elements with a central CACGTG sequence (21, 22). However, Myc homodimers bind this sequence with low affinity and whether they exist in cells is doubtful. Instead, Myc heterodimerizes with a second HLH/LZ protein termed Max (21, 23, 24, 25). The heterodimeric complex is a potent activator of transcription from E-box elements (26, 27, 28). Detailed mutational analysis has left no serious doubt that complex formation with Max and transcriptional activation are critically required for most, if not all biological properties of the Myc protein (29, 30).

In contrast to Myc, Max is a stable protein and is expressed at relatively constant levels in both proliferating and quiescent cells. In resting cells, in which Myc is not expressed, Max forms homodimers (23). As Max, in contrast to Myc, lacks transcriptional activation domains, Max homodimers act as passive repressors of transcription (31). Thus, the contrasting biochemical properties together with the contrasting patterns of expression of Myc and Max provide a model as to how the growth regulation of E-box containing genes might be achieved (for review, see 32).

Recently, a novel class of proteins has been identified that associate with Max, not with Myc. This proteins, termed Mad 1-4 (with Mad 2 also being called Mxi-1) are active repressors of transcription and often accumulate in differentiating cells (33, 34, 35). They exert their repressive effect by recruiting a repressor protein-mouse sin3 into a heterodimeric Mad/Max complex (36, 37). Thus, E-box containing promoters may be strongly repressed by Mad/Max complexes in differentiated cells. Not surprisingly, Mad proteins strongly antagonise Myc transforming abilities and can inhibit growth-factor induced cell proliferation (e.g. 38) (B. Lüscher, personal communication).

Several genes have been identified that are targets for transcriptional activation by Myc *in vivo*. These encode ornithine decarboxylase, a rate limiting enzyme of polyamine biosynthesis (39, 40), prothymosin-α, an acidic chromatin protein which is essential for cell proliferation (41, 42), bendless, an ubiquitin-conjugating enzyme, an RNA helicase (C. Grandori and R. Eisenman, personal communication) and cdc25A, a phosphatase involved in activation of (most likely) cyclin E/cdk2 complexes (D. Beach, personal communication). In these genes, E-boxes have been identified that mediate activation by Myc. Thus, target genes of Myc presumably enclose both proteins generally required for cell proliferation (ODC and prothymosin-α) and direct components of the cell cycle machinery (cdc25A). However, the analysis of biological effects of Myc on the cell cycle strongly suggests that at least some of the critical target genes have not been identified (see below).

One central question of Myc biology concerns the issue of specificity as several proteins are known that recognise identical DNA elements and that are also potent activators of transcription. These include USF, a relatively abundant protein present in many cells, and TFE-3, a protein involved in B-cell differentiation. Neither protein has any documented effect on cell proliferation and transformation. So if Myc is indeed an oncogene because it activates from E-box containing promoters, then mechanisms must exist that discriminate between these closely related transcription factors. Such mechanisms have been identified for both prothymosin-α and for ornithine decarboxylase: it appears that a distal location of the E-box, protein-protein interactions with repressor proteins binding to adjacent sites (43) and, in the case of ODC, the ability to bind co-operatively to two adjacent E-boxes are critical for specificity (M. Timmers, personal communication). Further, some E-boxes which are bound by Myc *in vivo* appear to have non consensus core sequences and this might add another factor to specificity (C. Grandori and R. Eisenman, personal communication, 44).

A final open question concerning the model of Myc function is whether Myc also acts directly as a transcriptional repressor. Numerous genes have been identified that are repressed in cells overexpressing Myc, including *c-myc* itself (45), cyclin D1 (46, 47) and several genes encoding cell surface proteins (which may be important in the escape of tumour cells from immune surveillance (e.g. 48). However, whether this reflects a direct repressive property of Myc or is an indirect consequence of cell cycle disturbances or transformation, is still somewhat open. Repression by Myc has been suggested to be caused by interference with CCAAT-binding factors (49, 50) or to be mediated by complex formation with proteins interacting at the start site of transcription (so-called initiator (Inr) elements), in particular YY-1 and TfII-I (51, 52). Conflicting results have been obtained as to whether mutations at Inr elements affect repression by Myc in transient transfection assays; thus this issue remains unresolved (49, 53).

CELL PROLIFERATION

Mice that express an IgH-myc chimera show a defined preneoplastic phenotype: whereas, as indicated, tumours that develop are clonal, these mice show a polyclonal expansion of pre-B cells (54). In the pre-B cell compartment the proportion of cycling cells is significantly higher. This is one of several indications that deregulated expression of myc may be sufficient to provide a strong mitogenic stimulus. Other indications come from tissue culture experiments that document a strongly reduced growth factor requirement for myc transformed cells to stay in the cell cycle (55). Indeed, in some extreme cases, the growth factor dependency for S-phase entry is essentially abrogated after ectopic expression of myc (56). For example, BALB/c-3T3 fibroblasts engineered to express a hormone-inducible allele of Myc (MycER chimera) enter S-phase upon addition of hormone to serum-deprived cells. The cells also progress through mitosis, thus the action of Myc is not restricted to a specific phase of the cell cycle (41).

In line with these observations, expression of the endogenous *c-myc* gene was found to be under tight control by external growth factors and to correlate closely with the proliferative status of the cell (e.g. 57). More precisely, *c-myc* belongs to the class of "immediate early genes" that are induced by growth factors in the absence of protein synthesis. In contrast to other genes of this class, expression of *c-myc* is not shut off after a few hours; it is instead maintained throughout the cell cycle at a somewhat reduced levels as long as cells proliferate (58, 59).

Numerous observations document this behaviour; however, the detailed biochemical pathways leading to activation are not as clear as for example for the c-fos gene. Recently, stimulation of *c-myc* expression in response to addition of platelet-derived growth factor (PDGF) was found to be sensitive to dominant negative alleles of src, defining kinases of the src-family as upstream regulators of *c-myc* expression (60). Interestingly, dominant negative alleles of ras, although they also interfered with PDGF-induced mitogenesis, did not interfere with induction of *c-myc*. but blocked *c-fos* induction by PDGF. Thus, addition of PDGF appears to induce two distinct pathways, one leading via src to the induction of *c-myc*, the other via ras to induction of *c-fos*. This notion is further supported by observations that ectopic expression of limiting amounts of *c-myc* rescued mitogenesis in cells blocked by dominant negative alleles of src, not in cells blocked by dominant negative alleles of ras.

The *c-myc* promoter is sensitive to induction by growth factors; it appears that several elements contribute to this regulation. In particular, a binding site for the transcription factor E2F has been implicated in growth factor-mediated induction of *c-myc* (61). In extracts from quiescent cells, this site is bound by E2F-p130 complexes and p130 can be phosphorylated in response to addition of growth factors even in the absence of protein synthesis (62); (S. Mittnacht, personal communication). However, several cyclin/cdk complexes are also able to activate *c-myc* expression from this site (see below). The picture is further complicated by several issues: for example, a quantitatively significant element of control of *c-myc* expression is provided by a regulated pausing step of the polymerase (at the end of the first exon) (for review, see 63). Also, so far no stably reintegrated *c-myc* promoter construct was expressed in transgenic mice (64), raising the possibility that *c-myc* expression *in vivo* is dictated by far upstream, yet unknown, regulatory elements.

Is Myc function required for cell proliferation? Disruption of both alleles of *c-myc* is lethal early in embryogenesis, demonstrating that myc function is required during embryogenesis; the same is true for N-Myc (65, 66, 67). In tissue culture, a number of antisense inhibition experiments document that inhibition of c-myc function can arrest cell proliferation of certain cell lines (e.g. 68). Further, microinjection of expression plasmids encoding Mad proteins inhibits growth-factor induced proliferation of rodent fibroblasts (B. Lüscher, personal communication). Taken together, these data show that Myc not only provides a strong mitogenic stimulus, but is also required for cell proliferation in at least some cells. Exceptions to this rule may exist: for example, PC12 cells have been found to have no intact Max gene (69). Thus, either, as yet unknown, partners of Myc exist or PC12 cells somehow bypass the requirement for Myc function. Also, a B-cell line expressing both bcl-2 and c-fos also apparently bypasses Myc function (70). Thus, pathways leading to proliferation may exist that do not absolutely require Myc function.

MYC-INDUCED CELL PROLIFERATION: MODELS

Several models have been suggested as to how Myc might induce cell cycle progression. Early suggestions focused on a direct role of Myc in DNA replication (71); although this remains a possibility, no strong biochemical evidence supporting a direct participation of Myc in replication has been presented.

Second, Myc has been suggested to function similarly to oncogenes of DNA tumour viruses which act at least in part by binding to and functionally sequestering the products of tumour suppressor genes. Similarly, it was suggested that Myc protein directly interacted with the retinoblastoma protein (72). This original suggestion could not be supported by *in vivo* interaction data. Also, in contrast to cells expressing nuclear oncogenes of DNA tumour viruses, inhibition of the cyclin D1/cdk4 kinase leads to an arrest of G1 progression after induction of Myc (73, 74). As this kinase is only required in cells expressing functional retinoblastoma protein, this suggest that the retinoblastoma protein is not sequestered in Myc expressing cells. However, these data do not immediately address the issue as to whether retinoblastoma protein is genetically upstream (i.e. that it controls Myc function) or downstream (i.e. that it is phosphorylated in response to activation by Myc) or both of Myc.

Recently, Myc has been found to form a complex with a related pocket protein, p107. p107 physically associates with transactivation domain of Myc and, in transient transfection experiments, expression of p107 represses transcriptional activation by Myc (75, 76). Interestingly, mutant alleles of Myc isolated from Burkitt lymphomas appear to be resistant to p107-mediated repression, although they still bind the protein (75). In support for a role of Myc in overcoming the negative effect of p107 on cell proliferation, ectopic expression of Myc rescues cells from p107 mediated growth arrest in SaOs-2 cells. Thus sequestering pocket proteins may be one mechanism by which Myc can stimulate cell proliferation. However, other biological effects have been suggested to be mediated by the Myc/p107 interaction (see below).

Third, Myc has been shown to act as an upstream regulator of cyclin-dependent kinases. This was demonstrated using cells that express conditional, hormone-inducible alleles of *c-myc*. Addition of hormone to resting cells expressing these chimeras led to a rapid induction of cyclin E/cdk2 kinase activity (74). This induction was sensitive to inhibition of transcription by actinomycin. Also, cells that expressed inducible alleles of Myc that were deficient for association with Max, transcriptional activation or DNA binding did not show increases in cyclin E-dependent kinase activity. Most likely therefore, induction of cyclin E/cdk2 kinase activity by Myc involved transcriptional activation of a target gene by Myc (74).

What might be this target gene? First, it did not appear to be cyclin E itself, as the increase in cyclin E dependent kinase activity preceded any later increase in cyclin E protein. Also, expression of cdk2 is unaffected by Myc under these conditions. Second, activation appeared not to be mediated by cdc25A, as treatment of cyclin E/cdk2 kinase complexes isolated from cells before and after activation of Myc revealed that cdc25A activity was indeed strongly limiting for full activation, even after induction of Myc (74). This is somewhat surprising as cdc25A has recently been shown to be a target for induction by Myc (David Beach, personal communication); most likely, an activation step in which cdc25A protein function is activated by phosphorylation by raf did not occur in serum-deprived cells after activation of Myc alone (77).

Instead, cyclin E/cdk2 complexes isolated before activation of Myc were found to contain the inhibitor protein p27, whereas activation of Myc led to a degradation of p27 and dissociation from cyclin E/cdk2 complexes (74). Recent experiments using specific inhibitors of the proteasome pathway suggest that Myc can lead to a partial activation of cyclin E/cdk2 in these cells even when degradation of p27 is blocked and p27-free complexes of cyclin E/cdk2 are formed under these conditions. Taken together the data show that Myc antagonises the action of p27 and suggest that Myc may interfere with the function of p27 not (only) by degradation, but also by hitherto unidentified mechanisms (P. Steiner and M.E., personal communication).

Strong evidence suggests that this is also observed *in vivo*. First, Myc-expressing cells have been shown to keep proliferating in the presence of p21, a related inhibitor (78). For example, cells that express both a MycER-chimera and a temperature-sensitive allele of p53 stayed in cycle and maintained elevated levels of cdk2 kinase activity even after shifting to the restrictive temperature (79). Activation of p21 expression by p53 was unaffected by Myc. More directly, cells expressing Myc were found to be resistant to elevated levels of both p27 and p21 delivered to cells by retroviral infection (Yaromir Flach and Bruno Amati, personal communication). Control cells were arrested by these levels of either p27 or p21.

Under the conditions of the experiment, ectopic expression of p27 did not affect the level of cyclin D1 dependent kinase; however, p27 inhibited cyclin E/cdk2 kinase and expression of cyclin A. Expression of Myc restored cyclin E/cdk2 kinase activity and expression of cyclin A, although p27 was not significantly degraded in cells expressing Myc. The conclusion is, similar to those reached above, that ectopic expression of Myc interferes with the cdk inhibitory function of p27.

The mechanism of this reaction is not yet clear. Biochemical evidence shows that cyclin E/cdk2 complexes isolated from Myc-expressing cells are sensitive to inhibition by p27 with similar affinities to those isolated from control cells, suggesting that neither cyclin E nor cdk2 are modified in such a manner as to become resistant to p27 (D. Müller, unpublished; J. Flach, personal communication). Thus it appears that p27 is modified either by covalent modification or by protein/protein association. Evidence from yeast shows that association of cdc28 with and inhibition by the FAR-1 inhibitor is regulated both by phosphorylation and degradation of FAR-1 (80, 81) and similar reactions may occur with p27. Whether Myc-induced degradation of p27 is simply a consequence of cell cycle progression and whether the recently discovered transcriptional activation of an ubc-gene by Myc (bendless; C. Grandori and R. Eisenman, personal communication) contributes to Myc-induced degradation of p27 has not been determined.

Is cyclin E/cdk2 kinase activity required for Myc-mediated cell cycle entry? Almost certainly the answer to this is yes. First, resistance to p27 is dose-dependent. Taking expression of cyclin A protein as a marker for cell cycle progression, microinjection of either p16, p21, p27, or of dominant-negative alleles of cdk2 all inhibited cell cycle induction by Myc (82). Also, infection with high-titers of retroviruses containing p27 is also dominant over Myc (B. Amati, personal communication). Further, addition of roscovitine, a specific inhibitor of cdk2, inhibits induction of cyclin A by Myc (82). Indeed, ectopic expression of cyclin E is sufficient to activate the cyclin A promoter. Activation of cyclin A expression both by Myc and by cyclin E occurs via the core of the cyclin A promoter and requires the integrity of two elements - termed CDE and CHR (83). The CDE is a potential binding site for the transcription factor E2F and E2F may well mediate activation by cyclin E (84, 85). Finally, activation of cyclin E/cdk2 kinase precedes activation of the cyclin A promoter by about three hours, whereas a Myc-induced increase in cyclin D1/cdk4 activity is significantly slower. Taken together, these data suggest that active cyclin E/cdk2 complexes mediate activation of cyclin A expression by Myc and thus suggest that Myc acts upstream of cyclin-dependent kinases to activate at least one downstream target gene in the cell cycle.

These findings also contrast cell cycle induction by Myc with that induced by E2F-1, a protein whose biological properties are similar to Myc: ectopic expression of E2F also induces both cell cycle entry and apoptosis in resting cells (e.g. 86). The similarity is not accidental, as E2F can act as an activator of *c-myc* expression (61) and Myc is an upstream regulator of E2F function (87). However, cell cycle induction by E2F differs in several key aspects from cell cycle induction by Myc: first, E2F induces cell cycle entry with no detectable cyclin E/cdk2 and cyclin D1/cdk4 kinase activity (88). In contrast, activation of Myc is followed by a rapid activation of cyclin E/cdk2 and a slower induction of cyclin D1/cdk4 kinase activity. Second, microinjection and transfection experiments show that neither inhibition of cyclin D1/cdk4 nor of cyclin E/cdk2 kinase strongly affects E2F-mediated S-phase entry. In particular, high expression of neither p16, nor of p21 or p27 inhibited E2F-mediated S-phase entry, in striking contrast to Myc-induced events (89). These experiments place E2F downstream of G1 cyclin-dependent kinases, presumably because over-expression of E2F generates free transcription factor even in the presence of non-phosphorylated retinoblastoma protein. In contrast, they further support the notion that Myc acts upstream of at least some G1 cyclin dependent kinases in stimulating cell cycle induction.

CYCLIN/CDK COMPLEXES AS UPSTREAM REGULATORS OF MYC

Two observations suggest that cyclin/cdk complexes can also act as upstream regulators of *c-myc* function and expression.

First, addition of CSF-1 to NIH-3T3 cells that have been engineered to express a CSF-1 receptor induced both expression of both *c-myc* and cyclin D1. Point mutations have been found in the carboxy-terminus of the CSF-1 receptor that abolish both CSF-mediated induction of *c-myc* and of cyclin D1 mRNAs. These cells fail to grow in soft agar in response to addition of CSF-1; ectopic expression of c-myc restores growth (90). Interestingly, ectopic expression of cyclin D1 not only restores growth in soft agar, but also expression of *c-myc* in these cells (73). In parallel experiments, ectopic expression of cyclin E failed to induce proliferation. These data imply that D-type cyclins can control *c-myc* expression and potentially function. This is further supported by observations that the *c-myc* promoter can be stimulated by ectopic expression of cyclin D1 via the E2F binding site (91).

Second, as indicated above, transcriptional activation by Myc is under negative control by the p107 pocket protein. Two observations suggest that this association may implicate cyclin/cdk complexes as upstream regulators of Myc function. First, p107 also negatively regulates transcriptional activation by E2F-4; during the cell cycle, p107 becomes phosphorylated by D1/cdk4 complexes and, in the case of E2F-4, phosphorylation of p107 releases transcriptionally active factor (92). Thus, if p107 acts in a similar fashion on Myc, D1/cdk4 complexes may act as positive effectors of Myc function. Second, p107 can target cyclin A/cdk2 complexes to Myc *in vitro*, leading to phosphorylation of Myc at sites that are also phosphorylated *in vivo* (93). Similar to models proposed for E2F-1 (94), association with cyclin A may negatively regulate Myc function during later phases of the cell cycle.

Taken together, these observations suggest models in which both expression of and transcriptional activation by Myc are restricted by the activity of cyclin dependent kinases. However, it should be added that some critical observations supporting similar models for E2F proteins (evidence for *in vivo* complexes between E2F-proteins and cyclin/cdk's complexes; availability of specific mutants allowing to test for the relevance of such interactions) have not yet been reported for Myc.

DO ONCOGENES COOPERATE TO ACTIVATE CYCLIN-DEPENDENT KINASES?

How can these findings be reconciled into a consistent model? It appears that cyclin/cdk complexes can act both upstream and downstream of Myc. As indicated, several findings suggest that cyclin D1/cdk4 complexes may act as upstream regulators of Myc function; in contrast, activation of D1/cdk4 kinase after induction of Myc is relatively slow and the extent of activation is low, suggesting that Myc may not be a physiologically significant factor in regulation of this kinase. In support of this notion, ectopic expression of Myc represses, not activates, expression of cyclin D1 in many instances (46, 47). Taken together, these data argue for a role of cyclin D1/cdk4 upstream of Myc, potentially via association of Myc with p107 or via stimulation of Myc expression. Second, rapid activation of cyclin E/cdk2 kinase and of cyclin A expression is observed after activation of Myc; activation of cyclin E/cdk2 is driven by degradation and functional inactivation of p27 and potentially by transcriptional activation of cdc25A. Thus, it appears, that both cyclin E and cyclin A/cdk2 complexes are downstream of activation by Myc.

We suggest, therefore, that Myc may serve to integrate signals form several cyclin/cdk pathways and may stimulate the commitment of cells to proliferation. Similar to models proposed for the control of cyclin E expression by the retinoblastoma protein (R. Herrera and R. Weinberg, personal communication), these data suggest a model in which Myc may act at the restriction point in G1 progression: activation of cyclin D1/cdk4 contributes to activation of Myc, which in turn contributes to activation of cyclin E/cdk2 kinase activity and cyclin A expression. Myc appears not to mainly act on cyclin E expression, but contributes to activation of cyclin E/cdk kinases by interfering with the function p27.

Such a model could also take into account a potential synergism with ras. Essentially two effects of oncogenic alleles of ras have been documented in terms of cell cycle regulation: First, activated alleles of ras enhance expression of the cyclin D1 gene via AP-1 and ets-dependent pathways (95). Second, the ras-activated protein kinase, raf, has been shown to bind to and activate cdc25A, a phosphatase involved in activation of G1 cyclin dependent kinases (77). In cells engineered to express oncogenic ras under the control of an MMTV-promoter, induction of ras stimulated expression of cyclin D1 and, to a lower extent, cyclins E and A (96). However, the kinase activity associated with these cyclins remained low as they remained bound to p27. Addition of growth factors decreased levels of p27 and induced kinase activity. Thus, effects of ras on cyclin expression and kinase activity appear to be largely complementary to those of Myc and may thus complement mitogenic effects of this oncogene.

APOPTOSIS

Ectopic expression of *c-myc* also induces or accelerates cell death in a number of cultured cell lines. This death often shows all the hallmarks of apoptosis (for reviews, see 17, 97). Antisense

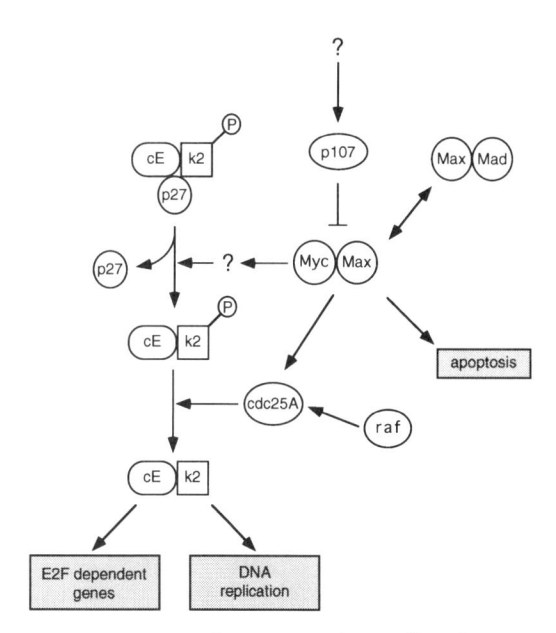

Figure 1. Role of *Myc* in G1 progression of mammalian cells.

imbalance of cell cycle progression. Second, apoptosis may reflect a physiological function of Myc and act as a safeguard against Myc-induced carcinogenesis; thus, mutations leading to activation or deregulated expression of Myc alone would be lethal for the affected cells and thereby reduce the risk of cancer for the whole organism.

Several observations argue against a conflict model of apoptosis. First, Myc-induced cell death appears to be independent of the position of a given cell in the cycle (56). Myc-induced apoptosis can occur both early in G1, in post-commitment, and even in post-DNA synthesis phases of the cell cycle (103). For example, activation of conditional alleles of Myc activates apoptosis in cells blocked in S/G2 by a topoisomerase II inhibitor etoposide. However, during these later phases, cells are independent of growth factors for progression through the remainder of the cycle. Therefore, it can be argued that no conflict between "stop" and "go" signals is apparent.

Second, Myc-induced apoptosis is most prominent upon serum deprivation as it is prevented by a number of cytokines that are, for example, present in serum (103). Individual cytokines can be either only mitogenic (e.g. EGF), only anti-apoptotic (e.g. insulin-like growth factor) or both (e.g. PDGF). Therefore the signalling pathways involved in both processes may be overlapping, but are not identical.

Third, a conflict model of apoptosis would predict that inhibition of Myc-induced proliferation would also protect cells from Myc-induced apoptosis. To test this, expression plasmids encoding inhibitors of cyclin-dependent kinases were microinjected into rat fibroblasts carrying a conditional allele of Myc (82). Under these experimental conditions, high amounts of inhibitors are expressed and Myc does not overcome them. As indicated above, Myc-induced expression of cyclin A (as a marker of cell cycle progression) was blocked completely under these conditions. However, Myc-induced apoptosis was unimpaired in these cells, suggesting that activation of cdk activity is not required for Myc to induce cell death.

Taken together, these data strongly suggest that the target genes via which Myc induces apoptosis are at least partly different from those involved in the control of cell proliferation. Targets in this apoptotic pathway may include ornithine decarboxylase; activation of this enzyme has been suggested to lead to reactive oxygen intermediates via excessive oxidation of polyamines (104). In support of this hypothesis, inhibition of ornithine decarboxylase by specific inhibitors blocks Myc-induced apoptosis in murine myeloid cells (104).

inhibition of Myc function also interferes with glucocorticoid-mediated death of T-lymphocytes, suggesting that Myc may also be involved in physiologically occurring forms of apoptosis (98). In culture, induction of apoptosis by Myc depends on its ability to heterodimerize with Max, activate transcription and bind to DNA, strongly suggesting that Myc induces apoptosis by activating a critical set of target genes (29, 56).

Myc-induced apoptosis shares many characteristics with other forms of apoptosis: for example, it does not occur in p53-deficient fibroblasts and is temperature-sensitive in cells expressing a temperature-sensitive allele of p53 (79, 99). Both observations argue for a role of p53 in this process; however, p53-independent forms of myc-induced apoptosis have also been found (100; T. Littlewood, personal communication). Further, Myc-induced apoptosis is blocked by expression of bcl-2 and is affected by inhibition of ICE-proteases; thus, many components of the general apoptotic machinery appear to be involved (101, 102).

The mechanisms by which Myc induces apoptosis are not known. In particular two models have been proposed to explain the relationship between Myc-induced apoptosis and cell proliferation (for a detailed discussion, see 17, 103): According to the first, apoptosis results from an abortive entry into the cell cycle that is induced when Myc alone is activated. In this conflict model, apoptosis arises because obligatory co-factors (e.g. cytokines stimulating Myc-independent signal transduction pathways) are missing and cells recognise an

Alternatively, Myc-induced pathways leading to transcriptional induction (105) and stabilisation of p53 (99) (as is also observed in cells expressing E1A (106)) may contribute to induction of apoptosis. However, it appears likely that at least some target genes leading to Myc-induced apoptosis are not yet identified.

ACKNOWLEDGEMENTS

We apologise to the numerous colleagues whose work has not been quoted due to the limited space available. Work in my laboratory in supported by grants from the Deutsche Forschungsgemeinschaft, the Human Frontiers of Science Organization and the Bundesministerium für Forschung und Technologie (BMFT).

REFERENCES

1. Meichle, A., Philipp, A., and Eilers, M. (1992) *Biochim. Biophys. Acta* **1114**, 129-46
2. Downs, K. M., Martin, G. R., and Bishop, J. M. (1989) *Genes Dev.* **3**, 860-869
3. Mugrauer, G., Alt, F. W., and Ekblom, P. (1988) *J. Cell Biol.* **107**, 1325-1335
4. Mugrauer, G., and Ekblom, P. (1991) *J. Cell Biol.* **112**, 13-25
5. Morgenbesser, S. D., Schreiber-Agus, N., Bidder, M., Mahon, K. A., Overbeek, P. A., Horner, J., and DePinho, R. A. (1995) *EMBO J.* **14**, 743-756
6. Berns, A., Breuer, M., Verbeek, S., and van, L. M. (1989) *Int. J. Cancer Suppl.* **4**, 22-5
7. Brodeur, G. M., Seeger, R. C., Schwab, M., Varmus, H. E., and Bishop, J. M. (1984) *Science* **224**, 1121-1124
8. Mäkelä, T. P., Saksela, K., Evan, G., and Alitalo, K. (1991) *EMBO J.* **10**, 1331-1335
9. Field, J. K., and Spandidos, D. A. (1990) *Anticancer Res.* **10**, 1-22
10. Land, H., Parada, L. F., and Weinberg, R. A. (1983) *Science* **222**, 771-778
11. Land, H., Chen, A. C., Morgenstern, J. P., Parada, L. F., and Weinberg, R. A. (1986) *Mol. Cell. Biol.* **6**, 1917-1925
12. Sinn, E., Muller, W., Pattengale, P., Tepler, I., Wallace, R., and Leder, P. (1987) *Cell* **49**, 465-475
13. Stewart, T. A., Pattengale, P. K., and Leder, P. (1984) *Cell* **38**, 627-637
14. Habets, G. G., Scholtes, E. H., Zuydgeest, D., van der Kammen, R. A., Stam, J. C., Berns, A., and Collard, J. G. (1994) *Cell* **77**, 537-549
15. van Lohuizen, M., Verbeek, S., Scheijen, B., Wientjens, E., van der Gulden, H., and Berns, A. (1991) *Cell* **65**, 737-752
16. Hunter, T. (1991) *Cell* **64**, 249-270
17. Harrington, E. A., Fanidi, A., and Evan, G. I. (1994) *Curr. Opin. Genet. Dev.* **4**, 120-129
18. Ridley, A. J., Paterson, H. F., Noble, M., and Land, H. (1988) *EMBO J.* **6**, 635-645
19. Landschulz, W. H., Johnson, P. F., and McKnight, S. L. (1988) *Science* **240**, 1759-1764
20. Murre, C., McCaw, P.S., and Baltimore, D. (1989) *Cell* **56**, 777-783
21. Blackwell, T. K., Kretzner, L., Blackwood, E. M., Eisenman, R. N., and Weintraub, H. (1990) *Science* **250**, 1149-1151
22. Prendergast, G. C., and Ziff, E. B. (1991) *Science* **251**, 186-189
23. Blackwood, E. M., Lüscher, B., and Eisenman, R. N. (1992) *Genes Dev.* **6**, 71-80
24. Blackwood, E. M., and Eisenman, R. N. (1991) *Science* **251**, 1211-1217
25. Prendergast, G. C., Lawe, D., and Ziff, E. B. (1991) *Cell* **65**, 395-407
26. Kretzner, L., Blackwood, E. M., and Eisenmann, R. N. (1992) *Nature* **359**, 426-429
27. Amati, B., Dalton, S., Brooks, M. W., Littlewood, T. D., Evan, G. I., and Land, H. (1992) *Nature* **359**, 423-426
28. Amin, C., Wagner, A. J., and Hay, N. (1993) *Mol. Cell. Biol.* **13**, 383-390
29. Amati, B., Littlewood, T. D., Evan, G. I., and Land, H. (1993) *EMBO J.* **13**, 5083-5087
30. Amati, B., Brooks, M. W., Levy, N., Littlewood, T. D., Evan, G. I., and Land, H. (1993) *Cell* **72**, 233-245
31. Kato, G., Lee, W. M. F., Chen, L., and Dang, C. V. (1992) *Genes Dev.* **6**, 81-92
32. Amati, B., and Land, H. (1994) *Curr. Opin. Genet. Dev.* **4**, 102-108
33. Zervos, A. S., Gyuris, J., and Brent, R. (1993) *Cell* **72**, 223-232
34. Ayer, D. E., Kretzner, L., and Eisenman, R. N. (1993) *Cell* **72**, 211-222
35. Hurlin, P. J., C. Queva, Koskinen, P. J., Steingrimsson, E., Ayer, D. E., Copeland, N. G., Jenkins, N. A., and Eisenman, R. N. (1995) *EMBO J.* **14**, 5646-5659
36. Ayer, D. E., Lawrence, Q. A., and Eisenman, R. N. (1995) *Cell* **80**, 767-776
37. Schreiber-Agus, N., Chin, L., Chen, K., Torres, R., Rao, G., Guida, P., Skoultchi, A. I., and DePinho, R. A. (1995) *Cell* **80**, 777-786
38. Cerni, C., Bousset, K., Seelos, C., Burkhardt, H., Henriksson, M., and Lüscher, B. (1995) *Oncogene* **11**, 587-596
39. Bello-Fernandez, C., Packham, G., and Cleveland, J. L. (1993) *Proc. Natl. Acad. Sci. USA* **90**, 7804-7808
40. Wagner, A. J., Meyers, C., Laimins, L. A., and Hay, N. (1993) *Cell Growth Differ.* **4**, 879-883
41. Eilers, M., Schirm, S., and Bishop, J. M. (1991) *EMBO J.* **10**, 133-141
42. Gaubatz, S., Meichle, A., and Eilers, M. (1994) *Mol. Cell. Biol.* **14**, 3853-3862
43. Desbarats, L., Gaubatz, S., and Eilers, M. (1996) *Genes Dev.* **10**, 447-460

44. Blackwell, T. K., and Weintraub, H. (1990) *Science* **250**, 1104-1110
45. Penn, L. J. Z., Brooks, M. W., Laufer, E. M., and Land, H. (1990) *EMBO J.* **9**, 113-121
46. Marhin, W. W., Hei, Y.-J., Chen, S., Jiang, Z., Gallie, B., Phillips, R. A., and Penn, L. Z. (1996) *Oncogene* **12**, 43-52
47. Philipp, A., Schneider, A., Väsrik, I., Finke, K., Xiong, Y., Beach, D., Alitalo, K., and Eilers, M. (1994) *Mol. Cell. Biol.* **14**, 4032-4043
48. Bernards, R., Dessain, S. K., and Weinberg, R. A. (1986) *Cell* **47**, 667-674
49. Antonson, P., Pray, M. G., Jacobsson, A., and Xanthopoulos, K. G. (1995) *Eur.J. Biochem.* **232**, 397-403
50. Yang, B.-S., Gilbert, J. D., and Freytag, S. O. (1993) *Mol. Cell. Biol.* **13**, 3093-3102
51. Shrivastava, A., Saleque, S., Kalpana, G. V., Artandi, S., Goff, S. P., and Calame, K. (1993) *Science* **262**, 1889-1892
52. Roy, A. L., Carruthers, C., Gutjahr, T., and Roeder, R. G. (1993) *Nature* **365**, 359-361
53. Li, L., Nerlov, C., Prendergast, G., MacGregor, D., and Ziff, E. B. (1994) *EMBO J.* **13**, 4070-4079
54. Langdon, W. Y., Harris, A. W., Cory, S., and Adams, J. M. (1986) *Cell* **47**, 11-18
55. Keath, E. J., Caimi, P. G., and Cole, M. D. (1984) *Cell* **39**, 339-348
56. Evan, G. I., Wyllie, A. H., Gilbert, C. S., Littlewood, T. D., Land, H., Brooks, M., Waters, C. M., Penn, L. Z., and Hancock, D. C. (1992) *Cell* **69**, 119-128
57. Kelly, K., Cochran, B. H., Stiles, C. D., and Leder, P. (1983) *Cell* **35**, 603-610
58. Thompson, C. B., Challoner, P. B., Neiman, P. E., and Groudine, M. (1985) *Nature* **314**, 363-366
59. Hann, S. R., Thompson, C. B., and Eisenman, R. E. (1985) *Nature* **314**, 366-369
60. Barone, M. V., and Courtneidge, S. A. (1995) *Nature* **387**, 509-512
61. Mudryj, M., Hiebert, S. W., and Nevins, J. R. (1990) *Embo J.* **9**, 2179-2184
62. Wolf, D. A., Hermeking, H., Albert, T., Herzinger, T., Kind, P., and Eick, D. (1995) *Oncogene* **10**, 2067-78
63. Eick, D., Wedel, A., and Heumann, H. (1994) *Trends Genet* **10**, 292-296
64. Lavenu, A., Pournin, S., Babinet, C., and Morello, D. (1994) *Oncogene* **9**, 527-536
65. Charron, J., Malynn, B. A., Fisher, P., Stewart, V., Jeannotte, L., Goff, S. P., Robertson, E. J., and Alt, F. W. (1992) *Genes Dev.* **6**, 2248-2257
66. Sawai, S., Shimono, A., Hanaoka, K., and Kondoh, H. (1991) *New Biol.* **3**, 861-869
67. Davis, A. C., Wims, M., Spotts, G. D., Hann, S. R., and Bradley, A. (1993) *Genes Dev.* **7**, 671-682
68. Heikkila, R., Schwab, G., Wickstrom, E., Loke, S. L., Pluznik, D. H., Watt, R., and Neckers, L. M. (1987) *Nature* **328**, 445-449
69. Hopewell, R., and Ziff, E. B. (1995) *Mol. Cell Biol.* **15**, 3470-3478
70. Miyazaki, T., Liu, Z. J., Kawahara, A., Minami, Y., Yamada, K., Tsujimoto, Y., Barsoumian, E. L., Permutter, R. M., and Taniguchi, T. (1995) *Cell* **81**, 223-231
71. Iguchi-Ariga, S. M. M., Itani, T., Kiji, Y., and Ariga, H. (1987) *EMBO J.* **6**, 2365-2371
72. Rustgi, A. K., Dyson, N., and Bernards, R. (1991) *Nature* **352**, 541-544
73. Roussel, M. F., Theodoras, A. M., Pagano, M., and Sherr, C. J. (1995) *Proc. Natl. Acad. Sci. USA* **92**, 6837-6841
74. Steiner, P., Philipp, A., Lukas, J., Godden-Kent, D., Pagano, M., Mittnacht, S., Bartek, J., and Eilers, M. (1995) *EMBO J.* **14**, 4814-4826
75. Gu, Y., Rosenblatt, J., and Morgan, D. O. (1992) *EMBO J.* **11**, 3995-4005
76. Beijersbergen, R. L., Hijmans, E. M., Zhu, L., and Bernards, R. (1994) *EMBO J.* **13**, 4080-4086
77. Galaktionov, K., Jessus, C., and Beach, D. (1995) *Genes Dev.* **9**, 1046-1058
78. Steinman, R. A., Hoffman, B., Iro, A., Guillouf, C., Liebermann, D. A., and El-Houseini, M. E. (1994) *Oncogene* **9**, 3389-3396
79. Wagner, A. J., Kokontis, J. M., and Hay, N. (1994) *Genes Dev.* **8**, 2817-2830
80. Peter, M., and Herskowitz, I. (1994) *Science* **265**, 1228-1231
81. Peter, M., Gartner, A., Horecka, J., Ammerer, G., and Herskowitz, I. (1993) *Cell* **73**, 747-760
82. Rudolph, B., Saffrich, R., Zwicker, J., Henglein, B., Müller, R., Ansorge, W., and Eilers, M. (1996) *EMBO J.*, in press
83. Zwicker, J., Lucibello, F. C., Wolfraim, L. A., Gross, C., Truss, M., Engeland, K., and Müller, R. (1995) *EMBO J.* **14**, 4514-4522
84. Dynlacht, B. D., Flores, O., Lees, J. A., and Harlow, E. (1994) *Genes Dev.* **8**, 1772-1786
85. Schulze, A., Zerfass, K., Spitkovsky, D., Middendorp, S., Berges, J., Helin, C., Janssen-Dürr, P., and Henglein, B. (1995) *Proc. Natl. Acad. Sci. USA* **92**, 11264-11268
86. Kowalik, T. F., DeGregori, J., Schwarz, J. K., and Nevins, J. R. (1995) *J. Virol.* **69**, 2491-2500
87. Jansen-Dürr, P., Meichle, A., Steiner, P., Pagano, M., Finke, K., Botz, J., Wessbecher, J., Draetta, G., and Eilers, M. (1993) *Proc. Natl. Acad. Sci. USA* **90**, 3685-3689
88. DeGregori, J., Kowalik, T., and Nevins, J. R. (1995) *Mol. Cell. Biol.* **15**, 4215-4224
89. DeGregori, J., leone, G., Ohtani, K., Miron, A., and Nevins, J. R. (1995) *Genes Dev.* **9**, 2873-2887
90. Roussel, M. F., Cleveland, J. L., Shurtleff, S. A., and Sherr, C. J. (1991) *Nature* **353**, 361-363

91. Oswald, F., Lovec, H., Möröy, T., and Lipp, M. (1994) *Oncogene* **9**, 2029-2036
92. Beijersbergen, R. L., Kerkhoven, R. M., Zhu, L., Carlee, L., Voorhoeve, P. M., and Bernards, R. (1994) *Genes Dev.* **15**, 2680-2690
93. Hoang, A. T., Lutterbach, B., Lewis, B. C., Yano, T., Chou, T. Y., Barrett, J. F., Raffeld, M., Hann, S. R., and Dang, C. V. (1995) *Mol. Cell. Biol.* **15**, 4031-4042
94. Krek, W., Ewen, M. E., Shirodkar, S., Arany, Z., Kaelin, W. G., Jr., and Livingston, D. M. (1994) *Cell* **78**, 161-172
95. Albanese, C., Johnson, J., Watanabe, G., Eklund, N., Vu, D., Arnold, A., and Pestell, R. G. (1995) *J. Biol. Chem.* **270**, 23589-23597
96. Winston, J. T., Coats, S. R., Wang, Y.-Z., and Pledger, W. J. (1996) *Oncogene* **12**, 127-134
97. Jacobson, M. D., and Evan, G. I. (1994) *Curr. Biol* **4**, 337-340
98. Shi, Y., Glynn, J. M., Guilbert, L. J., Clotter, T. G., Bissonnette, R. I., and Green, D. R. (1992) *Science* **257**, 212-214
99. Hermeking, H., and Eick, D. (1994) *Science* **265**, 2091-2093
100. Bennett, M. R., Evan, G. I., and Schwartz, S. M. (1995) *Circ Res* **77**, 266-273
101. Fanidi, A., Harrington, E. A., and Evan, G. I. (1992) *Nature* **359**, 554-556
102. Wagner, A. J., Small, M. B., and Hay, N. (1993) *Mol. Cell. Biol.* **13**, 2432-2440
103. Harrington, E. A., Bennett, M. R., Fanidi, A., and Evan, G. I. (1994) *EMBO J.* **13**, 3286-3295
104. Packham, G., and Cleveland, J. L. (1994) *Mol. Cell Biol.* **14**, 5741-5747
105. Reisman, D., Elkind, N. B., Roy, B., Beamon, J., and Rotter, V. (1993) *Cell Growth Differ.* **4**, 57-65
106. Lowe, S. W., and Ruley, H. E. (1993) *Genes Dev.* **7**, 535-545

chapter 8
DNA replication licensing factor

James P. J. Chong[1] and J. Julian Blow

DNA Replication Control Laboratory, ICRF Clare Hall Laboratories,
South Mimms, Herts., EN6 3LD, United Kingdom
[1]To whom correspondence should be addressed

DNA Replication Licensing Factor (RLF) is an essential activity required to restrict the duplication of genomic DNA to precisely once per cell cycle. Recent fractionation of RLF activity from *Xenopus* egg extracts has resulted in the identification of two essential components, RLF-B and RLF-M. RLF-M has been purified to homogeneity and has been shown to consist of a complex of proteins in the MCM/P1 family. RLF-B is still unidentified, but possible candidates for this activity have been identified in yeast. Elucidation of the RLF mechanism will provide important insights into the way that chromosome replication is controlled.

INTRODUCTION TO LICENSING FACTOR

In order for a cell to undergo cell division and generate two daughters it must first copy its entire genome exactly once. This is of the utmost importance, since if any stretch of DNA is left unreplicated then information is lost and mitosis will not proceed successfully. On the other hand, should any stretch of DNA be copied more than once problems are also likely to occur through recombination and gene dosage effects.

The term Licensing Factor was first coined by Blow and Laskey in 1988 (1). It was used to describe an activity which ensured that DNA replicated only once in each cell cycle. This hypothetical factor was required to only bind unreplicated DNA and thus distinguish it from the already replicated regions of the genome. Part of its specificity for unreplicated DNA was due to its inability to cross the nuclear envelope. The Licensing Factor model was based on observations made in a cell-free replication system made from eggs from the South African Clawed Toad, *Xenopus laevis* (1-3). Previous work with this system had demonstrated that cell-free extracts made by the low-speed centrifugation of *Xenopus* eggs will assemble any given double-stranded DNA template into nuclei either using demembranated *Xenopus* sperm nuclei (2, 4) or purified double-stranded DNA (2, 5, 6). The system requires the DNA to be surrounded by an intact envelope of membrane before initiation of replication can occur (3, 6-9). Once this is in place, complete chromosomal DNA replication will take place under apparently normal cell cycle control.

Chromatin assembled into intact nuclei replicate only once in each *in vitro* cell cycle (1, 2, 10). Even if replicated ("G2") nuclei were isolated and transferred to fresh extract, no re-replication occurred (1). However, if the nuclear envelope of isolated G2 nuclei was permeabilised prior to transfer to fresh extract, re-replication then occurred (1, 11, 12). These observations led to the suggestion of a "Replication Licensing Factor" (RLF) which could bind to unreplicated chromatin only during mitosis (in the absence of a nuclear envelope). Unbound RLF was excluded from the nucleus and would only support a single replication event, after which it was inactivated (Figure 1). This resulted in the whole genome being replicated precisely once; each initiation event and consequent elongation required the presence and subsequent inactivation of RLF, which ensured that initiation could not re-occur on replicated DNA (1).

THE HUNT FOR LICENSING FACTOR

Recently, several different functional assays for licensing factor have been developed (13-15). Synchronised G2 cell culture nuclei will replicate in *Xenopus* extract only if they have been treated with lysolecithin in order to permeabilise the nuclear envelope (12, 13). These permeabilised G2 nuclei could then be incubated with protein fractions (potentially containing licensing factor) before resealing the nuclei using a membrane fraction prepared from low-speed *Xenopus* egg extracts. Coverley et al (13) were able to demonstrate that the licensing event was not due to the release of an inhibitory factor which had accumulated in nuclei during S-phase, as permeabilised G2 nuclei which were not treated with extract before being resealed did not re-replicate. This demonstrated that licensing depended on the nuclear entry of a positive initiation activity which cannot cross the nuclear envelope of intact G2 cells.,

Treatment of *Xenopus* extracts with certain protein kinase inhibitors during mitosis prevents the activation of RLF without affecting other mechanisms required for the extract to perform

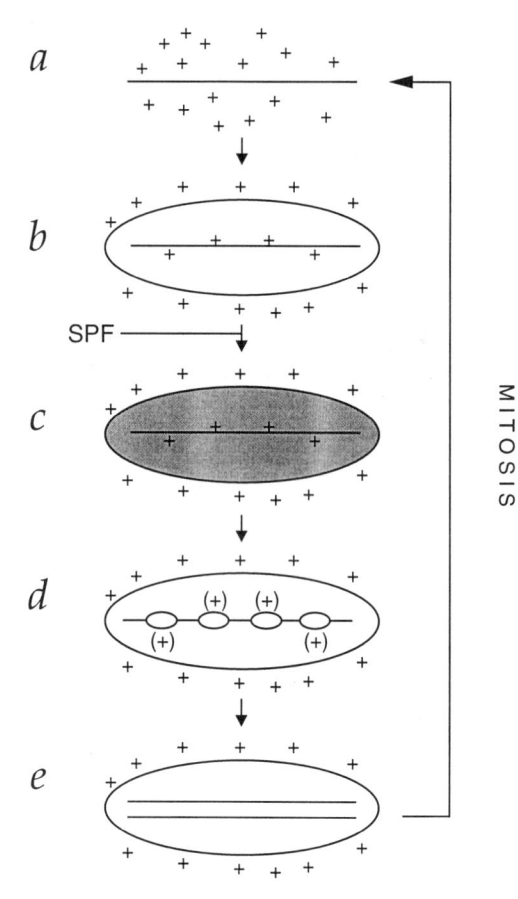

Figure 1. The original Licensing Factor model proposed by Blow and Laskey in 1988 (1). (*a*) Licensing Factor (denoted by "+") is free in the cytoplasm to bind to chromatin. (*b*) Unbound licensing factor is excluded from the nucleus as the nuclear envelope is assembled. (*c*) Initiation occurs at licensed sites. (*d*) Licensing factor is inactivated by initiation. (*e*) Replicated DNA cannot re-replicate due to the exclusion of licensing factor by the nuclear envelope. Nuclear envelope breakdown at mitosis allows licensing of the DNA to occur for the subsequent cell cycle. Reproduced with permission from Blow, J.J. and Laskey, R.A. (1988) *Nature* 332, 546-548. Macmillan Magazines Limited.

replication such as nuclear envelope assembly, initiation (in the presence of chromatin-bound RLF) and elongation (14, 15). Using one of these inhibitors, 6-dimethylaminopurine (6-DMAP), it was possible to develop an activity assay for RLF and establish a purification protocol (16). During the purification process RLF separates into two components, RLF-B and RLF-M. Both components are required for licensing together with the presence of hydrolysable ATP (16). RLF-M was purified to apparent homogeneity by following its activity in chromatographic fractions supplemented with crude RLF-B. Purification resulted in the identification of RLF-M as a complex consisting of several polypeptides of between 115 and 92kDa. Immunoblotting showed that the RLF-M complex contained the *Xenopus* Mcm3 homologue (16).

Xenopus Mcm3, a member of the MCM/P1 family (see below) was also shown to be bound to unreplicated chromatin in G1, but not to replicated chromatin in G2 (16-19). Quantification of Mcm3 bound to chromatin showed a decrease through S-phase which approximately corresponded to the amount of replicated DNA. Mcm3 was shown to be essential for DNA replication by immunodepletion of Xenopus extracts with anti-Mcm3 antibodies (17, 18). These immunodepleted extracts could be rescued either by anti-Mcm3 immunoprecipitated material (17, 18) or the purified RLF-M complex (16). Further, permeabilised nuclei from G1, but not G2 cells, which contained chromatin-bound Mcm3 were capable of replicating in an Mcm3-depleted extract (18). These results conform to the licensing factor model.

MCMS IN YEAST

The MCM/P1 family of proteins was first identified in a screen for mutants defective in the maintenance of episomes in the yeast *Saccharomyces cerevisiae*. These "minichromosome maintenance" (MCM) mutants lost stable circular plasmids containing a centromere and a replication origin (ARS) (20). Some of these mutants were shown to be defective due to their inability to activate specific replication origins (21-23). It was also shown that a shift to the non-permissive temperature of an *mcm3* mutant during S-phase prevented mitosis. Three of the genes (*MCM2, 3* and *5*) had highly homologous sequences. Other genes required for DNA replication that were identified in a cell division cycle (*CDC*) screen were shown to be members of the MCM/P1 family of genes. *CDC46* is identical to *MCM5* (24), whilst *CDC54* and *CDC47* represented new members of the family.

Observations concerning the interaction of these genes showed that *CDC46*, (allelic to *MCM5* (24)), could suppress either the *cdc45* or the *cdc54* mutations in an allele-specific manner (25). *CDC47* was also shown to suppress a *cdc45* mutation. In addition, *CDC46* and *CDC54* showed allele-specific lethality, as did *CDC47* and *CDC54*, and *CDC46* and *CDC47* (Figure 2) (26). *MCM2* and *MCM3* were also shown to be synthetically lethal (27). The *mcm3-1* mutant could be partially rescued by the over-production of Mcm2, but *mcm2-1* could not be rescued by Mcm3. Specificity for different ARSs was also seen in the *mcm2-1* and *mcm3-2* mutants (24, 27, 28). One possible explanation for these genetic interactions is that the MCM/P1 proteins function together in a large complex. Data consistent with this interpretation has been produced in a number of organisms (see below).

Cdc46 was also found to show variation in its subcellular localisation throughout the cell cycle.

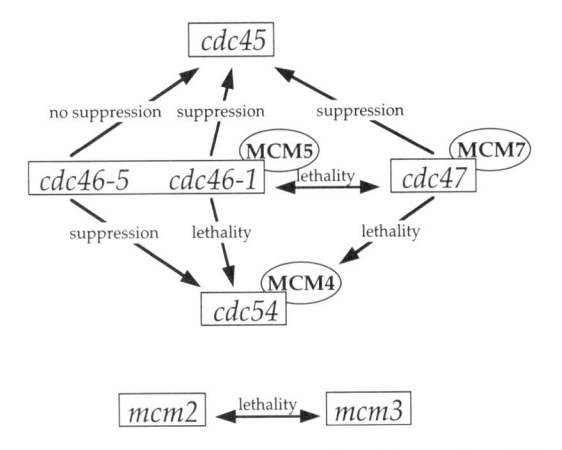

Figure 2. Summary of genetic interactions of *S. cerevisiae* CDC genes, adapted from (26). *CDC46* shows allele-specific suppression of *cdc54* and *cdc45*. *cdc47* could also suppress *cdc45*. In general, the combination of two mutant MCM proteins produced synthetic lethality. The mechanism of interaction between *CDC45* and MCM proteins has not yet been determined. With permission from Hennessy, K.M. et al. *Genes Dev.* 1991, 5, 958-69.

Immunofluorescence of Cdc46 showed accumulation of the protein within the nucleus up until early in S-phase (29), whereupon immunofluorescence disappeared. Nuclear immunofluorescence did not reappear until mitosis (Figure 3). This cyclic behaviour is analogous to that predicted for RLF, making Cdc46 a possible licensing factor candidate. Nuclear accumulation of Mcm2, Mcm3 and Cdc47 has also been observed (28, 30). Interestingly, in cells held in early S-phase by a hydroxyurea arrest, partial nuclear Mcm2 staining was observed to a much greater extent than that of Mcm3 (28), an observation consistent with Mcm2 and Mcm3 affecting different origins of replication. Yan et al (28) demonstrated that the MCM/P1 proteins were present in the cytoplasm of cycling cells, but could not be detected in this location by immunofluorescence. They also demonstrated the presence of a chromatin-bound fraction of MCM/P1 proteins in an asynchronous population by DNaseI digestion (see later and reference (31)). These results are consistent with the cyclic chromatin-association of Mcm3 in *Xenopus*.

MCM/P1 proteins have also been identified out of various screens in *Schizosaccharomyces pombe* and the mutants again show replication defect phenotypes. *cdc19/nda1* and *cdc21* mutants show synthetic lethality (32). *cdc21* mutants show a defect in the mitotic maintenance of some plasmids, as is seen in *S. cerevisiae* for *mcm2* and *mcm3* (33). In addition, *mis5+* has been shown to encode a novel member of the MCM/P1 family not yet found in *Saccharomyces cerevisiae* (34), although homologues have been found in other organisms (35).

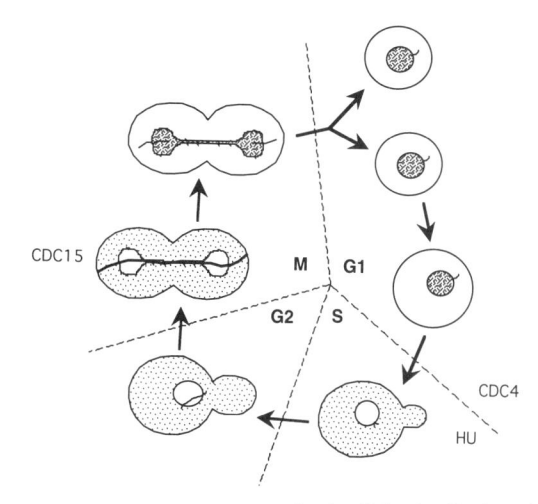

Figure 3. Scheme of variation of sub-cellular localisation of CDC46 in the yeast cell cycle, as judged by immunofluorescence. CDC46 is seen to accumulate in the nucleus of the cell during late mitosis ("M"). It remains there through G1, and disappears during S-phase ("S"). CDC46 remains cytoplasmic (as detected by immunoblotting) throughout G2 and early M-phases. Reproduced from Blow, J. (1995) in *Cell Cycle Control* (Hutchinson, C. and Glover, D.M. eds.) IRL Press. By permission of Oxford University Press.

MCM HOMOLOGUES IN HIGHER EUKARYOTES: THE MCM/P1 FAMILY

Since their first description in budding yeast, *MCM* homologues have been identified in many other organisms including fission yeast, insects, plants, amphibians and mammals (33, 36-39). The first mammalian homologues were actually isolated when polyclonal antibodies, raised to a protein co-fractionating with DNA polymerase-α in calf thymus, were used to probe human and mouse cDNA expression libraries. These *MCM* homologues were called "nuclear protein P1" (39). Sequence analysis of the MCM/P1 family members identified to date shows a very highly conserved region towards the C-terminus and a completely conserved MCM-box "IDEFDKM" (40). Other small motifs are also completely conserved, including a putative ATP-binding domain, but no structure/function information is available to date. We have performed an alignment of the known MCM/P1 family members and constructed a phylogenetic tree which suggests that the proteins fall into 6 groups, (which have been designated as MCM2-MCM7). In this classification each organism has only a single protein in each group (41, 42) (Figure 4). Recently, a great quantity of data has been published on the MCM/P1 family proteins, including the identification of many homologues, all of which can be assigned to the 6 groups described above. It thus seems likely that MCM/P1 proteins play a conserved role in the replication licensing system in all eukaryotic organisms.

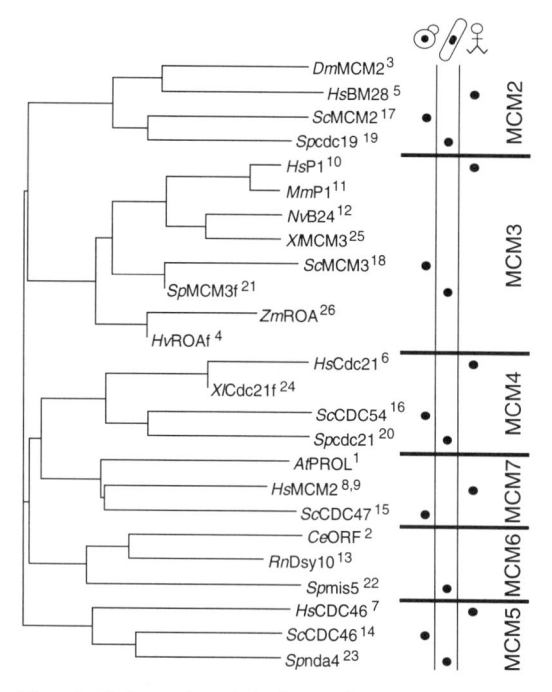

Figure 4. Phylogenetic analysis of currently sequenced MCM/P1 genes taken from (41). The MCM/P1 proteins form 6 groups (MCM2-MCM7) within a highly conserved family. Organisms are abbreviated: Dm, *Drosophila melanogaster*; Hs, *Homo sapiens*; Sc *Saccharomyces cerevisiae*; Sp, *Schizosaccharomyces pombe*; Mm, *Mus musculus*; Nv, *Notophthalmus viridescens*; Xl, *Xenopus laevis*; Zm, *Zea mays*; Hv, *Hordeum vulgare*; At, *Arabidopsis thaliana*; Ce, *Caenorhabditis elegans*; Rn, *Rattus norvegicus*. Protein fragments are indicated in the figure by the suffix 'f'. Accession numbers (shown as superscripts in the figure) are as follows (P=Swiss-Prot, otherwise Genbank). Reproduced from Chong, JPJ. et al. (1996), *Trends Biochem. Sci.* **21**, 102-107 with permission of Elsevier Trends Journals.

1	L39954	2	P34647	3	L42762
4	Z29369	5	X67334	6	X74794
7	X74795	8	D28480	9	X74796
10	P25205	11	P25206	12	X78322
13	U17565	14	P29496	15	P38132
16	P30665	17	P29469	18	P24279
19	P40377	20	P29458	21	P30666
22	D31960	23	P41389	24	P30664
25	U26057/D38074	26	Z29368		

MCM/P1 COMPLEXES

Data in *Xenopus* suggests that the MCM/P1 proteins may be found as an active complex. Purification of the RLF-M fraction revealed a complex of at least 3 polypeptides which cross-reacted with Mcm2, Mcm3 and Mcm5 antibodies and had a molecular weight of approximately 600kDa on gel filtration (16). Immunoprecipitation of *Xenopus* egg extracts using anti-Mcm3 antibodies also showed, in addition to *Xl*Mcm3, the presence of *Xl*Mcm2, *Xl*Mcm5 and *Xl*Cdc21 (*Xl*Mcm4) bands, and at least one other band which cross-reacts with an antibody raised to the conserved MCM/P1 peptide (19). It therefore seems likely that all 6 MCM/P1 family members are complexed in some way. Exactly what these interactions are still remains to be precisely determined.

Two distinct MCM/P1 complexes of about 600kDa have been identified in the fruit fly *Drosophila melanogaster*. One of these complexes contains *Dm*Cdc46 (*Dm*Mcm5) while the other contains *Dm*Mcm2 and Dpa (*Dm*Mcm4) (36). This is consistent with earlier data on human Cdc46 (*Hs*Mcm5) and P1 (*Hs*Mcm3) forming a complex of approximately 500kDa, although other proteins in this complex have not been identified (43). A similar interaction between Cdc46 and P1 homologues has also been seen in mouse cells, where the *Mm*Cdc21 (*Mm*Mcm4) protein has also been excluded from the complex (44). All this data suggests that the MCM/P1 proteins form at least two distinct complexes *in vivo*; one containing Mcm3 and Mcm5 homologues, and the other containing Mcm2 and Mcm4 homologues. Data concerning the association of Mcm6 and Mcm7 group homologues in the above complexes is obviously of great interest. It is likely that some protein-protein interactions are more favourable than others, and this has been demonstrated to a degree by all the workers in this field by the variation in their results due to using different salt concentrations in the isolation of the complexes.

MCM/P1 PHOSPHORYLATION AND NUCLEAR LOCALISATION

Certain MCM/P1 proteins appear to undergo changes in phosphorylation state as the cell cycle progresses. Mouse P1 (*Mm*MCM3) was shown to be present in cultured cell nuclei in at least two different phosphorylation states. One hypophosphorylated (or "underphosphorylated") form was found to be tightly associated with chromatin, while the second, hyperphosphorylated form appeared in the nucleus throughout S-phase, and was much less tightly associated with the chromatin. These two forms could be distinguished by their susceptibility to detergent extraction (the underphosphorylated form was detergent resistant) (31). Kimura *et al* (31) also observed that *Mm*P1 was removed first from the euchromatin regions and then the heterochromatin regions, correlating with the replication of these regions during S-phase. Other groups have produced similar results for murine P1 (45) and human P1 (46). Recent work has shown similar nuclear localisation and resistance to detergent extraction for the murine Cdc21 (*Mm*MCM4) and CDC46 (*Mm*MCM5) proteins (44).

The human BM28 (*Hs*Mcm2) protein has also been shown to be localised to the nucleus. Anti-BM28 antibodies were microinjected into cells synchronised either in G1 or S-phase of the cell cycle. Following microinjection, G1 cells were unable to progress into S-phase, and S-phase cells were unable to progress through mitosis, suggesting a two-fold requirement for BM28 during the cell cycle - both to enter S-phase and to perform cell division (47). Microinjection of murine P1 antibodies also resulted in an inhibition of DNA synthesis (31). BM28 has also been shown to be variably phosphorylated during the cell cycle. As is seen for murine P1, hyperphosphorylation occurs during S-phase, again in conjunction with loss of tight chromatin association (48). Similarly, human Cdc47 (*Hs*Mcm7) has been shown to dissociate from chromatin during S-phase (49) although no data on phosphorylation states is currently available.

Using an antisense oligonucleotide to the first 18 nucleotides of the human CDC47 cDNA sequence, Fujita *et al* (50) have also shown a requirement for this protein in S-phase progression. A more detailed study of *Xenopus* Cdc21 (*Xl*MCM4) suggests that the phosphorylation of MCM proteins throughout the cell cycle may be even more complicated (51). Cdc21 binds chromatin in an underphosphorylated state which is then partially phosphorylated in early S-phase, before dissociating from the chromatin. Finally, Cdc21 is hyperphosphorylated at the onset of mitosis (51). Phosphorylation data on MCM5-MCM7 group proteins has not been published at the present time. It is possible that formation or binding of the MCM/P1 protein complexes described previously could be regulated by the phosphorylation changes outlined here.

CDC6/cdc18+:
POSSIBLE CANDIDATES FOR RLF-B

Functional work in *Xenopus* has demonstrated that RLF consists of two separable components which are both required for licensing of DNA for replication, RLF-M (the MCM/P1 complex) and RLF-B (16). Licensing occurs when Mcms are bound to chromatin. However, the incubation of RLF-M and chromatin is not sufficient for the Mcms to bind and license chromatin. The second fraction, RLF-B is required at the same time as RLF-M to bind the Mcms to chromatin, thus facilitating its licensing. Hydrolysable ATP was also required to drive the licensing reaction (16). RLF-B has not yet been identified, but it is possible to speculate on likely components of this activity.

One possible candidate for RLF-B is the *S. cerevisiae CDC6* gene. *CDC6* is an essential gene which is required for DNA replication (52). *CDC6* has been cloned and shown to encode a 58kDa protein (53, 54). Purified Cdc6 protein will bind both ATP and GTP and shows significant ATP/GTPase activity (55). *CDC6* was shown to be periodically transcribed, with transcripts reaching a peak prior to initiation of S-phase (55, 56). As well as being required for correct DNA replication, *CDC6* appears to have a second role. This was demonstrated by constitutive expression of *CDC6* which was shown to delay the onset of mitosis in both *S. cerevisiae* and *S. pombe*, apparently by acting through a pathway which inhibits the activation of mitotic kinases (57). Further data to support this mechanism has been produced from work in *S. pombe* where the *CDC6* homologue, *cdc18+* demonstrates a similar effect and interaction with mitotic kinases, as shown by the effects seen by the over-expression of the kinase inhibitors Rum1 and SIC1 (58).

cdc6 mutants have also been shown to have a high minichromosome loss rate which can be overcome by the inclusion of additional ARS elements onto the minichromosomes. This suggests that *CDC6* may act directly at origins of replication (59). However, Zwerschke *et al.* (55) were unable to show any helicase or DNA binding activity of Cdc6. Their data do show a burst of transcription of *CDC6* on exit from G0, which correlates with work suggesting that *de novo* synthesis of Cdc6 may be important for S-phase progression (60, 61).

Cdc6 has recently been shown to be a multicopy suppressor of the *orc5-1* mutant, as well as physically interacting with the purified ORC complex (62). *In vivo* a *cdc6ts* strain causes the loss of the pre-replicative complex (61) (see below). Interestingly, *CDC6* and *cdc18+* show strong homology to the C-terminus of *ORC1* and the mating-type transcriptional silencer *SIR3*. Homologues of *CDC6* in higher eukaryotes remain to be identified.

ORC AND THE PRE-REPLICATIVE COMPLEX: POTENTIAL INTERACTION WITH RLF

The origin recognition complex (ORC) was first identified in *S. cerevisiae* as a complex which bound to origins of replication (63). Homologues of ORC have also been identified in higher eukaryotes, suggesting that ORC is conserved in metazoa (64-66). *In vivo* and *in vitro* footprinting studies have demonstrated the presence of an ORC-specific footprint at origins of replication (63, 67). However, cell cycle studies show that there are two distinct footprints at yeast origins, which vary before and after replication has occurred. In the post-replicative state, genomic footprints of replication origins closely resemble those seen *in vitro* with purified ORC (68). From anaphase to the start of S-phase, a larger footprint is seen over replication origins extending for approximately 50 extra base-pairs 3' to the ORC-specific footprint. The presence of this larger footprint suggests that a larger complex of proteins is assembled onto ORC during late mitosis and is removed as initiation of DNA replication occurs.

The appearance of the pre-replicative footprint in G1 coincides with the time when MCMs bind to chromatin and suggests that the formation of a "pre-replicative complex" on replication origins may be related to the licensing event. Formation of the pre-replicative complex is dependent on *CDC6*. Studies using a temperature-sensitive *CDC6* mutant show that the pre-replicative complex will not form, or is lost, at the non-permissive temperature (61). The pre-replicative complex is also lost when cells enter quiescence (68). Presumably this complex has to be reassembled before DNA replication can occur, and this may be facilitated by the burst of *CDC6* transcription observed on exit from quiescence (55, 60). The presence of *CDC6/cdc18+* also appears to inhibit entry into mitosis (58, 60), and so the presence of this protein in pre-replicative complexes may serve as a good restraint on attempts to perform mitosis before replication is complete. The parallels between *CDC6* being required for the assembly of a pre-replicative complex, and that of RLF-B being required to assemble RLF-M onto chromatin before replication can occur, are very striking, and suggest *CDC6* as a good candidate for the RLF-B activity.

OVERALL CONTROL OF LICENSING FACTOR: THE NUCLEAR ENVELOPE

The original Licensing Factor model postulated an essential role for the nuclear envelope in the control of licensing DNA for replication. It suggested that the nuclear envelope prevented the entry of further licensing factor, and hence avoided the possibility of re-replication of the DNA in a single cell cycle (1). Further evidence for a role for the nuclear envelope was presented using synchronised tissue culture cell nuclei which were added to *Xenopus* eggs or egg extracts with and without nuclear envelope permeabilisation (11, 12). Here it was shown that G1 HeLa nuclei did not require nuclear envelope permeabilisation for efficient replication in extracts, but that G2 nuclei did. The technology from these experiments was then developed into an assay for functional licensing factor by resealing the nuclear envelope of G2 nuclei after incubating them with fractions containing licensing factor (13).

While MCM proteins are obviously required for the once-and-only-once replication of the DNA in the cell cycle, and appear to be a good marker for unreplicated DNA, their ability to cross the nuclear envelope does not fit the licensing factor model. In *Xenopus*, Mcm3 appears to be imported into the nucleus at all stages of the cell cycle, possibly as part of a larger RLF-M complex. Strikingly, nuclei assembled in the absence of *Xl*Mcm3 could subsequently import *Xl*Mcm3 and initiate DNA replication (18, 19). In contrast, a recombinant fusion *Xl*Mcm3 protein could not cross the nuclear envelope (17). However, although endogenous *Xl*Mcm3 will cross the nuclear envelope of G2 nuclei, these nuclei would still not replicate as the nuclear MCM/P1 protein would not bind to chromatin in G2 nuclei (19). This data is consistent with immunofluorescence studies in mammalian cells showing that MCM/P1 proteins are nuclear throughout the cell division cycle (44, 48), and suggests, in the case of the recombinant protein, that the correct conformation of MCM/P1 proteins is important.

One obvious explanation for the lack of MCM/P1 binding in intact G2 nuclei is that RLF-B, which is required for RLF-M to associate with chromatin, is unable to cross the nuclear envelope in its active form. Indeed, Madine *et al* demonstrate the presence of a second "loading factor" activity, which is unable to cross the nuclear envelope and is required in order for MCM/P1 proteins to bind the chromatin of G2 nuclei (19). This "loading factor" seems likely to be synonymous with RLF-B (16, 19).

A SUMMARY OF THE LICENSING FACTOR MODEL

The binding of the RLF-M complex of MCM/P1 proteins to chromatin is mediated via RLF-B activity. RLF-B is either consumed in the process of binding the MCM/P1 proteins to chromatin, or RLF-B is inactivated by the time the signal for initiation of replication occurs. It seems likely that the replication of a specific stretch of DNA is controlled by the presence of chromatin-bound MCM/P1 proteins *cis* to the stretch of DNA to be copied. Initiation of replication, or passage of a replication fork, causes bound MCM/P1 proteins to be released from the chromatin. In the absence of RLF-B, the released RLF-M complex is unable to rebind chromatin. Thus re-replication of replicated regions of DNA cannot occur until RLF-B can again gain access to chromatin which occurs after mitosis (Figure 5). It is also possible that the hyperphosphorylation of MCM/P1 proteins which coincides with their release from chromatin during S-phase may further inhibit their ability to reassociate with chromatin.

PROSPECTS

The recent identification of the MCM/P1 family of proteins as the RLF-M component of replication licensing factor has enabled us to integrate much of the information concerning this family. However, many problems are still unanswered. What controls the variation in licensing activity through the cell cycle? What is the identity of the essential RLF-B component and is it related to *CDC6*? How do RLF-B and RLF-M interact? How do the licensing proteins interact with replication origins and do they require the presence of ORC? Elucidation of these questions should yield further insights into the role of RLF in the control of initiation of DNA replication.

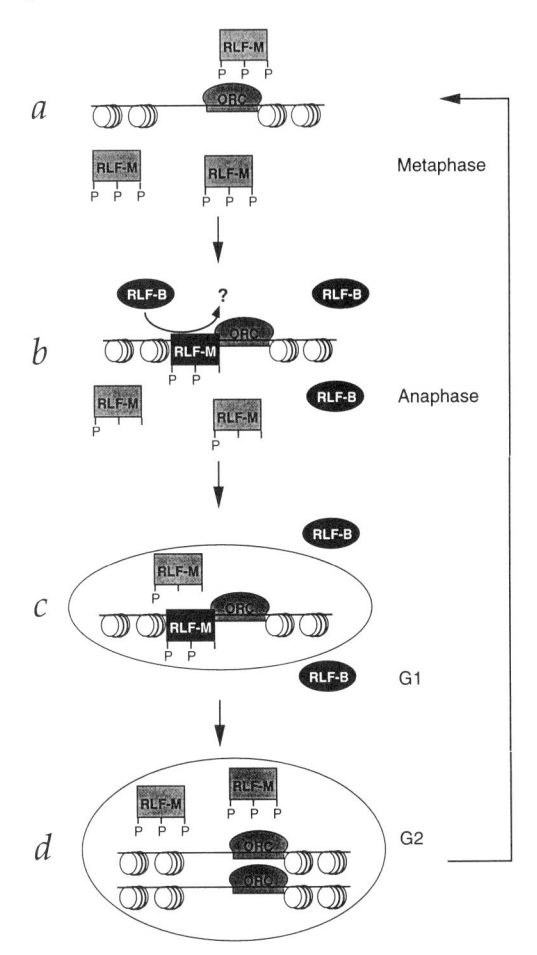

Figure 5. Summary of the known mechanism of Replication Licensing Factor. (*a*) Origins of DNA replication are marked by Origin Recognition Complex (ORC), probably throughout the cell cycle. Licensing Factor consists of two components, RLF-B and RLF-M. (*b*) Activation of RLF-B at the metaphase-anaphase transition results in the binding (and phosphorylation?) of RLF-M to chromatin. (*c*) RLF-B cannot cross the nuclear envelope, unbound RLF-M cannot rebind chromatin in the absence of RLF-B. (*d*) RLF-M is released from the chromatin as DNA replication occurs ensuring that the DNA is not re-replicated in a single cell cycle. Hyperphosphorylation of specific MCM/P1 proteins also occurs at this stage. The exact significance of this is not known. MCM/P1 proteins return to their hypophosphorylated form during mitosis (*a*).

REFERENCES

1. Blow, J.J. and Laskey, R.A. (1988) *Nature* **332**, 546-8.
2. Blow, J.J. and Laskey, R.A. (1986) *Cell* **47**, 577-87.
3. Blow, J.J. and Watson, J.V. (1987) *EMBO J.* **6**, 1997-2002.
4. Lohka, M.J. and Masui, Y. (1983) *Science* **220**, 719-721.
5. Newmeyer, D.D., Lucocq, J.M., Burglin, T.R. and De Robertis, E.M. (1986) *EMBO J.* **5**, 501-510.
6. Newport, J. (1987) *Cell* **48**, 205-217.
7. Sheehan, M.A., Mills, A.D., Sleeman, A.M., Laskey, R.A. and Blow, J.J. (1988) *J. Cell Biol.* **106**, 1-12.
8. Blow, J.J., Sheehan, M.A., Watson, J.V. and Laskey, R.A. (1989) *J. Cell Sci. Suppl.* **12**, 183-95.
9. Blow, J.J. and Sleeman, A.M. (1990) *J. Cell Sci.* **95**, 383-391.
10. Blow, J.J., Dilworth, S.M., Dingwall, C., Mills, A.D. and Laskey, R.A. (1987) *Philos. Trans. R. Soc. Lond. B. Biol. Sci.* **317**, 483-94.
11. De Roeper, A., Smith, J.A., Watt, R.A. and Barry, J.M. (1977) *Nature* **265**, 469-470.
12. Leno, G.H., Downes, C.S. and Laskey, R.A. (1992) *Cell* **69**, 151-8.
13. Coverley, D., Downes, C.S., Romanowski, P. and Laskey, R.A. (1993) *J. Cell Biol.* **122**, 985-92.
14. Blow, J.J. (1993) *J Cell Biol* **122**, 993-1002.
15. Kubota, Y. and Takisawa, H. (1993) *J. Cell Biol.* **123**, 1321-1331.
16. Chong, J.P., Mahbubani, H.M., Khoo, C.Y. and Blow, J.J. (1995) *Nature* **375**, 418-21.
17. Kubota, Y., Mimura, S., Nishimoto, S., Takisawa, H. and Nojima, H. (1995) *Cell* **81**, 601-9.
18. Madine, M.A., Khoo, C.Y., Mills, A.D. and Laskey, R.A. (1995) *Nature* **375**, 421-4.
19. Madine, M.A., Khoo, C.-Y., Mills, A.D., Musahl, C. and Laskey, R.A. (1995) *Curr. Biol.* **5**, 1270-1279.
20. Maine, G.T., Sinha, P. and Tye, B.K. (1984) *Genetics* **106**, 365-385.
21. Sinha, P., Chang, V. and Tye, B.-K. (1986) *J. Mol. Biol.* **192**, 805-814.
22. Gibson, S.I., Surosky, R.T. and Tye, B.K. (1990) *Mol. Cell Biol.* **10**, 5707-20.
23. Ray, A., Roy, N., Maitra, M. and Sinha, P. (1994) *Curr. Genet.* **26**, 403-409.
24. Chen, Y., Hennessy, K.M., Botstein, D. and Tye, B.K. (1992) *Proc. Natl. Acad. Sci. USA* **89**, 10459-63.
25. Moir, D., Stewart, S.E., Osmond, B.C. and Botstein, D. (1982) *Genetics* **100**, 547-63.
26. Hennessy, K.M., Lee, A., Chen, E. and Botstein, D. (1991) *Genes Dev.* **5**, 958-69.

27. Yan, H., Gibson, S. and Tye, B.K. (1991) *Genes Dev.* **5**, 944-57.
28. Yan, H., Merchant, A.M. and Tye, B.K. (1993) *Genes Dev.* **7**, 2149-60.
29. Hennessy, K.M., Clark, C.D. and Botstein, D. (1990) *Genes Dev.* **4**, 2252-63.
30. Dalton, S. and Whitbread, L. (1995) *Proc. Natl. Acad. Sci. USA* **92**, 2514-8.
31. Kimura, H., Nozaki, N. and Sugimoto, K. (1994) *EMBO J.* **13**, 4311-20.
32. Forsburg, S.L. and Nurse, P. (1994) *J. Cell Sci.* **107**, 2779-88.
33. Coxon, A., Maundrell, K. and Kearsey, S.E. (1992) *Nucl. Acids Res.* **20**, 5571-7.
34. Takahashi, K., Yamada, H. and Yanagida, M. (1994) *Mol. Biol. Cell* **5**, 1145-58.
35. Sykes, D.E. and Weiser, M.M. (1995) *Gene* **163**, 243-247.
36. Su, T.T., Feger, G. and O'Farrell, P.H. (1996) *Mol. Biol. Cell* **7**, 319-329.
37. Springer, P.S., McCombie, W.R., Sundaresan, V. and Martienssen, R.A. (1995) *Science* **268**, 877-80.
38. Bucci, S., Ragghianti, M., Nardi, I., Bellini, M., Mancino, G. and Lacroix, J.C. (1993) *Int. J. Dev. Biol.* **37**, 509-17.
39. Thömmes, P., Fett, R., Schray, B., Burkhart, R., Barnes, M., Kennedy, C., Brown, N.C. and Knippers, R. (1992) *Nucl. Acids Res.* **20**, 1069-74.
40. Koonin, E.V. (1993) *Nucl. Acids Res.* **21**, 2541-2547.
41. Chong, J.P.J., Thömmes, P. and Blow, J.J. (1996) *Trends Biochem. Sci.* **21**, 102-107.
42. Kearsey, S.E., Maiorano, D., Holmes, E.C. and Todorov, I.T. (1996) *Bioessays* **18**, 183-190.
43. Burkhart, R., Schulte, D., Hu, D., Musahl, C., Gohring, F. and Knippers, R. (1995) *Eur. J. Biochem.* **228**, 431-8.
44. Kimura, H., Takizawa, N., Nozaki, N. and Sugimoto, K. (1995) *Nucl. Acids Res.* **23**, 2097-104.
45. Starborg, M., Brundell, E., Gell, K., Larsson, C., White, I., Daneholt, B. and Hoog, C. (1995) *J. Cell Sci.* **108**, 927-34.
46. Schulte, D., Burkhart, R., Musahl, C., Hu, B., Schlatterer, C., Hameister, H. and Knippers, R. (1995) *J. Cell Sci.* **108**, 1381-9.
47. Todorov, I.T., Pepperkok, R., Philipova, R.N., Kearsey, S.E., Ansorge, W. and Werner, D. (1994) *J. Cell Sci.* **107**, 253-65.
48. Todorov, I.T., Attaran, A. and Kearsey, S.E. (1995) *J. Cell Biol.* **129**, 1433-45.
49. Fujita, M., Kiyono, T., Hayashi, Y. and Ishibashi, M. (1996) *J. Biol. Chem.* **271**, 4349-4354.
50. Fujita, M., Kiyono, T., Hayashi, Y. and Ishibashi, M. (1996) *Biochem. Biophys. Res. Comm.* **219**, 604-607.
51. Coué, M., Kearsey, S.E. and Méchali, M. (1996) *EMBO J.* **15**, 1085-1097.
52. Hartwell, L.H. (1976) *J. Mol. Biol.* **104**, 803-817.
53. Lisziewicz, J., Godany, A., Agoston, D.V. and Kuntzel, H. (1988) *Nucl. Acids Res.* **16**, 11507-20.
54. Zhou, C., Huang, S.H. and Jong, A.Y. (1989) *J. Biol. Chem.* **264**, 9022-9.
55. Zwerschke, W., Rottjakob, H.W. and Kuntzel, H. (1994) *J. Biol. Chem.* **269**, 23351-6.
56. Zhou, C. and Jong, A. (1990) *J. Biol. Chem.* **265**, 19904-9.
57. Bueno, A. and Russell, P. (1992) *EMBO J.* **11**, 2167-76.
58. Jallepalli, P.V. and Kelly, T.J. (1996) *Genes Dev.* **10**, 541-552.
59. Hogan, E. and Koshland, D. (1992) *Proc. Natl. Acad. Sci. USA* **89**, 3098-102.
60. Piatti, S., Lengauer, C. and Nasmyth, K. (1995) *EMBO J.* **14**, 3788-99.
61. Cocker, J.H., Piatti, S., Santocanale, C., Nasmyth, K. and Diffley, J.F.X. (1996) *Nature* **379**, 180-182.
62. Liang, C., Weinreich, M. and Stillman, B. (1995) *Cell* **81**, 667-76.
63. Bell, S.P. and Stillman, B. (1992) *Nature* **357**, 128-134.
64. Gavin, K.A., Hidaka, M. and Stillman, B. (1995) *Science* **270**, 1667-1671.
65. Gossen, M., Pak, D.T.S., Hansen, S.K., Acharya, J.K. and Botchan, M.R. (1995) *Science* **270**, 1674-1677.
66. Carpenter, P.B., Mueller, P.R. and Dunphy, W.G. (1996) *Nature* **379**, 357-360.
67. Diffley, J.F.X. and Cocker, J.H. (1992) *Nature* **357**, 169-172.
68. Diffley, J.F., Cocker, J.H., Dowell, S.J. and Rowley, A. (1994) *Cell* **78**, 303-16.

chapter 9
Tyrosine kinases wee1 and mik1 as effectors of DNA replication checkpoint control

Jérôme Tourret*‡ and Frank McKeon*1

*Department of Cell Biology, Harvard Medical School
240 Longwood Avenue, Boston, MA USA 02115
‡École Normale Supérieure
45 rue d'Ulm, 75005 PARIS, FRANCE
1 To whom correspondence should be addressed

Cell cycle studies have revealed mechanisms that prevent cell division if DNA fails to be completely replicated or sustains damage[1-3]. Here we focus on the evidence from yeast genetics that the wee1 and mik1 tyrosine kinases cooperate in the inhibitory phosphorylation of cdc2p, and the possibility that these kinases function in pathways that ensure the integrity of the genome prior to cell division. We also review the progress in cloning and analysing wee1-like tyrosine kinases from higher eukaryotes, and the evidence for and against their functioning in ensuring DNA replication prior to mitosis. Finally, we discuss the genes involved in these feedback controls and suggest that wee1p and mik1p might be the ultimate effectors that prevent mitosis when a checkpoint is triggered.

INTRODUCTION

The S and G2 phases of the cell cycle are particularly important because they represent the last opportunity for a cell to complete DNA replication or survey defects to the genome before chromosomes are irreparably transmitted to daughter cells. It is therefore not surprising that intricate controls, in the form of signal transduction pathways, act to prevent the onset of M phase of the cell cycle until the completion of DNA replication. Extensive genetic and biochemical evidence exists for the role of cdc2p kinase activity in the G2/M transition, and the notion that the G2 phase can be extended by suppressing this activity by activating or suppressing the immediate regulators of cdc2p, including wee1p, mik1p, and cdc25p[2, 4]. Wee1 and mik1 encode tyrosine kinases which act in a coordinate manner to suppress cdc2p activity by phosphorylating Y15 within the ATP binding site of the cdc2p kinase domain[4-9]. Conversely, cdc25p is a tyrosine phosphatase whose function appears to activate cdc2p by removing this inhibitory phosphate on Y15[10]. Despite this detailed knowledge of the immediate regulators of the activity of the cdc2p kinase, less is understood how this regulatory scheme is affected by sensors of the state of DNA replication and genome integrity during S phase. However, additional genetic screens have identified several categories of genes which function to integrate DNA replication with progression through G2 into mitosis. One class of genes includes CDC6 in budding yeast, and cdc18, cut5, and cdt1 in fission yeast[11-16]. These genes are essential for both DNA replication and the maintenance of G2 arrest in the absence of DNA replication. It is likely that they perform two roles by somehow functioning in the initiation process at replication complexes, and in generating signals that retard progression through G2. Other genes identified appear important only when DNA replication is blocked or when DNA is damaged by exogenous agents during S phase. These include RAD9, RAD17, RAD24, MEC1, MEC2, and MEC3 in budding yeast[17, 18] and rad1, rad3, rad9, rad17, hus1, and hus2 in fission yeast[19, 20]. It is likely that these genes play important roles in the signal transduction events that report incomplete DNA replication or damage to the cell cycle machinery. Despite this monumental progress in dissecting the players involved in ensuring the completion of DNA replication prior to cell division, coherent signal transduction pathways have yet to emerge that would explain the exquisite control of the G2 phase of the cell cycle. Moreover, it is now clear that mutations in mammalian homologs of some of these control proteins are associated with pre-disposition to cancers, suggesting that the unravelling of these pathways will lead to new strategies for controlling malignant cells[21]. In this review we will discuss the function and regulation of the wee1 and mik1 tyrosine kinases, their potential roles as effectors for components of the checkpoint pathways, and highlight efforts to integrate control of replication and the G2/M transition.

WEE1 AND MIK1 ARE REQUIRED TO MAINTAIN THE DNA REPLICATION CHECKPOINT

Wee mutants of fission yeast apparently uncouple cell division from cell growth, and therefore divide at a small cell size. Despite this small cell size, DNA replication is completed prior to division in these strains[22, 23]. In contrast, strains overexpressing cdc25p in a wee1- background fail to complete DNA replication at the time of cell division, resulting in a lethal event termed a mitotic catastrophe[24]. The complementing *wee1* gene was shown to encode a protein kinase which could prevent the onset of mitosis in a dose-dependent manner[25]. Significantly, cdc2p kinase activity, which is required for the G2/M transition, was shown to be regulated by tyrosine phosphorylation at Y15[5, 26]. In agreement with the genetic schemes, *wee1* and *cdc25* were shown to encode a tyrosine kinase and tyrosine phosphatase, respectively, and which see Y15 of cdc2p as the preferred substrate[6, 8, 9, 27]. Probably the most telling demonstration of the importance of Y15 phosphorylation on cdc2p for cell cycle coordination in fission yeast was the observation that a cdc2Y15F mutant underwent mitotic catastrophes, albeit only 10% per generation[5]. The ability of cdc2Y15F strains to enter mitosis prematurely was inconsistent, however, with the *wee1* loss of function phenotype, which simply divides at a small cell size with completely replicated DNA. Moreover, it was subsequently observed that interphase cdc2p in the temperature-sensitive wee1-50 strains was phosphorylated on tyrosine 15 even at non-permissive temperature[28], suggesting additional tyrosine kinases target cdc2p. The resolution to these inconsistencies came from a selection of genes for their ability to rescue a cdc2-3w, wee1-50 strain from mitotic catastrophes[4] (cdc2-3w displays cdc25-independent activation). This selection revealed the *mik1* gene which encoded a kinase most related to *wee1*. Further genetic analysis revealed why *mik1* remained cryptic and yet played such a central role in cdc2p regulation at the G2/M transition. A de-letion of the *mik1* gene was shown to be without phenotype, unlike the wee phenotype resulting from the *wee1* deletion. Therefore the *mik1* gene is not essential when wee1 activity is present. However, profound levels of mitotic catastrophes result in a Dmik1/wee1-50 strain at the nonpermissive temperature, even when no other defects or blocks to DNA replication were present. Thus it appears that cdc2 activity must be restrained throughout S phase by the combined activity of *wee1* and *mik1* to prevent the premature onset of mitosis before the completion of DNA replication. Furthermore, these results indicate that no other kinases in the cell are sufficient for restraining cdc2 activity. Figure 1 shows the cooperation between *wee1* and *mik1* in inactivating cdc2p.

Despite the dramatic effects of a simultaneous loss of function of both *wee1* and *mik1* activity in fission yeast, the generality of this scheme for controlling cdc2 activity at the G2/M transition was brought into question because the budding yeast strains with cdc28Y19F mutations appear to arrest normally in response to incomplete DNA replication or DNA damage[29, 30]. Thus despite the analogous tyrosine phosphorylation of Y19 during the budding yeast cell cycle, other mechanisms, possibly including the transcription regulation of the mitotic cyclins (*CLBs*), may suppress mitosis in the presence of incomplete DNA replication. It should be noted here that the Wee1-like kinase in S. cerevisiae, *SWE1*, can bind and suppress the activity of the cdc28Y19F mutant *in vitro*, suggesting a *SWE1*-dependent regulation in the absence of tyrosine phosphorylation[31]. Regardless, in fission yeast and higher eukaryote[7, 32, 33] an essential feature of the coordination between DNA replication and cell division appears to be the state of tyrosine phosphorylation of the conserved cdc2 kinase. While discrete pathways connecting the state of DNA replication and the activity of the cdc2 kinase are unknown, genetic approaches have discovered a myriad of factors that play important, if functionally obscure roles in this process.

GENES REQUIRED FOR BOTH DNA REPLICATION AND THE G2 CHECKPOINT

CDC6, *cdc18*, *orp1*, *orp2*, *cdt1*, and *cut5* represent an important class of genes for the coordination of DNA replication and cell division because they are required for both S phase and G2. In fact, loss-of-function phenotypes include not only a failure to initiate DNA replication but a complete bypassing of G2, resulting in an aberrant mitosis marked by the attempted segregation of unreplicated chromatids[11-16, 32, 33]. *CDC6* of budding yeast and *cdc18* and *orp1* of fission yeast share considerable sequence homology in their C-terminal domains which show general homology with ATPases generally found in multiprotein complexes. Recent genetic and biochemical evidence indicate that CDC6 and cdc18 directly interact with components of the Origin Recognition Complex, *ORC* and *orp*, respectively[12, 32, 34]. The overexpression phenotypes of *cdc18* in fission yeast are equally impressive and informative[35, 36]. These cells are prevented from entering mitosis and instead continue a re-replication process that leads to a massive accumulation of DNA in enlarged nuclei. The expression of *CDC6* in fission yeast leads to a very similar phenotype of arrest prior to cell division and massive re-replication (Wolf and McKeon, unpublished observations). Interestingly, *CDC6* appeared in a screen of budding yeast genes that would suppress mitotic catastrophes in cdc2-3w/wee1-50 fission yeast strains[11]. Thus *CDC6* was able to suppress mitosis long enough for these cells to complete DNA replication prior to

wee1 ⊣
       cdc2 → Mitosis delayed
mik1 ⊣

Figure 1. While S. pombe *wee1* and *mik1* cooperate in inactivating cdc2, *wee1* plays a predominant role.

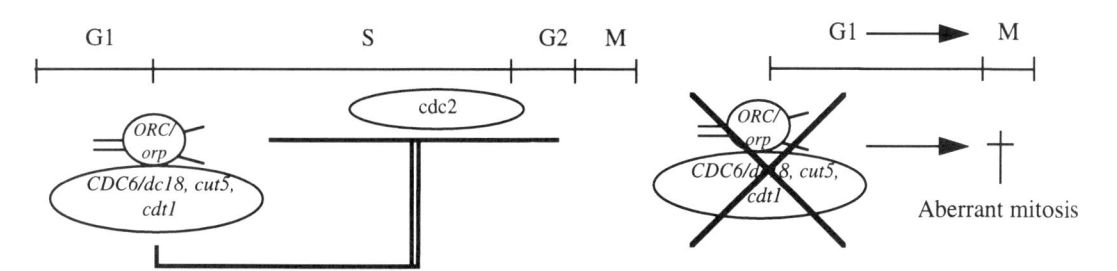

Figure 2. *S. cerevisiae* CDC6 or *S. pombe* cdc18, cut5, orp1, orp2 and cdt1 suppress cdc2 activity throughout the S and G2 phases. Loss of function of any of these genes results in mitosis immediately after G1.

mitosis. However, further analysis showed that *CDC6* was unable to suppress mitotic catastrophes in Dmik1/wee1-50 strains, suggesting that the G2 arrest imparted by *CDC6* was dependent, directly or indirectly, on the presence of these tyrosine kinases. The similar dual functions of *cdt1, cut5, orp1,* and *orp2* for the initiation of DNA replication and the suppression of mitosis suggest that these proteins might serve as proteins at the replication complex that initiate signals affecting cdc2 activity during S phase. The common loss-of-function phenotype of these genes underscores the importance of suppressing cdc2 activity throughout the process of DNA replication and, at least in the case of *CDC6*, the possibility that this is mediated through the combined actions of *wee1* and *mik1*. Figure 2 depicts the role of *CDC6* or *cdc18, cdt1, orp1, orp2,* and *cut5* in the replication checkpoint.

GENES WHICH SIGNAL INCOMPLETE DNA REPLICATION AND DAMAGE TO G2 ARREST MECHANISM

Genetic screens in budding and fission yeast have firmly established the presence of genes whose function appears to be in signalling the presence of DNA damage and incomplete DNA replication in S phase but are otherwise dispensable for either repair or synthesis reactions. In *S. cerevisiae*, such genes include *RAD9, RAD17, RAD24,* and *MEC1, MEC2,* and *MEC3* [17, 37]. In *S. pombe*, such screens have yielded *rad1, rad3, rad9,* and *rad17*, as well as *hus 1-5* [19, 20, 38, 39]. These fission yeast genes were originally selected for either radiation sensitivity or sensitivity to hydroxyurea, which blocks DNA replication through inhibition of ribonucleotide reductase. Thus the rad and hus mutants show increased sensitivity to either UV and ionising radiation or blocking of DNA replication because they continue past G2 to a lethal mitosis. Despite the separate screens employed, most of the genes identified show concomitant sensitivity to both radiation and prevention of DNA replication, suggesting that both of these events ultimately trigger a common signal transduction pathway to prevent the onset of mitosis. Another feature of these genes is that most of them show no phenotype at the restrictive temperature provided no exogenous radiation damage or DNA synthesis inhibitors are present. This feature is significant in that it suggests that these genes function in checkpoint pathways rather than in more mechanical processes such as repair or DNA replication itself. It should be noted that these genes are probably not essential components of pathways mediating cell cycle arrest by *CDC6, cdc18, cdt1, orp1-2,* and *cut5* described above, which apparently are constitutively activated during S phase to forestall mitosis. Regardless, the final mediators of G2 arrest by the *rad* and *hus* genes remain unclear at present. It is probably significant that most of the *rad* and *hus* mutations are synthetically lethal in wee1-50 strains at the non-permissive temperature, suggesting that these genes function in pathways that ultimately suppress cdc2 kinase activity[19]. One of the few exceptions is the hus4-16 allele, which maps so close to the wee1 gene that it may represent a wee1 allele itself. Furthermore, hus4-16 does not display any synthetic lethality with wee1-50. This would be consistent with the increased sensitivity of wee1-50 strains to DNA damage by irradiation[38]. It is formally possible then that the *rad* and *hus* genes function, in part, by suppressing cdc2 kinase activity through both wee1p and mik1p, and that the loss of one of these tyrosine kinases, coupled with the coincident loss of one of the *rad* or *hus* genes, impairs the total tyrosine kinase activity to the point where cdc2 activity drives a cell to unrestrained mitosis. While other mechanisms for inhibiting cdc2 kinase activity as well as other means of suppressing mitosis may also be involved in the G2 checkpoint, we will focus on the critical questions remaining for the control of the wee1 and mik1 tyrosine kinases during the cell cycle. Figure 3 gives a representation of possible pathways underlying DNA replication and repair checkpoints.

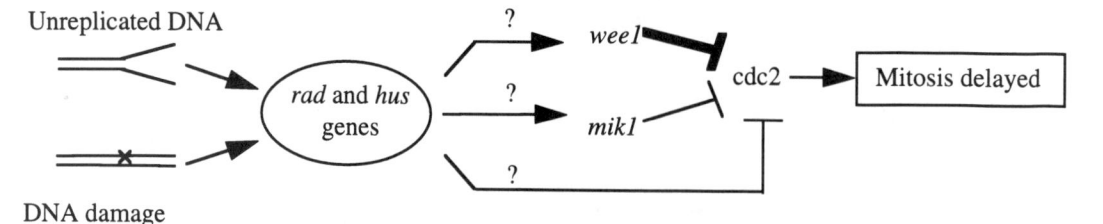

Figure 3. Both unreplicated and damaged DNA checkpoints are mediated by *rad* and *hus* genes. The ultimate result is an inactivation of cdc2. Three different effectors may be involved: wee1, mik1, and/or wee1/mik1 independent-effectors.

REGULATION OF WEE1 AND MIK1 TYROSINE KINASE ACTIVITY

While yeast genetics has provided an elegant scheme of the immediate regulators of the cdc2 protein kinase, less progress has been made towards extending this network to understanding the regulators of the regulators, information necessary to decipher the signal transduction pathways coordinating DNA replication with mitosis. A notable exception is the discovery of a negative regulator of *wee1* made through two independent efforts[40, 41]. These studies uncovered the *Nim1/Cdr1* gene which encodes a protein kinase which, when overexpressed, results in a wee phenotype consistent with the ability of *nim1* to suppress wee1 activity. The nim1 null phenotype results in an elongated cell at the time of division, again consistent with its role in suppressing wee1 activity. While homologs of the nim1/cdr1 kinase have yet to be described, considerable progress towards understanding interactions between nim1p and wee1p have resulted from expressing these proteins in sf9 insect cells and *Xenopus* extracts confirm the ability of nim1p to directly inhibit wee1p through phosphorylation of its C-terminal, catalytic domain[42-44]. However, when *S. pombe* wee1p alone is expressed in *Xenopus* extracts, no similar inhibitory phosphorylation events directed to the C-terminus of wee1 is apparent. In contrast, the inhibitory phosphorylation on *wee1* occurs on the N-terminal domain[44]. While the cdc2 protein kinase can also phosphorylate wee1p, cdc2p-depleted mitotic extracts also show inhibitory Wee1p phosphorylation[44]. In analysing the recently cloned *Xenopus* Wee1 homolog[45] in *Xenopus* cell cycle extracts, it is clear that Wee1 is tightly regulated by phosphorylation. In interphase extracts, exogenous *Xenopus* wee1p is underphosphorylated and very active towards the cdc2-cyclin B substrate. As these extracts are moved to mitosis, wee1p becomes hyperphosphorylated and shows very little activity towards cdc2-cyclin B. Similar results have been obtained by analysis of human Wee1 in cultured cells [46, 47]. Wee1 activity is apparently increased during S phase and in G2 in parallel with its expression levels. At mitosis, Hu-Wee1 appears to be hyperphosphorylated and possibly degraded, two processes that appear to contribute to its loss of activity.

Probably the most significant finding of the various analyses of Wee1 activity in higher eukaryotes was that unreplicated DNA or arresting cells in S phase does not result in an increase in Wee1 activity [46, 47]. This finding is in contrast with the significantly elevated tyrosine kinase activity directed against cdc2p in *Xenopus* extracts in which the DNA checkpoint had been activated with unreplicated DNA[48]. Cdc2p phosphorylation was also significantly elevated in irradiated human keratinocytes [49]. This discrepancy may be explained, on one level, by differences in the analysis of Wee1 activity in the various studies. Alternatively, these discrepancies may signal something more fundamental about our assumptions of Wee1-like functions in higher eukaryotes. For instance, the most detailed genetic analysis of these tyrosine kinases has been performed in the fission yeast *S. pombe*, where *wee1* appears to play a major role in cdc2 tyrosine phosphorylation and *mik1* acts as an important backup to *wee1* [4]. Surprisingly, the extensive search for *wee1/mik1* homologs in higher eukaryotes has yielded only one close relative in each species examined. Table 1 summarises some features of the known wee1-like proteins. The human and *Drosophila* Wee1-like tyrosine kinases were originally cloned by complementation of *S. pombe* strains (OP-cdc25/wee1-50 and cdc2-3w/wee1-50, respectively) that undergoes mitotic catastrophes at the non-permissive temperature[50, 51]. Both of these selections yielded Wee1-related proteins that were truncated versions of the full-length cDNAs such that the N-terminal domain was deleted and only the C-terminal catalytic domain expressed. The fact that this selection isolated human and *Drosophila* clones that were truncated almost at the identical homologous sites suggested that a full-length protein would be less efficient at rescuing the OPcdc25/wee1-50 strain. While the function of the N-terminal domain of Wee1 and the Wee1-like proteins remains obscure, the apparently constitutive activity of the human N-terminal deletion Wee1-like protein (p50-Wee1) indicates that the N-terminus acts as an auto inhibitory domain. To obtain a complementing Wee1 protein might therefore require specific modifications of the N-terminus to activate tyrosine kinase activity, a process that has evolved to the point where *S. pombe* fails to activate a full-length human Wee1-like

CHAPTER 9/ WEE1 AND MIK1 TYROSINE KINASES

Table 1. Features of several Wee1-like molecules in various organisms.

Wee1/Mik1	Cloning strategy	Mol. mass	Specificity	Comments	Regulation	Ref.
S. pombe wee1	Complementation of a wee⁻ strain	107kD	S/T, Y		Neg. regulation by *nim1/cdr1*; loss-of function has short G2	6,8,9,22, 23,25,40, 41,42,43, 44,54
S. pombe mik1	Rescues cdc2-3w/wee1-50 strain from mitotic catastrophes	66kD	Y only	ΔMik1/wee1-50: profound mitotic catastrophes	ΔMik1 has no phenotype in a wee1+ background	4
S. cerevisiae SWE1	PCR based on *S. pombe wee1/mik1*	92kD	Y	swe1p binds and inhibits cdc28F19p *in vitro*	Δswe1 no phenotype	29,30,31
Hu-Wee1	Complementation of cdc25-OP/wee1-50 *S. pombe* strain	50kD (95kD)	Y only	C-terminal catalytic domain rescues, full-length is 95kD	S phase: active, hypophos.; mitosis: hyperphos. degraded(?), inactive	46,47,50, 55
Drosophila Dwee1	Complementation of a cdc2-3w/wee1-50 *pombe* strain	69kD	Y only	C-terminal catalytic domain rescues; full-length is 69kD		51
Xenopus Wee1	PCR based on the most conserved region of *S pombe wee1*, *S p. mik1* and Hu-Wee1	62kD	Y only		Hypophosphorylated (68kD) and active in interphase extracts, hyper-phosphorylated (75kD) and inactive during mitosis	45
Xenopus Myt1	PCR based on the most conserved region of *S pombe wee1*, *S p. mik1* and Hu-Wee1	62kD	S/T, Y	Has a trans-membrane domain; less related to *pombe wee1* and *mik1* than the others	Hypophosphorylated and active in interphase extracts, hyper-phosphorylated and inactive during mitosis	53

protein. Considering the apparently exacting requirements for higher eukaryotic Wee1 complementation of *S. pombe*, the isolation of a single cDNA from both human and *Drosophila* may in fact be an underestimate of the Wee1-like kinases in higher cells. In light of this possibility it is interesting that it has been argued that the human Wee1 protein is somewhat more similar to *S. pombe mik1* than *S. pombe wee1* [47]. This notion is supported by the observation that *S. pombe wee1* is a dual specificity serine/threonine and tyrosine kinase while the human Wee1-like protein appears relatively specific for tyrosine residues, as does *S. pombe mik1 in vitro*[52]. Recently, an additional Wee1-like protein was cloned from *Xenopus* libraries and show to phosphorylated both T14 and Y15 of cdc2 when introduced into *Xenopus* extracts [53]. Interestingly, *S. pombe wee1* was recently shown to be responsible for both T14 and Y15[54]. Moreover, this *Xenopus* Wee1-like molecule appears to undergo similar inhibitory phosphorylation in mitotic extracts as experienced by either *S. pombe wee1* or *Xenopus* Wee1. However, this protein, termed Myt1p, has an unusual transmembrane domain not seen in either fission yeast *wee1* or *mik1* and shows considerably less similarity to these proteins than does the human Wee1-like protein. It remains to be determined whether other Wee1-like proteins exist in higher eukaryotes or whether the Wee1/Mik1 functions in fission yeast have somewhat been dispersed in evolution to the Wee1-like protein and another that specifically phosphorylates T14[7]. The surprisingly passive response of the human and *Xenopus* Wee1-like proteins to agents which would activate the DNA replication checkpoint suggest that these molecules play no obvious role in G2 arrest in these situations and that other means, either through other Wee1 isotypes or other pathways altogether mediate the G2/M checkpoint.

CONCLUSIONS

We have summarised the evidence, primarily from studies in fission yeast of tyrosine kinases that suppress cdc2 activity, that *wee1* and *mik1* play a role in mediating the DNA damage and replication checkpoints during S and G2 phases of the cell cycle. However, the analysis of the regulation of Wee1-like tyrosine kinases from higher eukaryotes in general fail to support their involvement in checkpoint control. While numerous agents and pathways independent of cdc2-specific tyrosine kinases may function in these checkpoints, it is likely that other Wee1-like homologs exist in higher eukaryotes that respond to signals reporting DNA damage and incomplete replication.

ACKNOWLEDGEMENTS

We thank our colleagues for helpful suggestions and Dieter Wolf for reading this manuscript. We gratefully acknowledge the support of the American Cancer Society, the NIH, and the Council for Tobacco Research.

REFERENCES

1. Hartwell, L.H. and Weinert, T.A. (1989) *Science* **246**, 629-634.
2. Enoch, T. and Nurse, P. (1991) *Cell* **65**, 921-923.
3. Murray, A.W. (1993) *Nature* **359**, 599-604.
4. Lundgren, K., Walworth, N., Booher, R., Dembski, M., Kirschner, M. and Beach, D. (1991) *Cell* **64**, 1111-1122.
5. Gould, K.L. and Nurse, P. (1989) *Nature* **342**, 39-45.
6. Featherstone, C. and Russell, P. (1991) *Nature* **349**, 808-811.
7. Krek, W. and Nigg, E.A. (1991) *Embo J* **10**, 3331-3341.
8. Parker, L.L. and Piwnica-Worms, H. (1992) *Science* **257**, 1955-1957.
9. Parker, L.L., Atherton-Fessler, S. and Piwnica-Worms, H. (1992) *Proc Natl Acad Sci U S A* **89**, 2917-2921.
10. Kumagai, A. and Dunphy, W.G. (1991) *Cell* **64**, 903-914.
11. Bueno, A. and Russell, P. (1992) *Embo J* **11**, 2167-2176.
12. Piatti, S., Lengauer, C. and Nasmyth, K. (1995) *Embo J* **14**, 3788-3799.
13. Kelly, T.J., Martin, G.S., Forsburg, S.L., Stephen, R.J., Russo, A. and Nurse, P. (1993) *Cell* **74**, 371-382.
14. Hofmann, J.F. and Beach, D. (1994) *Embo J* **13**, 425-434.
15. Saka, Y., Fantes, P., Sutani, T., McInerny, C., Creanor, J. and Yanagida, M. (1994) *Embo J* **13**, 5319-5329.
16. Saka, Y., Fantes, P. and Yanagida, M. (1994) *J Cell Sci Suppl 1994;18:57-61*
17. Weinert, T.A., Kiser, G.L. and Hartwell, L.H. (1994) *Genes Dev* **8**, 652-665.
18. Paulovich, A.G. (1995) *Cell* **82**, 841-847.
19. Enoch, T., Carr, A.M. and Nurse, P. (1992) *Genes Dev* **6**, 2035-2046.
20. al-Khodairy, F. and Carr, A.M. (1992) *Embo J* **11**, 1343-1350.
21. Hartwell, L.H. and Kastan, M.B. (1994) *Science* **266**, 1821-1828.
22. Nurse, P. (1975) *Nature* **256**, 457-451.
23. Nurse, P. and Thuriaux, P. (1980) *Genetics* **96**, 627-637.
24. Russell, P. and Nurse, P. (1986) *Cell* **45**, 145-153.
25. Russell, P. and Nurse, P. (1987) *Cell* **49**, 559-567.
26. Draetta, G., Piwnica-Worms, H., Morrison, D., Druker, B., Roberts, T. and Beach, D. (1988) *Nature* **336**, 738-744.
27. Kumagai, A. and Dunphy, W.G. (1992) *Cell* **70**, 139-151.
28. Gould, K.L., Moreno, S., Tonks, N.K. and Nurse, P. (1990) *Science* **250**, 1573-1576.
29. Sorger, P.K. and Murray, A.W. (1992) *Nature* **355**, 365-368.
30. Amon, A., Surana, U., Muroff, I. and Nasmyth, K. (1992) *Nature* **355**, 368-371.
31. Booher, R.N., Deshaies, R.J. and Kirschner, M.W. (1993) *Embo J* **12**, 3417-3426.
32. Leatherwood, J., Lopez-Girona, A. and Russell, P. (1996) *Nature* **379**, 360-363.
33. Muzi-Falconi, M. and Kelly, T.J. (1995) *Proceedings of the National Academy of Sciences of the United States of America* **92**, 12475-12479.
34. Liang, C., Weinreich, M. and Stillman, B. (1995) *Cell* **81**, 667-676.
35. Nishitani, H. and Nurse, P. (1995) *Cell* **83**, 397-405.
36. Muzi-Falconi, M., Brown, G.W. and Kelly, T.J. (1996) *Proceedings of the National Academy of Sciences of the United States of America* **93**, 1566-1570.
37. Weinert, T.A. and Hartwell, L.H. (1993) *Genetics* **134**, 63-80.
38. Rowley, R., Hudson, J. and Young, P.G. (1992) *Nature* **356**, 353-355.
39. Rowley, R., Subramani, S. and Young, P.G. (1992) *Embo J* **11**, 1335-1342.
40. Russell, P. and Nurse, P. (1987) *Cell* **49**, 569-576.
41. Feilotter, H., Nurse, P. and Young, P.G. (1991) *Genetics* **127**, 309-318.

42. Coleman, T.R., Tang, Z. and Dunphy, W.G. (1993) *Cell* **72**, 919-929.
43. Parker, L.L., Walter, S.A., Young, P.G. and Piwnica-Worms, H. (1993) *Nature* **363**, 736-738.
44. Tang, Z., Coleman, T.R. and Dunphy, W.G. (1993) *Embo J* **12**, 3427-3436.
45. Mueller, P.R., Coleman, T.R. and Dunphy, W.G. (1995) *Mol Biol Cell* **6**, 119-134.
46. McGowan, C.H. and Russell, P. (1995) *Embo J* **14**, 2166-2175.
47. Watanabe, N., Broome, M. and Hunter, T. (1995) *Embo J* **14**, 1878-1891.
48. Smythe, C. and Newport, J.W. (1992) *Cell* **68**, 787-797.
49. Herzinger, T., Funk, J.O., Hillmer, K., Eick, D., Wolf, D.A. and Kind, P. (1995) *Oncogene* **11**, 2151-2156.
50. Igarashi, M., Nagata, A., Jinno, S., Suto, K. and Okayama, H. (1991) *Nature* **353**, 80-83.
51. Campbell, S.D., Sprenger, F., Edgar, B.A. and O'Farrell, P.H. (1995) *Mol Biol Cell* **6**, 1333-1347.
52. McGowan, C.H. and Russell, P. (1993) *Embo J* **12**, 75-85.
53. Mueller, P.R., Coleman, T.R., Kumagai, A. and Dunphy, W.G. (1995) *Science* **270**, 86-90.
54. Den Haese, G.J., Walworth, N., Carr, A.M. and Gould, K.L. (1995) *Mol Biol Cell* **6**, 371-385.
55. Baldin, V., Ducommun, B. (1995) *J Cell Sci* **108**, 2425-2432.

chapter 10
Regulation of Cdc2 activity by phosphorylation at T14/Y15

Lynne D. Berry and Kathleen L. Gould[1]

Howard Hughes Medical Institute, Department of Cell Biology,
Vanderbilt University, Nashville, TN 37212 USA
[1]To whom correspondence should be addressed

The highly conserved Cdc2 serine/threonine kinase plays a central role in cell cycle progression. Although Cdc2 levels remain constant throughout the cell cycle, Cdc2 kinase activity peaks at the G2/M boundary, in order to drive entry into mitosis. In the model organism *Schizosaccharomyces pombe*, potentially active Cdc2/Cdc13 kinase complex accumulates throughout the S and G2 phases of the cell cycle. This complex, however, is maintained in an inactive state by Wee1/Mik1-mediated phosphorylation at Y15 (and, possibly, T14). At the G2/M boundary, the Cdc25 protein phosphatase is activated to dephosphorylate the Cdc2/Cdc13 complex, resulting in abrupt activation of Cdc2 kinase activity and entry into mitosis.

CDC2 AS A KEY REGULATOR OF CELL CYCLE PROGRESSION

Early genetic screens identified the *cdc2* gene product as a key regulator of cell cycle progression in the fission yeast *Schizosaccharomyces pombe*. Two types of mutants mapped to the *cdc2* locus: wee mutants (1,2,3,4,5), and temperature sensitive cell division cycle (cdc[ts]) mutants (6). *cdc2*[ts] mutant cells, when incubated at restrictive temperature, arrest progression through the cell cycle but continue to elongate; wee mutants, on the other hand, undergo mitosis and cell division at reduced cell size, resulting in the observed "wee" phenotype. That different mutations in a single gene product could give rise to either cell cycle arrest or acceleration of the cell cycle suggested that *cdc2* must play a central role in driving cell cycle progression. Indeed, *cdc2* activity is required at both of two major cell cycle control points (7,8,9): Start, at which cells commit to progression through the mitotic cell cycle; and the G2/M transition, at which cells initiate the processes of chromosome condensation, mitotic spindle assembly, and actin ring formation, processes which culminate in mitosis and, subsequently, cytokinesis.

The *cdc2* gene product plays a critical role in cell cycle regulation not only in *S. pombe*, but in all other eukaryotic organisms, as well. Since the initial cloning of *cdc2* from fission yeast (10), *cdc2* homologs have been identified and cloned in all other eukaryotic systems studied, including budding yeast (CDC28, in fact cloned prior to *S. pombe cdc2*; 9,10); starfish (11); frog (12,13); and human (14,15,16,17,18,19,20). Across the eukaryotic lineage, Cdc2 homologs demonstrate a high degree of conservation not only at the protein sequence level, but at the functional level, as well; both human *cdc2* (19) and budding yeast CDC28 (10) rescue *S. pombe cdc2*[ts] mutants.

cdc2 encodes a 34 kDa protein serine/threonine kinase (21,22). *In vivo*, in all systems studied, the kinase activity of Cdc2 peaks at the G2/M boundary, while Cdc2 protein levels remain constant throughout the cell cycle (11,15,22,23,24,25,26,27,28,29,30; Figure 1). The peak of Cdc2 activity at the G2/M boundary is consistent with the demonstrated requirement for *cdc2* gene activity at G2/M (7). If Cdc2 protein levels remain constant throughout the cell cycle, however, by what mechanism is Cdc2 activity regulated, in order to drive orderly progression of cell cycle events?

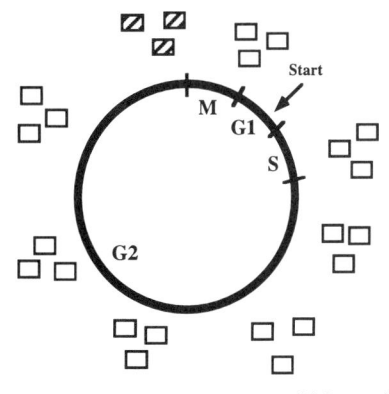

Figure 1. Schematic diagram representing Cdc2 protein levels and kinase activity through the *S. pombe* cell cycle. Cdc2 kinase activity peaks at the G2/M boundary, while Cdc2 protein levels remain constant throughout the cell cycle.

Key: ☐ , inactive Cdc2;

▨ , active Cdc2 kinase.

REGULATION OF CDC2 KINASE ACTIVITY: PHOSPHORYLATION AT T167 AND T14/Y15

Phosphorylation of key conserved residues within the Cdc2 molecule has emerged as an essential means of regulating Cdc2 activity. In vivo, S. pombe Cdc2 is phosphorylated at both threonine 167 (T167; 31) and tyrosine 15 (Y15; 32). Phosphorylation at T167 (T161 in higher eukaryotes) is essential for activation of the Cdc2 kinase (30,31). In contrast, Y15 phosphorylation plays a negative regulatory role, as demonstrated by the phenotype of the cdc2-F15 mutant. In cdc2-F15, the structurally similar but non-phosphorylatable amino acid phenylalanine replaces Y15. Cells expressing cdc2-F15 traverse the G2/M boundary prematurely, displaying cytological abnormalities and the wee phenotype characteristic of premature entry into mitosis. (32). Consistent with a periodic role in preventing premature activation of the Cdc2 kinase, Y15 phosphorylation of Cdc2 varies through the cell cycle: Y15 phosphorylation begins to accumulate after the execution of cdc10 function at Start; steadily increases during S phase and G2; then abruptly decreases at the G2/M boundary, coincident with the abrupt activation of Cdc2 and initiation of mitosis (17,24,25,27,29,32,33,34,35,36; Figure 2)).

Although Y15 functions as the primary site of inhibitory phosphorylation in S. pombe Cdc2, inhibitory phosphorylation events at both Y15 and the adjacent T14 occur in higher eukaryotes (30,37,38,39). Indeed, under certain circumstances, S. pombe Cdc2 may also be phosphorylated at T14; the physiological significance of such phosphorylation, however, remains unclear (40). Both Y15 and T14 are located in the ATP-binding domain of Cdc2, the kinase-conserved GXGXXG motif (41). This localization has prompted speculation that phosphorylation of T14/Y15 may inhibit Cdc2 activity by blocking the binding or utilization of ATP. Three-dimensional modeling of Cdc2, however, suggests that Y15 phosphorylation is more likely to disrupt a substrate recognition site than to prevent ATP binding (42). Moreover, Y15 phosphorylation of Cdc2 does not block the specific binding of 5'-p-fluorosulfonylbenzoyladenosine (FSBA), an ATP analog (43). Despite the apparent ability of Y15-phosphorylated Cdc2 to bind ATP, tyrosine-phosphorylated Cdc2 fails to generate free $^{32}P_i$ or ADP in kinase assays performed in the presence of [gamma-^{32}P]ATP. Thus, Y15 phosphorylation, rather than blocking ATP binding, may alter the positioning of the gamma phosphate of ATP, rendering it inaccessible for cleavage by Cdc2 and subsequent transfer to substrate (43).

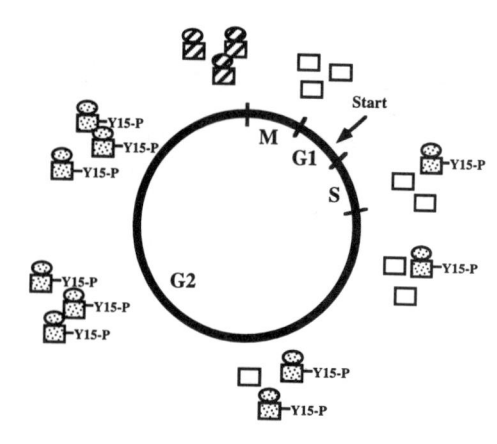

Figure 2. Inhibitory phosphorylation of Cdc2 at Y15 varies through the cell cycle; accumulation of potentially active Cdc2/Cdc13 kinase complex parallels accumulation of Y15 phosphorylation. Prior to Start, neither Cdc2/Cdc13 nor tyrosine-phosphorylated Cdc2 are present in rapidly growing S. pombe cells; after passing Start, cells begin to accumulate both Cdc2/Cdc13 and Cdc2 phosphotyrosine; Cdc2/Cdc13 as well as Cdc2 phosphotyrosine levels peak in late G2; Cdc2 phosphotyrosine disappears as cells enter mitosis; Cdc2/Cdc13 levels drop as Cdc13 is destroyed at the end of mitosis. Cdc2/Cdc13 and Cdc2 phosphotyrosine levels remain low throughout the subsequent G1 phase of the cell cycle, until cells once again pass Start.

Key: □, inactive monomeric Cdc2;
Y15-P, Y15-phosphorylated (inactive) Cdc2/Cdc13 complex;
, active Cdc2/Cdc13 complex.

T14/Y15 PHOSPHORYLATION: THE MOLECULAR PLAYERS

A number of molecular players involved in the regulatory phosphorylation of Cdc2 at T14/Y15 have been identified. In S. pombe, these players include the products of the cdc13, wee1, mik1, and cdr1/nim1 genes.

cdc13

cdc13, like cdc2, was originally identified as a gene essential for normal cell cycle progression (6). cdc13 was cloned (44, 45) and shown to encode a B-type cyclin, one of a large family of proteins whose abundance and association with CDKs oscillate through the cell cycle, thus contributing to orderly cell cycle progression (see Chapter 15, this volume). In S. pombe, activity of Cdc2 at the G2/M boundary depends upon the association of Cdc2 with Cdc13 (44,45,46). Cyclin B binding, however, also promotes Y15 phosphorylation of Cdc2 (34,47,48). Indeed, cell-cycle dependent accumulation of the Cdc2/Cyclin B complex in S. pombe parallels accumulation of phosphotyrosine on Cdc2 (36; Figure 2): Prior to start, neither the Cdc2/Cdc13 complex nor tyrosine-phosphorylated Cdc2 are present at detectable levels in rapidly growing S. pombe cells; after passing Start, cells begin to accumulate both Cdc2/Cdc13 and Cdc2 phosphotyrosine; both Cdc2/Cdc13 and Cdc2 phosphotyrosine levels peak in late G2; Cdc2 phosphotyrosine disappears as cells enter mitosis; Cdc2/Cdc13 levels drop as Cdc13 is destroyed at the end of mitosis. Cdc2/Cdc13 and Cdc2 phosphotyrosine levels remain low throughout the subsequent G1 phase of the cell cycle, until cells once again pass through Start (36). That Cdc13 binding stimulates Y15-phosphorylation of Cdc2 suggests a role for Y15 phosphorylation in maintaining the potentially active Cdc2/Cdc13 complex in an inactive state. Thus, the Cdc2/Cdc13 complex accumulates gradually throughout the S and G2 phases of the cell cycle; abrupt entry into mitosis, however, occurs only upon abrupt Y15-dephosphorylation of Cdc2, after cells have completed DNA replication (in S) and accumulated sufficient cell mass to undergo viable cell division (primarily in G2).

wee1

Genetic evidence first suggested a role for the *wee1* gene product as an inhibitor of mitosis. Specifically, loss of function alleles of *wee1* give rise to the wee phenotype characteristic of cells which undergo premature entry into mitosis (1,2,3). Consistent with the wee phenotype of loss of function alleles of *wee1*, overexpression of *wee1* delays entry into mitosis, resulting in elongation of cells (49). *wee1* was cloned, and shown to encode a protein kinase homolog (49). Might *wee1* encode the Cdc2 tyrosine kinase? Evidence accumulated slowly. First, Wee1 protein kinase activity was demonstrated. *In vitro*, Wee1 autophosphorylates, as well as phosphorylating exogenous peptide substrates, on both serine/threonine and tyrosine residues (50). Thus, *wee1* encodes a dual-specificity (serine/threonine and tyrosine) kinase. And indeed, *S.pombe* Wee1 is capable of phosphorylating human Cdc2 on Y15 (47; Figure 3)). *In vitro*, Wee1-mediated Y15 phosphorylation of Cdc2 is cyclin-dependent, and results in inhibition of Cdc2/cyclin kinase activity, as measured by the ability of Cdc2 to phosphorylate the exogenous substrate histone H1 (47). Recent evidence implicates *S. pombe* Wee1 in the *in vivo* phosphorylation of Cdc2 T14, as well. T14 phosphorylation during an S phase block (imposed by incubation of S phase mutants at restrictive temperature) is observed in the presence of wild type *wee1*, but not in a strain deleted for the *wee1* gene. Additionally, overexpression of *wee1* results in a high stoichiometry of phosphorylation at T14 (40).

Putative *wee1* homologs have been identified and cloned in a number of other eukaryotic systems, including budding yeast (51), frog (52), fly (53), and human (54,55). *In vitro*, human Wee1 (WEE1Hu) autophosphorylates on tyrosine, but not on serine/threonine residues (56). Likewise, *in vitro* kinase assays demonstrate that WEE1Hu phosphorylates human Cdc2/Cyclin B complex on Y15, but not on T14 (56,57). Y15 phosphorylation of human Cdc2 by WEE1Hu results in inhibition of Cdc2, as demonstrated by decreased *in vitro* Cdc2 kinase activity toward histone H1 (56), as well as by the failure of kinase-dead WEE1Hu to inhibit the onset of mitosis *in vivo* (57). The behavior of *Xenopus* Wee1 mirrors that of WEE1Hu. That is to say, *in vitro*, *Xenopus* Wee1 mediates the cyclin-dependent phosphorylation of Cdc2 exclusively on Y15. Addition of Wee1 protein to *Xenopus* cell cycle extracts delays entry into mitosis, with a concomitant increase in tyrosine phosphorylation of Cdc2 (52). Budding yeast Swe1, like WEE1Hu and *Xenopus* Wee1, tyrosine-phosphorylates and inhibits activity of Cdc28, the budding yeast homolog of Cdc2. Interestingly, Swe1 phosphorylates and inhibits Cdc28 bound to Clb2, a G2-specific cyclin, but not Cdc28 bound to Cln2, a G1-specific cyclin (51). Thus, in *S. cerevisiae*, as in *S. pombe*, tyrosine-phosphorylation may play a G2-specific role in the inhibition of Cdc28 activity. In *S. pombe*, as described above, Y15 phosphorylation accumulates only after cells have passed Start; prior to Start, Cdc2 lacks G2/M-specific kinase activity not because of Y15 phosphorylation, but rather because of low levels of Cdc13 (36). [The *rum1* gene product, a G1-specific inhibitor of the Cdc2/Cdc13 mitotic kinase, also plays a critical role as a negative regulator of any Cdc2/Cdc13 complex present in the G1 cell (58,59). For a full discussion of Rum1, see Chapter 4.]

Regulation of Wee1 activity

Genetic analyses first identified the *nim1* (new inducer of mitosis; 60)/*cdr1* (changed divsion response; 61) gene product as a dose-dependent mitotic inducer (60). Epistasis studies further suggested that Nim1/Cdr1 positively regulates mitosis by negatively regulating Wee1. *nim1* mutant cells delay entry into mitosis, resulting in an elongated cell phenotype; *wee1* mutant cells enter mitosis prematurely, resulting in a wee phenotype. The double mutant, *nim1*- *wee1*-, is wee. Thus, *wee1*- is epistatic to *nim1*-, suggesting that *wee1* acts downstream of *nim1* (60,61). *nim1*/*cdr1* was cloned, and shown to encode a protein kinase homolog (61,62). Does Nim1/Cdr1 negatively regulate Wee1 via phophorylation? Both *in vitro* and in insect cell expression systems, Cdr1/Nim1 autophosphorylates on serine/threonine and tyrosine residues; Cdr1/Nim1 also directly phosphorylates Wee1 on serine/threonine residues, resulting in a decreased ability of Wee1 to phosphorylate Cdc2 (63,64,65). Moreover, levels of Nim1/Cdr1 activity parallel levels of Wee1 phosphorylation *in vivo*. Serine/threonine phosphorylation of Wee1 is greatest (4X) in cells overexpressing *nim1*/*cdr1*; intermediate (2X) in wild type (*nim1*+/*cdr1*+) cells; and least (1X) in cells deleted for *nim1*/*cdr1* (64). Thus, Nim1/Cdr1 appears to promote mitosis by phosphorylating and inhibiting Wee1 (Figure 3). As discussed in greater detail elsewhere in this work (Volume 1, Chapter 16), the regulation by Nim1/Cdr1 of Wee1-mediated Cdc2 tyrosine phosphorylation may provide a link between nutritional sensing and cell cycle control (61,66).

Nim1/Cdr1 homologs have not been identified in other organisms. Nevertheless, recombinant *S. pombe* Wee1, when introduced into *Xenopus* egg extracts, oscillates between an interphase hypophosphorylated form with high Cdc2-directed kinase activity, and a mitotic hyperphosphorylated form with low Cdc2-directed kinase activity (67). Egg extract-mediated phosphorylation of Wee1 occurs in the N-terminal portion of the molecule (67), in contrast to the C terminal-localized phosphorylation mediated by Nim1/Cdr1 *in vitro* (65). Thus, in *Xenopus*, Wee1 phosphorylation state and corresponding activity are subject to cell cycle regulation. This regulation, however, may not involve a Nim1/Cdr1 homolog.

mik1

Despite the well-established role of Wee1 in the tyrosine phosphorylation of Cdc2, *S. pombe wee1* mutants do not fail to accumulate phosphotyrosine on Cdc2 (68) Cdc2 tyrosine phosphorylation even in the absence of Wee1 function suggests that, *in vivo*, Wee1 does not play a unique role as the Cdc2 tyrosine kinase. Indeed, the *mik1* gene encodes a protein kinase homolog which cooperates with Wee1 in the phosphorylation of Cdc2 at Y15 (69,70). *S. pombe* cells deleted for *mik1* display no apparent phenotype. In the absence of both Wee1 and Mik1 function, however, Cdc2 tyrosine phosphorylation disappears, and cells undergo "mitotic catastrophe," or lethal premature entry into mitosis (69). Consistent with the *in vivo* evidence of a role for Mik1 in the tyrosine phosphorylation of Cdc2, Mik1 immunoprecipitated from insect cells or from *S. pombe* cell lysates directly phosphorylates Cdc2 on Y15 *in vitro* (70). Thus, in *S. pombe*, inhibitory phosphorylation of Cdc2 depends upon the activity of at least two different kinases, Wee1 and Mik1 (Figure 3). Comparison of single mutant phenotypes, however, suggests that Wee1 plays the major role in regulation of Cdc2 activity. Loss of *wee1* function in a *mik1+* background results in premature entry into mitosis, while loss of *mik1* function in a *wee1+* background has no apparent effect on cell cycle progression.

myt1

Phosphorylation of Cdc2 occurs at T14 as well as at Y15 in higher eukaryotes and, under certain circumstances, in *S. pombe*. *In vitro*, however, *S. pombe* Wee1, WEE1Hu, *Xenopus* Wee1, *Drosophila*

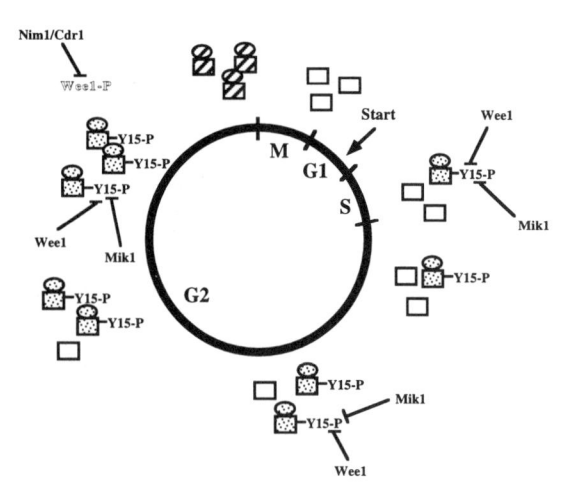

Figure 3. The dual specificity (S/T and Y) kinase Wee1, in cooperation with the Mik1 protein kinase, phosphorylates Cdc2/Cdc13 on Y15, in order to maintain the potentially active Cdc2/Cdc13 kinase in an inactive state during S and G2. The Nim1/Cdr1 kinase phosphorylates and inhibits Wee1.

Key: □, inactive monomeric Cdc2;
Y15-P, Y15-phosphorylated (inactive) Cdc2/Cdc13 complex;
, active Cdc2/Cdc13 complex;
Wee1, active Wee1 kinase;
Mik1, active Mik1 kinase;
Wee1-P, Wee1 kinase inactivated by phosphorylation;
Nim1/Cdr1, Nim1/Cdr1 kinase.

Wee1 (Dwee1), and Swe1 phosphorylate Cdc2 exclusively on tyrosine. What, then mediates T14 phosphorylation? The recently cloned *myt1* (from *Xenopus*) encodes a Wee1 homolog capable of *in vitro* phosphorylation of Cdc2/CyclinB on both T14 and Y15 (71). Such a dual-specificity Cdc2 kinase has been described only in *Xenopus*, although as yet unidentified homologs may well exist in other systems.

cdc25

Wee1 and Mik1 tyrosine phosphorylate and inhibit Cdc2. What mediates tyrosine dephosphorylation of Cdc2, with consequent activation of the Cdc2 kinase and entry into mitosis? Again, genetics led the way in identifying the *cdc25* gene product as a dose dependent activator of mitosis, acting in opposition to Wee1. *cdc25*, like *cdc2* and *cdc13*, plays an essential role in cell cycle progression; cells lacking *cdc25* gene function arrest in G2 (6,72). Cell cycle arrest in the *cdc25* mutant, however, occurs only in the presence of functional *wee1*; the *wee1 cdc25* double mutant is viable (49,72,73). Suppression of the *cdc25* mutant phenotype in a *wee1-* background suggests that the *cdc25* mutant in a *wee1+* background arrests in G2 because of a failure to counteract *wee1*-mediated inhibition of mitosis. In other words, Cdc25

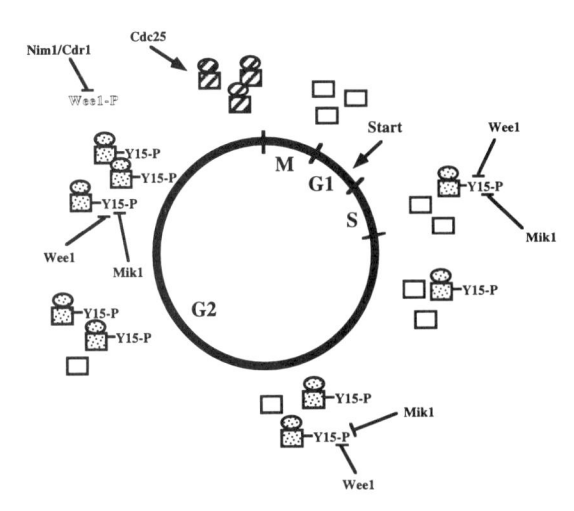

Figure 4. At the G2/M boundary, Cdc25 dephosphorylates Cdc2/Cdc13 on Y15, activating the kinase complex to induce entry into mitosis.

Key: ☐, inactive monomeric Cdc2;
  Y15-P, Y15-phosphorylated (inactive) Cdc2/Cdc13 complex;
  , active Cdc2/Cdc13 complex;
  Wee1, active Wee1 kinase;
  Mik1, active Mik1 kinase;
  Wee1-P, Wee1 kinase inactivated by phosphorylation;
  Nim1/Cdr1, Nim1/Cdr1 kinase;
  Cdc25, Cdc25 protein phosphatase.

functions antagonistically to Wee1 in order to induce entry into mitosis. Specifically, Cdc25 may dephosphorylate Cdc2 on tyrosine 15. The predicted amino acid sequence of Cdc25 (72), however, bears only limited homology to known protein phosphatases (74,75). Nevertheless, both *in vivo* and *in vitro* studies provide compelling evidence that *cdc25* encodes the Cdc2 tyrosine phosphatase. First, expression in *S. pombe* of either *cdc2*-F15 (32), or a human protein tyrosine phosphatase capable of dephosphorylating Cdc2 on Y15 (68), suppresses the cell cycle arrest phenotype of the *cdc25* mutant. Moreover, *in vitro*, recombinant Cdc25 directly tyrosine dephosphorylates and activates Cdc2 (75,76,77,78,79,80). In sum, evidence strongly suggests that the *cdc25* gene product induces entry into mitosis by dephosphorylating Cdc2 on Y15, thus activating Cdc2 mitotic kinase activity (Figure 4).

PHYSIOLOGICAL ROLE OF T14/Y15 PHOSPHORYLATION: CELL CYCLE CHECKPOINTS

That *wee1* mutants enter mitosis at approximately half the size of wild type cells suggests that Wee1 plays a role in coupling the onset of mitosis to the accumulation of cell mass. That is to say, Wee1-mediated phosphorylation of Y15 acts as a cell size checkpoint, delaying entry into mitosis until cells have attained a critical mass (2,81,82).

Y15 phosphorylation also plays a role in the DNA synthesis checkpoint, as demonstrated in *S. pombe* by overexpression of Cdc25 in the presence of hydroxyurea, as well as by the behavior of the *wee1⁻ mik1⁻* double mutant. Wild type *S. pombe* cells blocked in S phase by treatment with hydroxyurea (an inhibitor of DNA synthesis) do not enter mitosis with wild type kinetics, but rather delay entry into mitosis until DNA synthesis is complete. Cells overexpressing Cdc25, however, fail to delay the cell cycle in response to hydroxyurea, proceeding into mitosis even in the presence of unreplicated DNA (83). Likewise, with loss of both Wee1 and Mik1 function and consequent loss of Cdc2 Y15 phosphorylation, cells enter mitosis prior to completion of S phase, undergoing mitotic catastrophe as they attempt to segregate incompletely replicated chromosomes (69).

Does Y15 phosphorylation act as a checkpoint in other eukaryotic systems? In budding yeast, Y19 (equivalent to *S. pombe* Cdc2 Y15) is not required for the cell size or DNA synthesis checkpoint. Mutations of Y19 which prevent phosphorylation do not cause premature entry into mitosis and do not abolish the dependence of mitosis on completion of DNA synthesis (38,84). Rather, Cdc28 Y19 plays a role in a cell cycle checkpoint unique to budding yeast, a cell morphogenesis checkpoint (85). In budding yeast, inhibition of bud formation delays entry into mitosis due to delayed activation of Cdc28. Inhibition of Cdc28 in the budding checkpoint depends upon Y19 phosphorylation, as well as upon delayed accumulation of Clb1 and Clb2, budding yeast B-type cyclins (85). Thus, the details of the Y15/Y19-mediated checkpoints differ between fission and budding yeast. Nevertheless, in budding yeast as in fission yeast, Y19 phosphorylation plays a key role in ensuring orderly progression of cell cycle events (i.e. budding before mitosis).

DIRECTIONS FOR FUTURE RESEARCH

Although T14/Y15 phosphorylation plays an undisputed role in negative regulation of Cdc2 kinase activity, the structural basis for this inhibition remains unexplained. Despite the location of T14/Y15 in the presumptive ATP binding domain of Cdc2 (41), phosphorylation does not interfere with the binding of ATP (42,43). How, then, does T14/Y15 phosphorylation inhibit Cdc2? Does T14/Y15 phophorylation interfere with the utilization of ATP, perhaps by repositioning the gamma phosphate of ATP and rendering it inaccesible for cleavage and/or transfer to substrate? Does T14/Y15 phosphorylation interfere with substrate recognition and/or binding? Or, does phosphorylation at T14/Y15 interfere with the catalytic activity of Cdc2 in some other manner? These questions remain to be answered.

Additional questions include those concerning the roles of Wee1, Mik1, and Myt1 as Cdc2 kinases. Specifically, why do eukaryotic systems utilize two different Cdc2-directed kinases: Wee1 and Mik1 in *S. pombe*; Wee1 and (presumably) Myt1 homologs in higher eukaryotes? In *S. pombe*, is Mik1 function strictly redundant in a *wee1+* background, or does Mik1 perform any unique function? If Mik1 performs functions unique from those of Wee1, why do *wee1* mutants undergo premature entry into mitosis, while *mik1* mutants appear to progress through the cell cycle in a wild-type manner? Likewise, in higher eukaryotes, what, if any, is the unique role of a tyrosine-specific kinase (the Wee1 homologs) in the background of a dual specificity kinase (Myt1 and presumptive homologs)?

Finally, questions remain concerning the regulation of Wee1 and Mik1. The identification of both Nim1/Cdr1-mediated (in *S. pombe*) and Nim1/Cdr1-independent (in *Xenopus*) phosphorylation and inhibition of Wee1 places Wee1 in a regulatory network of kinases ultimately impinging on Cdc2 Y15 phosphorylation to control cell cycle progression. What other regulatory networks might feed into the Wee1 and/or Mik1 pathways to influence Y15 phosphorylation? In S. pombe, the *wis1* gene product has been identified as a dosage dependent inducer of mitosis (86) with homology to the map kinase kinase sub-family of serine/threonine protein kinases (87,88,89,90). Might Wis1 act through Wee1 to regulate the onset of mitosis (Warbrick and Fantes, 1991)? If so, Wee1-mediated phosphorylation of Cdc2 Y15 (Y19) may ultimately integrate signals from pathways as divergent as those involved in the sensing of internal cues, such as completion of DNA replication, accumulation of critical cell mass, or bud emergence; as well as those involved in sensing environmental cues, such as nutritional state or osmotic stress. Insight into the pathways impinging upon Wee1 and/or Mik1 may also provide insight into the differential functions of these kinases. Wee1 and Mik1 may integrate unique signals from independent pathways in order to coordinate T14/Y15 phosphorylation and, consequently, cell cycle timing, with the widest possible range of internal and external cues.

ACKNOWLEDGEMENTS

L.D.B is supported by a National Science Foundation Graduate Research Fellowship. K.L.G. is an assistant investigator of the Howard Hughes Medical Institute.

REFERENCES

1. Thuriaux P, Nurse P, Carter B (1978) *Mol. Gen. Genet.* **161**, 215-220.
2. Nurse P, Thuriaux P (1980) *Genetics* **96**, 627-637.
3. Fantes PA (1981) *J. Bacteriol.* **146**, 746-754.
4. Carr, AM, MacNeill, SA, Hayles, J, and Nurse, P (1989) *Mol. Gen. Genet.* **218**, 41-49.
5. MacNeill, SA and Nurse, P (1989) *Curr. Genet.* **16**, 1-6.
6. Nurse P, Thuriaux P, Nasmyth, K (1976) *Mol. Gen. Genet.* **146**, 167-178.
7. Nurse P, Bissett Y (1981) *Nature* **292**, 558-560.
8. Piggott JR, Rai R, and Carter BLA (1982) *Nature* **298**, 391-393.
9. Reed SI, Wittenberg C (1990) *Proc. Natl. Acad. Sci. USA* **87**, 5697-5701.
10. Beach, D, Durkacz, B, Nurse, P (1982) *Nature* **300**, 140-142.
11. Labbe JC, Lee MG, Nurse P, Picard A, Doree M (1988) *Nature* **335**, 251-254.
12. Dunphy WG, Brizuela L, Beach D, Newport J (1988) *Cell* **54**, 423-431.
13. Gautier J, Norbury C, Lohka M, Nurse P, Maller J (1988) *Cell* **54**, 433-439.
14. Arion D, Meijer L, Brizuela L, Beach D (1988) *Cell* **55**, 371-378.
15. Draetta G, Beach D (1988) *Cell* **54**, 17-26.
16. Draetta G, Brizuela L, Potashkin J, Beach D (1987) *Cell* **50**, 319-325.
17. Draetta G, Piwnica-Worms H, Morrison D, Druker B, Roberts, T, Beach D (1988) *Nature* **336**, 738-743.
18. Draetta G, Luca F, Westendorf J, Brizuela L, Ruderman J, Beach D (1989) *Cell* **56**, 829-838.
19. Lee MG, Nurse, P (1987) *Nature* **327**, 31-35.
20. Lee MG, Norbury CJ, Spurr, NK, Nurse P (1988) *Nature* **333**, 676-679.
21. Hindley, J, Phear, GA (1984) *Gene* **31**, 129-134.
22. Simanis V, Nurse P (1986) *Cell* **45**, 261-268.
23. Mendenhall MD, Jones CA, Reed SI (1987) *Cell* **50**, 927-935.
24. Booher RN, Alfa CE, Hyams JS, Beach DH (1989) *Cell* **58**, 485-497.
25. Dunphy WG, Newport JW (1989) *Cell* **58**, 181-191.
26. Felix MA, Pines J, Hunt T, Karsenti E, (1989) *EMBO J.* **8**, 3059-3069.
27. Gautier J, Matsukawa T, Nurse P, Maller J (1989) *Nature* **339**, 626-629.
28. Moreno S, Hayles J, Nurse P (1989) *Cell* **58**, 361-372.
29. Pondaven P, Meijer L, Beach D (1990) *Genes Dev.* **4**, 9-17.
30. Krek W, Nigg EA (1991) *EMBO J.* **10**, 305-316.
31. Gould KL, Moreno S, Owen DJ, Sazer S, Nurse P (1991) *EMBO J.* **10**, 3297-3309.

32. Gould KL, Nurse P (1989) *Nature* **342**, 39-45.
33. Morla AO, Draetta G, Beach D, Wang JYJ (1989) *Cell* **58**, 193-203.
34. Solomon MJ, Glotzer M, Lee TH, Philippe M, Kirschner MW (1990) *Cell* **63**, 1013-1024.
35. Norbury C, Blow J, Nurse P (1991) *EMBO J.* **10**, 3321-3329.
36. Hayles J, Nurse P (1995) *EMBO J.* **14**, 2760-2771.
37. Krek W, Nigg EA (1991) *EMBO J.* **10**, 1331-3341.
38. Amon A, Surana U, Muroff I, Nasmyth K (1992) *Nature* **355**, 368-372.
39. Solomon MJ, Lee T, Kirschner MW (1992) *Mol. Biol. Cell* **3**, 13-27.
40. Den Haese GJ, Walworth N, Carr AM, Gould KL (1995) *Mol. Biol. Cell* **6**, 371-385.
41. Hanks S (1991) *Curr. Opin. Structural Biol.* **1**, 369-383.
42. Marcote MJ, Knighton DR, Basi G, Sowadski JM, Brambilla P, Draetta G, Taylor SS (1993) *Mol. Cell Biol.* **13**, 5122-5131.
43. Atherton-Fessler S, Parker LL, Geahlen RL, Piwnica-Worms H (1993) *Mol. Cell Biol.* **13**, 1675-1685.
44. Hagan I, Hayles J, Nurse P (1988) *J. Cell Sci.* **91**, 587-595.
45. Booher R, Beach D (1988) *EMBO J.* **7**, 2321-2327.
46. Booher R, Beach D (1987) *EMBO J.* **6**, 177-182.
47. Parker LL, Atherton-Fessler S, Lee MS, Ogg S, Kalk JL, Swenson KI, Piwnica-Worms H (1991) *EMBO J.* **10**, 1255-1263.
48. Meijer L, Azzi L, Wang JYJ (1991) *EMBO J.* **10**, 1545-1554.
49. Russell P, Nurse P (1987) *Cell* **49**, 559-567.
50. Featherstone C, Russell P (1991) *Nature* **349**, 808-811.
51. Booher RN, Deshaies RJ, Kirschner MW (1993) *EMBO J.* **12**, 3417-3426.
52. Mueller PR, Coleman TR, Dunphy WG (1995) *Mol. Biol. Cell* **6**, 119-134.
53. Campbell, SD, Sprenger F, Edgar, BA, O'Farrell PH (1995) *Mol. Biol. Cell.* **6**, 1333-1347.
54. Igarashi M, Nagata A, Jinno S, Suto K, Okayama H (1991) *Nature* **353**, 80.
55. Watanabe N, Broome M, Hunter T (1995) *EMBO J.* **14**, 1878-1891.
56. Parker LL, Piwnica-Worms H (1992) *Science* **257**, 1955-1957.
57. McGowan CH, Russell P (1993) *EMBO J.* **12**, 75-85.
58. Moreno S, Nurse P (1994) *Nature* **367**, 236-242.
59. Correa-Bordes J, Nurse P (1995) *Cell* **83**, 1001-1009.
60. Russell P, Nurse P (1987) *Cell* **49**, 569-576.
61. Young PG, Fantes PA (1987) *J. Cell Sci.* **88**, 295.
62. Feilotter H, Nurse P, Young PG (1991) *Genetics* **127**, 1-10.
63. Parker LL, Walter SA, Young PG, Piwnica-Worms H (1993) *Nature* **363**, 736-738.
64. Wu L, Russell P (1993) *Nature* **363**, 738-741.
65. Coleman TR, Tang Z, Dunphy WG (1993) *Cell* **72**, 919-929.
66. Fantes PA, Warbrick E, Hughes DA, MacNeill, SA (1991) *Cold Spring Harbor Symp. on Quant. Biol.* **56**, 605-611.
67. Tang Z, Coleman TR, Dunphy WG (1993) *EMBO J.* **12**, 3427-3436.
68. Gould KL, Moreno S, Tonks NK, Nurse P (1990) *Science* **250**, 1573-1576.
69. Lundgren K, Walworth N, Booher R, Dembski M, Kirschner M, Beach D (1991) *Cell* **64**, 1111-1122.
70. Lee MS, Enoch T, Piwnica-Worms H (1994) *J. Biol. Chem.* **269**, 30530-30537.
71. Mueller PR, Coleman TR, Kumagai A, Dunphy WG (1995) *Science* **270**, 86-90.
72. Russell P, Nurse P (1986) *Cell* **45**, 145-153.
73. Fantes P (1979) *Nature* **279**, 428-430.
74. Moreno S, Nurse, P (1991) *Nature* **351**, 194.
75. Gautier J, Solomon MJ, Booher RN, Bazan JF, Kirschner MW (1991) *Cell* **67**, 197-211.
76. Dunphy WG, Kumagai A (1991) *Cell* **67**, 189-196.
77. Kumagai A, Dunphy WG (1991) *Cell* **64**, 903-941.
78. Lee MS, Ogg S, Xu M, Parker LL, Donoghue DJ, Maller JL, Piwnica-Worms H (1992) *Mol. Biol. Cell* **3**, 73-84.
79. Millar JBA, McGowan CH, Lenaers G, Jones R, Russell P (1991) *EMBO J.* **10**, 4301-4309.
80. Strausfeld U, Labbe JC, Fesquet D, Cavadore JC, Picard A, Sadhu K, Russell P, Doree M (1991) *Nature* **351**, 242-245.
81. Nurse P (1975) *Nature* **256**, 547-551.
82. Fantes PA, Nurse P (1978) *Exp. Cell Res.* **115**, 317-329.
83. Enoch T, Nurse P (1990) *Cell* **60**, 665-673.
84. Sorger PK, Murray AW (1992) *Nature* **355**, 365-368.
85. Lew DJ, Reed SI (1995) *J. Cell Biol.* **129**, 739-749.
86. Warbrick E, Fantes PA (1991) *EMBO J.* **10**, 4291-4299.
87. Kosako H, Nishida E, Gotoh Y (1993) *EMBO J.* **12**, 787-794.
88. Millar JB, Buck V, Wilkinson MG (1995) *Genes Dev.* **9**, 2117-2130.
89. Shiozaki K, Russell P (1995) *EMBO J.* **14**, 492-502.
90. Shiozaki K, Russell P (1995) *Nature* **378**, 739-743.

chapter 11
The family of polo-like kinases

Roy M. Golsteyn[1], Heidi A. Lane[2], Kirsten E. Mundt[2], Lionel Arnaud[3] and Erich A. Nigg[3]*

[1]Institute Curie, Section de Recherche, 26, Rue d'Ulm, F-75231 Paris, France
[2]Swiss Institute for Experimental Cancer Research, 155 Chemin des Boveresses,
CH-1066 Epalinges, Switzerland
[3]Dept. of Molecular Biology, Sciences II, University of Geneva, 30, Quai Ernest-Ansermet
CH-1211 Geneva, Switzerland
*To whom correspondence should be addressed

Here we discuss members of a new family of serine/threonine protein kinases with a likely role in cell cycle control. These kinases are referred to as polo-like kinases, after the prototypic founding member of the family, the *polo* gene product of *Drosophila melanogaster*. The polo kinase was originally identified in mutants that display abnormal mitotic spindle organization. Subsequently, potential homologues of *Drosophila* polo have been identified in yeasts (Cdc5p in *Saccharomyces cerevisiae*; plo1$^+$ in *Schizosaccharmoyces pombe*) and in mammals (polo-like kinase 1; Plk1). Genetic and biochemical studies suggest that polo, Cdc5p and plo1$^+$ may be required for mitotic spindle organization and, possibly, for cytokinesis. Likewise, the patterns of expression, activity and subcellular localization of Plk1 strongly suggest that this mammalian kinase functions also during mitosis, possibly in spindle assembly and function. In addition to Plk1, however, more distantly related members of the polo-like kinase family have been identified in mammalian cells, and the available data are consistent with the idea that some of these may act earlier in the cell cycle, possibly during G1. If this hypothesis is correct, different members of the polo-like kinase family would act at several points during the cell cycle, reminiscent of the behaviour of Cdk/cyclin complexes.

INTRODUCTION

Protein phosphorylation regulates many key events and transitions in the eukaryotic cell cycle. Accordingly, the identification of cell cycle regulatory protein kinases is of great interest for understanding cell cycle progression. During mitosis, a dramatic increase in protein phosphorylation occurs [1, 2]. This increase correlates with a profound restructuring of the cell, including breakdown of the nuclear envelope, condensation of interphase chromatin to chromosomes, and the assembly of a mitotic spindle [3]. Through the application of antibodies that specifically recognize phosphorylated epitopes, mitosis-specific phosphoproteins have been detected on many mitotic structures [4-10] and biochemical analyses strongly indicate that phosphorylation regulates multiple mitotic events [11, 12]. For instance, centrosome nucleation in *Xenopus* egg extracts is blocked by the addition of protein kinase inhibitors [13], and phosphatase treatment reduces the nucleating capacity of centrosomes *in vitro* [7].

One of the best characterized mitotic protein kinases is the p34cdc2/cyclin B complex known as Cdc2 kinase [14, 15]. The list of putative Cdc2 substrates continues to expand [16], and some likely relevant substrates are associated with the mitotic spindle [17, 18]. Of particular interest in the present context, Cdc2 kinase has been implicated in mediating the changes in microtubule dynamics and increases in centrosome nucleation activity that are required for bipolar spindle formation [19-21]. However, although the key role of Cdc2 kinase in triggering mitosis is well established, there is no doubt that other protein kinases are also important for mitotic progression. One prominent example for this is provided by a newly emerging family of kinases that are structurally related to the 65 kDa *polo* gene product of *Drosophila*. Members of this family have now been identified in several species, including yeasts and human. Some of these may perform functions during mitosis, similar to *Drosophila* polo, whereas others may play roles at other stages of the cell cycle. The structural and functional properties of polo-like kinases constitute the focus of this review.

STRUCTURAL COMPARISON OF POLO-LIKE KINASES

In the absence of detailed functional information, it may be appropriate to group polo-like kinases by structural criteria. Human and murine polo-like kinase 1 (Plk1, also called Plk), budding yeast Cdc5p and fission yeast plo1$^+$ all share a substantial degree of sequence similarity with *Drosophila* polo, both inside and outside of the N-terminally located catalytic domain (Figures

A

```
              ▼▼▼▼▼
Plk1 Human  53  YVRGRFLGKG GFAKCFEISD ADTKEVFAGK IVPKSLLLKP HQREKMSMEI SIHRSLAHQH VVGFHGFFED NDFVFVVLEL CRRRSLLELH  142
Plk1 Mouse  53  YIRGRFLGKG GFAKCFEISD ADTKEVFAGK IVPKSLLLKP HQREKMSMEI SIHRSLAHQH VVGFHDFFED SDFVFVVLEL CRRRSLLELH  142
Polo        25  YKRMRFFGKG GFAKCYEIID VETDDVFAGK IVSKKLMIKH NQKEKTAQEI TIHRSLNHPN IVKFHNYFED SQNIYIVLEL CKKRSMMELH  114
plo1+       41  YTRYDCIGEG GFARCFRVKD .NYGNIYAAK VIAKRSLQND KTKLKLFGEI KVHQSMSHPN IVGFICFED STNIYLILEL CEHKSLMELL  129
Cdc5p       82  YHRGHFLGEG GFARCFQIKD .DSGEIFAAK TVAKASIKSE KTRRKLLSEI QIHKSMSHPN IVQFICFED DSNVYILLEI CPNGSLMELL  170
Consensus       Y-R----G-G GFA-C----D ------A-K ---K------ ----K--EI --H-S--H-- -V-F---FED -------LE- C---S--EL-
Kinase Domain         I          II         III                              IV                          V

Plk1 Human      KRRKALTEPE ARYYLRQIVL GCQYLHRNRV IHRDLKLGNL FLNEDLEVKI GDFGLATKVE YDGERKKTLC GTPNYIAPEV L..SKKGHSF  230
Plk1 Mouse      KRRKALTEPE ARYYLRQIVL GCQYLHRNQV IHRDLKLGNL FLNEDLEVKI GDFGLATKVE YEGERKKTLC GTPNYIAPEV L..SKKGHSF  230
Polo            KRRKSITEFE CRYYIYQIIQ GVKYLHDNRI IHRDLKLGNL FLNDLLHVKI GDFGLATRIE YEGERKKTLC GTANYIAPEI L..TKKGHSF  202
plo1+           KRRKQLTEPE VRYLMMQILG ALKYMHKKRV IHRDLKLGNI MLDESNNVKI GDFGLAALLM NESERKMTIC GTPNYIAPEI LFNSKEGHSF  218
Cdc5p           KRRKVLTEPE VRFFTTQICG AIKYMHSRRV IHRDLKLGNI FFDSNYNLKI GDFGLAAVLA NESERKYTIC GTPNYIAPEV LMGKHSGHSF  260
Consensus       --RK--TE-E -R------ ---Y-H---- IHRDLKLGN- -------KI GDFGLA----  ---ERK-T-C GT-NYIAPE- L----GHSF
Kinase Domain              VIA        VIB                 VII                          VIII

Plk1 Human      EVDVWSIGCI MYTLLVGKPP FETSCLKETY LRIKKNEYSI PKH..INPVA ASLIQKMLQT DPTARPTINE LLNDEFF  305
Plk1 Mouse      EVDVWSIGCI MYTLLVGKPP FETSCLKETY LRIKKNEYSI PKH..INPVA ASLIQKMLQT DPTARPTIHE LLNDEFF  305
Polo       /    EVDIWSIGCV MYTLLVGQPP FETKTLKDTY SKIKKCEYRV PSY..LRKPA ADMVIAMLQP NPESRPAIGQ LNFEFL  277
plo1+           EVDLWSAGVV MYYLLIGKPP RKIKANSYSF PSNVDISAEA KDLISSLLTH DPSIRPSIDD IVDHEFF  296
Cdc5p           EVDIWSLGVM LYALLIGKPP FQARDVNTIY ERIKCRDFSF PRDKPISDEG KILIRDILSL DPIERPSLTE IMDYVWF  337
Consensus       EVD-WS-G-- -Y-LL-G-PP F-------- Y --IK------ P--------- -------L-- -P--RP---- -------
Kinase Domain              IX                X              IK                                   XI
```

B

```
Plk1 Human 405  ...CIPIFWV SKWVDYSDKY GLGYQLCDNS VGVLFNDSTR LILYNDGDSL QYIERDGTES YLT....VSS HPNSLMKKIT LLKYFRNYMS  487
Plk1 Mouse 405  ...CIPIFWV SKWVDYSDKY GLGYQLCDNS VGVLFNDSTR LILYNDGDSL QYIERDGTES YLT....VSS HPNSLMKKIT LLNYFRNYMS  487
Polo       385  ...AQPLFWI SKWVDYSDKY GFGYQLCDEG IGVMFNDTTK LILLPNQINV HFIDKDGKET YMT....TTD YCKSLDKKMK LLSYFKRYMI  468
plo1+      489  ....EPVLFI TKWVDYSNKY GLGYQLSDES VGVHFNDDTS LLFSADEEVV EYAHLPKDTE IKPYIYPASK VPESIRSKLQ LLKHFKSYMG  574
Cdc5p      505  LPKIKHPMIV TKWVDYSNKH GFSYQLSTED IGVLFNNGTT VLRLADAEEF WYISYDDREG WVASHYLLSE KPRELSRHLE VVDFFAKYMK  594
Consensus       --------- -KWVDYS-K- G--YQL---- -GV-FN--T- --------- -------- --------- ----F--YM-

Plk1 Human      EHLLKAGA.. ...NITPREGD ELARLPYLRT WFRTRSAIIL HLSNGSVQIN FFQDHTKLIL CPLMAAVTYI DEKRDFRTYR LSLLEEYG..  571
Plk1 Mouse      EHLLKAGR.. ...NITPREGD ELARLPYLRT WFRTRSAIIL HLSNGTVQIN FFQDHTKLIL CPLMAAVTYI NEKRDFQTYR LSLLEEYG..  571
Polo            EHLVKAGA.. ..NNVNIESD QISRMPHLHS WFRTTCAVVM HLTNGSVQLN .FSDHMKLIL CPPRMSAITYM DQEKNFRTYR FSTIVENG..  551
plo1+           QNLSK..AVQ DESFEKPKNS TSNTMLFMQH YLRTRQAIMF RLSNGIFQFN .FLDHRKVVI SSTARKIIVL DKERERVELP LQEASAFSE.  660
Cdc5p           ANLSRVSTFG REEYHKDD.. ......VFLRR YTRYKPFVMF ELSDGTFQFN .FKDHHKMAI SDGGKLVTYI SPSHESTTYP LVEVLKYGEI  676
Consensus       --L------- --------- ------R--- -L--G--Q-N -F-DH-K--- --------- --------- --------

Plk1 Human      ...CCKELAS RLRYARTMVD KLLSSRSASN RLKAS         603
Plk1 Mouse      ...CCKELAS RLRYARTMVD KLLSSRSASN RLKAS         603
Polo            ...VSKDLYQ KIRYAQEKLR KMLEKMFT.. .....         576
plo1+           ......DLRS RLKYIRETLE SWASKMEVS. .....         683
Cdc5p           PGYPESNFRE KLTLIKEGLK QKSTIVTVD. .....         705
Consensus       --------- --------- ---------
```

C

```
               ▼▼▼▼▼
MmPlk1    53  YIRGRFLGKG GFAKCFEISD ADTKEVFAGK IVPKSLLLKP HQKEKMSMEI SIHRSLAHQH VVGFHDFFED SDFVFVVLEL CRRRSLLELH  141
MmSnk     79  YCRGKVLGKG GFAKCYEMTD LTNNKVYAAK IIPHSRVAKP HQREKIDKEI ELHRLLHHKH VVQFYHYFED KENIYILLEY CSRRSMAHIL  168
MmFnk     63  YTKGRLLGKG GFAKCYEATD TESGIAYAVK VIPQSRVAKP HQREKILNEI IVRFSHHFED ADNIYIFLEL LLNYFRNYMS  152
Consensus     Y--G---LGKG GFA-C-E--D ------A-K --P-S---KP HQ-EK---EI --HR-L-H-H -V-F---FED -------LE- C-R-S-----
MmSak-a/b 12  FKVGNLLGKG SFAGVYRAES IHTGLEVAIK MIDKKAMYKA GMVQRVQNEV KIHCQLKHPS VLELYNYFED NNYVYLVLEM CHNGEMNRYL  101
Kinase Domain         I          II          III                           IV                          V

MmPlk1        KRR.KALTEP EARYYLRQIV LGCQYLHRNQ VIHRDLKLGN LFLNEDLEVK IGDFGLATKV EYEGERKKTI CGTPNYIAPE VLSKKGHSFE  231
MmSnk         KAR.KVLTEP FVRYYLRQIV SGLKYLHEQE ILHRDLKLGN FFINEAMELK VGDFGLAARL EPLEHRRRTI CGTPNYLSPE VLNKQGHGCE  257
MmFnk         KAR.HTLLEP EVRYYLRQIL SGLKYLHQRG ILHRDLKLGN FFITDNMELK VGDFGLAARL EPPEQRRKTI CGTPNYVAPE VLLRQGHGPE  241
Consensus     K-R.--L-EP E-RYYLRQI- -G--YLH--- --HRDLKLGN F--E--K --GDFGLAARL E----R--T- CGTPNY--PE VL---GH--E
MmSak-a/b     KNRMKPFSER EARHFMHQII TGMLYLHSHG ILHRDLTLSN ILLTRNMNIK IADFGLATQL NMPHEKHYTL CGTPNYISPE IATRSAHGLE  191
Kinase Domain           VIA        VIB                  VII                          VIII

MmPlk1        VDVWSIGCIM YTLLVGKPPF ETSCLKETYL RIKKNEYSIP KHINPVAASL IQKMLQTDPT ARPTIHELLN DEFF     305
MmSnk         SDIWALGCVM YTMLLGRPPF ETTNLKETYR CIREARYTMP SSLLAPAKHL IASMLSKNPE DRPSLDDIIR HDFF     331
MmFnk         ADVWSLGCVM YTLLCGSPPF ETADLKETYR CIKQVHYTLP ASLSLPARQL LAAILRASPR DRPSIEQILR HDFF     315
Consensus     -D-W--GC-M YT-L-G-PPF ET--LKETY- -I----Y--P ------A--L ----L---P- -RP------ --FF
MmSak-a/b     SDIWSLGCMS YTLLIGRPPF DTDTVKNTLN KVVLADEYEMP AFLSREAQDL IHQLLRRNPA DRLSLSSVLD HPFM     265
Kinase Domain           IX                X                                            XI
```

D

```
MmPlk1   405  CIP...... IFWVSKWVDY SDKYGLGYQL CDNSVGVLFN DSTRLILYND GDSLQYIERD GTESYLTVSS HPNSLMKKIT LLNYFRNYMS  487
MmSnk    488  CIPKEQLSTS FQWVTKWVDY SNKYGFGYQL SDHTVGVLFN NGAHMSLLPD KKTVHYYAEL GQCVSFPATD APEQFISQVT VLKYFSHYME  577
MmFnk    437  FAPLAQ.PEP LVWVSKWVDY SNKFGFGYQL SRRVAVLFN DGTHMALSAN RKTVHYNPTS TKHFSFSMGS VPRALQPQLG ILRYFASYME  525
Consensus     --P------ -WV-KWVDY S-K-GFGYQL ----V-VLFN ------L-- -----Y---- --------- -P-------- -L-YF--YM-

MmPlk1        EHLLKAGRNI TPREGDELAR LPYLRTWFRT RSAIILHLSN GTVQINFFQD HTKLILCPLM AA..VTYINE KRDFQTYRLS LLEEYGCCKE  575
MmSnk         ENLMDGG.DL PSVTDIRRP. RLYLLQWLKS DKALMMLFND GTFQVNFYHD HTKIICNQSS EEYLLTYINE DRISTTFRLT TLLMSGCSLE  665
MmFnk         QHLMKGG.DL PSVEEAEVPA PPLLLQWVKT DQALLMLFSD GTVQVNFYGD HTKLILSG.W EPLLVTFVAR NRSACTYLAS HLRQLGCSPD  613
Consensus     --L--G--- --------- ---L--W--- -A------- GT-Q-NF-D HTK-I---- -----T---- -R------- -L---GC---

MmPlk1        LASRLRYART MVDKLLSSRS AS          603
MmSnk         LKNRMEYALN M...LLQRCN            682
MmFnk         LRQRLRYALR ....LLRDQS PA          631
Consensus     L-R--YA-- ----LL---- --
```

1A and B). Furthermore, as discussed in more detail below, polo, Plk1, Cdc5p and plo1+ may all be important for regulating some aspect(s) of mitosis, and may therefore be considered collectively as "mitotic Plks". Whether these kinases represent functional homologues in a strict sense is not entirely clear, since no successful complementation across species has yet been achieved.

Within the catalytic domains of mitotic Plks (Figure 1A), residues essential for the structure of protein kinases are conserved, as expected [22]. However, a few features appear to be characteristic for mitotic Plks, identifying them as a distinct kinase family. In particular, the ATP binding region of subdomain I displays the motif GxGGxAxC (residues 60 - 67 in the human sequence) instead of the highly conserved consensus GxGxxGxV [23]. Furthermore, the mitotic Plks share high sequence identity in subdomain VIII, a region that is typically highly conserved among members of the same family [23]. Immediately C-terminal to the catalytic domain (i.e. over a region corresponding to approximately 200 amino acids in the case of human Plk1) the above kinases display variable lengths and no obvious sequence similarities. Importantly, however, significant sequence conservation can be observed further downstream. In particular, there is a motif comprising approximately 30 amino acids (Figure 1B, black bar), hereafter referred to as "the polo-box", which may constitute a signature for Plks [24-26]. The function of this polo-box is presently unknown.

In addition to Plk1, mammalian cells were found to express other polo-like kinases (Figures 1C and 1D), less closely related to *Drosophila* polo. These were termed Snk (serum-induced kinase; [27]), Fnk (FGF-induced kinase; [28]) and Sak-a/b (Snk/Plk-akin kinase; [29]). However, we emphasize that mammalian Plk1 shares a higher level of sequence similarity with *Drosophila* polo than with other mammalian polo-like kinases. In the case of Snk and Fnk, sequence similarity to Plk1 and *Drosophila* polo extends beyond the catalytic domain (Figures 1C and 1D), but Sak-a and Sak-b lack the polo-box. Thus, by structural criteria alone, it is difficult to decide whether Sak-a/b should be considered as bonafide members of the polo-like kinase family. As described below, experimental data suggest that Snk, Fnk and Sak-a/b may all function in relation to cell proliferation [27-29]. Of particular interest, the available evidence indicates that Snk and Fnk may not function during mitosis but earlier in the cell cycle. If this possibility can be corroborated by further experiments, it would imply that Snk and Fnk may constitute a class of "interphasic Plks", functionally related, but distinct from the "mitotic Plks" (i.e. Plk1 and its homologues in other species).

DROSOPHILA POLO

The first and founding member of the polo-like kinase family has been identified through the analysis of mitotic mutants in *Drosophila melanogaster* [30]. Depending upon the severity of the allele, mutations in the *polo* gene were either lethal (*polo²*) or resulted in arrest at either embryonic or late larval stages of development (*polo¹*) (30, 47). Detailed inspection of homozygous *polo¹* larval brains revealed an array of mitotic abnormalities that could be explained by aberrations in mitotic spindle formation. These abnormalities included misalignment of chromosomes on a bipolar spindle, bipolar spindles with one pole larger than the other, and monopolar spindles. The spindle defects, although not immediately lethal for cells, produced numerous polyploid cells which ultimately resulted in the death of the larvae or, rarely, a sterile adult. The embryos produced by the few homozygous females able to reach maturity subsequently became populated with highly branched, unorganised microtubule structures. The aberrant spindles in these embryos were correlated with the absence of centrosomal staining by Bx63 antibody, a marker for centrosomes, suggesting that the *polo* kinase might be required for the proper organisation of centrosomes.

Analysis of polo kinase expression in embryos revealed that *polo* transcripts were localised to regions rich in dividing cells, consistent with the view that the function of this kinase is related to cell proliferation. More specifically, a role in mitosis was suggested from direct measurements of polo kinase activity, performed with extracts from single embryos undergoing synchronous syncytial mitoses [31]. In these experiments, the timing of polo kinase activation at mitosis was determined

---

Figure 1. Sequence alignments of mitotic polo-like kinases from mammals, *Drosophila* and yeasts. Source of data: [24-26, 32, 38-40, 47].
A. The catalytic domains of Plk1 (human and mouse), polo (*Drosophila*), plo1+ (*S. pombe*), and Cdc5p (*S. cerevisiae*) are compared using GCG version 7 software and printed in single letter amino acid code. Numbering of amino acids is taken from sequences in original papers. Amino acids common to all sequences are listed in the consensus sequence, and kinase subdomains, as described by Hanks and Hunter (1995) [23], are indicated. Arrowheads mark an unusual GxGGxAxC motif in subdomain I.
B. Alignment of the C-terminal domains of the same polo-like kinases listed in A. The alignment begins at amino acid 405 relative to the human sequence, and the approximate position of the polo-box signature [24-26] is indicated by a bar.
C. Alignment of the catalytic domains of murine Plk1, Snk, Fnk and Sak-a/b. Sources of data [27-29]. Note that the latter kinase does not display the typical GxGGxAxC motif in subdomain I (arrowheads).
D. Alignment of the C-terminal domains of the murine kinases Plk1, Snk and Fnk. Note the conservation of a polo-box (indicated by a bar).

with reference to Cdc2 kinase activity and microscopic observations of chromosomal behaviour. Using casein as a preferred *in vitro* substrate, polo kinase activity was found to peak after Cdc2 kinase, suggesting a role in late stages of mitosis, i.e. anaphase and/or telophase.

MITOTIC POLO-LIKE KINASES IN YEASTS: CDC5p OF *SACCHAROMYCES CEREVISIAE*

When the budding yeast *Saccharomyces cerevisiae CDC5* gene was cloned, it was found to encode a 81 kDa serine/threonine protein kinase with striking sequence similarity to *Drosophila polo* (40% identity within the catalytic domain) [32]. Immunoprecipitation of Cdc5 protein after overexpression in wild type cells enabled the detection of a casein kinase activity, which was not detected with a catalytically-inactive mutant. Consistent with a mitotic role, *CDC5* mRNA expression was maximal at the G2/M-phase border, indicating that protein levels might also increase at a corresponding time. It is not known if Cdc5p is the functional homologue of *Drosophila* polo, but the phenotype of *cdc5* mutants does reveal some similarities to the mitotic phenotypes described for polo. *Cdc5* was originally classified as a late mitotic mutant, since cells did not complete mitosis and, as a result, retained one elongated nucleus [33, 34]. *CDC5* function was also required for meiotic divisions, and temperature-sensitive *cdc5* mutants arrested at either the first or second meiosis, depending upon the timing of shift-up to the restrictive temperature [35]. *Cdc5* mutant cells induced to arrest in meiosis I displayed separated spindle pole bodies, but microtubules were not interconnected as normally observed for wild-type cells. In addition, arrest in meiosis II resulted in the formation of diploid spores (dyads) instead of normal haploid spores. Examination of *cdc5* mutants by electron microscopy revealed that spindle elongation did not occur, resulting in inadequate separation of the haploid nuclei in most (but not all) cases. Not only did chromosomal segregation fail, but *cdc5* mutants also exhibited "mixed meiotic segregation" in that diploid spores were commonly found to contain chromosomes from both reductional and equational divisions [36]. These results suggested, therefore, that *CDC5* is required for the correct orientation of chromosomes on spindle fibres.

It is intriguing that the *CDC5* gene was actually isolated as a multicopy suppressor of *dbf4* [32], a gene required for S-phase [37]. Furthermore, when transcribed from multicopy plasmids, *CDC5* also suppressed temperature sensitive mutations in *cdc15*, *cdc20* and *dbf2* mutants, all of which encode protein kinases. The physiological significance of these findings remains to be explained.

MITOTIC POLO-LIKE KINASES IN YEASTS: PLO1+ OF *SCHIZOSACCHAROMYCES POMBE*

Recently, a polo-like kinase has also been isolated from the fission yeast *Schizosaccharomyces pombe*, using PCR protocols [38]. The gene, named *plo1+*, encodes a 77 kDa protein kinase that shares about 40% sequence identity with the polo and Cdc5p kinases (50-60% within the catalytic domain). The *plo1+* gene is essential, and disruption caused cells to arrest either in mitosis, with highly condensed chromosomes associated with monopolar spindle structures, or after nuclear division as single cells with two nuclei and no septa [38]. Further examination of the latter cells revealed defects in actin ring formation and in the deposition of septal material. These two processes are necessary for cytokinesis in fission yeast, and their absence in *plo1+* disrupted cells suggests a role for *plo1+* in early cytokinesis, in addition to an apparent role in spindle organisation. In complementary experiments, overexpression of *plo1+* also induced mitotic defects. Again, cells were arrested in the cell cycle with either highly condensed chromosomes associated with monopolar spindle structures or with multiple septa in the absence of mitosis. Indeed, premature septum formation could be induced by *plo1+* overexpression in both G1 and G2 cells. In wild-type fission yeast, septum formation and spindle assembly are believed to occur at about the same time. It is conceivable, therefore, that plo1+ kinase activity may peak at mitosis, and that plo1+ kinase may phosphorylate substrates required for both spindle formation and/or function, as well as for cytokinesis. However, no direct measurements of the timing of plo1+ kinase activity have been reported.

MAMMALIAN POLO-LIKE KINASES

As judged by the criterion of structural similarity to *Drosophila* polo, the polo-like kinase family appears to comprise multiple members. This is true at least for mammals, but may apply to other species as well. Hence, it is attractive to postulate that different members of the polo-like kinase family may perform specific functions during the cell cycle and/or during development. Although it is tempting to classify Plk1 as a "mitotic kinase" and kinases such as Snk or Fnk as "interphasic kinases", the precise function of each of these enzymes remains to be determined. Moreover, although the three kinases display sequence hallmarks that appear characteristic of the polo-like kinase family, they also show substantial sequence diversity between themselves (Figures 1C and 1D). It is possible, therefore, that the classification of polo-like kinases may have to be refined as the functions of individual members are clarified, and, perhaps, additional family members discovered.

Polo-like kinase 1 (Plk1)

Evidence that mammalian Plk1 functions during mitosis is supported by experiments describing changes in mRNA levels, protein levels, enzymatic activity, and intracellular localisation during different phases of the cell cycle. Indeed, Plk1 transcripts are only detected in actively growing cultured cells [25, 26, 39-41], embryonic tissues [24], or adult tissues that are composed of actively dividing cells [24-26, 39, 40]. Furthermore, Plk1 mRNA is absent or greatly reduced in serum starved NIH 3T3 cells or A431 cells, and, interestingly, is deadenylated upon differentiation of MELC cells [39]. Conversely, when quiescent cultured cells are induced to re-enter the cell cycle by stimulation with serum, or when human lymphocytes are treated with the mitogen phytohemagglutinin, Plk1 mRNA levels accumulate drastically [39-41]. Plk1 transcript levels also fluctuate during cell cycle progression, being highest in cells collected after a mitotic block with the drug nocodazole, greatly reduced or absent in G1 and low in S-phase cells [39, 41]. On the basis of nuclear run-off assays, it was suggested that transcription rates of Plk1 mRNA may remain constant during the cell cycle and that fluctuations in mRNA levels may therefore be regulated post-transcriptionally [39]. The recent identification of the core promotor regions of both the murine and human Plk1 gene may help in testing this proposition and may contribute to further studies on the regulation of Plk1 mRNA expression [42].

Plk1 protein levels also fluctuate during the cell cycle, being highest in G2 and mitotic cells, and lowest in G1 cells [25, 41, 43]. Plk1 enzymatic activity, measured *in vitro* after immunoprecipitation, also peaks during mitosis. The kinetics of Plk1 activation at the G2/M transition roughly parallel that of p34cdc2/cyclin B, but the inactivation of Plk1 upon exit from mitosis may occur after the inactivation of p34cdc2/cyclin B [41, 44]. Although Plk1 protein levels are highest during the period of maximal activity, the increase in protein level at G2/M is insufficient to account for the observed drastic increase in activity. These results suggest that Plk1 activity is also regulated post-translationally. In this respect, mitosis-specific phosphorylation of Plk1 *in vivo* has been detected both by direct [^{32}P]-phosphate labelling, and by visualisation of a form with reduced mobility on SDS-PAGE gels [43]. Treatment of active Plk1 from mitotic cells with phosphatase resulted in greatly reduced Plk1 activity, suggesting that phosphorylation is required for kinase activity. Whether this phosphorylation is due to "autophosphorylation" or phosphorylation by another protein kinase remains to be determined. Further characterisation and identification of the mitotic phosphorylation sites will help to resolve this question.

During the period of its highest kinase activity, Plk1 undergoes striking changes in intracellular localisation [44]. The protein, which is mainly cytoplasmic in interphase cells, accumulates at the centrosomes as cells go into prophase. The association with spindle poles persists until the metaphase/anaphase transition at which point centrosomal staining is lost and the protein collects at the equatorial microtubule overlap zone (the "midzone") [25, 41, 44]. Ultimately, Plk1 is seen adjacent to the midbody in the intracellular bridge of daughter cells. Considering that the midbody is shed shortly after cell division, the accumulation of Plk1 in postmitotic bridges may contribute, at least in part, to reduce Plk1 protein levels during the early stages of the subsequent cell cycle. The mechanisms that regulate Plk1 localisation at different stages of mitosis have not been determined. One interesting possibility is that phosphorylation events may control not only the activity of Plk1 but also its localisation, as has recently been demonstrated for the human kinesin-like motor protein HsEg5 [17]. Conversely, localisation may influence Plk1 activity and/or substrate specificity. In this context, it is interesting that an association between Plk1 and MKLP-1, a kinesin-related mitotic motor protein, has recently been reported [41]. MKLP-1 may also be a substrate of Plk1, as suggested by *in vitro* phosphorylation of MKLP-1 in Plk1 immunoprecipitates. The regulation of motor protein function by Plk1 phosphorylation would be in keeping with the putative mitotic spindle function of Plk1. However, whether Plk1 and MKLP-1 interact directly remains unclear, and the physiological significance of the reported interaction remains to be determined.

Studies that directly address the function of mammalian Plk1 remain scarce. One study has led to the suggestion that Plk1 may play a role in S-phase events [26]. This proposal is based upon the results of sense and antisense Plk mRNA microinjection experiments in NIH 3T3 cells. Microinjection of *in vitro* transcribed Plk sense mRNA was reported to induce quiescent cells to enter S-phase, as measured by tritiated thymidine incorporation, whereas antisense RNA blocked thymidine incorporation in growing cells. At this time, it is not obvious how to reconcile the suggested G1/S role for Plk1 with the observations that Plk mRNA is almost completely absent in G1 cells, that Plk1 mRNA and protein only accumulate once DNA synthesis is well under way, and that Plk1 kinase activity peaks in G2/M. One is left to wonder,

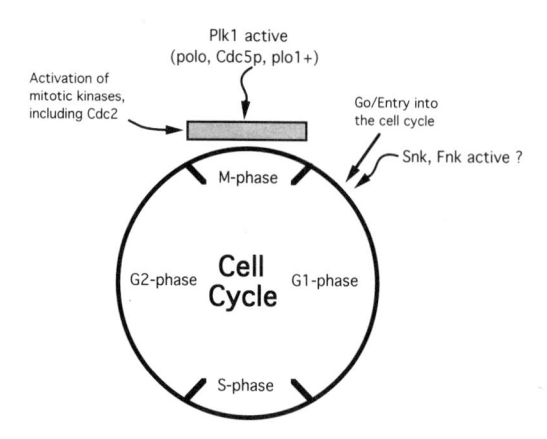

Figure 2. A tentative scheme summarizing the functions of polo-like kinases.
The activities of mitotic polo-like kinases (Plk1, polo, Cdc5p and plo1+) and of putative interphasic family members, Snk and Fnk, are shown relative to the phases of the cell cycle. Sak-a and Sak-b are not included in this scheme. For the sake of simplicity, Plk1, polo, Cdc5p and plo1+ are being considered as functional homologues. However, we emphasize that this point has not been rigorously proven. Thus, both the precise timing of periodic activation as well as the precise functions of the various kinases may vary among the corresponding species.

therefore, whether Plk1 related kinases, such as those described below, might account for the apparent G1/S function that has been attributed to Plk1 [26]. Recent antibody-microinjection experiments lend no support for a role of human Plk1 in DNA synthesis; instead, these studies unequivocally demonstrate a requirement for Plk1 during mitosis (H. Lane and E.A. Nigg, manuscript in preparation).

Serum-inducible kinase (Snk)

The Snk transcript was identified as an mRNA that showed strong induction upon serum stimulation or phorbol-ester treatment of NIH 3T3 cells [27]. Accordingly, Snk was classified as an immediate early gene. Sequencing of a Snk cDNA revealed a 78 kDa serine/threonine kinase. Although Snk is a murine kinase, it is as divergent from mammalian Plk1 as it is from Drosophila polo [27]. Snk mRNA appears to persist for only a short period after serum induction and cell cycle entry, and the short half-life of this mRNA may be related to the presence of several adenine-uracil rich elements (AREs) in the 3' untranslated trailer [45]. These data suggest that Snk is more likely to perform a function early in the cell cycle than during mitosis. In support of this view, preliminary data suggests that Snk may function in relation to mitogenic signalling pathways that lead to the induction of c-*myc* and c-*fos*. [27].

FGF-inducible kinase (Fnk)

Fnk cDNA was isolated using a differential display approach for identifying genes induced by fibroblast growth factor (FGF) treatment of NIH 3T3 cells. Fnk encodes a 70 kDa serine/threonine protein kinase that displays a comparable extent of sequence similarity to both Drosophila polo and mammalian Plk1. Its closest relative is Snk, with which it shares about 50% sequence identity [28]. Analysis of Fnk mRNA levels revealed that this kinase also belongs to the immediate-early gene family, in that expression levels were maximal shortly after stimulation with FGF-1, phorbol esters and serum, and expression was independent of *de novo* protein synthesis. Fnk mRNA was only transiently expressed, levels being greatly reduced 8 hours post stimulation, suggesting that the cognate protein is required for early G1 signal transduction. These properties are reminiscent of Snk, yet, the two kinases share little or no sequence similarity beyond the catalytic domain and the polo-box. Analysis of mouse tissues revealed strong Fnk mRNA expression in tissues of the newborn, whereas in adult tissues a strong signal was observed only in skin, but not in other tissues with a high proliferative index [24].

Snk/Plk-akin kinase (Sak-a and Sak-b)

The Sak-a and Sak-b cDNAs were isolated serendipitously in the course of a search for genes regulating sialylation [29]. Analysis of cDNAs isolated from a murine lymphoid library revealed two open reading frames for Sak-a (925 amino acids) and Sak-b (464 amino acids). These kinases are identical for the first 416 amino acids, including the protein kinase catalytic domain, but they differ in their noncatalytic C-terminal ends. Sak-a and Sak-b may arise as alternatively spliced mRNAs from a single Sak gene, as they diverge after an AG dinucleotide representative of a splice donor site [46]. The C-termini of Sak-a and Sak-b show no obvious similarity to each other or to the polo-like kinases described above [29]. The N-terminal catalytic domain common to both Sak-a and Sak-b shares significant sequence sim-ilarity with Drosophila polo (42%) and murine Snk and Plk1 (41% and 37% respectively), but it displays a GxGxxAxV motif (instead of the GxGGxAxC characteristic of Plks) in subdomain I (Figure 1C). Also, both Sak-a and Sak-b lack the polo-box. Consequently, Sak-a/b should probably be considered as distant members of the family.

Analysis of Sak mRNA expression has given some insight into the possible role of the these kinases. By Northern blotting, Sak mRNA expression was found to be highest in adult testes, and localisation of Sak mRNA by *in situ* hybridisation revealed high expression in regions of tissues undergoing cell division [29]. These data are suggestive, therefore, of a role for Sak-a/b in cell proliferation. Further evidence for such a role was gained from antisense experiments in CHO cells. Expression of Sak-a antisense constructs greatly reduced CHO colony formation, whereas sense constructs had no effect. Although it remains to be proven that the observed block on cell growth was due solely to ablation of Sak-a mRNA, these data, as well at the expression analyses, suggest that Sak kinases may be involved in mitotic and meiotic processes.

CONCLUSION AND PERSPECTIVE

A family of polo-like kinases may be defined by the criterion that its members display structural similarity to the polo kinase of *Drosophila melanogaster*. Putative homologs of *Drosophila* polo have been identified in yeasts and mammals, and genetic, biochemical and immunocytochemical studies support the view that *Drosophila* polo, budding yeast Cdc5p, fission yeast plo1$^+$ and mammalian Plk1 all function in mitosis (Figure 2). With Snk and Fnk, two additional members of the mammalian polo-like kinase family have been identified, and it is possible that homologues of Snk and Fnk remain to be discovered in lower eukaryotes. On the basis of their expression, Snk and Fnk are not expected to function during mitosis. Instead, they may play important roles in signal transduction at early stages of the cell cycle (Figure 2). Finally, the more distantly related kinases Sak-a/b may also play a role in cell proliferation. However, to what extent these latter kinases functionally resemble Plk1, Snk or Fnk remains to be determined.

It is attractive to view polo-like kinases as a newly emerging family of cell cycle regulatory enzymes. However, so far, cell cycle roles have been directly established only for polo, the founding member of the family, and its likely homologs in other species (Cdc5p, plo1$^+$ and Plk1). Also, comparatively little definitive information is available on the precise physiological role of any of the kinases described in this review. Nevertheless, future work on polo-like kinases is likely to uncover novel and fascinating aspects of cell cycle regulation. Of particular importance, future studies should aim at understanding how polo-like kinases are regulated during the cell cycle, and at identifying their physiological substrates.

ACKNOWLEDGEMENTS

RMG was supported by grants from the Fondation pour la Recherche Medicale and Association pour la Recherche sur le Cancer (France) during the preparation of this manuscript. HAL acknowledges a postdoctoral fellowship from the Schering Research Foundation, Berlin. Work in EAN's laboratory was supported by the Swiss National Science Foundation.

REFERENCES

1. Maller, J.L., Wu, M. and Gerhart, J.C. (1977) *Dev. Biol.* **58**, 295-312.
2. Karsenti, E., Bravo, R. and Kirschner, M.W. (1987) *Dev. Biol.* **119**, 442-453.
3. McIntosh, J.R. and Koonce, M.P. (1989) *Science* **246**, 622-628.
4. Vandré, D.D., Davis, F.M., Rao, P.N. and Borisy, G.G. (1984) *Proc. Natl. Acad. Sci.* **81**, 4439-4443.
5. Vandré, D.D., Davis, F.M., Rao, P.N. and Borisy, G.G. (1986) *Eur. J. Cell Biol* **41**, 72-81.
6. Kuriyama, R. (1989) *Cell Motil. Cytoskeleton* **12**, 90-103.
7. Centonze, V.E. and Borisy, G.G. (1990) *J. Cell Sci.* **95**, 405-411.
8. Vandré, D.D., Centonze, V.E., Peloquin, J., Tombes, R.M. and Borisy, G.G. (1991) *J. Cell Sci.* **98**, 577-588.
9. Vandré, D.D. and Burry, R.W. (1992) *J. Histochem. Cytochem.* **40**, 1837-1847.
10. Gorbsky, G.J. and Ricketts, W.A. (1993) *J. Cell Biol.* **122**, 1311-1321.
11. Snyder, J.A. and McIntosh, J.R. (1975) *J. Cell Biol.* **67**, 744-760.
12. Kuriyama, R. and Borisy, G.G. (1981) *J. Cell Biol.* **91**, 822-826.
13. Ohta, K., Shiina, N., Okumura, E., Hisanaga, S.-I., Kishimoto, T., Endo, S., Gotoh, Y., Nishida, E. and Sakai, H. (1993) *J. Cell Sci.* **104**, 125-137.
14. Nurse, P. (1990) *Nature* **344**, 503-508.
15. King, R.W., Jackson, P.K. and Kirschner, M.W. (1994) *Cell* **79**, 563-571.
16. Nigg, E.A. (1995) *BioEssay* **17**, 471-480.
17. Blangy, A., Lane, H.A., d'Hérin, P., Harper, M., Kress, M. and Nigg, E.A. (1995) *Cell* **83**, 1159-1169.
18. Liao, H., Li, G. and Yen, T.J. (1994) *Science* **265**, 394-398.
19. Verde, F., Labbé, J.-C., Dorée, M. and Karsenti, E. (1990) *Nature* **343**, 233-238.
20. Karsenti, E. (1991) *Seminars Cell Biol.* **2**, 251-260.
21. Buendia, B., Draetta, G. and Karsenti, E. (1992) *J. Cell Biol.* **116**, 1431-1442.
22. Hanks, S. (1991) *Curr. Op. Struc. Biol.* **1**, 369-383.
23. Hanks, S.K. and Hunter, T. (1995) *FASEB J.* **9**, 576-596.
24. Clay, F.J., McEwen, S.J., Bertanocello, I., Wilks, A.F. and Dunn, A.R. (1993) *Proc. Natl. Acad. Sci.* **90**, 4882-4886.
25. Golsteyn, R.M., Schultz, S.J., Bartek, J., Ziemiecki, A., Ried, T. and Nigg, E.A. (1994) *J. Cell Sci.* **107**, 1509-1517.
26. Hamanaka, R., Maloid, S., Smith, M.R., O'Connell, C.D., Longo, D.L. and Ferris, D.K. (1994) *Cell Grow. Diff.* **5**, 249-257.

27. Simmons, D.L., Neel, B.G., Stevens, R., Evett, G. and Erikson, R.L. (1992) *Mol. Cell. Biol.* **12**, 4164-4169.
28. Donohue, P.J., Alberts, G.F., Guo, Y. and Winkles, J.A. (1995) *J. Biol. Chem.* **270**, 10351-10357.
29. Fode, C., Motro, B., Yousefi, S., Heffernan, M. and Dennis, J.W. (1994) *Proc. Natl. Acad. Sci.* **91**, 6388-6392.
30. Sunkel, C.E. and Glover, D.M. (1988) *J. Cell Sci.* **89**, 25-38.
31. Fenton, B. and Glover, D.M. (1993) *Nature* **363**, 637-640.
32. Kitada, K., Johnson, A.L., Johnston, L.H. and Sugino, A. (1993) *Mol. Cell. Biol.* **13**, 4445-4457.
33. Hartwell, L.H., Mortimer, R.K., Culotti, J. and Culotti, M. (1973) *Genetics* **74**, 267-286.
34. Byers, B. and Goetsch, L. (1974) *Cold Spring Harbor Symp. Quant. Biol.* **38**, 123-131.
35. Schild, D. and Byers, B. (1980) *Genetics* **96**, 859-876.
36. Sharon, G. and Simchem, G. (1990) *Genetics* **125**, 475-485.
37. Chapman, J.W. and Johnston, L.H. (1989) *Exp. Cell Res.* **180**, 419-428.
38. Ohkura, H., Hagan, I.M. and Glover, D.M. (1995) *Gene Dev.* **9**, 1059-1073.
39. Lake, R.J. and Jelinek, W.R. (1993) *Mol. Cell. Biol.* **13**, 7793-7801.
40. Holtrich, U., Wolf, G., Bräuninger, A., Karn, T., Böhme, B., Rübsamen-Waigmann, H. and Strebhardt, K. (1994) *Proc. Natl. Acad. Sci.* **91**, 1736-1740.
41. Lee, K.S., Yuan, Y.O., Kuriyama, R. and Erikson, R.L. (1995) *Mol. Cell. Biol.* **15**, 7143-7151.
42. Bräuninger, A., Strebhardt, K. and Rübsamen-Waigmann, H. (1995) *Oncogene* **11**, 1793-1800.
43. Hamanaka, R., Smith, M.R., O'Connor, P.M., Maloid, S., Mihalic, K., Spivak, J.L., Longo, D.L. and Ferris, D.K. (1995) *J. Biol. Chem.* **270**, 21086-21091.
44. Golsteyn, R.M., Mundt, K.E., Fry, A.M. and Nigg, E.A. (1995) *J. Cell Biol.* **129**, 1617-1628.
45. Chen, C.A. and Shyu, A. (1995) *Trend Biochem. Sci.* **20**, 465-470.
46. Padgett, R.A., Grabowski, P.J., Konarska, M.M., Seiler, S. and Sharp, P.A., in *Annual Review of Biochemistry.* 1986, p. 1119-1150.
47. Llamazares, S., Moreira, A., Tavares, A., Girdham, C., Spruce, B., Gonzalez, C., Karess, R., Glover, D. and Sunkel, C. (1991) *Genes and Dev.* **5**, 2153-2165.

chapter 12
Ubiquitin-dependent proteolysis and cell cycle control in yeast

Kristin T. Chun, Neal Mathias and Mark G. Goebl[1]

Department of Biochemistry and Molecular Biology
and The Walther Oncology Center, 635 Barnhill Drive, Medical Sciences Building,
Indiana University School of Medicine, Indianapolis, IN 46202-5122, USA
[1]To whom correspondence should be addressed

Genetic and biochemical data indicate that ubiquitin-mediated proteolysis is involved in the regulated turnover of proteins required for controlling cell cycle progression. In general, mutations in some genes that encode proteins involved in the ubiquitin pathway cause cell cycle defects and affect the turnover of cell cycle regulatory proteins. Furthermore, some cell cycle regulatory proteins are short-lived, ubiquitinated, and degraded by the ubiquitin pathway. This review will examine how the ubiquitin pathway plays a role in regulating progression from the G1 to the S phase of the cell cycle, as well as the G2 to M phase transition.

THE UBIQUITIN PATHWAY

Many of the details of the ubiquitin system have been reviewed previously (1,2,3oncentrate on the components of the system in the yeast, *Saccharomyces cerevisiae*. Ubiquitin is a small, highly conserved polypeptide of 76 amino acids (see above reviews) that is one of the most abundant proteins in the cell. Yeast ubiquitin varies by only two or three amino acids from the ubiquitin of plants and animals, respectively. In yeast, this polypeptide is encoded by four genes, *UBI1*, *UBI2*, *UBI3*, and *UBI4* (4). All four genes encode identical ubiquitin proteins, but in each case, proteolytic processing of the gene product is required to release ubiquitin. *UBI1* and *UBI2* encode identical fusion proteins consisting of ubiquitin and a COOH-terminal tail, while *UBI3* codes for a fusion between ubiquitin and a distinct COOH-terminal tail. The tail portions of these two fusion proteins are ribosomal proteins important for ribosome assembly (5), although the significance of their fusion to ubiquitin remains unclear. Finally, *UBI4* encodes a polyubiquitin protein which must be processed to generate multiple, free ubiquitin molecules (4).

Ubiquitin is found attached to other cellular proteins as a post-translational modification. While attachment of a single ubiquitin has been described, a better understood event is the attachment of a polyubiquitin chain. The initial attachment to the substrate protein occurs via the formation of an isopeptide linkage between a lysine on the substrate protein and the COOH-terminal glycine of ubiquitin. This attached ubiquitin is then itself ubiquitinated in the same manner by a second ubiquitin and so on. The best described site of modification on ubiquitin itself is lysine 48 (6); however, ubiquitin chains involving lysine 29 and lysine 63 have also been described *in vivo* (7, 8). While the ability to form ubiquitin chains through lysine 48 is essential for viability in yeast, the loss of the other lysines as sites of polyubiquitination can be tolerated (8). Recently, mutants were isolated which are specifically defective in the ability to form ubiquitin chains on lysine 29 or lysine 48 (9).

The attachment of ubiquitin proceeds via several steps (summarised in Figure 1). First, in an ATP-dependent manner, ubiquitin is activated by the formation of a thiolester linkage between the COOH-terminal glycine of ubiquitin and a cysteine on a ubiquitin-activating enzyme (E1). In yeast, this enzyme is encoded by the essential *UBA1* gene (10), although a second essential gene, *UBA2*, encodes a product highly related to that of *UBA1* (11). Second, the ubiquitin-activating enzyme transfers ubiquitin, again via a thiolester linkage, to a member of a family of ubiquitin-conjugating enzymes (E2's). In yeast, at least 10 genes encoding putative ubiquitin-conjugating enzymes have been identified, including *CDC34* (*UBC3*), *UBC4*, and *UBC9*, which are relevant to this review (see below). At least some E2's can directly transfer ubiquitin to its substrate by forming an isopeptide linkage between the COOH-terminal glycine of ubiquitin and the epsilon amine of a lysine residue within the substrate (12, 13). Alternatively, some enzymes, including Ubc4p and Ubc5p, transfer ubiquitin to yet another protein known as a ubiquitin ligase (E3) (14, 15). In the same manner as the E2 ligase, the E3 then transfers the ubiquitin to a substrate protein. While the best characterised E3 ubiquitin ligase is the E6 associated protein known to target the ubiquitination of p53 via the HPV E6 protein, several yeast proteins (including Rsp5p, Tom1p, Yjr036p, and Ykl010p) are also likely members of this protein family. These proteins may only represent one class of E3 ubiquitin ligases, since

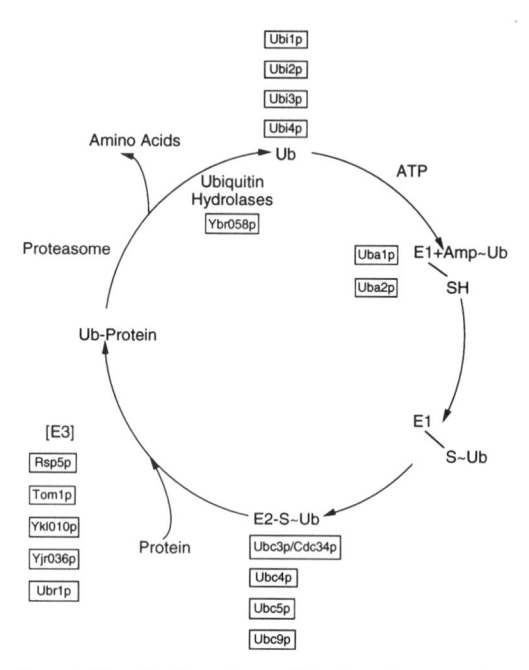

Figure 1. The ubiquitin pathway. Ubiquitin is first attached to a ubiquitin-activating enzyme (E1), Uba1p or Uba2p. The E1 transfers this molecule to one of a family of ubiquitin-conjugating enzymes (E2's). Ubc3p (Cdc34p), Ubc4p, Ubc5p, and Ubc9p are E2's that play a role in the cell division cycle. The E2 either transfers the ubiquitin to a substrate protein or to a ubiquitin ligase (E3). Several putative E3's (Rsp5p, Tom1p, Ykl010p, Yjr036p, and Ubr1p) are listed. In either case, the attachment of a poly-ubiquitin chain targets the substrate protein for degradation by the proteasome. The ubiquitin chain is removed from the substrate protein, and the component ubiquitin molecules are released, perhaps by ubiquitin hydrolases, to be reused.

the Ubr1 protein may have a similar function, although it is structurally distinct from the other E3's (16, 17, 18).

The formation of a ubiquitin chain on a substrate protein targets that protein for degradation by a high molecular weight protease known as the 26S proteasome (for review, see 19). This proteasome can be separated into a 20S catalytic core and a regulatory component, PA700. The structure for the *Thermoplasma acidophilum* 20S proteasome has recently been determined, and it has a barrel-like structure composed of 14 copies each of two different subunits, α and β (20). These 28 subunits are arranged into four stacked rings, two inner rings of 7 β subunits each and two outer rings of 7 α subunits each. In yeast, 14 distinct 20S proteasome subunits have been identified (reviewed in 21, Table 1), and they appear to be α- or β-like in structure, which suggests that the basic structure of the eukaryotic 20S proteasome will be very similar to that of *T. acidophilum*. These subunits all range from 20-32kD in size. A deletion of any one of the genes encoding these various subunits (except that encoding Y13) is lethal.

While the 20S proteasome can degrade denatured proteins into small peptides, only when combined with the PA700 component does it degrade proteins in a ubiquitin-dependent manner (for review, see 19). The activity of this 26S proteasome is also ATP-dependent. The mammalian PA700 has on the order of 15 subunits, many with identifiable yeast homologs (26, Table 2). While the precise workings of PA700 are not clear, some functions can be assigned to several of the subunits. Several of the subunits resemble ATPases and are presumably involved in the energy-dependent activation of the 26S proteasome (26, Table 2). The S5 component, which is highly homologous to the predicted translation product of the yeast ORF *YHR200*, binds polyubiquitin and is presumably necessary for the 26S proteasome's ability to recognise polyubiquitinated proteins (27). Many of the yeast genes encoding the predicted subunits of PA700 are also essential (Table 2).

The 26S proteasome does not degrade the polyubiquitin attached to the targeted proteins (28); rather, the ubiquitin is recycled and can be re-utilised in other ubiquitinating events. The best candidates for the enzymes that recycle ubiquitin are a large family of proteins (over 13 members identified in yeast) known as ubiquitin hydrolases. The many members of this family suggest a high degree of complexity in substrate recognition or regulation of this step. In higher cells, a specific ubiquitin hydrolase, isopeptidase T, has been described which can disassemble ubiquitin chains. Its closest yeast relative is Ybr058p (28).

Genetic and biochemical data indicate that ubiquitin-mediated degradation is involved in the regulated turnover of proteins required for controlling cell cycle progression. A list of these cell cycle and ubiquitin-related genes and where in the cell cycle each seems to be important appears in Figure 2. In general, mutations in genes that encode proteins involved in the ubiquitin pathway cause cell cycle defects and affect the turnover of cell cycle regulatory proteins. Furthermore, some cell cycle regulatory proteins are short-lived, ubiquitinated, and degraded by the ubiquitin pathway. This review will examine how the ubiquitin pathway plays a role in regulating progression from the G1 to the S phase of the cell cycle, as well as through the G2 to M phase transition.

G1 AND S PHASE

Ubiquitin-mediated degradation is required for progression through the G1 to S phase transition of the cell cycle. It appears that the G1 cyclins (Cln1p, Cln2p, and Cln3p), as well as the cyclin-dependent kinase inhibitor, Sic1p, are degraded by the ubiquitin system. In particular, three genes (*CDC4*, *CDC34*,

CHAPTER 12/ UBIQUITIN PATHWAY AND CELL CYCLE CONTROL IN YEAST

Table 1. **Subunits of the 20S proteasome (adapted from Hilt and Wolf, 1995).** The known subunits of the yeast 20S proteasome are listed. Alternate protein names appear in parentheses. α and β refer to the *T. acidophilum* subunit-type that each protein most resembles.

Yeast Protein	Type	Length (amino acids)	Predicted Mass	Essential	Accession Number
Prs1p (YC1)	α	288	31KD	yes	M55436/ J05616/ M37208
Prs2p (Scl1p/YC7/Y8)	α	252	28KD	yes	M55440/ J05616/ M37209
Y7	α	250	27KD	yes	M63640
Y13	α	258	29KD	no	M63851
Pup2p	α	260	29KD	yes	X64918/ S47209
Pre5p	α	234	26KD	yes	L34347
Pre6p	α	254	28KD	yes	L34348
Pre1p	β	198	22KD	yes	X56812
Pre2p (Prg1p)	β	287	32KD	yes	X68662
Pre3p (Yer012p)	β	194	21KD	yes	X78991
Pre4 p (Yfr050p)	β	266	29KD	yes	X68663
Prs3p (Ybl041p)	β	242	27KD	yes	D00845
Pup1p	β	261	28KD	yes	X61189
Pup3p (Yer094p)	β	205	23KD	yes	M88470

Table 2. **Yeast homologs of the PA700 subunits.** The yeast proteins that are likely homologs of the designated subunits of the mammalian PA700 complex are listed.
[a] These proteins contain an ATPase domain.
[b] These proteins contain a polyubiquitin binding domain.
"NA" indicates the information is not available.

Yeast Protein	PA700 Subunit	Length (amino acids)	Predicted Mass	Essential	Accession Number
Sen3p (22)/Yil075p	S1	945	104KD	yes	L06321
Yhr027p	S2	993	110KD	NA	U10399
Yer021p/Sun2p	S3	523	60KD	NA	D78023/ U18778
Yta5p (23)/Yhs4p [a]	S4	437	49KD	yes	X81070
Yhr200p/Sun1p [b]	S5	268	30KD	NA	U00030/ D78022
Yta2p/ Ynt1p [a]	S6	428	48KD	yes	U06229/ X73570
Yta3p/Cim5p (24)/Ykl145p [a]	S7	467	52KD	yes	Z28145/ X73571
Cim3p/ Sug1p [a]	S8/S9	405	45KD	yes	X66400
S10 Homolog	S10	429	49KD	NA	U32445
None Identified	S11				
None Identified	S12				
Nin1p (25)	S13/S14	257	30KD	yes	D10515
None Identified	S15				

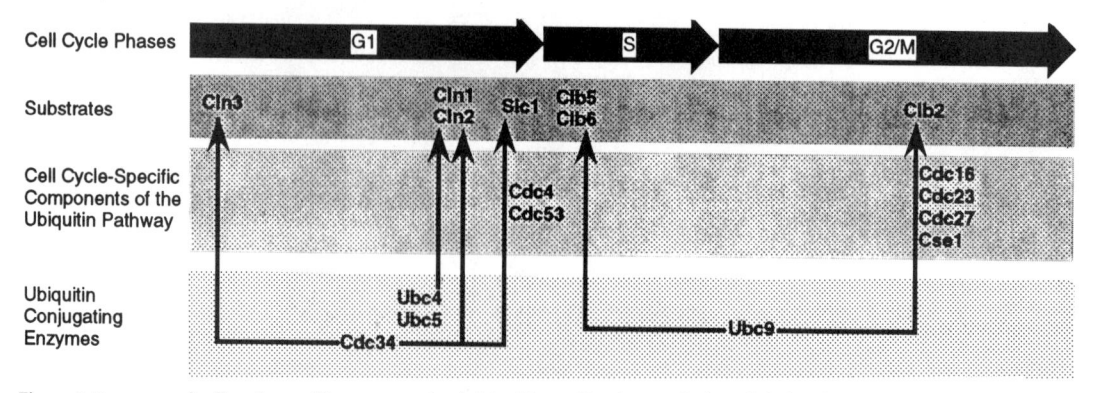

Figure 2. Summary of cell cycle specific components of ubiquitin-mediated proteolysis and their substrates. Ubiquitin-conjugating enzymes and other components of the ubiquitin pathway, all of which are required to degrade yeast cell cycle regulatory proteins are displayed. The heavy black lines connect substrates with the enzymes involved in their regulated turnover. Whereas some of the connections are supported by substantial evidence, others are more speculative.

and *CDC53*) play a role in the protein degradation events necessary for the G1 to S phase transition.

G1 Cyclins: Cln3p

Although the G1 cyclins are functionally redundant, it appears that Cln3p functions quite distinctly from Cln1p and Cln2p. Cln3p is an upstream activator for a variety of proteins, including Cln1p and Cln2p, and unlike these other two cyclins, its patterns of oscillation are less pronounced. Mutant yeast strains with a non-degradable form of Cln3p yield a detectable phenotype, which in two independent screens, led to the identification of *CLN3*. In one screen to identify genes that control Start, mutants displaying a small cell size phenotype were isolated by zonal centrifugation (29). One mutant isolated in this manner, termed *whi-1*, initiates bud formation when the parent cell is only half the size of a wild type cell at the same stage. A second screen was designed to find mutants resistant to mating factor, and it identified a mutant containing *daf1-1*, in which Start also occurs in small cells (30). Cloning of the wild type genes revealed that the mutations are allelic (30, 31). Furthermore, sequencing showed that both mutations are similar. In each, a premature termination codon results in the expression of a truncated protein that lacks COOH-terminal sequences (discussed below).

A detailed sequence comparison of wild type *WHI1*, now called *CLN3*, showed that its encoded amino acid sequence is homologous to that of other cyclins in clam, sea urchin, and *Schizosaccharomyces pombe* (31). Concurrently, Rogers *et al.* (1986) identified sequences that are common to rapidly degraded proteins (32). A comparison of the amino acid sequences of proteins whose half-lives are less than two hours revealed regions rich in proline (P), glutamic acid (E), serine (S), and threonine (T). Such "PEST" sequences are located in the COOH-terminus of Cln3p (31). It was thus proposed that the truncated Cln3p is stabilised compared to the wild type version, because it lacks the de-stabilising PEST sequences.

This prediction was proved correct when antisera to Cln3p became available. Although this antisera is not able to detect Cln3p in extracts from yeast containing the wild type allele of *CLN3*, it did detect Cln3p in extracts from yeast containing the mutant allele that encodes the PEST-less Cln3p (33). The fusion of a *CLN3* allele encoding an HA epitope-tagged version of Cln3p to a galactose inducible promoter resulted in overexpression of the chimera and also eased detection (using anti-HA monoclonal antibodies) of Cln3p. The half-life of HA-Cln3p is only about ten minutes, while the half-life of the truncated, PEST-less Cln3p is approximately two hours (34).

In strains with either the *daf1-1* or *whi-1* mutations (which encode only the first 397 and 403 amino acids of Cln3p, respectively), the mutated proteins accumulate and therefore appear to be stabilised (30, 31). Next, a comprehensive study was carried out to identify more precisely the Cln3p COOH-terminal sequences that are required for its rapid turnover. Yaglom *et al.* (1995) constructed a series of nested deletions at the 3' end of *CLN3* and introduced these truncated genes for the wild type, endogenous alleles (35). The effects of these deletions on the steady state abundance of the proteins and their half-lives, as well as the phenotypic effects of altered Cln3p stability, such as changes in cell volume, were then measured. This allowed a direct correlation between stability and *in vivo* activity. They observed that an increase in Cln3p stability is associated with a decrease in cell volume. This work identified a region from amino acids 404-580, the COOH-terminus of Cln3p, that if deleted, stabilises the protein and confers a small cell phenotype. Deletions that extend beyond this tail region result in a large cell phenotype. This COOH-terminal tail of Cln3p contains five PEST regions as defined by Rogers *et al* (32). As each PEST sequence is deleted, the protein's stability increases. Although analysis of internal deletions that disrupt one or two PEST sequences identified a portion of PEST region 2, between amino acids 449-483, as the sequence that contributes the greatest effect on Cln3p instability, it was concluded that all 5 regions are needed for efficient degradation.

The signal for Cln3p degradation was shown to be also transferable. The half-life of a β-galactosidase fusion protein containing amino acids 404-580 of Cln3p (Cln3-β-Galp) is 30 minutes (highly unstable yet not as unstable as wild type Cln3p) suggesting that this region is not the sole contributor to Cln3p degradation (35). Further analysis of this chimera revealed that a minimal region between amino acids 404-488 can confer instability upon β-galactosidase and, presumably, Cln3p.

What pathway is used to degrade Cln3p, and what enzyme(s) targets Cln3p for degradation? If ubiquination of Cln3-β-Galp occurs, then its degradation is likely to depend upon the proteasome (1, 2, 36). It was thus reasoned that if Cln3p is degraded in a ubiquitin-dependent manner, conditional mutations in two genes that encode the essential proteins, Pre1p and Pre4p, of the proteasome 20S complex (Table 1) would inhibit the degradation of Cln3-β-Galp. Indeed, the Cln3-β-Galp fusion is significantly more stable in a *pre1 pre4* mutant strain (35). Furthermore, in the presence of mutated ubiquitin molecules that fail to produce poly-ubiquitin chains, Cln3-β-Galp is also stabilised.

Tyers *et al.* (1992) established a link between Cln3p, an unstable protein active at Start, and the ubiquitin conjugating enzyme, Cdc34p, which is required for cell cycle progression shortly after Start (34, 37, 12). They compared the levels of Cln3p immunoprecipitated from strains containing either the wild type allele for *CDC34* or a temperature sensitive allele, *cdc34-2*. They found that the total levels of Cln3p isolated from these different strains do not change dramatically, even in extracts from *cdc34-2* cells incubated at the nonpermissive temperature. However, the authors revealed that Cln3p can exist as a phosphoprotein and showed that, relative to wild type cells, phosphorylated Cln3p accumulates in *cdc34-2* cells at the nonpermissive temperature. In addition, the kinase activity of Cln3p-associated Cdc28p isolated from *cdc34-2* cells is much higher than that isolated from wild type cells. Thus, Cdc34p may not be responsible for the turnover of Cln3p *per se*, but it may be required to degrade phosphorylated (and presumably active) forms of Cln3p. In addition, the associated kinase activity of a PEST-less Cln3p isolated from *cdc34-2* cells is also greater than that isolated from wild type *CDC34* cells. This mutant allele of *CDC34* also affects the response to mating pheromone. As discussed previously, *DAF1* was identified by a mutation that confers resistance to mating factor, presumably because the mutant Cln3p is stabilised. Even when Cln3p is overexpressed in a *GAL-CLN3* strain, these cells arrest in the presence of mating pheromone. However, a *GAL-CLN3 cdc34-2* strain is resistant to mating pheromone. Taken together, these results suggest that if active Cln3p is not degraded at the appropriate time during the cell division cycle, it becomes misregulated.

A more complete survey to determine which ubiquitin conjugating enzymes affect the stability of Cln3-β-Galp fusion protein determined that mutant alleles of *UBC1*, *UBC2*, *UBC6*, *UBC7*, *UBC9* or *UBR1* do not increase the stability of this fusion protein. However, the half-life of the Cln3-β-Galp fusion protein increases in strains with a mutant allele of *CDC34* or in those with mutant alleles of both *UBC4* and *UBC5* (35). Ubc4p and Ubc5p are redundant proteins (36). Thus it was proposed that Cdc34p, Ubc4p, and Ubc5p may act together for the efficient degradation of Cln3p. However, a measure of Cln3p stability in a strain mutant for all three *UBC* genes has not been reported. Furthermore, there may be still other enzymes that target Cln3p for degradation.

Thus, the sequences that contribute to the instability of Cln3p and the enzymes that target Cln3p for degradation are beginning to be revealed. However, what signals are required for these components to recognise each other? The observation that the abundance of phosphorylated Cln3p increases in *cdc34* mutants (34, 35) indicates that in such strains, phosphorylated Cln3p is stabilised. Furthermore, deletions that stabilise Cln3p remove not only PEST sequences but also remove potential phosphorylation sites in the COOH-terminal tail. Therefore, the role of phosphorylation for degradation was explored by Yaglom *et al.* (35). Previous studies showed that Cln3p is phosphorylated by Cdc28p, and in strains containing a temperature sensitive allele of *CDC28*, *cdc28-4*, the stability of Cln3-β-Galp is significantly increased. Only one site on the COOH-terminal tail, serine 468, satisfies the stringent consensus of a Cdc28p phosphorylation site (38). However, mutating this site increases the stability of Cln3p significantly but has only a minor effect on the fusion protein.

What is the function of Cln3p degradation? It is known that Cln3p acts at the top of a cascade and is an upstream activator of a variety of cyclins (39) that accumulate in response to a positive feedback loop. Little oscillation of Cln3p abundance or associated kinase activity is detected throughout the cell cycle. Stabilisation of Cln3p causes a small cell size, perhaps indicative of a premature Start. Therefore its degradation may be triggered by general growing conditions such as nutrient availability and protein synthesis.

To summarise, Cln3p is degraded by the ubiquitin pathway and may be the substrate of three Ubc enzymes, Cdc34p, Ubc4p, and Ubc5p. Some of the sequences required for the instability of Cln3p have been identified, but the possibility of additional sequences exists. Phosphorylation stimulates Cln3p degradation, and Cdc28p might phosphorylate residues in the COOH-terminal tail of Cln3p to target it for degradation.

G1 Cyclins: Cln1p and Cln2p

The relevance of Cln1p and Cln2p regulated degradation is more readily inferred than that of Cln3p, since the abundance of Cln1p and Cln2p clearly oscillate during the cell division cycle (39, 40). Cln1p and Cln2p accumulate in late G1, just prior to bud emergence, and are rapidly lost, but then they reappear for the next cycle. This fluctuation in abundance corresponds with their associated kinase activity. The accumulation is made possible by a positive feedback mechanism (39), but why and how are these molecules degraded? Cln1p and Cln2p display similar characteristics and high sequence homology (41), so they will be discussed together.

Artificial increases of either Cln1p or Cln2p (by overexpression of either *CLN1* or *CLN2* from a *GAL* promoter) cause an aberrant, elongated (yet small) cell morphology (41, 42, 43). Considering such detrimental effects, as well as the presence of PEST sequences in the COOH-terminal tail, it was hypothesised that these molecules may be subject to negative regulation, in part, by degradation. A half-life measurement for Cln2p of less than 15 minutes shows that this molecule is highly unstable (40).

Deshaies *et al.* (1995) set out to determine the mechanism responsible for the rapid turnover of Cln1p and Cln2p (44). To investigate the degradation of Cln2p, the fate of *in vitro* translated *CLN2* added to yeast extracts was monitored. A fraction of the Cln2p appeared as a high molecular weight smear, so it was hypothesised that, as observed for some other labile proteins, the presence of these high molecular weight species is due to the addition of ubiquitin to Cln2p. Three lines of evidence support this contention. First, the addition of purified ubiquitin to the same yeast extract stimulates the formation of the Cln2p high molecular weight species. Secondly, the addition of a ubiquitin methylated at the site required for chain elongation, thus inhibiting this reaction, decreases the size of the modified Cln2p produced. Finally, this high molecular weight Cln2p is selectively immuno-precipitated by anti-ubiquitin antibodies.

Since the disappearance of Cln2p coincides with the requirement for Cdc34p function, experiments were carried out to uncover any link between these two events (44). Interestingly, the overexpression of Cln2p in a *cdc34-2* strain enhances the *cdc34* mutant phenotype. Comparing the amount of Cln2p modification in yeast extracts, the abundance of the high molecular weight Cln2p was reduced in extracts from cells with a *cdc34-2* allele compared to extracts from wild type cells. Furthermore, the addition of purified Cdc34p to mutant extracts stimulated ubiquitination of Cln2p. Evidence concerning Cln3p degradation suggests that other Ubc's are involved, yet the addition of one of these enzymes, Ubc4p, has no effect on restoring the amount of multi-ubiquinated Cln2p in a *cdc34* mutant. Is the degradation of Cln2p solely the responsibility of Cdc34p? Deshaies *et al.* suggest that this is so (44).

Salama *et al.*, (1994) analysed the COOH-terminal portion of Cln2p and its role in determining Cln2p stability (46). Contradicting the previously established models, they found the half-life of Cln2p in cells arrested at S or M phase to be the same and identical to that in unarrested cells. They also showed that in cells with a *cdc28* ts or *cdc34* ts mutant allele, and at the permissive or nonpermissive temperature, the half-life of Cln2p is the same.

Previously, we discussed the relevance of phosphorylation as a means of signalling Cln3p destruction. Both Cln1p and Cln2p are also found as phosphoproteins. In fact, *in vitro*, the phosphorylation of Cln2p by yeast extracts is sufficient to cause an 18kD shift on SDS-PAGE gels (44). Similar to Cln3p, Cdc28p was shown to be responsible for this modification, since incubation of Cln2p in extracts from a *cdc28* mutant strain produces severely reduced levels of hyperphosphorylation, as well as reduced levels of ubiquitinated Cln2p. The addition of Cdc28p protein to these extracts restores the levels of phosphorylated Cln2p and ubiquitinated Cln2p to levels comparable to that found in a strain wild type for *CDC28*. Furthermore, the amount of hyperphosphorylated Cln2p is greater when incubated with extracts from *cdc34* cells than when incubated with those from wild type cells. These data suggest that Cln2p is phosphorylated by Cdc28p and that this phosphorylation signals the ubiquitination of Cln2p by Cdc34p. The half-life of Cln2p in *cdc28* ts extracts is 8 minutes versus 20 minutes in *cdc34* ts extracts. Compared to the half-life of Cln2p in wild type extracts, 5 minutes, these half-lives are still extremely short, indicating that perhaps phosphorylation by Cdc28p is not the critical signal and/or that Cdc34p may not be the only enzyme responsible for Cln2p ubiquitination. This raises the possibility that Cln2p, and presumably Cln1p, are metabolised by a variety of different pathways.

There are a number of physiological roles for Cln1p and Cln2p degradation. The destruction of Cln1p and Cln2p is required for proper regulation of the cell cycle. It provides a means by which Cdc28p is rid of one set of binding partners and thus becomes available for another set of binding partners required for the next stage of the cycle. But what is required for such a process to take place? Alternatively, what conditions have to prevail in the cell that disallows Cln1p or Cln2p degradation? This was the basis for a clever genetic screen conducted by Barral and colleagues (46). They selected for mutants that contain stable Cln1p as reported by the fusion protein encoded by *CLN1-lacZ*. In this screen, mutants that were both blue and had the aberrant elongated cell morphology indicated stabilised or overexpressed Cln1p (42, 43). In a mutant identified by this screen, the half-lives for both Cln1p and Cln2p are at least 30 minutes. In the wild type strain, the half-lives of these proteins are only about 5 minutes. As expected, the abundance of Cln1p and Cln2p is significantly higher in the mutant strain than in the wild type; however, Cln3p is not affected. The mutant phenotype is complemented by the glucose repression resistance gene, *GRR1* (47, 48, 49). *GRR1* encodes a 135kD protein of low abundance and unknown biochemical activity. One role for *GRR1* could be to coordinate the timing of Start with the availability of nutrients by regulating cyclin proteolysis. In the presence of glucose, activation of the *GRR1* pathway would allow two events to occur. First, it would stimulate the transcription of various nutrient transporter genes, and second, it would activate the degradation of Cln1p and Cln2p and, consequently, delay Start. However, in the absence of glucose, the expression of glucose transporters would be down regulated, and cyclin abundance would accumulate more freely, leading to an early Start. This fits nicely with an old observation that cells growing slowly perform Start at a smaller size than cells growing rapidly (37).

In addition, the abundance of Cln1p was shown to decrease when cells were shifted from a poor carbon source to a rich one (50). However, the abundance of Cln2p does not change. Cells were also shown to pause briefly in G1, and Start was reset to a slightly larger cell size. Cln1p proteolysis must have been rapidly activated for the cells to rid themselves of existing Cln1p. Thus, G1 cyclin stability is not only dependent on the stage of the cycle but also on the environment in which the cell resides.

Sic1p Degradation Mediated by Cdc34p

A phosphoprotein called p40 was shown to bind tightly to Cdc28p (51). This complex has no associated kinase activity, since p40 is a potent inhibitor of the kinase (52). The p40 is the product of the *SIC1* gene (53) and is also known as Sic1p. After establishing that S phase initiation is directly dependent on Clb5p- and Clb6p-associated kinase activity, Schwob *et al.* (1994) postulated the presence of an inhibitor that prevents S phase (54). They identified this inhibitor as being Sic1p. The level of *SIC1* transcripts stays constant during the cell cycle, but the abundance of HA epitope tagged Sic1p fluctuates. The abundance of this protein increases during G1, drops prior to S phase, and is absent during S phase and G2. The levels then increase as cells exit mitosis, thus allowing the protein to be present for the next G1. Apparently, in order for S phase to begin and for the Clb5p/Clb6p kinases to be activated, Sic1p must be destroyed. Clb5p can carry out S phase in the absence of all other Clbs, but in the absence of all six B-type cyclins, the resulting terminal morphology is identical to that displayed by *cdc34* mutants: DNA replication does not take place, bud growth is hyperpolarised, and most cells have undergone bud emergence several times (54). In *cdc34* mutants at the nonpermissive temperature, no kinase activity can be detected for the Clb5p-Cdc28p complex, while activity is detected at the permissive temperature. Furthermore, Sic1p levels remain high in *cdc34* mutants at the nonpermissive temperature. These data suggest that the drop of Sic1p level prior to S phase is dependent on Cdc34p activity.

The observation that Sic1p appears to migrate in SDS-PAGE gels as a doublet raised the possibility that it could exist as a phosphoprotein. In addition, the putative phosphorylated form of Sic1p is more abundant in *cdc34* mutants (54). Could the Clns be involved in signalling Sic1p degradation through phosphorylation? Direct evidence is not available. However this scenario would ensure the correct ordering of the cycle, with Clb-associated kinases becoming active after Cln-associated kinases.

Transition into S Phase

The three genes (*CDC4*, *CDC34*, and *CDC53*) that will be described in this section are specifically required for entry into S phase. *CDC34* encodes a ubiquitin conjugating enzyme (11) that tags its specific substrates with ubiquitin and thus targets them for degradation via the proteasome. *CDC4* encodes a protein that has no known biochemical activity yet contains multiple β-transducin-like repeats (55). A recent description of *CDC53* shows that, based on its predicted amino acid sequence, Cdc53p also cannot be assigned an obvious biochemical function (56). However, it does bear homology with gene products involved in the cell cycle of the worm, *Caenorhabditis elegans* (57). Cells with a temperature sensitive allele of any one of these genes have the same terminal morphology: arrest at the G1 to S phase transition, no DNA replication, spindle pole bodies duplicated but not separated, and the repeated formation of elongated or hyperpolarised buds. Such a striking similarity in terminal phenotype suggests that these three genes are involved in the same pathway. Sic1p function and degradation are similarly affected in *cdc4*, *cdc34*, and *cdc53* mutants (54). For instance, the kinase activity of Clb5p-Cdc28p is undetectable in *cdc4* and *cdc53* mutants, as well as in *cdc34* mutants. Mutant *cdc34* cells also deleted for *SIC1* pass through S phase but fail to undergo nuclear division and arrest with a large bud. The identical phenotype is observed in *sic1 cdc4* and *sic1 cdc53* double mutants. Interestingly, the *sic1* mutation rescues the hyperpolarised bud growth displayed in *cdc4*, *cdc34* and *cdc53* mutants. Hyperpolarisation is indicative of overactive Cln1p or Cln2p, but it is not clear whether these Clns are present in these double mutants or whether they have they been disposed of by an alternate pathway.

Mutant alleles of *CDC4*, *CDC34*, and *CDC53* also display synthetic lethality with each other (56). That is, at the permissive temperature, the combination of any two of the respective temperature sensitive mutations in the same cell is lethal. Furthermore, the overexpression of *CDC53* appears to stabilise temperature sensitive Cdc4p and Cdc34p. The co-purification of the three proteins suggests that they form a complex that acts at the same point in the pathway.

The function of Cdc34p is clear, but the roles of Cdc4p and Cdc53p are unknown. Since Cdc34p has been linked to the degradation of a variety of substrates, Cdc4p and Cdc53p may aid the recognition of these substrates. Such a function would make these proteins functionally similar to the E3's.

We have discussed evidence that the potential substrates of this complex may be phosphorylated and that phosphorylation might act as a signal for ubiquination. It is possible that the complex would be active throughout the cell cycle and would only recognise phosphorylated substrates. However, Cdc34p is also present in the cell as a phosphoprotein (58), hinting that phosphorylation may also regulate the complex activity. Cdc34p can ubiquinate itself *in vitro* (59) and *in vivo* (58), so as an additional means of regulation, phosphorylation of Cdc34p might signal its own destruction.

G2 AND M PHASE

In addition to the G1 to S phase transition, ubiquitin-mediated degradation is also required for progression from G2 to M phase. Several genes have been identified that, when mutated, cause cells to arrest during this period of the cell cycle and in addition, play a role in protein turnover, specifically in the ubiquitin-mediated degradation pathway. Often, these mutant cells arrest after the completion of DNA replication, but before the completion of chromosome segregation and nuclear division. At least one set of essential targets of this pathway are the mitotic cyclins.

Genes That Encode Elements of the Proteasome are Required for Mitosis

A mutation in *NIN1*, which encodes a yeast homolog of the S14 component of PA700 (Table 2), causes a defect in cell cycle progression. At the restrictive temperature, most *nin1-1* ts mutant cells arrest at G2 (60). However, 20-30% of these cells arrest as unbudded cells, suggesting an earlier arrest. If cells are synchronized with hydroxyurea at S phase, and then shifted to the restrictive temperature, some cells are delayed during S phase, but they all eventually arrest at G2 (25). Apparently, *NIN1* is required for the transition from G2 to M phase. In addition, a mutant allele of *CDC28* (*cdc28-1N*, which causes a G2/M arrest at the nonpermissive temperature (61)) and *nin1-1* are synthetically lethal, indicating a potential interaction between their gene products. In a *nin1-1* mutant, mitotic cyclin associated-Cdc28p histone H1 kinase activity fails to increase during the cell cycle, which suggests that *NIN1* and the proteasome might play a role in the degradation of a mitotic cyclin dependent kinase inhibitor (25). Alternatively, if *nin1-1* cells are first synchronised with mating pheromone at late G1 and then shifted to the restrictive temperature, these *nin1-1* cells arrest before S phase with 1N DNA. Since Sic1p is the only known essential G1/S target, these results implicate *NIN1* in the degradation of Sic1p and support the idea that the proteasome and ubiquitin-mediated degradation are required to dispose of Sic1p at the G1 to S phase transition. Apparently, *NIN1* is somehow required to activate Cdc28p at both the G1 to S phase and the G2 to M phase transitions. Without Nin1p, the inhibitors are never inactivated, and Cdc28p kinase is never activated.

Analysis of crude cell extracts by glycerol density gradient centrifugation revealed that *NIN1* protein copurifies with proteasome activity (25). Furthermore, antibodies against Nin1p crossreact with a 31 kD subunit of the mammalian proteasome regulatory subunit, PA700. Finally, in *nin1-1* mutant cells, antibodies against polyubiquitin chains detect high molecular weight proteins of various sizes (presumably polyubiquitinated proteins that have been marked for degradation) that are not detected in wild type cells. All of these results are consistent with *NIN1* encoding a regulatory component of the proteasome, which is somehow required for activating Cdc28 kinase activity and for normal cell cycle progression through both the G1 to S and G2 to M phase transitions.

Two other probable proteasome components also play a role in mitosis. *CIM3* and *CIM5* (co-lethal in mitosis) were identified by mutant alleles that, in combination with *cdc28-1N*, cause lethality (synthetic lethality) (24). In the presence of a wild type allele of *CDC28*, the phenotype of each mutant allele, *cim3-1* or *cim5-1*, is temperature sensitive and most cells arrest with a large bud, a single nucleus of replicated DNA, and a short spindle, characteristics of a G2 or M phase arrest. These arrested cells have greater than wild type levels of Cdc28p-associated histone H1 kinase activity, and even at 30°C, they accumulate greater than normal levels of the mitotic cyclins, Clb2p and Clb3p.

CIM3 is identical to *SUG1* (suppressor of *gal4*), which was identified in a screen for suppressors of a mutant allele of *GAL4* (62). The predicted amino acid sequences of Cim3p and Cim5p are 40% identical to each other, and Cim5p is 70% identical to the product of the human *MSS1* gene (24). Mss1p was identified as a modulater of HIV Tat-mediated transactivation (63), and is also a component (S7, see Table 2) of the human 26S proteasome (64). *CIM5* and *MSS1* appear to be homologs, since the *MSS1* cDNA complements a *cim5* mutation (24). Therefore, Cim3p and Cim5p appear to be components of the yeast proteasome. Consistent with this, they are required for certain ubiquitin-mediated degradation events. In either *cim3* or *cim5* mutants, an artificial substrate of non-N-end rule ubiquitin-mediated degradation (ubiquitin-pro-β-galactosidase) is stabilised. However, substrates known to be degraded by the N-end rule pathway are not stabilised in these strains. Finally, in a nearly homogenous preparation of *Drosophila melanogaster* 26S proteasome, antibodies against Cim3p and antibodies against Cim5p recognise a 43 kD and a 48 kD protein, respectively. These two *Drosophila* proteins are components of the regulatory complex of the proteasome, PA700. Therefore, it appears that *CIM3* and *CIM5* encode proteasome subunits that are required to degrade positive effectors of Cdc28p kinase activity, possibly Clb2p and Clb3p, and presumably, the failure of *cim3* and *cim5* mutants to inactivate this activity prevents progression into M phase. Since *NIN1*, *CIM3*, and *CIM5* all appear to encode components of the proteasome regulatory complex, their protein products might help direct the substrate specificity of the proteasome, in this case specificity for positive or negative regulators of Cdc28p kinase activity.

PRG1 (*PRE2*) encodes a 20S proteasome-related protein that is essential (Table 1), and *prg1* mutants are defective in a chymotrypsin-like activity (65). At the restrictive temperature, conditional mutants divide 2.6-4.3 times slower than wild type and die after about 8 hours. At the permissive temperature, a large fraction of the cells have a single nucleus and a short spindle, suggesting they have successfully executed the early steps of spindle formation, but they then pause later in the cell cycle. The nucleus is usually near the neck between the mother and daughter cells, or it is stretched between them. Some cells lack a nucleus, but most have 2N or greater DNA content. There also appears to be a higher than wild type rate of chromosome loss in these cells. All of these observations suggest *PRG1* is required for chromosome segregation. In addition, a deletion of *CLB2* suppresses both the temperature sensitive growth defect of a *prg1-4* ts mutation as well as the lethality of a *PRG1* deletion, further indicating that the proteasome degrades Clb2p.

Mitotic Cyclin Degradation

The studies concerning *CIM3*, *CIM5*, and *PRG1* all suggest that, as for the G1 cyclins, mitotic cyclins are also degraded by the ubiquitin pathway. The mitotic cyclins (B-type cyclins) are encoded by at least four different genes (*CLB1*, *CLB2*, *CLB3*, and *CLB4*). Mutants that lack all four of these cyclins are inviable (66, 67). However, expression of only *CLB2* is sufficient for survival, so it appears that *CLB2* is the most important for mitosis (66). During the cell division cycle, transcripts from *CLB2* are absent during G1, their level then rises as cells enter G2, peaks just before anaphase, and then drops as cells exit from mitosis (66-69). Clb2p levels and the associated Cdc28p kinase activity approximate this pattern (68). In addition, changes in the turnover rate of the encoded protein play an important role in regulating abundance. The turnover of Clb2p is rapid at the end of mitosis, and the protein remains unstable throughout G1 ($t_{1/2} < 1$ minute) until the kinase activity of Cdc28p associated with the G1 cyclins is activated, and Clb2p becomes stabilised ($t_{1/2} \geq 10$ minutes) (70). This change in turnover rate is dependent upon the presence of the destruction box of Clb2p. When G1 cyclin genes are expressed, Clb2p produced from a *CLB2* gene fused to a heterologous *GAL* promoter begins to accumulate, but an interruption of G1 cyclin gene expression causes Clb2p levels to decline. Apparently, the accumulation of G1 cyclins results in the stabilisation of Clb2p.

Along with the observations described above, several lines of evidence indicate that the degradation of some mitotic cyclins, especially Clb2p, is mediated by ubiquitin. *UBC9* encodes a ubiquitin conjugating enzyme, and mutations in this gene cause cell cycle defects (71). At the nonpermissive temperature, a temperature sensitive mutation, *ubc9-1*, causes most cells to arrest with a large bud and a single nucleus of 2N DNA at the neck between the mother and daughter cells, as well as a short, pre-anaphase spindle. Presumably, the arrest occurs before sister-chromatid separation. The absence of *RAD9*, which prevents cells with incompletely replicated DNA from dividing, does not affect the arrest phenotype of *ubc9-1* cells. In addition, when *ubc9-1* cells are released from their cell cycle block, they proceed to divide once more in the presence of hydroxyurea, which interferes with DNA synthesis. These results indicate that the *ubc9-1* block occurs after the *RAD9* and hydroxyurea execution points and that *UBC9* is required after S phase, sometime during the G2 to M phase transition.

Furthermore, genetic evidence suggests that *UBC9* is involved in the degradation of both Clb2p and Clb5p (71). Normally, Clb2p is relatively stable during S and M phases and very short-lived before Start. In wild type cells, Clb2p is barely detectable during G1, but this protein accumulates during G1 in cells with a *ubc9-1* mutation. Clb5p is normally short-lived during the cell cycle. When cells are blocked in G1, its half-life is 5-10 minutes, but its half-life is lengthened to 15-20 minutes during S or M phase. However, in a *ubc9-1* mutant, Clb5p becomes stabilised with a half-life of about 1 hour during M phase. Moreover, overproduction of Clb5p is tolerated by wild type cells, but this is toxic in *ubc9-1* cells, as well as in cells with a mutant proteasome subunit (*pre1-1* cells).

Additionally, other genes involved in mitotic cyclin degradation have been identified. In a screen for mutants that, during G1, hyperaccumulate a Clb2p-β-galactosidase fusion protein, mutant alleles of three genes (*CDC16*, *CDC23*, and *CSE1*) were identified (72). At the nonpermissive temperature, in strains with a conditional allele of *CDC16*, *CDC23*, or *CSE1*, the half-life of Clb2p becomes ≥25 minutes (72). Since the degradation of several other unstable proteins is unaffected in these mutant strains, it appears that these three genes do not encode general components of a proteolytic pathway. Each is essential, and, additionally, certain mutant alleles of *CSE1* cause a defect in chromosome segregation (73). Lamb *et al.* (1994) demonstrated that Cdc16p, Cdc23p, and another protein, Cdc27p, form a complex *in vivo* and *in vitro* (74). A mutation in any one of these three genes causes cells to arrest with a large bud, replicated DNA, a single nucleus at or near the neck between the mother and daughter cells, and a short mitotic spindle, typical of a G2 or M phase arrest (74). Relating these genes to the ubiquitin pathway, mutant alleles of *CDC23* and *UBC4* are synthetically lethal (72). At the permissive temperature, a mutation in either *CDC23* or *UBC4* has no obvious effect on cell viability, but under the same conditions, the combination of these two mutations causes cells to arrest with large buds after 1-3 divisions. However, unlike the G1 cyclins, Clb2p degradation does not appear to involve *CDC34*. At the nonpermissive temperature, a conditional mutation in *CDC34* does not cause the accumulation of Clb2p in G1 arrested cells (72).

Clb2p proteolysis seems to require Cdc23p and, to a lesser extent, Cdc16p (72). Mutations in *CLB2* and *CDC23* are also synthetically lethal. At the nonpermissive temperature, a conditional mutation in *CDC23* causes cells to arrest with stable Clb2p. However, under the same conditions, a mutation in *CDC16* causes cells to arrest as well, but over time, these cells slowly enter G1, and *CLB2* protein levels slowly decline.

These results observed in yeast agree with what is known about mitotic cyclin, cyclin B, degradation in more complex eukaryotes. It has been shown in *Xenopus* oocyte extracts that the sea urchin cyclin B is degraded by the ubiquitin pathway, and that the "destruction box" sequence (amino acids 42-RAALGNISEN-50) targets cyclin B for degradation (75). Recently, investigators used this *in vitro* system to identify components from mitotic *Xenopus* oocyte extracts that are sufficient to ubiquitinate an N-terminal fragment of the sea urchin cyclin B (76). These components include Ubc4p and homologs of the yeast Cdc27p and Cdc16p proteins.

Cdc27p function appears to be required for anaphase to proceed. HeLa cells injected with anti-Cdc27p antibodies arrest with apparently normal spindles and aligned chromosomes, indicating that Cdc27p is not required for metaphase but is required for the onset of anaphase (77). If these antibodies are injected during either prophase or metaphase, the cells arrest after metaphase and before chromosome segregation. However, injecting the antibodies into cells that have reached anaphase fails to cause an arrest. Because of its probable role in the onset of anaphase, the 20S complex containing Cdc27p, as well as Cdc16p and Cdc23p, is called the anaphase promoting complex (76).

In yeast, the *CDC20* gene product appears to be required for proper chromosome segregation. A mutation in *CDC20* causes cells to arrest with a large bud, replicated DNA in a single nucleus at the neck between the mother and daughter cells, and a short spindle (37, 78, 79). At the semi-permissive temperature, these cells exhibit a rate of chromosome loss that is 100 times that of wild type cells (80). The deduced amino acid sequence for *CDC20* is 32% identical to that of the *D. melanogaster* gene, fizzy (*fzy*) (81). In *Drosophila*, *fzy* appears to be required for normal progression through metaphase or anaphase and perhaps for cyclin degradation. In embryos, maternally derived *fzy* gene product is present, but after it disappears, embryonic protein is present in actively dividing tissues and declines in tissues where cell division is ending. In the dorsal epidermal region of later stage embryos, most cells have stopped dividing, and few stain positively for either cyclin A or cyclin B. However, in *fzy*⁻ mutant embryos, many more cells stain positively for cyclins A and B and many of these cells appear to be arrested at metaphase. It is possible that a defect in cyclin turnover causes this abnormal accumulation of cyclins, and wild type *fzy* protein is involved in targeting mitotic cyclins for degradation.

Since mutations in *CDC20* and *fzy* both cause a defect during mitosis, and the corresponding wild type gene products share significant sequence homology, it is possible that each is involved in regulating the turnover of mitotic cyclins. In addition, the sequence similarity between Cdc20p and the *fzy* protein is greatest in a region predicted to encode WD-40 repeats (81), which are also found in Cdc4p (55). As described earlier, it is likely that Cdc4p forms a complex with Cdc34p and Cdc53p, and this complex seems to be involved in the turnover of cell cycle progression regulatory proteins required during the G1 to S phase transition (G1 cyclins and Sic1p). Again, perhaps analogous to Cdc4p, Cdc20p might be involved in the turnover of mitotic cyclins.

Taken together, these results indicate that a complex containing Cdc16p, Cdc23p, and Cdc27p is required for the ubiquitin-dependent proteolysis of mitotic cyclins and perhaps of another protein that must be degraded before sister chromatid separation at anaphase can proceed. Throughout this review we have discussed the substrates, signals, and enzymes involved in the rapid turnover of important cell cycle regulatory components. These proteins are required during various periods of the cycle, and in the absence of their rapid destruction, cell cycle progression either arrests or is defective. Ubiquitin-mediated degradation provides an efficient means to target the rapid destruction of regulatory proteins that have completed their function, and in doing so, maintains the proper order of cell cycle events by ensuring that before a subsequent step in the cell cycle begins, a previous step ends.

ACKNOWLEDGEMENTS

We thank Edward Kipreos for sharing unpublished results and Prianto Moeljadi for expert technical assistance. This work was supported by National Institutes of Health grant GM-45460 to M.G.G. and a postdoctoral fellowship from the American Cancer Society to K.T.C.

REFERENCES

1. Finley, D., and Chau, V. (1991) *Annu. Rev. Cell Biol.* **7**, 25-69.
2. Hershko, A., and Ciechanover, A. (1992) *Annu. Rev. of Biochem.* **61**, 761-807.
3. Jentsch, S. (1992) *Annu. Rev. of Genet.* **26**, 179-207.
4. Özkaynak, E., Finley, D., Solomon, M.J., and Varshavsky, A. (1987) *EMBO J.* **6**, 1429-1439.
5. Finley, D., Bartel, B., and Varshavsky, A. (1989) *Nature* **338**, 394-401.
6. Chau, V., Tobias, J.W., Bachmair, A., Marriott, D., Ecker, D.J., Gonda, D.K., and Varshavsky, A. (1989) *Science* **243**, 1576-1583.
7. Arnason, T., and Ellison, M.J. (1994) *Mol. Cell. Biol.* **14**, 7876-7883.
8. Spence, J., Sadis, S., Haas, A.L., and Finley, D. (1995) *Mol. Cell. Biol.* **15**, 1265-1273.
9. Johnson, E.S., Ma, P.C.M., Ota, I.M., and Varshavsky, A. (1995) *J. Biol. Chem.* **270**, 17442-17456.
10. McGrath, J.P., Jentsch, S., and Varshavsky, A. (1991) *EMBO J.* **10**, 227-236.
11. Dohmen, R.J., Stappen, R., McGrath, J.P., Forrová, Kolarov, J., Goffeau, A., and Varshavsky, A. (1995) *J. Biol. Chem.* **270**, 18099-18109.
12. Goebl, M.G., Yochem, J., Jentsch, S., McGrath, J.P., Varshavsky, A., and Byers, B. (1988) *Science* **241**, 1331-1335.
13. Jentsch, S., and McGrath, J.P., and Varshavsky, A. (1987) *Nature* **329**, 131-134.
14. Scheffner, M., Nuber, U., and Huibregtse, J.M. (1995) *Nature* **373**, 81-83.
15. Huibregtse, J.M., Scheffner, M., Beaudenon, S., and Howley, P.M. (1995) *Proc. Natl. Acad. Sci. U.S.A.* **92**, 2563-2567.
16. Dohmen, R.J., Madura, K., Bartel, B., and Varshavsky, A. (1991) *Proc. Natl. Acad. Sci. U.S.A.* **88**, 7351-7355.
17. Sharon, G., Raboy, B., Parag, H.A., Dimitrovsky, D., and Kulka, R.G. (1991) *J. Biol. Chem.* **266**, 15890-15894.
18. Sung, P., Berleth, E., Pickart, C., Prakash, S., and Prakash, L. (1991) *EMBO J.* **10**, 2187-2193.
19. Hochstrasser, M. (1995) *Curr. Opin. Cell Biol.* **7**, 215-223.
20. Löwe, J., Stock, D., Jap, B., Zwickl, P., Baumeister, W., and Hubert, R. (1995) *Science* **268**, 533-539.
21. Hilt, W. and Wolf, D.H. (1995) *Mol. Biol. Rep.* **21**, 3-10.
22. DeMarini, D.J., Papa, F.R., Swaminathan, S., Ursic, D., Rasmussen, T.P., Culbertson, M.R., Hochstrasser, M. (1995) *Mol. Cell. Biol.* **15**: 6311-6321.
23. Schnall, R., Mannhaupt, G., Stucka, R., Tauer, R., Ehnley, S., Schwarzloss, C., Vetter, I., and Feldmann, H. (1994) *Yeast* **10**: 1141-1155.
24. Ghislain, M., Udvardy, A., and Mann, C. (1993) *Nature* **266**: 358-362.
25. Kominami, K., DeMartino, G. N., Moomaw, C. R., Slaughter, C. A., Shimbara, N., Fujimuro, M., Yokosawa, H., Hisamatsu, H., Tanahashi, N., Shimizu, Y., Tanaka, K., and Toh-e, A. (1995) *EMBO J.* **14**, 3105-15.
26. Dubiel, W., Ferrell, K., and Rechsteiner, M. (1995) *Mol. Biol. Rep.* **21**, 27-34.

27. Deveraux, Q., Ustrell, V., Pickart, C., and Rechsteiner, M. (1994) *J. Biol. Chem.* **269**, 7059-7061.
28. Wilkinson, K.D., Tashayev, V.L., O'Connor, L.B., Larsen, C.N., Kasperek, E., and Pickart, C.M. (1995) *Biochemistry* **34**, 14535-14546.
29. Sudbery, P. E., Goodey, A. R., and Carter, B. L. A. (1980) *Nature* **288**, 401-404.
30. Cross, F. R. Mol. (1988) *Mol. Cell. Biol.* **8**, 4675-4684.
31. Nash, R., Tokiwa, G., Anand, S., Erickson, K. and Futcher, A. B. (1988) *EMBO J*. **7**, 4335-4346.
32. Rogers, S., Wells, R. and Rechsteiner, M. (1986) *Science* **234**, 364-368.
33. Cross, F. C. and Blake, C. M. (1993) *Mol. Cell. Biol.* **13**, 3266-3271.
34. Tyers, M., Tokiwa, G., Nash, R. and Futcher, B. (1992) *EMBO J.* **11**, 1773-1784.
35. Yaglom, J., Linskens, M. H. K., Sadis, S., Rubin, D. M., Futcher, B. and Finley, D. (1995) *Mol. Cell. Biol.* **15**, 731-741.
36. Seufert, W. and Jentsch, S. (1992) *EMBO J*. **11**, 3077-3080.
37. Pringle, J. R. and Hartwell, L. H. (1981) in *The Molecular Biology of the Yeast Saccharomyces* (Strathern, J. D., Jones, E. W. and Broach, J. R., eds.) pp 97-142 Cold Spring Harbor Laboratory Press, Cold Spring Harbor, NY.
38. Moreno, S. and Nurse, P. (1990) *Cell* **61**, 549-551.
39. Tyers, M., Tokiwa, G. and Futcher, B. (1993) *EMBO J*. **12**, 1955-1968.
40. Wittenberg, C., Sugimoto, K. and Reed, S. (1990) *Cell*, **62** 225-237.
41. Hadwiger, J. A., Wittenberg, C., Richardson, H. E., De Barros Lopes, M. and Reed, S. (1989) *Proc. Natl. Acad. Sci. U.S.A.* **86**, 6255-6259.
42. Lew, D. and Reed, S. I. (1993) *J. Cell Biol.* **120**, 1305-1320.
43. Richardson, H. E., Wittenberg, C., Cross, F. and Reed, S. I. (1989) *Cell* **59**, 1127-1133.
44. Deshaies, R. J., Chau, V. and Kirschner, M. (1995) *EMBO J.* **14**, 303-312.
45. Salama, S. R., Hendricks, K. B. and Thorner, J. (1994) *Mol. Cell. Biol.* **14**, 7953-7966.
46. Barral, Y., Jentsch, S. and Mann, C. (1995) *Genes Dev.* **9**, 399-409.
47. Bailey, R. B., and Woodward, A. (1984) *Mol. Gen. Genet.* **193**, 507-512.
48. Flick J. S. and Johnston, M. (1991) *Mol. Cell. Biol.* **11**, 5101-5112.
49. Gancedo, J. M. (1992) *Eur. J. Biochem.* **206**, 297-313.
50. Tokiwa, G., Tyers, M., Volpe, T. and Futcher, B. (1994) *Nature* **371**, 342-345.
51. Reed, S. I., Hadwiger, J. A. and Lorinz, A. T. (1985) *Proc. Natl. Acad. Sci. U.S.A.* **82**, 4055-4059.
52. Mendenhall, M. D. (1993) *Science* **259**, 216-219.
53. Nugroho, T. T. and Mendenhall, M. D. (1994) *Mol. Cell. Biol.* **14**, 3320-3328.
54. Schwob, E., Bohm, T., Mendenhall, M. D. and Nasmyth, K. (1994) *Cell* **79**, 233-244.
55. Yochem, J. and Byers, B. J. (1987) *J. Mol. Biol.* **195**, 223-245.
56. Mathias, N., Johnson, S. L., Winey, M., Adams, A. E., Goetsch, L., Pringle, J. R., Byers, B. and Goebl, M. submitted.
57. Kipreos, E.T., Lander, L.E., Wing, J.P., He, W.W., and Hedgecock, E.M. submitted.
58. Goebl, M. G., Goetsch, L. and Byers, B. (1994) *Mol. Cell. Biol*. **14**, 3022-3029.
59. Banerjee, A., Gregori, L., Xu, Y. and Chau, V. (1993) *J. Biol. Chem.* **268**, 5668-5675.
60. Nisogi, H., Kominami, K., Tanaka, K., and Toh-e, A. (1992) *Exp. Cell Res.* **200**, 48-57.
61. Piggot, J.R., Rai, R., and Carter, B.L.A. (1982) *Nature* **298**, 391-393.
62. Swaffield, J. C., Bromberg, J. F., and Johnston, S. A. (1992) *Nature* **357**, 698-700
63. Shibuya, H., Irie, K., Ninomiya-Tsuji, J., Goebl, M., Taniguchi, T., and Matsumoto, K. (1992) *Nature* **357**, 700-702.
64. Dubiel, W., Ferrel, K., and Rechsteiner, M. (1993) *FEBS Lett.* **323**, 276-278.
65. Friedman, H., and Snyder, M. (1994) *Proc. Natl. Acad. Sci. U.S.A.* **91**, 2031-2035.
66. Richardson, H., Lew, D. J., Henze, M., Sugimoto, K., and Reed, S. I. (1992) Genes Dev. **6**, 2021-2034.
67. Fitch, I., Dahmann, C., Surana, U., Amon, A., Nasmyth, K., Goetsch, L., Byers, B., and Futcher, B. (1992) *Mol. Cell. Biol.* **3**, 805-818.
68. Surana, U., Amon, A., Dowzer, C., McGrew, J., Byers, B., and Nasmyth, K. (1993) *EMBO J.* **12**, 1969-1978.
69. Surana, U., Robitsch, H., Price, C., Schuster, T., Fitch, I., Futcher, A. B., and Nasmyth, K. (1991) *Cell* **65**, 145-161.
70. Amon, A., Irniger, S. and Nasymth, K. (1994) *Cell* **77**, 1037-1050.
71. Seufert, W., Futcher, B., and Jentsch, S. (1995) *Nature* **373**, 78-81.
72. Irniger, S., Piatti, S., Michaelis, C., and Nasmyth, K. (1995) *Cell* **81**, 269-278.
73. Xiao, Z., McGrew, J. T., Schroeder, A. J., and Fitzgerald-Hayes, M. (1993) *Mol. Cell. Biol.* **13**, 4691-4702.
74. Lamb, J. R., Michaud, W. A., Sikorski, R. S., and Hieter, P. A. (1994) *EMBO J.* **13**, 4321-4328.
75. Glotzer, M., Murray, A. W., and Kirschner, M. W. (1991) *Nature* **349**, 132-138.
76. King, R. W., Peters, J. M., Tugendreich, S., Rolfe, M., Hieter, P., and Kirschner, M. W. (1995) *Cell* **81**, 279-288.
77. Tugendreich, S., Tomkiel, J., Earnshaw, W., and Hieter, P. (1995) *Cell* **81**, 261-268.
78. Byers, B., and Goetsch, L. (1974) *Cold Spring Harbor Symp. Quant. Biol.* **38**, 123-131.

79. Hartwell, L. H., Mortimer, R. K., Culotti, J., and Culotti, M. (1973) *Genetics* **74,** 267-286.
80. Hartwell, L. H., and Smith, D. (1985) *Genetics* **110,** 381-395.
81. Dawson, I. A., Roth, S., and Artavanis-Tsakonas, S. (1995) *J. Cell Biol.* **129,** 725-737.

chapter 13
Suc1: cdc2 affinity reagent or essential cdk adaptor protein?

Lee Vogel[1] and Blandine Baratte

CNRS - Station Biologique, BP 74, 29682 Roscoff cedex, France.
[1] To whom correspondence should be addressed

CKS proteins, for which the original member, p13^{suc1}, was identified as a suppressor of cdc2 alleles in *S. Pombe*, have long served as a reagent for the purification of p34^{cdc2}, whereas their biological function has remained elusive. Apparently conflicting data derived from different model systems may indicate a diversity of function for these proteins. Several new observations in yeast and *Xenopus* egg extracts together with new structural information tends to enhance the hypothesis that CKS proteins function to alter the activity of cdc2 at several important points in the cell cycle. Here we review previous observations and recent data that suggest CKS proteins serve as adaptor proteins that modify the functions of cdc2 throughout the cell cycle.

INTRODUCTION

The reader is either familiar with cell cycle control enzymes or is now aware, through reading of other chapters in this book, that the cell growth and division cycle is precisely controlled by a multi-component enzyme circuit. Many of the component proteins in this circuit are physically associated with one another either as regulatory and catalytic subunits of a single enzyme or as enzyme and substrate. As will be detailed below, another category of interaction that must be included is an association between an enzyme and a co-factor of ill-defined function.

Recent lessons from other cellular signalling mechanisms indicate that eukaryotic cellular control circuits are often composed of multi-enzyme complexes in which physical association and/or co-localization is promoted by highly specific protein association domains. Often these domains are repeated in several different proteins within the same control circuit. (e.g. Rel/dorsal homologies and ankyrin repeats in Nf-κB and related proteins (6)). This may also involve small proteins with two distinct binding domains capable of bridging two proteins in an enzyme complex.(e.g. α and β-γ subunits of G-protein complexes (25), SH2, SH3 domains in molecules such as GRB2 involved in tyrosine kinase signalling pathways (38,43)). The following will examine the thesis that a similar concept can be observed within the molecular mechanisms governing control of the cellular division cycle.

As for many questions pertaining to cell cycle regulation, this particular question originates from a mutant phenotype of *Schizosaccharomycys pombe*. According to the original description Suc1$^+$ was isolated as a gene capable of suppressing several temperature sensitive alleles of cdc2 in *S. pombe* (24). Shortly thereafter, the protein coding sequence of Suc1$^+$ was identified (26), allowing a more detailed analysis of Suc1 function. A clear phenotype was observed for *S. pombe* strains strongly overexpressing Suc1$^+$ (26), with elongated cells suggesting a delay in cell cycle progression. The subsequent isolation of a close Suc1$^+$ homologue, Cks1, as a suppressor of Cdc28 in *Saccharomyces cerevisiae* provided the second genetic clue that these genes were intimately involved in the cell cycle control program (22). The *S. cerevisiae* protein has also stolen some of the limelight from *S. pombe* as Suc1 homologues have become known generally as CKS proteins (for CDC28 Kinase Subunit). The Suc1 protein has been shown to form a tight physical association with cdc2 (1,9), a property extensively exploited for cdc2 purification. Yet, other than providing a useful function as a purification reagent, the precise molecular function of CKS proteins remains a subject of considerable speculation.

p13^{suc1} THEN AND NOW

Much of the speculation concerning the function of CKS proteins has focused on regulation of cdk's. It is therefore pertinent to recall some of the key features of cdk regulation. For many question pertaining to biochemistry of the cell cycle, the logical starting point is amphibian and invertebrate eggs and oocytes. At present Xenopus eggs provide, arguably, the best experimental model for biochemical studies of the cell division cycle. Interphase extracts from Xenopus eggs are very sensitive to exogenously added active MPF activity. The quantity of active MPF activity required to induce cell cycle progression is much less than the endogenous level of MPF observed in vitro in M-phase extracts. This suggests that the addition of active MPF may serve to activate the endogenous MPF pool (15). What then is inhibiting the endogenous MPF in Xenopus eggs? Cyclins are not the limiting factor as cyclin accumulation occurs throughout interphase (31,34). Furthermore, by mid-interphase in sea urchin eggs enough cyclin has accumulated that cycloheximide addition does not block entry into mitosis (47). By extrapolation, other unidentified protein factors are also in sufficient supply, at least by mid-interphase in sea urchin eggs. From these and other data, it became clear early on that the factor limiting activation of MPF was likely to be a postranslational modification. This, of course, opens a Pandora box of complicated possibilities. Within the context of regulatory phosphorylation events, the current state of knowledge for the cell cycle can be briefly summarised by the following. The ATP binding site of cdc2 contains a tyrosine at position 15 that negatively regulates kinase activity when phosphorylated (17). An adjacent threonine 14 site is also phosphorylated in higher eukaryotes (35). Contrary to the tyrosine 15 and threonine 14 sites, cdc2 and cdk2 contain threonine residues at positions 161 and 167 respectively, whose phosphorylation is required for activation (21,35,39). The crystal structure of cdk2 has revealed that this phosphorylation site resides on a helix loop (T-loop) that occludes the substrate binding site (11). Although cyclin A binding is not required for Thr167 phosphorylation of cdk2 (12), structural studies have shown that cyclin A induces conformational changes that expose the substrate binding cleft and position the T-loop for potentially more efficient phosphorylation (27). In contrast to cdk2, cdc2 appears to require cyclin B association prior to phosphorylation of Thr161 (14). Regardless of the order of events, it is likely that cyclin binding stabilises the active conformation, suppressing dephosphorylation and down-regulation of the kinase activity. Given that $p13^{suc1}$ forms a tight association with $p34^{cdc2}$, is has been suggested that inhibition of cdc2 kinase activity could arise through a physical masking of regulatory phosphorylation sites (17,24,28,36). Theoretically such an inhibitory action of $p13^{suc1}$ could be accounted for by restricting access of phosphatases to the phosphorylated Tyr15/Thr14 sites. Similarly, $p13^{suc1}$ binding could manifest an inhibitory activity by restricting the access of CAK to the non-phosphorylated Thr161/167 site. Consistent with the first model, it has been observed that $p13^{suc1}$ blocks the entry into mitosis of Xenopus egg extracts (16,17). Additionally, $p13^{suc1}$ was observed to inhibit Xenopus extract cdc2 kinase activity by interfering with the dephosphorylation of tyrosine 15 (17). An inhibitory function of $p13^{suc1}$ was also observed at mitosis in mammalian cells (40). Although this represents an intriguing model, these results are ambiguous for two reasons: first, inhibition of cdc2 requires high concentrations of $p13^{suc1}$ both in Xenopus extracts and in microinjected mammalian cells. Second, microinjection of $p13^{suc1}$ into mammalian cells produces a cell cycle arrest and cell morphology that are indistinguishable from the cell cycle arrested morphology observed when $p13^{suc1}$ is depleted by microinjection of anti-$p13^{suc1}$ antibodies (40). Furthermore, recent structural information from CksHs1-cdk2 co-crystals demonstrates that CksHs1 interacts with the C-terminal lobe of cdk2, at considerable distance from Tyr15 (8). Consequently, a direct inhibitory function of $p13^{suc1}$ appears unlikely.

AND WHAT ABOUT YEAST ?

The picture of $p13^{suc1}$ function derived from fission yeast experiments is somewhat different from that of Xenopus. In fact, based solely on the cell cycle defects of CKS mutants, the role of this protein in both fission yeast and budding yeast is consistent with a cdc2 activating function. However, analysis of cdc2 kinase activity in several CKS deletion mutants shows that absence of CKS protein results in increased cdc2 kinase activity (4,32,46). Consistent with this, $p13^{suc1}$ had no effect on the S. pombe in vitro cdc2 protein kinase assay (32). As previously mentioned, cells overexpressing $Suc1^+$ demonstrate an elongated morphology suggesting a delay in cell cycle progression (26). Initial experiments with $p13^{suc1}$ deletions clearly demonstrated the essential nature of this protein for cell viability, but the complicated morphology of arrested cells precluded the designation of $p13^{suc1}$ function to a particular cell cycle stage (23). A subsequent, more detailed investigation of the $p13^{suc1}$ deleted phenotype using anti-tubulin staining revealed that $p13^{suc1}$ appears to function late in mitosis near the anaphase/telophase transition (32). Surprisingly, the initiation of G2 cdc2 kinase activity in $p13^{suc1}$ deleted mutants occurs at the same time as for wild type cells. The difference for $p13^{suc1}$ deleted cells is seen in the atypical prolongation and augmentation of cdc2 kinase activity at the end of G2 instead of the normal G2 phasic peak of cdc2 kinase activity. What do these results tell us? Primarily that the use of different model systems is unlikely to result in a consensus on CKS function. Whereas Xenopus eggs may require a CkS protein to delay the onset of maturation, the function of $p13^{suc1}$ in S. pombe appears to be downstream of cdc2 activation as an indirect inactivator of cdc2 later in mitosis.

In addition to functional differences between Xenopus eggs and yeast, there is also evidence that CKS proteins can function independently at different phases of the cell cycle. In S. cerevisiae, temperature sensitive mutants of Cks1 cause asynchronously dividing cells to pause in G1, then arrest with a 2N complement of DNA in G2 when shifted to a partially restrictive temperature (46). If cells were first synchronised in the presence of mating pheromone then shifted to a completely restrictive temperature, no DNA replication was observed. This tight block in G1 indicates an absolute requirement for Cks1 at the G1-S boundary. In contrast, cells prearrested in S-phase with hydroxyurea were capable of completing DNA synthesis at the restrictive temperature following hydroxyurea release. These cells then arrested in G2 with an extended morphology. Regardless of the stage of presynchronisation and cell cycle block induced by shifting to the restrictive temperature, Cks1 mutants were unable to undergo bud formation. This defect in bud formation is interesting in light of the observation that both commitment to S-phase and bud formation are thought to be driven by the G1 cyclin activated form of cdc28 (30,42). In contrast, the transition from G2 to M is mediated by the cyclin B activated form of cdc28 (20,41,45). Together these observations indicate that Cks1 interacts with different forms of cdc2 performing diverse functions throughout the cell cycle. Consistent with this hypothesis are results obtained in *S. pombe* with $p13^{suc1}$ under the control of a thiamine-conditional promoter (4). As observed for *S. cerevisiae*, conditional suppression of $p13^{suc1}$ resulted in an increase in cdc2 kinase activity. The increased kinase activity appeared to be due to an increased synthesis of the B-type cyclin cdc13, while the level of cdc2 synthesis remained constant. A smaller increase in the activity of cdc2-cig1 kinase complex was also detected without a detectable increase in the synthesis of the cig1 cyclin. Unlike the *S. cerevisiae* system, no effect on G1 progression was observed for *S. pombe* $p13^{suc1}$ mutants under restrictive conditions. Conversely, a heterogeneous morphology of blocked cells, with a majority having condensed chromosomes, indicated a requirement for $p13^{suc1}$ during M-phase.

FROM STRUCTURE TO FUNCTION

Being a small protein, expressed well in recombinant systems, and relatively easy to purify, the CKS family of proteins have enjoyed considerable recent success in solution and crystal structure definition. The first of the CKS structures to be determined was for CKShs2 and was rather surprising in that it formed a hexamer derived from three interlocking dimers (36). However, the subunit structure of CKShs2 is described as an equilibrium between monomers, dimers and hexamers, where the hexameric form is dependant upon non-physiological pH and divalent cation concentrations. Given the ability of human and fission yeast CKS homologues to complement the cks1 deletion in *S. cerevisiae*, conservation of higher order structures is also expected. Consequently, it is surprising that hexamers have not been detected for CKShs1 (2), $p13^{suc1}$ (18), or the cks homologue from *Physarum* (5). The single fold monomeric form appears to be the common denominator among the cks homologues whose structure has been determined. However, while the $p13^{suc1}$ structure was initially determined as a monomer, a conformational switch leading to a b-strand exchanged dimer similar to CKShs2 was subsequently observed (7), indicating the possible functional significance of the dimeric form of CKS homologues. Finally, the recent co-crystal structure of cdk2 with CKShs1 has eliminated much of the speculation on binding sites and significance of CKS higher order structure (8). This latter work describes the interaction of monomeric CKShs1 with cdk2 and demonstrates that the CKS residues important for binding are sterically inaccessible for the b-interchanged dimer fold. It is suggested, however, that the dimeric form may be physiologically relevant for self-regulation of CKS binding (8). Another important aspect of the cdk2-CKS co-crystal structure as well as a prior publication on the structure of CKShs1, is the structural similarity between CKShs1 and the cdk2 N-lobe domain. The N-lobe domain of cdk2 is involved in ATP binding (11). The binding of phosphate and phosphate analogues by Ckshs1, the structural similarity with the cdk2 N-lobe and the structural conservation of the phosphate binding residues (Lys11,Arg20,Ser51, Trp54, Arg71) in Ckshs1, together support the functional importance of this site. Furthermore, the cdk2-CKShs1 co-crystal structure places the Ckshs1 phosphate binding pocket within a solvent exposed surface opposite to the cdk binding surface. Consequently, with two opposed binding surfaces, CKS homologues have the physical requirements to act as adaptors between cdk's and substrates or additional effector molecules. Finally, as mentioned, the position of CKS binding on cdk2 is sufficiently distant from the regulatory tyr-15 site to indicate that CKS binding does not interfere with dephosphorylation (17,24,28). This has also been confirmed by *in vitro* dephosphorylation of cks bound cdc2 by recombinant cdc25 phosphatase (3). Curiously, dephosphorylation of cdc2/cyclin B by cdc25 is inhibited by addition of the *Xenopus* CKS homologue (37). However, this assay appears to utilise cdc2/cyclin B purified with fission yeast suc1-agarose then eluted with a high concentration of *Xenopus* CKS homologue. Consequently, cdc2/cyclin B is already bound stoichiometrically with CKS protein and dephosphorylation by cdc25 is not inhibited. It is only upon adding additional *Xenopus* CKS protein that inhibition of cdc25 is observed. This leaves the possibility that the molar excess of *Xenopus* CKS protein (or another factor in the *Xenopus* CKS preparation) is directly or indirectly inhibiting cdc25.

```
           M                                     I YS   Y  D     EYRHV  LP
CKS1       MYHHYHA FQGRKLTDQERARVLEF  QDSIHYSPRYSDDNYEYRHVMLP
Suc1       M SK SG VPRLLTASERERLEPFID QIHYSPRYADDEYEYRHVMLP
CksPv      MSAR_____QIYYSDKYFDEDFEYRHVMLP
SF-P9      MSQK_____DIYYSDKYYDDVYEYRHVMLP
CksHs1     MSHK_____QIYYSDKYDDEEFEYRHVMLP
CksHs2     MAHK_____QIYYSDKYFDEHYEYRHVMLP
Xe-P9      MSYK_____NIYYSDKYTDEHFEYRHVMLP
Cks-phy    MPRD_____TIQYSEKYYDDKFEYRHVILP
Lmm-Cks1   MPAKPAQDFFSLDANGQREALIIIKKLQCKILYSDKYYDDMFEYRHVILP
                   10        20        30        40        50

                              E EWR   G   QS GW   Y  H PEPH
CKS1       KAMLKVIPSDYFNSEVGTLRILTEDEWRGLGITQSLGWEHYECHAPEPHI
Suc1       KAMLKAIPTDYFNPETGTLRILQEEEWRGLGITQSLGWEMYEVHVPEPHI
CksPv      KDIAKMVPKN_____HLMSEAEWRSIGVQQSHGWIHYMKHEPEPHI
SF-P9      KSIAKMVSKD_____RTMSEDEWRGIGVQQSQGWVHYMSHKPEPHI
CksHs1     KDIAKLVPKT_____HLMSESEWRNLGVQQSQGWVHYMIHEPEPHI
CksHs2     RELSKQVPKT_____HLMSEEEWRRLGVQQSLGWVHYMIHEPEPHI
Xe-P9      KELAKQVPKT_____HLMSEEEWRRLGVQQSLGWVHYMIHEPEPHI
Cks-phy    PDVAKEIPKN_____RLLSEGEWRGLGVQQSQGWVHYALHRPEPHI
Lmm-Cks1   KDLARLVPTS_____RLMSEMEWRQLGVQQSQGWVHYMIHKPEPHV
                   60        70        80        90       100

                     F
CKS1       LLFKRPLNYEAELRAATAAAQQQQQQQQQQQQQQQQHQTQSISNDMQVPPQIS
Suc1       LLFKREKDYQ   MK FS   QQRGG
CksPv      LLFRRKVTGQ
SF-P9      AV RI
CksHs1     LLFRRPL_____PKK_____PKK
CksHs2     LLFRRPL_____PKD_____QQK
Xe-P9      LLFRRPL_____PKD_____QQK
Cks-phy    LLFRREVPNPAASLSHNP
Lmm-Cks1   LLFKRPRT
                   110       120
```

Figure 1. Alignment of amino acid sequences of CKS1(*S. cerevisiae*) (22), suc1 (*S. pombe*) (26), CkxPv (*p. vulgata*) (10), CksHs1/2 (*H. sapiens*) (41), Xe-P9 (*Xenopus*) (37), Cks-phy (*Physarum polycephalum*) (5), and Lmm-Cks1 (*Leishmania mexicana*) (33). Amino acids that are conserved between the nine proteins are indicated above the CKS1 sequence. Deletions occurring in higher eucaryotes are underscored.

FUNCTIONAL DIVERSITY

Data from diverse models concerning the role of CKS proteins in regulation of the cell cycle is often conflicting or ambiguous. Even in the two different yeast species there is disagreement as to whether CKS proteins function strictly in G1-S progression or in G2-M as well. Given the high degree of sequence conservation (fig. 1) among the CKS proteins from evolutionary distant organisms it is difficult to envision substantial species differences in the role of these proteins. We prefer to adhere to the basic principle that has emerged since the first purification and characterisation of MPF; fastidious conservation of the principal mechanisms governing the cell cycle. Within this concept we see the differences and ambiguities concerning CKS function as resulting from a combination of both different experimental parameters and model systems. The interpretation of results from a rapidly dividing single cellular organism (yeast) will naturally be difficult to align with those of an invertebrate oocyte designed to precisely arrest in metaphase or undergo scincitial divisions following fertilisation. This problem is magnified for CKS proteins where there is no direct parameter to follow (phosphorylation, enzyme activity or degradation) as there is for cdk's, phosphatases and cyclins. Nevertheless, a pattern is emerging and together with structural complex information we can begin to speculate a little more seriously into the true cellular function of this interesting little cdc2 affinity reagent.

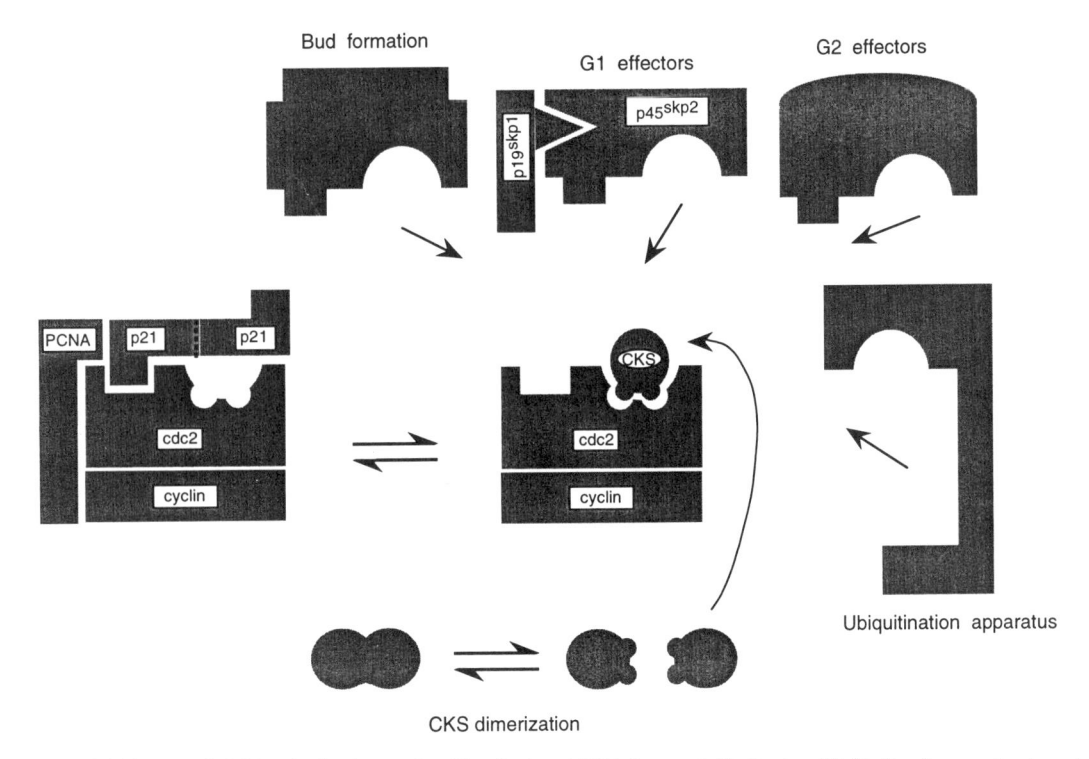

Figure 2. Modification of cdc2 function by cks proteins. Dimerization of CKShs1 prevents binding to cdk2 (8). Also, in normal quiescent fibroblasts the endogenous inhibitor p21 (binding at greater than unit stiochiometry) and/or PCNA may interfere with binding of monomeric CKS to cdk2/cyclin-A (48). Binding of monomeric CKS proteins to cdc2 (also cdk2 and cdk3) provides an additional binding surface which, together with binding of different cyclins, may diversify the function of cdk's for phosphorylation and/or interactions with various substrates or effectors. These may include effectors of bud formation, G1-S transition which may also include effectors of cellular transformation such as p19^{skp1} and p45^{skp2} (48), G2-M transition and ubiquitination.

Cellular and biochemical data for CKS proteins indicate disparity in two general areas. First, a conflict with respect to an activating or an inhibitory function. Second, an ambiguity as to the cell cycle phase for which CKS proteins have a regulatory role. If we adhere to the principle of functional conservation, then globally these results indicate a diversity of function for CKS proteins. Just as the functional diversity of cdk's is emerging, we should expect that these enzyme complexes will require the means to facilitate such functional differences. One manner in which to impose functional differences is to regulate the levels of required co-factors. While virtually identical at the primary structural level, CKShs1 and CKShs2 nevertheless show striking difference in their 3-dimensional form (2), an indication of functional differences. Furthermore, human CKShs1 mRNA levels oscillate with the cell cycle, showing peaks in G1 and G2, whereas CKShs2 transcripts have one peak in G2 (41).

Another means of enhancing functional diversity of cdk's is through association with so-called adaptor molecules, proteins that either alter an existing binding domain or provide an additional binding surface. A role for CKS proteins in the modification of substrate specificity for cdc2 has been suggested (41,46). It has even been suggested that p13^{suc1} could be a phosphotyrosine binding protein with structural similarities to the phosphotyrosine binding SH2 domain of p60^{c-src} (19). However, the only experimental evidence for a modification of substrate specificity is the p13^{suc1} induced attenuation of *in vitro* p34^{cdc2} kinase activity toward intermediate filament proteins (29). Therein lies a critical problem not just for the functional characterisation of CKS proteins but for the entire cell cycle field, a serious lack of identified *in vivo* substrates for cdk's. Obviously, without a knowledge of *in vivo* substrates it will be impossible to verify a substrate targeting role for CKS proteins.

Clearly, CKS proteins play a role in the regulation of the cell cycle and likely function at several phases. If these proteins are involved in the normal regulation of the cell cycle then there will likely also be examples where they are involved in deregulation. Normal human fibroblasts contain cdk2 complexed with cyclin A, p21 and PCNA (49). In transformed cells this complex is replaced by cdk2-cyclin A associated with p9$^{CkShs1/hs2}$ and two other proteins required for G1-S transition, p19^{Skp1}

and p45^{Skp2} (48). This suggests a role for CKS proteins in cell cycle deregulation associated with cellular transformation. Possibly, CKS proteins are required before cdk2-cyclin A can bind the Skp proteins. Consistent with this idea is the observation that when cell division is inhibited by TGF-β, a TGF-β specific down regulation of CKShs1 transcripts is also observed (44).

Yet another targeting role that may require the binding of CKS proteins to the cdk complex, is the process of dissociation and destruction of cyclins. Recent evidence has indicated that binding of cks to cdc2-cyclin B in *Xenopus* extracts is required for the ubiquitination and eventual destruction of cyclin B that marks the end of metaphase (37).

We have seen that CKS proteins appear to participate with cdk's in divers functional pathways of cell cycle regulation (fig. 2). These include control points involving transition of the G1-S boundary, transition of the G2-M boundary, mechanisms involved in *S. cerivisiae* bud formation and mechanisms for cyclin degradation. Each of these pathways certainly involves complexing with different substrate and effector molecules. With the aid of recent structural information we have advanced the hypothesis that CKS proteins serve as adaptor proteins to augment the functional diversity of cdk's (cdc2, cdk2 and cdk3). While a paucity of known cdk substrates makes it difficult to verify a substrate modification function for CKS proteins, a few carefully designed experiments should be able to address the question of whether or not CKS proteins can function as adaptors. For example, we have seen that deletion of p13^{suc1} in *S. pombe* results in a dramatic increase in the level of the B-type cyclin cdc13 (4). It is also apparent that a feedback control loop may exist between cdc2/cyclin-B and the regulation of cyclin-B transcription/translation (13). If this feedback control is regulated by cdc2/cyclin-B mediated phosphorylation of an effector protein whose interaction with cdc2 is mediated by CKS proteins, then it should be possible to demonstrate an interaction between these proteins. An interaction trap worth trying would be to subject the various CKS proteins to a two-hybrid screen.

ACKNOWLEDGEMENTS

Lee Vogel was supported by a fellowship from the "Fondation pour la Recherche Médicale". A portion of this work was also supported by a grant from the "Association pour la Recherche sur le Cancer" (ARC 6268) and from the "Région Bretagne".

REFERENCES

1. Arion, D., Meijer, L., Brizuela, L. and Beach, D. (1988) *Cell*,55,371-378.
2. Arvai, A.S., Bourne, Y., Hickey, M.J. and Tainer, J.A. (1995) *JMolBiol*,249,835-842.
3. Barette, B., Meijer, L., Galaktionov, K. and Beach, D. (1992) *Anticaner Res*,12,873-880.
4. Basi, G. and Draetta, G. (1995) *MolCell Biol*,15,2028-2036.
5. Birck, C., Raynaud-Messina, B. and Samama, J.P. (1995) *FEBS Lett*,363,145-150.
6. Blank, V., Kourilsky, P. and Israel, A. (1992) *TIBS*,17,135-140.
7. Bourne, Y., Avrai, A.S., Bernstein, S.L., Watson, M.H., Reed, S.I., Endicott, J.E., Nobel, M., Johnson, L.N. and Tainer, J.A. (1995) *Proc Natl Acad Sci*,92,10232-10236.
8. Bourne, Y., Watson, M.H., Hickey, M.J., Holmes, W., Rocque, W., Reed, S.I. and Tainer, J.A. (1996) *Cell*,84,863-874.
9. Brizuela, L., Draetta, G. and Beach, D. (1987) *EMBO J*,6,3507-3514.
10. Colas, P., Serras, F. and Van Loon, A.E. (1993) *IntJDevBiol*,37,589-594.
11. Debondt, H.L., Rosenblatt, J., Jancarik, J., Jones, H.D., Morgan, D.O. and Kim, S-H (1993) *Nature*,363,595-602.
12. Desai, D.Y.Gu and Morgan, D.O. (1992) *MolBiolCell*,3,571-582.
13. Dirick, L., Bohm, T. and Nasmyth, K. (1995) *EMBO*, 14, 4803-4813.
14. Ducommun, B., Brambilla, P., Felix, M.A., Franza, B.R.,Jr., Karsenti, E. and Draetta, G. (1991) *EMBO J*,10,3311-3319.
15. Dunphy, W.G. and Newport, J.W. (1988) *JCell Biol*,106,2047-2056.
16. Dunphy, W.G. and Newport, J.W. (1988) *Cell*,55,925-928.
17. Dunphy, W.G. and Newport, J.W. (1989) *Cell*,58,181-191.
18. Endicott, J.A., Noble, M.E., Garman, E.F., Brown, N., Rasmussen, B., Nurse, P. and Johnson, L.N. (1995) *EMBO J*,14,1004-1014.
19. Endicott, J.A. and Nurse, P. (1995) *Structure*,3,321-325.
20. Fitch, I., Dahmann, C., Surana, U., Amon, K., Nasmyth, K., Goetsch, L., Byers, B. and Futcher, B. (1992) *MolBiolCell*,3,805-818.
21. Gould, K.L., Moreno, S., Owens, D.J., Sazer, S. and Nurse, P. (1991) *EMBO J*,10,3297-3309.
22. Hadwiger, J.A., Wittenberg, C., Mendenhall, M.D. and Reed, S.I. (1989) *MolCell Biol*,9,2034-2041.
23. Hayles, J., Aves, S. and Nurse, P. (1986) *EMBO*,5,3373.
24. Hayles, J., Beach, D., Durkacz, B. and Nurse, P. (1986) *MolGenGenet*,202,291-293.
25. Helper, J.R. and Gilman, A.G. (1992) *TIBS*,17,383-387.

26. Hindley, J., Phear, G., Stein, M. and Beach, D. (1987) *MolCellBiol*,**7**,504-511.
27. Jeffrey, P.D., Russo, A.A., Polyak, K., Gibbs, E., Hurwitz, J., Massague, J. and Pavletich, N.P. (1995) *Nature*,**376**,313-320.
28. Jessus, C., Ducommun, B. and Beach, D. (1990) *FEBS Lett*,**266**,4-8.
29. Kusubata, M., Tokui, T., Matsuoka, Y., Okumura, E., Tachibana, K., Hisanaga, S., Kishimoto, T., Yasuda, H., Kamijo, M., Ohba, Y. and et al, (1992) *J Biol Chem*, **267**, 20937-20942.
30. Lew, D.J. and Reed, S.I. (1993) *J Cell Biol*,**120**,1305-1320.
31. Minshull, J., Blow, J.J. and Hunt, T. (1989) *Cell*,**56**,947-956.
32. Moreno, S., Hayles, J. and Nurse, P. (1989) *Cell*,**58**,361-372.
33. Mottram, J.C. and Grant, K.M. (1996) *BiochemJ*,**316**,833-839.
34. Murray, A.W. and Kirschner, M.W. (1989) *Nature*,**339**,275-280.
35. Norbury, C., Blow, J. and Nurse, P. (1991) *EMBO J*,**10**,3321-3329.
36. Parge, H.E., Arvai, A.S., Murtari, D.J., Reed, S.I. and Tainer, J.A. (1993) *Science*, **262**,387-395.
37. Patra, D. and Dunphy, W.G. (1996) *Genes & Dev*, **82**, 915-925.
38. Pazin, M. and Williams, L.T. (1992) *TIBS*,**17**,374-378.
39. Poon, R.Y., Yamashita, K., Adamczewski, J.P., Hunt, T. and Shuttleworth, J. (1993) *EMBO J*,**12**,3123-3132.
40. Riabowol, K., Draetta, G., Brizuela, L., Vandre, D. and Beach, D. (1989) *Cell*,**57**,393-401.
41. Richardson, H.E., Stueland, C.S., Thomas, J., Russell, P. and Reed, S.I. (1990) *Genes Dev*,**4**,1332-1344.
42. Richardson, H.E., Wittenberg, C., Cross, F. and Reed, S.I. (1989) *Cell*,**59**,1127-1133.
43. Schlessinger, J. (1993) *TIBS*,**18**,273-275.
44. Simon, K.E., Cha, H.H. and Firestone, G.L. (1995) *Cell Growth & Diff*,**6**,1261-1269.
45. Surana, U., Robitsch, H., Price, C., Schuster, T., Fitch, I., Futcher, A.B. and Nasmyth, K. (1991) *Cell*,**65**,145-161.
46. Tang, Y. and Reed, S.I. (1993) *Genes Dev*,**7**,822-832.
47. Wagenaar, E.B. (1983) *ExpCell Res*, **144**, 393-403.
48. Zhang, H., Kobayashi, R., Galakitiionov, K. and Beach, D. (1995) *Cell*,**82**,915-925.
49. Zhang, H., Xiong, Y. and Beach, D. (1993) *MolCellBiol*,**4**,897-906.

chapter 14
Structural basis for chemical inhibition of CDK2

Sung-Hou Kim[1], Ursula Schulze-Gahmen, Jeroen Brandsen and Walter Filgueira de Azevedo, Jr.

Department of Chemistry and E.O. Lawrence Berkeley National Laboratory
University of California, Berkeley, California 94720 USA
[1]To whom correspondence should be addressed

The central role of cyclin-dependent kinases (CDKs) in cell cycle regulation makes them a promising target for discovering small inhibitory molecules that can modify the degree of cell proliferation. The three-dimensional structure of CDK2 provides a structural foundation for understanding the mechanisms of activation and inhibition of CDK2 and for the discovery of inhibitors. In this article five structures of human CDK2 are summarised: apoprotein, ATP complex, olomoucine complex, isopentenyladenine complex, and des-chloro-flavopiridol complex.

INTRODUCTION

Cell cycle progression is tightly controlled by the activity of cyclin-dependent kinases (CDKs) (1-3). CDKs remain inactive by themselves, and activation requires binding to cyclins, a diverse family of proteins whose levels oscillate during the cell cycle, and phosphorylation by CDK-activating kinase (CAK) on a specific threonine residue (4, 5). In addition to the positive regulatory proteins such as cyclins and CAK, many negative regulatory proteins (CDK inhibitory proteins, CKIs) have been discovered (6-8) such as p16 (9), p21 (10-13) and p28 (14). Since deregulation of cyclins and/or alteration or absence of CKIs have been associated with many cancers, there is strong interest in chemical inhibitors of CDKs that could play an important role in the discovery of a new family of antitumour agents or antiproliferative drugs.

Since ATP is the authentic cofactor of CDK2 it can be considered as a "lead compound" for discovery of CDK2 inhibitors. However, there are two major concerns: adenine containing compounds are common ligands for many enzymes in cells, thus, any adenine derivatives may inhibit many enzymes in the cells; second, any highly charged groups such as phosphates in ATP will prevent uptake by the cells. Our studies provide the structural basis for overcoming these two difficulties by appropriate modification of a common base, such as adenine, to endow specificity and cell uptake of the derivatives.

MATERIALS AND METHODS

Materials
Human CDK2 was prepared as described in reference (15). Briefly, Sf9 insect cells were infected with baculovirus containing human CDK2 gene. The supernatant of cell lysate was loaded over a DEAE Sepharose column followed by an S-Sepharose column. The flow-through was loaded onto an ATP-affinity column and eluted by NaCl salt gradient. The purified protein is fully functional in that it binds to cyclin A, the resulting complex can be fully activated when incubated with partially purified human CAK, and phosphorylated CDK2-cyclin A complex can phosphorylate histon H1.

Olomoucine was obtained from Dr. Laurent Meijer of CNRS (Roscoff, France) and can be purchased from Calbiochem (La Jolla, CA), and isopentenyadenine and ATP were purchased from Sigma (St. Louis, MO). Des-chloro-flavopiridol is derived by chemical synthesis from a parent structure obtained from *Dysoxylum binectariferum* (16), a plant native to India. Des-chloro-flavopiridol was provided by Dr. Peter Worland of Mitotix (Cambridge, MA) who obtained it from Dr. Harold Sedlacek, Behringwerke, Marburg, Germany.

Inhibitor soaking and co-crystals
Recombinant CDK2 was crystallised as below (17). A CDK2 solution was concentrated to 10 mg/ml by dialysis against 20 mM Hepes buffer (pH 7.4) and 1 mM EDTA. Sitting drops were equilibrated by vapour diffusion at 4°C against reservoirs containing 200 mM Hepes buffer at pH 7.4. Diamond and wedge-shaped crystals appeared after 2 to 4 days. A gradual increase of the Hepes buffer concentration in the reservoirs (up to 800 mM) produced crystals with average dimensions of about 0.2 mm x 0.3 mm x 0.3 mm. After CDK2 crystals had formed, small amounts of powdered chemical inhibitors were added to the crystallisation drops. All inhibitor complexes were obtained by this soaking procedure (18, 19), except for the olomoucine complex, which was obtained by co-crystallisation of a precomplexed solution of CDK2 and olomoucine. Crystals were soaked usually for 48 hours before data collection. Crystallisation of CDK2 apoprotein was performed using the sparse matrix method (20). Co-crystals of

Table 1. Crystallographic parameters and refinement statistics.

	APO	ATP	Olomoucine	Isopentenyl-adenine	Des-chloro-flavopiridol
Space group	$P2_12_12_1$	$P2_12_12_1$	$P2_12_12_1$	$P2_12_12_1$	$P2_12_12_1$
Cell dimensions (Å)	a=73.12	a=72.82	a=73.77	a=73.25	a=71.30
	b=72.72	b=72.66	b=72.55	b=72.51	b=72.03
	c=54.25	c=54.07	c=54.06	c=54.03	c=53.70
Unique reflections	27,034	22,636	15,044	25,371	11,430
Completeness of data (Å)	∞-1.8 98.6%	∞-1.9 97.3%	∞-2.5 99.0%	∞-2.0 96.4%	∞-2.0 91.7%
	1.9-1.8 97.1%	2.0-1.9 92.8%	2.5-2.2 97.0%	2.0-1.8 84.0%	
R_{sym}[1] (%)	6.1	6.7	8.4	8.3	4.9
Resolution (Å)	8.00-1.8	8.00-1.9	8.00-2.2	8.00-1.8	8.00-2.33
R_{value}[2]	0.18	0.18	0.19	0.20	20.3
R_{free}[3]	0.25	0.27	0.27	0.27	28.8
$B_{average}$[4] (Å^2)	31	38	31	30	32.35
Deviations observed: rmsd from ideal bond length (Å)	0.011	0.012	0.013	0.011	0.012
rmsd from ideal bond angle (°)	1.61	1.78	1.70	1.76	1.7
No. of water molecules	180	108	76	99	84

[1] $R_{sym} = 100 \Sigma \Sigma | I(h) - \langle I(h) \rangle | / \Sigma I(h)$, with $I(h)$, observed intensity and $\langle I(h) \rangle$, mean intensity of reflection h over all measurement of $I(h)$.
[2] $R_{value} = 100 \Sigma |F_o - F_c| / \Sigma (F_o)$, the sums being taken over all reflections with $F/s(F) > 2$ cut off.
[3] $R_{free} = R_{value}$ for 10% of the data, which were not included during crystallographic refinement.
[4] $B_{average}$ = Average B values for all non-hydrogen atoms.

CDK2-olomoucine were obtained under the same conditions as apoenzyme.

X-ray crystallographic studies

Relevant crystallographic parameters and refinement parameters are given in Table 1. X-ray diffraction data were collected using a Rigaku RAXIS-II imaging plate area detector and processed with the RAXIS data processing software. The crystal structure of the apoprotein (17) was determined by the multiple isomorphous replacement method. The flavopiridol complex structure was solved by molecular replacement method using the apo structure as the probe. For other complexes the apoenzyme was the starting model for refinement against complex data. All the structures were refined using X-PLOR 3.0 (21). The refinement statistics are given also in Table 1.

RESULTS

Domain structure and conformation of CDK2

The CDK2 structure is almost identical in CDK-apoenzyme (17), ATP complex (17), olomoucine complex (18), isopentenyladenine complex (18), and des-chloro-flavopiridol complex (19), and very similar to that in CDK2-cyclin A complex (22), except in five regions (see below). The enzyme is folded into the bilobal structure typical for most of protein kinases, with the smaller N-terminal domain consisting predominantly of β-sheet structure and the larger C-terminal domain consisting primarily of α-helices (Fig. 1). There are no significant differences in the domain orientations between the ligand-enzyme complexes and the apoenzyme. In all cases, except CDK2-cyclin A complex, electron density is weak in two regions in the enzyme, spanning residues 36-47 which links the N-terminal domain and "PSTAIR" or cyclin recognition helix and residues 150-164 of the "T-loop" containing the activating phosphorylation site (Fig. 2). All inhibitors and ATP bind in the deep cleft between the two domains. As shown in Fig. 1 most of the conserved residues are located either around ATP-binding pocket or cyclin A binding side of CDK2.

Conformational differences between active and inactive form of CDK2 and a mechanism of activation

Comparison of the structure of CDK2 in ATP complex (17) and the ternary complex of CDK2 with ATP and cyclin (22) shows that there are five significant conformational differences (fig. 2): (1) the two domains are more open in the ternary complex, thus, pushing the phosphate-loop (P-loop) toward the N-terminal domain; (2) the conformation of the triphosphate of ATP and surrounding residues are different; (3) a short α-helix becomes a β-strand in the "α/β transition box" on cyclin A binding; (4) the T-loop containing Thr160, which gets phosphorylated for CDK2 to be fully activated, reorients completely; and (5) "PSTAIR or cyclin binding helix" changes its location and orientation. These differences are schematically shown in Fig. 2.

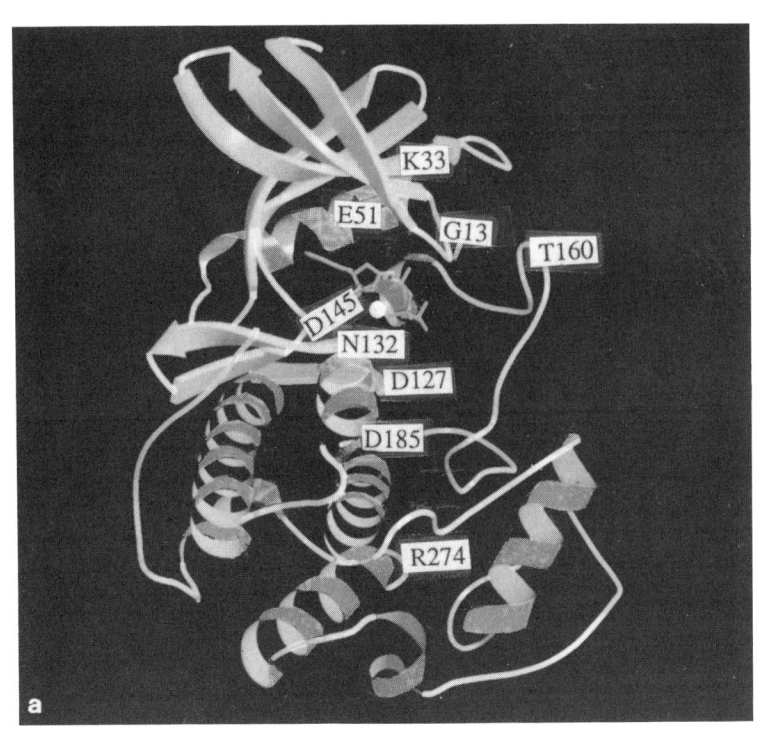

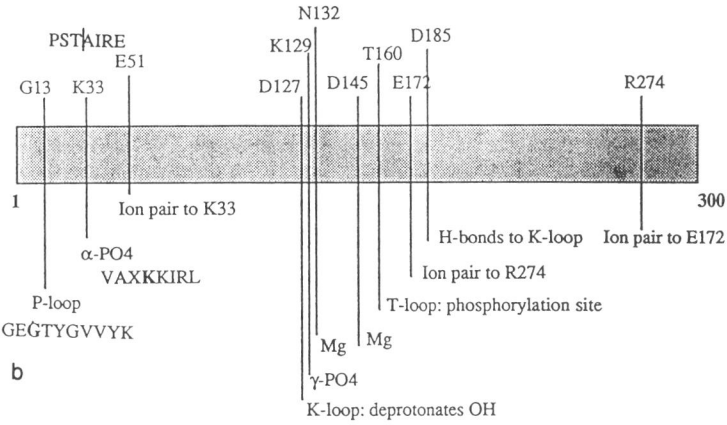

Figure 1. (a, top) The backbone structure of human CDK2. The β-strands and α-helices are represented by arrowed ribbons and helical coils, respectively, and ATP in a stick model and Mg^{++} ion as a white ball. All the connectors between secondary structures are shown as coils. (b, bottom) The conserved residues among CDKs and the structural bases for their conservation as interpreted based on the three-dimensional structure are indicated.

The conformational changes of the residues surrounding ATP, in turn, change the conformation of the triphosphate of ATP so that the scissile bond between β– and γ–phosphates now becomes "in line" with the direction of attacking hydroxyl oxygen of the substrates of CDK2, the necessary alignment for ATP hydrolysis.

Interaction between CDK2 and ligands
ATP

The binding site for ATP as well as the inhibitors is located between the N- and C-terminal domains (Fig. 3a), and the binding of these ligands are associated with conformational changes of surrounding residues (Fig. 3b). The refinement of the CDK2 apoenzyme and ATP complex to 1.8 Å and 1.9 Å, respectively, resulted in very accurate protein models as indicated by low R-values and good stereochemistry. The electron density for the refined CDK2-ATP complex shows clear density for the ATP molecule with all three phosphates, a Mg^{2+} ion, and a few water molecules in the binding pocket. The five-membered ribose ring is puckered into a C2'-*endo* envelope (17). As can be expected from the

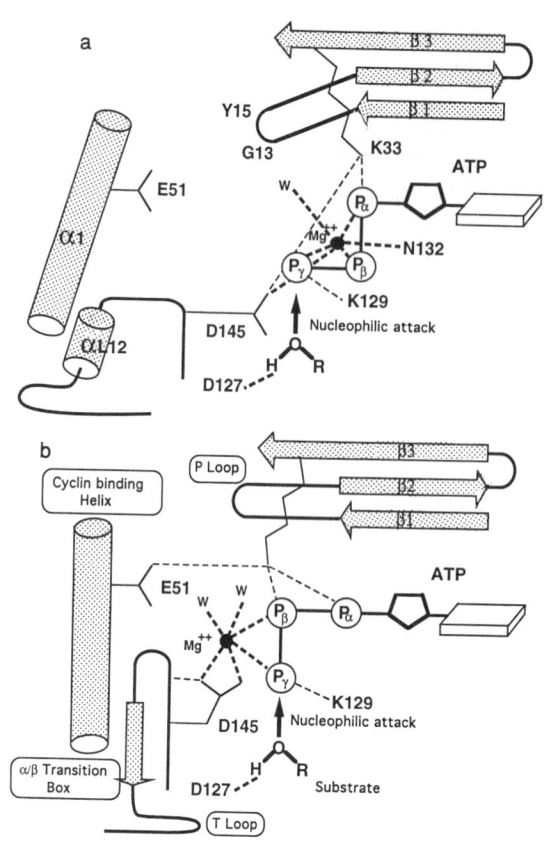

Figure 2. Schematic drawing of the five regions that have different conformations between (a) CDK-ATP binary complex and (b) CDK2-ATP-cyclin A ternary complex structures. They are: triphosphate of ATP, P-loop, cyclin binding helix, α/β transition box, and T-loop.

relative hydrophilicity of the ATP molecule, most hydrogen bonds and salt bridges are formed with the phosphate moieties, while the adenine base and ribose are involved in only three hydrogen bonds each. (Fig 4a). Three water molecules in the binding pocket are also mediating CDK2-ATP interactions. The hexacoordinated Mg^{2+} ion is bound by one oxygen from each phosphate, and one from Asn^{132} and Asp^{145} side-chains and one water.

Adenine derivative inhibitors: olomoucine and isopentenyladenine

Kinase assay of adenine derivatives revealed that two adenine derivatives have inhibitory properties of CDC2 and CDK2, with IC_{50} of 10 μM for olomoucine and 50 μM for isopentenyladenine. Olomoucine inhibits cell cycle kinases (except CDK4) quite specifically, and isopentanyladenine (IPA) inhibits both cell cycle kinases and non-cell cycle kinases (23). The crystal structures of CDK2-IPA complex and CDK2-olomoucine complex have been determined to high resolution (1.8 Å and 2.2 Å, respectively), with good stereochemistry (Table 1). As in the apoenzyme and in the CDK2-ATP complex, electron density is weak for two regions in the enzyme spanning residues 37-43 of interdomain connector and 153-163 of the T-loop containing the activating Thr^{160} phosphorylation site (17).

Both IPA and olomoucine are adenine derivatives (Fig. 5). Although the purine rings bind roughly in the same area of the binding cleft as the adenine ring of ATP, the relative orientation of each purine ring with respect to the protein is different for all three ligands (ATP, IPA, and olomoucine) as shown in Fig. 6. This is probably due to the different size of the substituent groups of the purine in the three ligands; the N6 amino group of the adenine ring is replaced by an isopentenylamino group in IPA and by a bulky benzylamino group in olomoucine. The complete atomic interactions between CDK2 and olomoucine are shown in Fig. 4b, and those between CDK2 and IPA in Fig. 4c.

A flavone inhibitor: des-chloro-flavopiridol

Previous studies have shown that flavopiridol, a novel flavonoid, and not an adenine derivative, can inhibit growth of breast and lung carcinoma cell lines and can inhibit CDK activity at the same nanomolar concentration range of the inhibitor (16). To understand how flavopiridol analogues bind to the ATP-pocket, we determined the X-ray structure of CDK2 in complex with the des-chloro-flavopiridol (DFP, Fig. 5d), and compared it with the X-ray structures of CDK2-ATP complex (17) and other inhibitor complexes of CDK2 (18).

DFP binds in the ATP-binding pocket, with the benzopyran ring occupying approximately the same region as the purine ring of ATP (Fig. 6). The two ring systems overlap in the same plane, but the benzopyran is rotated about 60° relative to the adenine in ATP, measured as the angle between the carbon-carbon bonds joining the two cycles in benzopyran and adenine rings, respectively. In this orientation, the O5 hydroxyl and the O4 of the inhibitor are close to the position of the N6 amino group and N1 in adenine, respectively. The piperidinyl ring partially occupies the α-phosphate pocket and is assigned to a chair conformation (although a boat conformation can not be ruled out in the current resolution map). The detailed interactions between DFP and CDK2 are shown in Fig. 4d.

Differences in protein side-chain conformations in the binding pocket

In contrast to the finding that cyclin A changes very little on CDK2 binding, the conformation of CDK2 appears to be much more adaptable as manifested by its changes on binding of ATP, olomoucine, IPA and DFP, as well as cyclin A (17, 18, 19, 24), To find out whether any inhibitor or ATP binding to CDK2 induces changes in side-chain

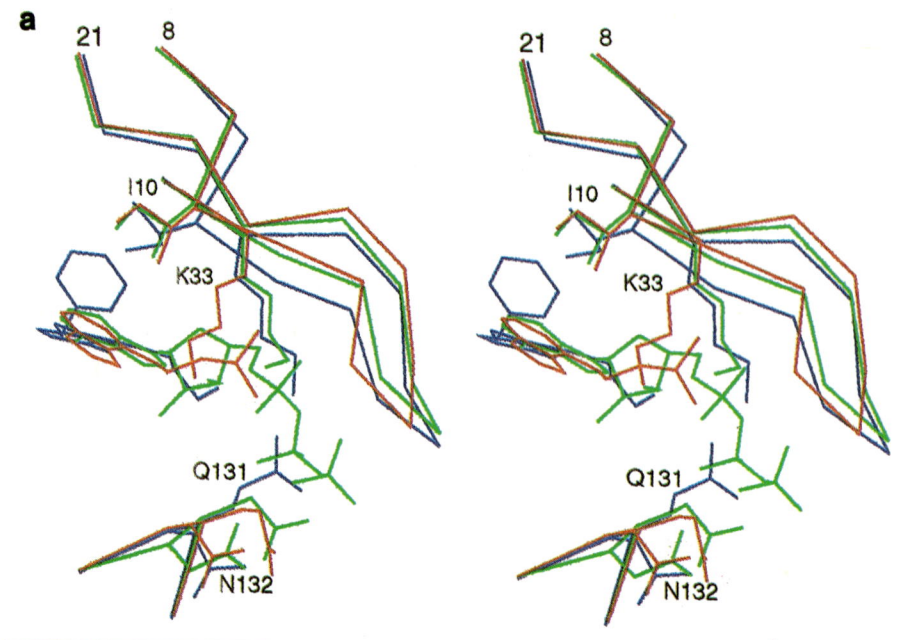

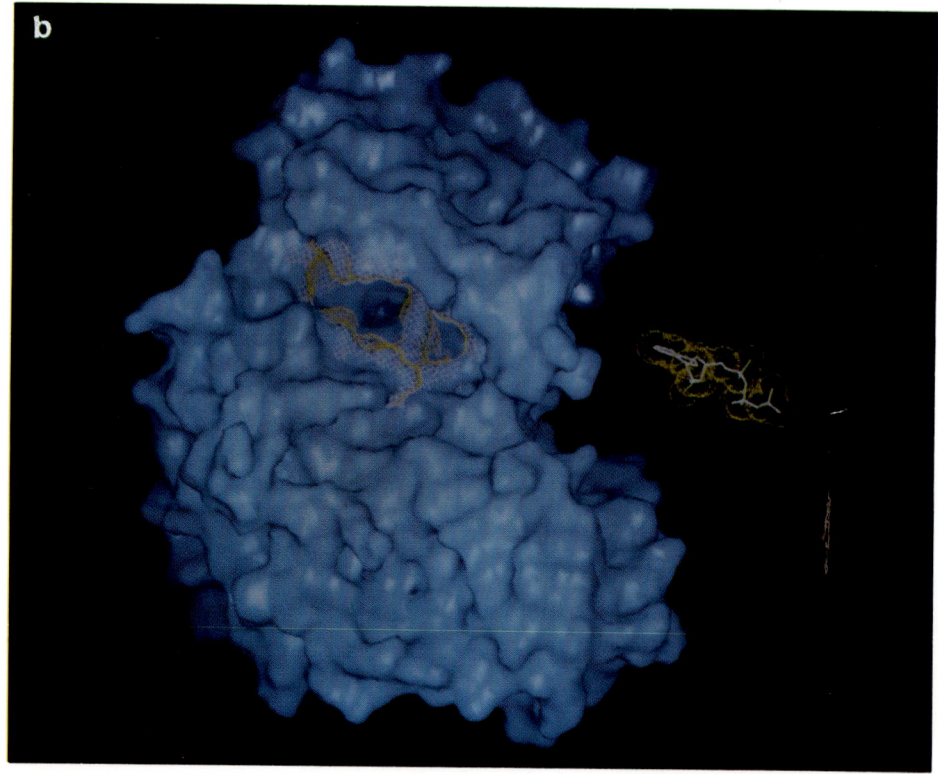

a

21 8
I10
K33
Q131
N132

21 8
I10
K33
Q131
N132

b

conformation in the binding pocket, we compared the binding-pockets of CDK2-ligand complexes and CDK2 apo protein after a superposition of the enzymes based on their Cα atoms (Fig. 3b). This superposition revealed the significant movement of few side-chains. In addition to small changes in the backbone of the P-loop, four residues change their side chain conformations most: Lys^{33}, Gln^{131}, Ile^{10}, and Asn^{132}. Lys^{33} and its equivalent Lys^{72} in cyclic AMP dependent protein kinase are important for ATP binding. In all ligand bound complexes the side-chain of this residue has moved away to accommodate the ligands except in the IPA complex, where the side chain conformation is the same as that of apoenzyme presumably because of the small size of the ligand. In the CDK2-ATP complex, Lys^{33} forms salt bridges with the α-phosphate of ATP and Asp^{145}, a residue involved in ATP-Mg^{2+} binding (17).

The side-chain of Gln^{131} is directed away from the binding pocket in the ATP- and IPA complex. In the olomoucine complex, however, the side-chain points into the binding pocket and forms van der Waals

Figure 3. (a) A stereo view of ATP binding pocket in the presence of ATP (green), olomoucine (blue), and isopentenyladenine (red). Protein residues with significant conformational differences are labelled. (b) Molecular surface of CDK2 and ATP binding pocket. ATP was moved out of the pocket for clarity.

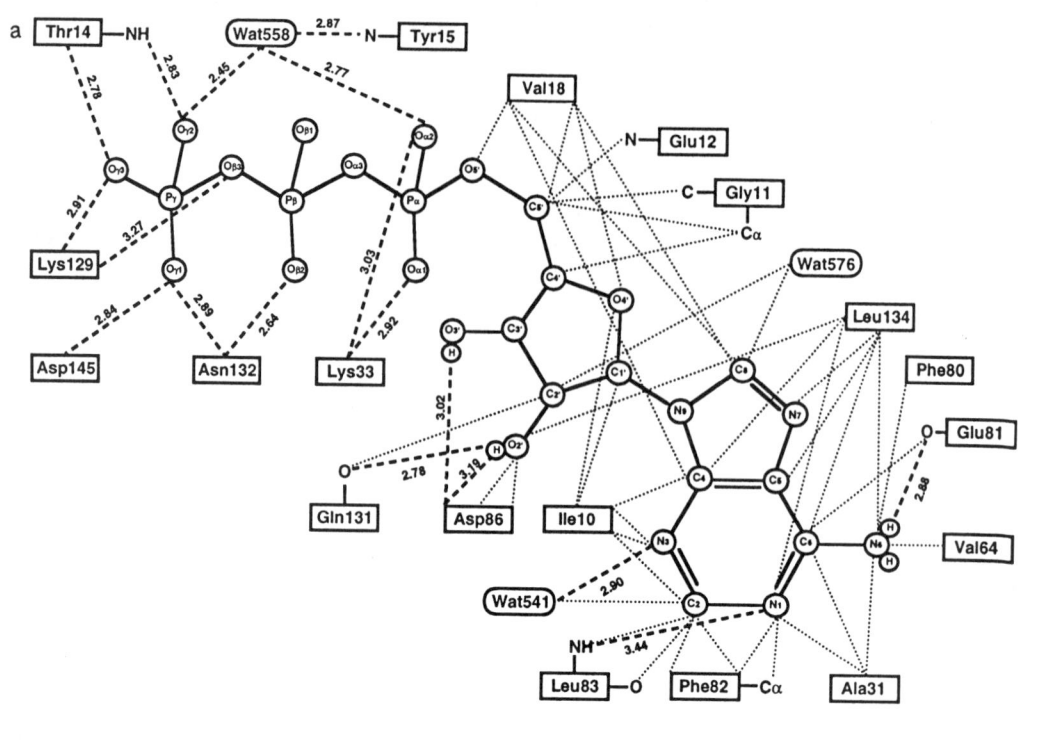

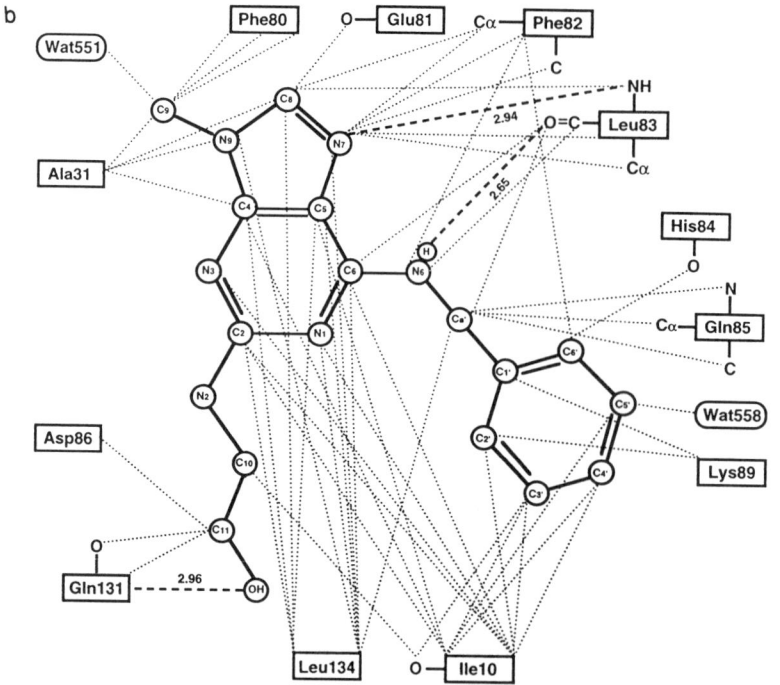

Figure 4.
All atomic interactions between CDK2 and (a) ATP, (b) olomoucine.

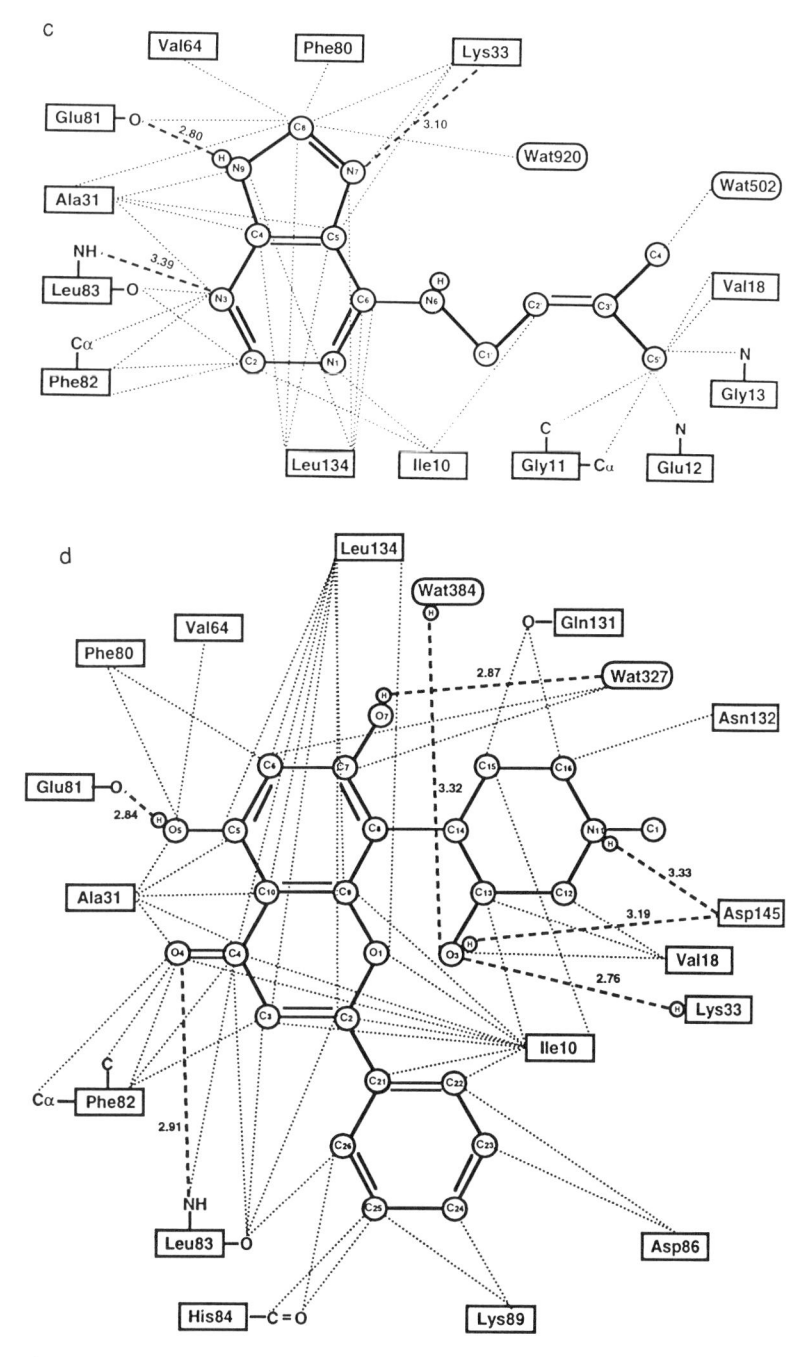

Figure 4. Continued
(c) isopentenyladenine and (d) des-chloro-flavopiridol. Contacts with protein side-chains are indicated by lines connecting to the respective residue box while interactions to main-chain atoms are shown as lines to the specific main-chain atoms indicated. Van der Waals contacts are indicated by dotted lines, and hydrogen bonds by thick broken lines. For the inhibitors all ligand contacts are shown, while for ATP, van der Waals contacts to phosphates were omitted for clarity.

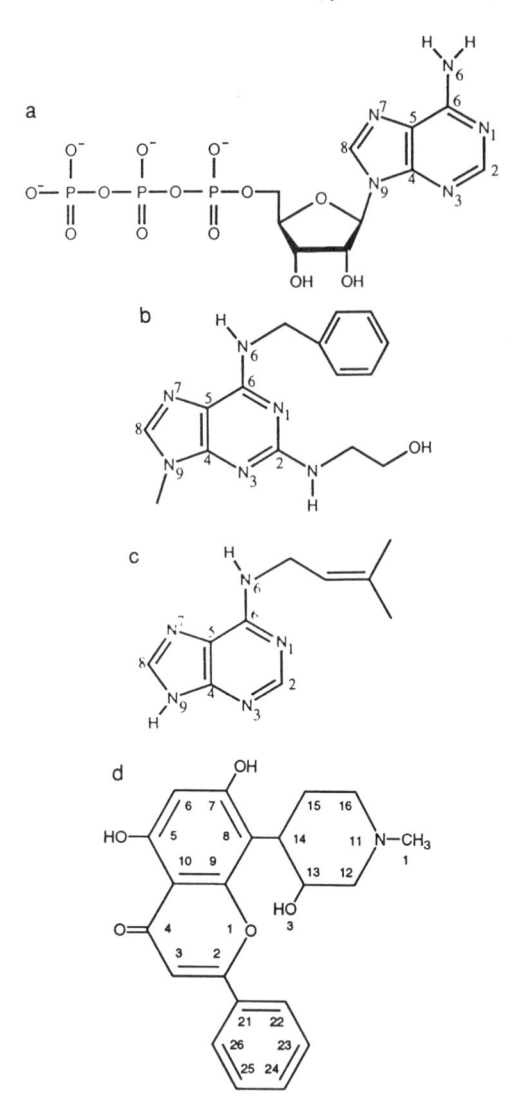

Figure 5. Chemical structures of four ligands of CDK2 discussed in this paper. (a) ATP, (b) olomoucine, (c) IPA and (d) des-chloro-flavopiridol.

contacts and a hydrogen bond with the hydroxyethylamino group of olomoucine. Ile[10] and Asn[132] make smaller conformational changes on the ligand binding. In summary, the observed side-chain differences in the binding pocket seem to contribute to an improved fit between protein and ligand molecules.

DISCUSSION

The high-resolution X-ray structures of human CDK2 apoenzyme and ATP complex as well as CDK2-cyclin A-ATP complex provide the structural basis for understanding the mechanism of activation of CDK2. Furthermore, the structures of the complexes between CDK2 with three inhibitors described here explain the specificity and potency of these inhibitors compared to other related derivatives, and suggest structure-based drug design and combinatorial library design that may enhance the chance of discovering better and improved inhibitors of CDK2 and other CDKs. Several interesting lessons were learned from these studies:

(a) The ATP binding pocket of CDK2 has the surprising capacity to bind varieties of flat aromatic moieties with some unpredictable conformational changes of residues in the pocket. This may be related to the relatively weak binding of the ligands ($K_d \sim \mu M$).

(b) Knowing the orientation of adenine of ATP in the ATP pocket did not allow us to predict how the adenine moiety of olomoucine and IPA would bind to CDK2. Experimental determination of IPA and olomoucine complexes made us realise that any "rational drug design" based on adenine position of ATP would have completely missed in predicting olomoucine and IPA as potential inhibitors.

(c) Specificity of each inhibitor for CDK2 over other kinase comes from the interaction between substitution groups outside of the aromatic scaffold and the peripheral surface of the ATP pocket. This observation dispels the conventional wisdom that "the common base such as adenine cannot be a good scaffold for inhibitor design because there are many proteins in cells that bind adenine derivatives." Our results clearly demonstrate that specificity can be created by appropriate modification of a common scaffold molecule.

(d) The knowledge of the structures of CDK2 complexes with the three inhibitors suggests a few new scaffolds to consider in structure-based drug design and combinatorial library design.

ACKNOWLEDGEMENTS

The work summarised here has been supported by a fellowship to (W.F.A.Jr.) from the Conselho Nacional de Pesquisas (Brazil), an American Cancer Society fellowship (U.S.-G.), and by grants from the Office of Health and Environmental Research, U.S. Department of Energy (DE-AC03-76SF00098) and National Institutes of Health. We thank Dr. Laurent Meijer for olomoucine, and Drs. Peter Worland and Harold Sedalcek for des-chloro-flavopiridol. We also gratefully acknowledge the funding of the research described here by Asahi Chemical Industry Co. Ltd., Japan.

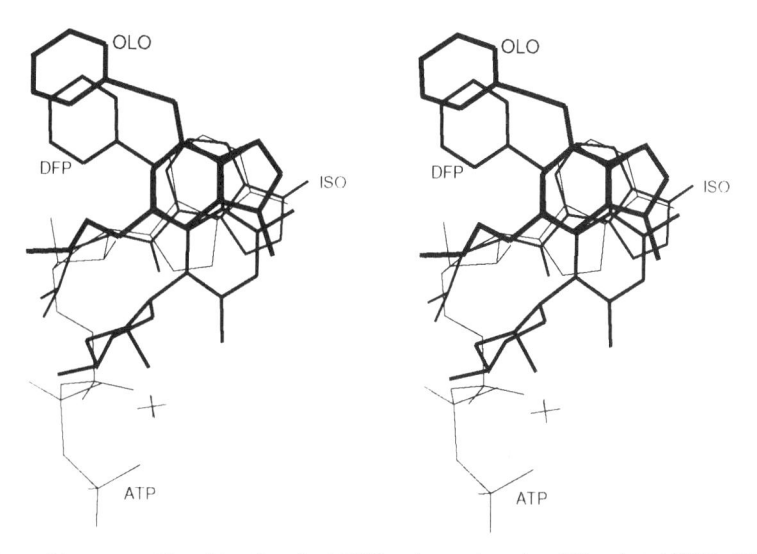

Figure 6. A stereo view of the superposition of four ligands of CDK2 as they are bound to ATP pocket of CDK2: ATP, IPA, olomoucine, and des-chloro-flavopiridol.

REFERENCES

1. Norbury, C. and Nurse, P. (1992) *Annu. Rev. Biochem.* **61**, 441-470.
2. Fang, F. and Newport, J. W. (1991) *Cell* **66**, 731-742.
3. Hunt, T. (1989) *Curr. Opin. Cell Biol.* **1**, 274-286.
4. Desai, D., Gu, Y., Morgan, D. O. (1992) *Mol. Biol. Cell* **3**, 571-582.
5. Gu, Y. Rosenblatt, J. and Morgan, D. O. (1992) *EMBO J.* **11**, 3995-4005.
6. Richardson, H. E., Stueland, C. S., Thomas, J. Russel, P. and Reed, S. I. (1990) *Genes Dev.* **4**, 1332-1344.
7. Serrano, M., Hannon, G. J. and Beach, D. (1993) *Nature,* **366**, 704-707.
8. Peter, M. and Herskowitz, I. (1994) *Cell* **79**, 181-184.
9. Nobori, T., Miura, K., Wu, D. J., Lois, A., Takabayashi, K. and Carson, D. A. (1994) *Nature,* **368,** 753-756.
10. Gu, Y., Turck, C. W. and Morgan, D. O. (1993) *Nature,* **366**, 707-710.
11. Xiong, Y., Hannon, G. J., Zhang, H., Casso, D., Kobayashi, R. and Beach, D. (1993) *Nature,* **366**, 701-704.
12. Harper, J. W., Adami, G. R., Wei, N., Keyomarsi, K. and Elledge, S. J.(1993) *Cell,* **75**, 805-816.
13. Dulic, V., Kaufmann, W. K., Wilson, S. J., Tisty, T. D., Lees, E., Harper, J. W. and Elledge, S. J. (1994) *Cell,* **76**, 1013-1023.
14. Hengst, L., Dulic, V., Slingerland, J. M., Lees, E. and Reed, S. I. (1994) *Proc. Natl. Acad. Sci. USA* **91**, 5291-5295.
15. Rosenblatt, J., De Bondt, H., Jancarik, J., Morgan, D.O. and Kim, S.-H. (1993) *J. Mol. Biol.* **230**, 1317-1319.
16. Losiewicz, M. D., Bradley, A. C., Kaur, G., Sausville, E. A. and Worland, P.J. (1994) *Biochem. Biophys. Res. Commun.* **201**, 589-595.: Kaur, G., Stetler-Stevenson, M., Seber, S., Wordland, P., Sedlacek, H., Myers, C., Czech, J., Naik, R. and Sausville, E. (1992) *J. Natl. Cancer Inst.* **84**, 1736-1740.
17. De Bondt, H. L., Rosenblatt, J., Jancarik, J., Jones, H. D., Morgan, D. O. and Kim, S.-H. (1993) *Nature,* **363**, 595-602.
18. Schulze-Gahmen, U. Brandsen, J. Jones, H. D., Morgan, D. O., Meijer, L., Vesely, J. and Kim, S.-H. (1995) *Proteins: Structure, Function, and Genetics* **22**, 378-391.
19. Azevedo, W.F.,Jr., Mueller-Diechmann, H.-J., Schulze-Gahmen, U., Worland, P.J., Sausville, E., and Kim, S.-H. (1996) *Proc. Natl. Acad. Sci. USA,* (in press).
20. Jancarik, J. and Kim, S.-H. (1991) *J. Appl. Cryst.* **24**, 409-411.
21. Brünger, A. T. (1991) *X-PLOR, a system for crystallography and NMR,* version 3.0, Yale University Press, New Haven, CT.
22. Jeffrey, P. D., Russo, A. A., Polyak, K., Gibbs, E., Hurwitz, J., Massagué, J.and Pavletich, N. P. (1995) *Nature,* **376**, 313-320.
23. Vesely, J., Havlicek, L., Strand, M., Blow, J.J., Donella-Deana, A., Pinna, L., Letham, D.S., Kato, J.-Y., Detivaud, L., Leclerc, S., and Meijer, L. (1994) *Eur. J. Biochem.* **224**: 771-786.
24. Owen, D. J., Noble, M. E., Garman, E. F., Papageorgiou, A. C., Johnson, L. N. (1995) *Structure* **3**, 467-482.

chapter 15
Apoptosis and the cell cycle

Rati Fotedar, Ludger Diederich and Arun Fotedar[*]

Institut de Biologie Structurale J. -P. Ebel, 41 ave. des Martyrs, 38027 Grenoble cedex 1, France
[*]La Jolla Institute for Allergy and Immunology, La Jolla, California 92037, USA

Apoptosis is a genetically controlled response by which eukaryotic cells undergo programmed cell death. This phenomenon plays a major role in developmental pathways (1), provides a homeostatic balance of cell populations, and is deregulated in many diseases including cancer. Control of cell number is determined by an intricate balance of cell death and cell proliferation. Accumulation of cells through suppression of death can contribute to cancer and to persistent viral infections, while excessive death can result in impaired development and in degenerative diseases. Identification of genes that control cell death, and understanding of the impact of apoptosis in both development and disease has advanced our knowledge of apoptosis in the past few years. There appears to be a linkage between apoptosis and cell cycle control mechanisms. Elucidating the mechanisms that link cell cycle control with apoptosis will be of key importance in understanding tumour progression and designing new models of effective tumour therapy.

REGULATORS OF APOPTOSIS: INDUCERS, EXECUTIONERS AND INHIBITORS OF APOPTOSIS

Apoptosis (2) is characterised by cytoskeletal disruption and decreased adhesion, condensation of cytoplasm and nucleus with accompanying cell shrinkage and membrane blebbing. Within the nucleus, the chromatin condenses along the nuclear periphery and the chromatin subsequently undergoes enzymatic degradation to form a characteristic DNA ladder when examined by gel electrophoresis (DNA fragmentation reviewed in 3). The dying cell does not leak its contents but instead fragments into membrane bound apoptotic bodies. The debris from dead cells is then rapidly phagocytosed and digested by macrophages. Apoptosis differs from necrosis, another form of cell death that results from acute cellular injury (reviewed in 4). During necrosis, cells swell and a loss of cell membrane results in the release of cytoplasmic contents, provoking an inflammatory response. In this aspect necrosis contrasts with apoptosis, where dying cells are cleared without evoking an inflammatory response.

In the past few years significant progress has been made in the identification of genes that regulate apoptosis (Table 1). Although apoptosis is triggered by many different and distinct signals in a cell type-specific manner, it appears that all these signals may converge to activate a common execution pathway.

Fas: A death receptor!
The immune system eliminates potentially autoreactive peripheral T lymphocytes and autoreactive B lymphocytes in bone marrow through apoptosis. Cytotoxic T cells eliminate virus infected cells through apoptosis. In these cases, killing can be triggered via molecules expressed on the surface of an activated cell which can either act to induce apoptosis in the same cell (suicide) or on target cell bearing a death receptor. The cell surface protein expressed on the target cell is Fas (also known as APO-1 or CD95) which acts as the receptor for Fas ligand (FasL) (reviewed in 5, 6). FasL is constitutively expressed in many cell types, but is induced upon activation in lymphocytes. When FasL binds to Fas, the target cell undergoes apoptosis (7-10).

Fas is expressed in a variety of tissues and cells (see 11 and references therein). Fas mRNA is expressed in mouse liver, heart, ovary, muscle and thymus but not in brain, bone marrow or spleen. Hepatocytes, endothelial cells, keratinocytes and epithelial cells in the intestine express Fas. Resting lymphocytes contain low or undetectable levels of Fas but greatly increase its level upon activation. Fas induced death is partially inhibited by overexpression of Bcl-2 (12, 13), and completely inhibited by the combination of Bcl-2 and Bag (14) although it appears that Bcl-2 and Fas regulate distinct pathways in lymphocytes (15). Fas mediated death involves an ICE like protease (15-19). The ICE family consists of cysteine proteases which share homology with the death inducing *ced-3* gene product of *Caenorhabditis elegans* (20; see below). Thymocytes from ICE[-/-] mice are resistant to Fas mediated apoptosis (21), while dexamethasone and γ irradiation induced death is normal (22).

Mice with autoimmune disorders caused by spontaneous mutations in Fas and FasL have been

Table 1. Regulators of Apoptosis.

Gene product	Inducers
TNF-R1	Cell surface receptor. Binds to TNFα and promotes apoptosis. Member of TNF-family of proteins which includes Fas. Shares homology with a conserved cytoplasmic region of Fas termed the death domain.
Fas/Apo/CD95	Cell surface receptor for Fas ligand (FasL). Promotes apoptosis when bound to FasL. Cells from ICE$^{-/-}$ mice are resistant to Fas mediated apoptosis. Apoptosis is inhibitable by CrmA. Fas expression is inducible by p53.
Death domain proteins	TRADD, FADD, RIP... interact with TNF-R1 or Fas via death domain. Transient expression in cells induces apoptosis. May convey the death-signal generated by TNF-R1 or Fas to the death machinery.
Reaper, Hid	*Drosophila* proteins that show homology to death domain receptors of Fas/TNF. Loss of gene suppresses all forms of programmed cell death in *Drosophila*. Death is inhibitable by baculovirus p35 protein or ICE protease inhibitory peptides
ICE family	Cysteine proteases that induce apoptosis. Homologues include *C. elegans* ced-3 gene product. Substrates include IL-1β, PARP, Gas2, UI-70. Physiological substrates unknown.
Ced-4	*C.elegans* gene product. No mammalian homolog. Ced-4 mediated death is inhibited by Ced-9, a homolog of Bcl-2.
Bcl-2 family	Bcl-2 is prototype of this family (see below), which is characterized by three conserved domains that mediate dimerization among the various members and are required for regulation of apoptosis. Bak, Bax, Bcl-x$_S$, Bad and Bik1/Nbk are mammalian representatives of this family. Bax, Bad and Bak bind to Bcl-2 and Bcl-x$_L$.
	Inhibitors
p35	Baculovirus (*Autographa californica*) protein which inhibits developmentally programmed cell death in *C. elegans* and *Drosophila*. Inhibits apoptosis in mammalian cells. Binds to ICE family members and inhibits their function.
CrmA	Protein encoded by cowpox virus. Blocks ICE-mediated apoptosis.
Dad1	Human and *C. elegans* dad-1 gene products suppress developmentally programmed cell death
survival factors	Growth factors and cytokines (e.g. IGF-1) that promote all survival. Mitogenic signalling by IGF-1 is inhibited by IGF-BP3, a p53 inducible gene.
Bcl-2 family	Members in mammalian cells include Bcl-2, Bcl-x$_L$, Mcl-1, A1. Bak can be an inhibitor or an inducer of apoptosis. T cells from Bcl-2$^{-/-}$ mice are resistant to glucocorticoids, radiation, anti-CD3 etc.. Bcl-2 and Bcl-x$_L$ inhibit p53 mediated cell death. Viral homolog E1B 19K is more effective than Bcl-2 in inhibiting Fas mediated death.

characterised (reviewed in 6, 23). These mice develop lymphoadenopathy and splenomegaly and produce autoantibodies and symptoms similar to systemic lupus erythematosus. The genetic defect in *lpr* mice has been identified as the insertion of a retroviral transposon into the second intron of the Fas gene, leading to premature transcript termination and reduced levels of Fas. Defective apoptosis in *lpr* mice results in aberrant depletion of mature T cells. Mice with mutations in the *gld* locus develop a syndrome indistinguishable from *lpr* mice. The defect in *gld* mice is a single amino acid change in FasL. Fas$^{-/-}$ mice exhibit hyperplasia of lymph nodes and of the spleen as a result of accumulation of T cells (24). In contrast, the thymus does not show any hyperplasia. Hyperplasia of liver in Fas$^{-/-}$ mice appears to be a consequence of lack of Fas mediated death.

Fas is a member of a growing family of receptors called TNF family. Other members of the family include TNF-R (Tumour Necrosis Factor receptor), CD30 (cell surface glycoprotein) and CD40 (B cell surface antigen)(see 6 for a review). Recently, like Fas and TNF receptors, CD30 was shown to be involved in cell death (25). While Fas and TNF mainly effect apoptosis in mature T cells, CD30 appears to effect apoptosis in thymocytes (25). CD30 deficient mice exhibit impaired negative selection (25). Genetic analysis in *Drosophila melanogaster* have led to the identification of a gene, *reaper*, that suppresses almost all forms of programmed death (26). *Reaper* shares limited homology with the "death" domain found in proteins of the TNF family including Fas (reviewed in 27). The death domain is a conserved region of the cytoplasmic domain of Fas and TNF-R1, which is critical for mediating cell death. The gene is expressed in cells destined to die and deletion of reaper abolishes all cell death in the embryo. *Reaper* appears to activate a death programme mediated by the ICE protease family, as death due to *reaper* can be prevented by ICE inhibitors such as a baculovirus p35 protein (28) or by specific protease inhibitory peptides (28, 29).

It is not known how the FasL-Fas interaction activates apoptosis. The conserved region of the cytoplasmic domain of Fas and TNF-R1, termed the death domain is critical for mediating cell death. This domain is involved in protein-protein interactions and a number of proteins that interact with this cytoplasmic domain of Fas have recently been identified (reviewed in 30, 31). These proteins include TRADD, FADD, RIP and FAF-1 (32) each of which, when transiently expressed in cells, can induce apoptosis. It is speculated that the death domain interactions with these ligands activate signal transduction pathways, with CAP1-4 (33) as prime candidates, that feed into the death machinery.

Bcl-2 family: The ying yang regulators!

The bcl-2 gene is a proto-oncogene which is constitutively activated upon translocation in human B cell follicular lymphomas. Unlike other proto-oncogenes, Bcl-2 extends cell survival (34) by preventing death in a variety of cell types in response to diverse apoptotic stimuli (reviewed in 35, 36). *C. elegans* gene, *ced-9*, which protects cells from programmed cell death is a functional homolog of the bcl-2 gene (37). Bcl-2 protein is localised to the mitochondria, endoplasmic reticulum and nuclear membrane (for a review see 35, 38). Bcl-2 is expressed widely in foetal tissues. In adult tissues, its expression becomes limited to lymphoid tissue, hematopoeitic precursors, and neurons in the peripheral nervous system (39, 40). Bcl-2$^{-/-}$ mice progress through development but with severe defects such as polycystic kidney disease, dramatic loss of mature B and T cells, distortion of the small intestine due to death of progenitor cells at the base of the crypt (41). These mice also exhibit hypopigmentation of hair, probably due to death of melanocytes. Bcl-2 expression in T cells leads to increased resistance to apoptosis induced by glucocorticoids, radiation, anti-CD3, phorbol myrstate acetate (PMA) and ionomycin (15, 42-44). Bcl-2 suppresses negative selection to a variable degree (see reviews 45, 46).

Bcl-2 is a prototype of a growing multi-gene family whose members in mammalian cells include Bcl-x, Mcl-1, Bax, Bak, Bad, A1 and Bik1. Bcl-2 family members interact with each other via conserved regions and these interactions are thought to regulate apoptosis (most recently reviewed in 47). Interestingly, the different Bcl-2 family members can act as either upregulators or downregulators of apoptosis. Bcl-x$_L$ like Bcl-2 is an inhibitor of apoptosis, and either protein can protect cells from p53-mediated apoptosis (48, 49). Proteins such as Bax, Bad, Bik1, that interact with Bcl-2 or Bcl-x$_L$, act to accelerate apoptosis. Bax and Bak can act as both inducers and inhibitors of apoptosis depending on cell type (50, 51). The outcome of the interaction among various members of the Bcl-2 family appear to depend on the ratio of the death promoter to the death suppressor.

The phenotypes of mice deficient for each of several Bcl-2 family members are now known. Bcl-x deficient mice die at day 13 of embryogenesis and death is proceeded by extensive apoptosis of neurons and hematopoietic cells (52). In contrast to bcl-2 deficient mice, in which there are defects in maturation of the peripheral T cell population, bcl-x$^{-/-}$ mice are deficient in immature T cells. Bax deficient mice are viable but display hyperplasia of thymocytes and B cells (51). In addition, the testis display atrophy and the male mice are infertile. Loss of bax therefore appears to result in hyperplasia or hypoplasia depending upon the cell type. Despite extensive work that has led to the identification of these proteins, establishment of protein-protein interactions, and analysis of the phenotypes of various knock-out mice, the biochemical function of Bcl-2 and related proteins remains to be established.

ICE-related proteases: The executioners!

In *Caenorhabditis elegans*, two genes essential for cell death (*ced*; <u>c</u>ell <u>d</u>eath abnormal), *ced-4* and *ced-3* have been identified. The function of *ced-4* remains unknown but *ced-3* codes for a protein that is 29% identical to human <u>i</u>nterleukin 1β <u>c</u>onverting <u>e</u>nzyme (ICE) (53). ICE is a cysteine protease that cleaves interleukin 1β (IL-1β) converting it into the mature form (54). Point mutations in the region homologous between ICE and *ced-3* abolish the ability of either gene product to induce apoptosis (53). The X-ray crystal structure of ICE reveals that residues essential for substrate specificity are conserved between *ced-3* and ICE (55, 56). Overexpression of either *ced-3* or ICE in mammalian cells induces apoptosis that is inhibitable by Bcl-2 (57). CrmA, a specific inhibitor of ICE encoded by Cowpox virus (58), inhibits death mediated by ICE proteases (57, 59, 60). CrmA also inhibits apoptosis induced by TNFα (61) and by Fas (16, 61). The baculovirus (*Autographa californica*) p35 protein inhibits developmentally programmed death in *C. elegans* (62) and *Drosophila* (63). Furthermore baculovirus p35 protein inhibits apoptosis in mammalian cell lines, and has been shown to bind with multiple members of the ICE family to inhibit their function (64, 65).

PCR cloning using the active site residues (QACRG) conserved between ICE and *ced-3* has lead to the identification of other ICE-related genes. These include: Nedd-2/Ich1, TX/ICE$_{rel}$-II/Ich-2, ICE$_{rel}$-III, Mch-2 and CPP32, indicating the existence of a multigene family (66 and references therein). It is not clear why there are so many members of the family. There may be selectivity of expression in different tissue and cell types or each enzyme may have a substrate preference.

ICE$^{-/-}$ mice develop normally without any developmental defects but exhibit defects with respect to Fas-mediated apoptosis (22). Other apoptotic responses, such as dexamethasone or irradiation induced apoptosis of thymocytes, occur normally in ICE$^{-/-}$ mice (21, 22). It is likely that multiple members of the ICE family lead to a redundancy so that loss of one member of the family may not significantly impact apoptosis or development. Together with the demonstrated ability of ICE-specific antisense oligonucleotides to reduce Fas mediated death, these observations suggest that ICE has an essential role in Fas dependent death pathway.

The physiological substrates of the ICE family of proteases remain elusive. Some members of this family can participate in self-amplifying cascades or they can activate other members of the family. For example, ICE can activate CPP32 (67) and TX can activate ICE (68). The substrates of ICE protease family include IL-1β (54), poly(ADP)ribose polymerase (PARP) (67, 69-72), nuclear lamin (73), U1-70 (74, 75), Gas 2 (76), DNA dependent protein kinase (77) and Protein Kinase C (78). How the cleavage of the substrates feeds into the death machinery remains an intriguing mystery. Further, it is not known if protease cleavage of any of the known substrates is essential for apoptosis. The substrates of the ICE protease family appear to be distributed throughout the cell and both cytoplasmic and nuclear proteolytic pathways may be necessary for apoptosis.

CELL CYCLE REGULATION

Recent evidence indicates that there is a link between apoptosis and the cell cycle. Loss of cell cycle checkpoint control or DNA damage leads to apoptosis. Coordination between the regulatory pathways controlling the cell cycle and apoptosis is therefore important. This section offers an overview of the cell cycle in order to provide a basis for understanding the molecular nature of this coordination.

Cyclin dependent kinases regulate progression through cell cycle

In higher eukaryotes, cell cycle progression is controlled by a family of protein kinases that share sequence similarity to cdc2 and are designated cyclin dependent kinases or cdk's (reviewed in 79). Microinjection studies have established that cdk2 regulates G1 to S phase transition (80, 81) whereas mammalian cdc2 regulates G2 to M transition (see 82). Overexpression of dominant negative mutants of cdk2 and cdk3 arrest the cells in G1, whereas cdc2 arrests cells in G2/M (83). In contrast, mutant cdk4, cdk5 and cdk6 do not have any effect on the cell cycle by this analysis.

The activity of cdk's is regulated in part by their association with distinct cyclins which are synthesised and subsequently destroyed at different times in the cell cycle. In mammalian cells, the key regulators of G1 progression are cyclins D and E while cyclin A functions later, during S phase (reviewed in 84). D type cyclins which associate with cdk4 or cdk6, are most closely linked to the regulation of the Restriction point (reviewed in 85). There are three D type cyclins (D1, D2 and D3) and the relative expression of the three cyclins may be tissue specific. The expression of D type cyclins is growth factor inducible and the kinase activity of the complex peaks during late G1. Cyclin E combines with cdk2 and this kinase activity peaks at the G1/S transition, after the D type cyclins but prior to A type cyclins. Overexpression of cyclin D (86, 87) or cyclin E (87) accelerates entry into S phase. The cdc2-cyclin B kinase is thought to activate a cascade of protein kinases during mitosis which directly act on proteins which control nuclear envelope breakdown and cytoskeletal rearrangement.

Regulation of cyclin dependent kinases

The timely activation of different cdk's is crucial to regulation of the cell cycle. For example, cdk2 is inactive when Thr 161 is dephosphorylated or Tyr14 and Thr15 are phosphorylated through the activation of the wee1/mik1 related kinases (reviewed in 89). Activation of cdk2 first requires association with cyclin and phosphorylation of a conserved Thr residue (Thr 161 in mammalian cells) by a cdk2 activating kinase, CAK. Interestingly, CAK has been shown to be comprised of a complex of MO15, a cdk homolog, with cyclin H. Final activation follows the de-phosphorylation of Thr 15 and Tyr 14 probably through the action of a protein phosphatase, cdc25 (reviewed in 89). Inactivation of cdks occurs through a decrease in cyclin levels, either due to a decrease in transcription of cyclin or to its specific proteolytic degradation.

Cyclin Dependent Kinase Inhibitors (CKIs)

Recently, it was discovered that cdk activity is modified by additional regulatory proteins (reviewed in 90-92). These proteins bind directly to either the cdk or the cdk-cyclin complex and inhibit their kinase activity. Interestingly, some of these proteins can inhibit previously activated cdks. Cdk inhibitors have also been described in yeast (reviewed in 90). In mammalian cells two distinct families of cyclin-cdk kinase inhibitors, p21 and p16, have been described. The p21 family includes structurally related proteins, $p21^{WAF-1/Cip1}$ (93-97), $p27^{Kip1}$ (98, 99), and $p57^{Kip2}$ (100, 101) all of which inhibit a variety of cdk-cyclin kinases by binding activated cyclin-cdk complexes. Deletional analysis of p21 reveals that the N-terminal of p21 binds to cdk2-cyclin complex and is sufficient to inhibit cdk-cyclin kinase activity *in vitro* (84, 102-107). N-terminal residues of p21, p27 and p57 share significant homology, and have been shown to bind cdk2 and inhibit cdk2-cyclin A kinase activity. p27, p21 and p57 differ in their C terminal regions. Although p21 expression is regulated by p53 upon DNA damage, p21 can also be expressed in a p53-independent manner (108, 109).

The p16 family is structurally unrelated to the p21 family. Its members, p14/p15 (110, 111), p16 (112), p18 (110) and p19 (113, 114) preferentially inhibit cdk4-cyclin D and cdk6-cyclin D kinases. These proteins inhibit kinase activity by binding to the cdk subunit and competing for the binding of the regulatory cyclin subunit to the catalytic cdk subunit.

Tumour suppressor genes and entry into S phase

The products of tumour suppressor genes, Rb and p53, act as negative regulators of cell growth (reviewed in 115). Rb (retinoblastoma protein) binds to and negatively regulates a subgroup of E2F transcription factors (E2F-1, -2, -3) whose function is to activate genes required for DNA replication. The phosphorylation of Rb at the G1 to S transition down-regulates its growth inhibitory effects. Rb is phosphorylated by cyclin D-cdk4 and cyclin D-cdk6, and a role has also been suggested for cyclin E-cdk2 in phosphorylation of Rb (115). E2F-1, thus relieved from inhibition, activates various genes associated with G1 to S phase progression including, cyclins D, E and A (116, 117). Regulation of transcription of cyclin genes by E2F-1 may thus control the progression of cells through G1 to S phase. Physiological growth inhibitory signals such as TGF-β, cyclic AMP and contact inhibition mediate G1 arrest by preventing phosphorylation of Rb. These signals activate a CKI that in turn can prevent the cdk-cyclin kinase from phosphorylating Rb (111, 118-121). Alternatively, as in the case of TGF-β, there is a dramatic reduction in the levels of cdk4 in some cell types (122).

p53 arrests cells in G1 following DNA damage and functions as a transcriptional activator of genes induced in response to DNA damage (reviewed in 123). The induced genes may be involved in cell growth delays or cell cycle checkpoints. An inhibitor of cdk activity, p21 (94-96) is inducible by p53 (93). Cell lines containing wild type p53 induce p21 upon irradiation while cells containing mutant p53 do not (93). These findings have led to the proposal of a model in which p21 acts as an effector of cell cycle arrest in response to the p53 checkpoint pathway (93, 124, 125). p21 is regarded as the primary mediator of p53-induced G1 growth arrest following DNA damage. G1 arrest is possibly mediated by binding of p21 to cdk2-cyclin E or A kinases, thereby decreasing the protein kinase activity required for G1 to S phase transition (124).

CELL CYCLE PROGRESSION, TUMOUR SUPPRESSORS AND APOPTOSIS

Cyclin dependent kinases and apoptosis

Activation of cyclin dependent kinases has been observed in many forms of apoptosis. Apoptosis induced in a T cell hybridoma by activation with anti-CD3 or concanavalin A occurs in G2/M phase and is accompanied by an inappropriate induction of cdc2 and cyclin B associated kinase activity (126). Apoptosis induced in these activated T cells is inhibited by antisense cyclin B oligonucleotides (126). Apoptosis induced in target cells by fragmentin-2, a granule serine protease produced by natural killer cells, is accompanied by activation of cdc2 dependent kinase activity, and apoptosis can be blocked by a peptide substrate of cdc2 dependent kinase activity (127). Ectopic expression of PITSLRE, a cdc2 related gene, induces apoptosis in Chinese hamster ovary cells (128), although PITSLRE has not been shown to be involved in cell cycle progression. Induction of apoptosis via the Fas pathway was shown to be associated with an increase in PITSLRE kinase activity that could be suppressed by serine protease inhibitors.

Apoptosis in HeLa cells blocked in S phase and treated with chemicals known to elicit premature mitosis is also accompanied by elevated levels of cdk2, cdc2 and cyclin A proteins as well as elevated kinase activity (129). Apoptosis induced by overexpression of *myc* is associated with elevated levels of cyclin A mRNA (130). Increase in cyclin A and E associated kinase activity has been observed in HIV-1 Tat induced apoptosis in T cells (131).

Activation of cyclin dependent kinases is however not a requirement for all forms of apoptotic death. Antisense cyclin B oligonucleotides prevent apoptosis in T cells activated by anti-CD3 but has no effect on dexamethasone induced apoptosis in the same cell type (126). Cdc2 kinase activity is not involved in apoptosis in fibroblasts following serum withdrawal (132) or apoptosis of thymocytes following treatment with etoposide or dexamethasone (133).

Transforming growth factor-β1: cell cycle arrest and apoptosis

TGF-β is a multifunctional cytokine involved in controlling cell cycle progression, cell differentiation and morphogenesis (reviewed in 134). In an inherited form of colon cancer known as hereditary nonpolyposis colorectal cancer (HNPCC), genetic instability caused by mismatch repair has been shown to lead to mutations in TGF-β (135). TGF-β1 inhibits proliferation of a variety of cells by preventing the phosphorylation of Rb leading to a G1 arrest (118). G1 cyclin dependent kinases regulate the phosphorylation of Rb (115). Cell cycle arrest by TGF-β1 may be mediated through inhibition of expression of G1 cyclins (136) or of cdk4 (122), with a resulting inhibition of cdk-cyclin kinase activity. The expression of cyclin kinase inhibitors, p15 (111) and p21 (137), is upregulated by TGF-β1. Recently, TGF-β1 has been shown to transcriptionally activate both p15 (138) and p21 (139).

While several investigators have shown that TGF-β1 induces apoptosis, the effect of TGF-β1 on

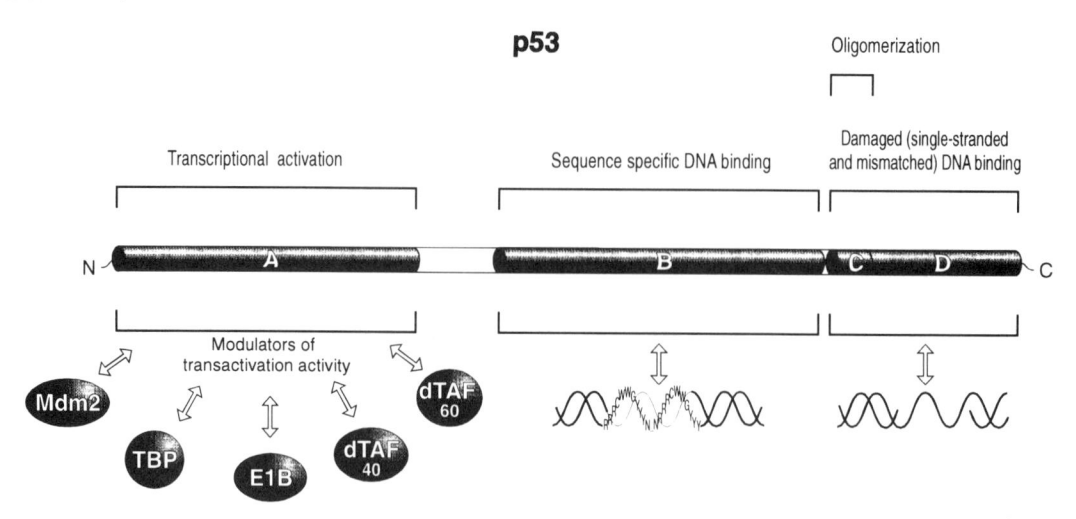

Figure 1. Functional domains of p53. The transcriptional activation domain of p53 binds to a number of proteins that regulate its function: Mdm2 (167), TBP; TATA binding protein (245), E1B; adenovirus E1B (214), dTAF40 and dTAF60; subunits of the transcription factor IID (246). p53 mutations found in human tumors map to the sequence specific DNA binding domain. Binding of p53 to mismatched and single stranded DNA via the C-terminus implicates p53 in DNA repair.

apoptosis appears to be variable (140-144). It is not clear whether the differences observed are due to cell type, or the state of transformation of the cell. We have found that TGF-β1 inhibits T cell death induced by activation via the T cell receptor (145). Since TGF-β1 is an inhibitor of cell cycle progression, and since we have shown that an increase of cyclin associated kinase activity in the same cell line is associated with activation induced cell death (126), we expect that TGF-β1 is preventing cell death by induction of cdk-cyclin kinase inhibitors (137). Recently, apoptosis mediated by anti-Fas antibody was shown to be partially inhibited in the presence of TGF-β1 (144).

p53 tumour suppressor: Cell cycle arrest and apoptosis

p53 tumour suppressor gene frequently undergoes loss of function mutations in human tumours (146). Transfection and overexpression of p53 in tumour cells lacking p53 function results in growth arrest. Its ability to induce cell cycle arrest has been long considered to be the mechanism by which p53 suppresses tumour cell growth. It now appears that the ability of p53 to promote apoptosis may be of singular importance to suppression of tumour growth.

Functions of p53

Levels of p53 are low in normal cells and the protein has a half life of 20-30 min. Following DNA damage, the protein is both stabilised and activated, leading to a significant increase in its abundance. Activation of p53 by DNA damage leads to either G1 arrest, allowing time for the damage to be repaired, or to apoptosis. The kinetics of induction of p53 protein and its ultimate abundance vary depending on the damaging agent.

p53 can activate or repress transcription of specific genes. p53 also senses damaged DNA and probably facilitates DNA repair. p53 has been shown to bind single-stranded DNA ends (147), and to exhibit DNA-DNA and RNA-DNA annealing activities (147-149). p53 has three functional domains, a central core (residues 90-290 of human p53) that binds DNA in a sequence specific manner, an N-terminal domain that activates transcription and a basic C-terminal region (residues 319-360) that facilitates tetramerization and binds to mismatched and single stranded DNA (Fig. 1) (reviewed in 150). The p53 mutations found in human tumours map largely to the central DNA binding domain. The 72 amino acids from the acidic N-terminal domain can alone activate transcription of a test gene. In addition, the function of this domain is regulated by interaction with Mdm2 and adenovirus E1B which repress its activity, or with dTAF60 and dTAF40 which activate it. It appears that the p53 binding to damaged DNA may activate the protein to regulate transcription of p53-specific DNA elements in target genes elsewhere (Fig. 2).

Among the various activities ascribed to p53, the ability of p53 to induce growth arrest is well correlated with its ability to function as a transcriptional activator (Fig. 2)(151, 152). p53 activates transcription of p21$^{WAF1/Cip1}$ (93), Gadd45 (153), cyclin G (154, 155), mdm2 (156), bax (157) and IGF-BP3 (158). Of substantial interest in the context of apoptosis control, p53 also induces transcription of the gene for Fas receptor (159).

CHAPTER 15/ APOPTOSIS AND CELL CYCLE

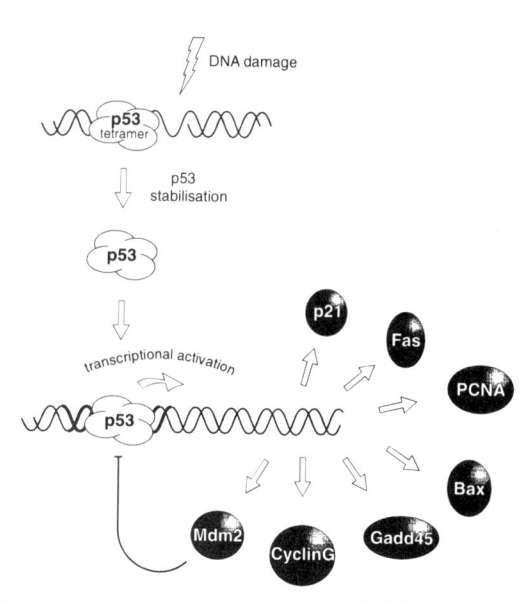

Figure 2. Transcriptional activation by p53 following DNA damage. The C-terminal domain of p53 normally masks the central sequence-specific DNA binding domain. Binding of p53 to damaged DNA through the C terminus activates the protein. p53 then binds to specific elements in target genes to activate transcription.

The best studied of the genes activated by p53 encodes p21, an inhibitor of cyclin dependent kinases (94-96). p21 contains two functional domains, an N-terminal domain that binds and inhibits cyclin-cdk kinase (84, 102-107) and a C-terminal domain that binds PCNA and inhibits its function (84, 102, 104, 106). Both the N-terminal domain (which binds cdk-cyclin complexes) (84, 102, 104, 105) and the C-terminal domain of p21 (which binds PCNA) can function independently to block cell cycle progression and inhibit cell growth albeit via different mechanism (84, 104). p21 inhibits the activity of G1 cyclin dependent kinases, cyclin D and cyclin E which normally phosphorylate Rb leading to its inactivation (Fig. 3). Hypophosphorylated Rb associates with and inactivates the E2F family of transcription factors which are responsible for transcription of growth related genes. Failure to release Rb from E2F leads to a G1 block. Mice that are $p21^{-/-}$ exhibit a partial failure to block in G1 (160,161). This may reflect redundancy of function, with other G1 cyclin de-pendent kinase inhibitors such as p16, p15, p27 and p57.

Insulin-like growth factor binding protein 3 (IGF-BP3) inhibits mitogenic signalling by insulin-like growth factor (IGF-1). The induction of IGF-BP3 gene expression by p53 suggests a role for this protein in growth inhibition (158). Gadd45 binds to PCNA (162, 163) and to p21 (164) and inhibits cell growth (165). The interaction between Gadd45, p21 and PCNA has been proposed to play a role in DNA repair (162). Mdm2 is a p53 inducible gene which acts as a negative regulator of p53 function (166, 167). Mdm2 binds to the N-terminus of p53 (168) and inhibits p53 dependent transcriptional activation (167). $Mdm2^{-/-}$ mice are not viable and die in embryogenesis. This suggests a failure to down-regulate p53 function, since mice deficient in both p53 and mdm2 are viable and develop normally (169, 170). Finally, cyclin G mRNA is elevated after activation of p53 but the functional significance of cyclin G induction is not known (154, 155, 171).

p53 also suppresses transcription of genes from specific promoters (172) which contain TATA box elements (173, 174). These include genes associated with proliferation such as c-*fos* (175) and PCNA (176, 177). Interestingly, p53 inhibits the expression of Bcl-2, an inhibitor of apoptosis (178).

p53 and apoptosis

Loss of p53 mediated apoptosis has been implicated in tumour progression both in mice (179) and in humans (180). Mice bearing an intact transgene for SV40 large T antigen, develop aggressive tumours as a result of sequestration of both Rb and p53, while mice expressing a truncated T antigen, capable of binding Rb but not p53, develop slow growing tumours (179). SV40 T antigen binds Rb directly and reverses its growth suppressive effects (181). The slow growth of tumours observed in this case is p53 dependent, as the same truncated T antigen expressed in $p53^{-/-}$ mice elicits aggressive tumours. The p53-dependent slow growth of tumours is accompanied by extensive apoptosis. Similarly, p53 mediated apoptosis has been shown to protect transformation of cells by E1A, another oncogenic viral product (182).

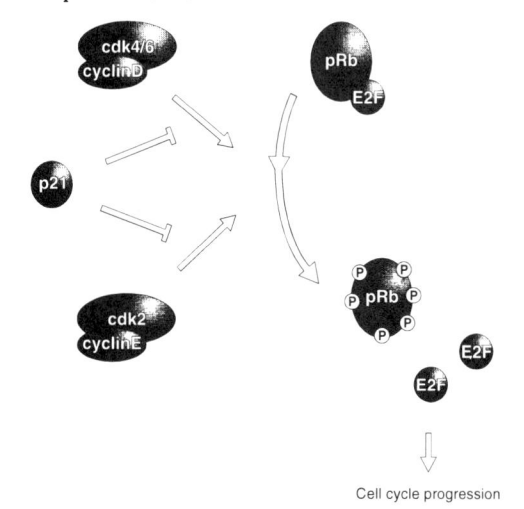

Figure 3. Regulation of G1 by Rb and p21. Hypophosphorylated Rb associates with and inactivates E2F family of transcription factors which are responsible for transcription of growth related genes and entry into S phase. p21 inhibits the activity of G1 cyclin dependent kinases, cyclin D and cyclin E which normally phosphorylate Rb leading to its inactivation. Failure to release Rb from E2F leads to a G1 block.

153

p53 induces apoptosis in skin (183, 184) and many cell types including thymocytes (185, 186) and intestinal epithelium cells (187, 188) in response to DNA damage. Cells of the spleen and of the thymus, and osteocytes, show dramatic accumulation of p53 after whole body irradiation (189). The induction of p53 correlates with apoptosis in spleen and thymus, but p53 does not induce apoptosis in osteocytes. Upon DNA damage different tissues thus respond differently to induction of p53. The intestinal cells and thymocytes in p53$^{-/-}$ mice are insensitive to irradiation induced death, demonstrating a critical role of p53 in this process (185, 186, 188). In contrast, glucocorticoid mediated apoptosis of thymocytes is p53-independent in these mice (185, 186). Mitogen activated mature T cells from p53 deficient mice undergo death that is independent of p53 (190) and dependent on Interferon regulatory factor-1 (IRF-1)(191).

p53 mediated apoptosis is not restricted to cells exposed to DNA damage, but also occurs in response to withdrawal of serum factors. Haematopoetic progenitors from p53$^{-/-}$ mice exhibit reduced apoptosis following deprivation of factors (192). Similar reduction of apoptotic death upon factor withdrawal has been observed in leukemic cells in which p53 is inactivated (193, 194). p53 is also involved in death due to metabolite imbalance (195) and after exposure to drugs such as PALA that do not damage DNA (196).

In model systems in which apoptosis is induced by E1A (197, 198), c-*myc* (199, 200) or E2F (201, 202), it has been found that apoptosis also requires p53. p53-dependent apoptosis induced by expression of proteins such as E1A, E2F and *myc* which drive cell cycle progression suggests that a conflict of signals, between growth induction and inhibition, triggers apoptosis. In this regard it is interesting that Rb, which blocks cell cycle progression, prevents p53 mediated apoptosis (201, 203).

Does p53 mediated apoptosis occur in G1 phase?
The ability of p53 to activate transcription and mediate G1 arrest has been correlated with apoptosis. On the other hand, transcriptional activation that leads to G1 arrest is not a pre-requisite for p53 mediated apoptosis (200, 203, 204). Cells from p21 deficient mice exhibit a defect in p53 induced G1 arrest but show a normal p53 mediated apoptosis response (160, 161). No clear consensus can be reached at present as to whether p53 mediated apoptosis requires G1 arrest. Evidence suggests that the apoptotic response is condition dependent, implying a complex relationship exists between these functions. p53-dependent apoptosis appears to occur in the G1 phase of the cell cycle in some cell types (205-208). Expression of p53 was shown to induce growth arrest followed by cell death in murine erythroleukemia cells transfected with a temperature sensitive mutant of murine p53 (p53 val-135), and this effect occurred specifically in cycling cells in G1 (205). Activation of p53 in a myeloblastic leukemia M1 cell line expressing a temperature sensitive mutant of murine p53 was similarly followed by apoptosis in G1 (206). Cisplatinin treated Epstein-Barr virus-immortalised human B lymphoblastoid cell lines (LCL) undergo p53 mediated apoptosis that occurs at the G1/S boundary (207). Irradiation of murine hematopoietic cell lines in the absence of IL-3 leads to cell death in S phase (209). In the presence of interleukin-3 (IL-3) these cells exhibit G1 arrest upon irradiation. p53 induced apoptosis in mouse T-lymphoma cells is preceded by G1 arrest (210). Cell cycle progression appears to be required for apoptosis as cells arrested earlier than G1 fail to apoptose despite accumulation of p53 (205-207). Ectopic expression of transcription factor E2F-1 overcomes p53 induced G1 arrest and leads to apoptosis, also confirming the requirement for cells to be cycling in order to undergo p53 mediated death (201, 202).

Although p53 plays a role in the G2/M checkpoint (211), p53 independent death appears to occur in G2/M. LCL cell lines containing mutant p53 incapable of transactivating do not undergo G1 arrest, but instead accumulate in the G2/M phase of the cell cycle and die (207). A human promyelocytic leukemia cell line, HL-60, which lacks p53 due to a large deletion in the gene, can be induced to undergo apoptosis upon X-irradiation, with death following a G2 arrest (208). Apoptosis induced by DNA damage in proliferating T cells derived from p53$^{-/-}$ mice, is accompanied by cell cycle arrest in both G1 and in G2 (190).

Choice between G1 arrest and apoptosis
The nature of the cell response to p53 activation, whether leading to G1 arrest or apoptosis, is a much debated subject. At present it seems that the outcome of activation rests primarily on the cell type in question and on environmental cues. Thymocytes undergo p53 dependent apoptosis upon irradiation (186, 187). In contrast, fibroblasts irradiated with similar doses undergo G1 arrest with increased p21 levels (212). Irradiation of murine hematopoietic cell lines in the presence of IL-3 induces G1 arrest while in the absence of IL-3 the irradiated cells undergo apoptosis (209). Apoptosis occurs despite induction of both p53 and p21. Inhibition of p53 mediated apoptosis by bcl-2 leads to a cell cycle arrest in G1 (206, 210). As cell cycle progression appears to be necessary for apoptosis (190, 205-207), it has been argued that the inability to arrest the cells in G1 leads to death. Conversely, if the cells arrest in G1 in response to the stimuli, apoptosis does not occur.

Trans-activation/transcription repressor function of p53 and apoptosis

p53 has been reported to mediate apoptosis in both a transcription-activation dependent and independent manner. p53 mediates apoptosis in the presence of inhibitors of protein synthesis and of transcriptional activation (204). Myc induced apoptosis can occur in the absence of new protein synthesis (200). A truncated p53 protein, containing the first 214 amino-terminal residues, is defective in transcription activation (203). Nevertheless, this protein retains the ability to mediate apoptosis. Conversely, a p53 mutant (p53175P) which is competent for transactivation is defective in inducing apoptosis (213). In contrast to these studies, mutations of residues 22 and 23 within the activation domain of p53 render it transcriptionally inactive, but still capable of binding to DNA in a sequence specific manner (214). In comparison to wild type p53 this transcriptionally defective mutant is defective in inducing apoptosis (215). A similar inability of transcriptionally defective p53 to induce apoptosis has been reported in HeLa cells in transient transfection assays (216). These studies together suggest that p53 can mediate apoptosis either in transcription-activation dependent or independent manner. The contribution of the transcription-activation function of p53 in apoptosis is made difficult on several accounts. Transcrip-tionally inactive mutants of p53 retain a selective capability to activate a subset of promoters (203). Further, the ever growing number of p53 responsive genes makes it difficult to test each of the genes to determine if their expression is affected by the mutations in question.

It should be noted that p53 can both activate and suppress transcription. The p53 mutant described by Sabbatini et al (215) is also defective in suppressing transcription. As another example, Bcl-2 and E1B-19 kDa protein both inhibit p53 mediated apoptosis. p53 mediated repression of promoters is inhibited in the presence of either protein (217). In contrast, the transactivation function of p53 is not changed. This raises the possibility that these proteins may exert their effect on apoptosis in part through their ability to inhibit p53-mediated repression of transcription. It is therefore likely that not only transcriptional activation of some genes but also the transcriptional suppression of others by p53 may be required for apoptosis.

Transcriptional regulation of genes involved in apoptosis

p53 has been recently shown to transcriptionally regulate genes whose products are involved in apoptosis. p53 can repress transcription of specific genes and has been reported to inhibit Bcl-2 expression (218). Bcl-2 is an inhibitor of apoptosis and has been shown to suppress p53 dependent apoptosis (48, 190, 219). A p53 dependent negative response element has been mapped within the Bcl-2 gene (178). Mice deficient in p53 exhibit increased levels of Bcl-2 and decreased levels of Bax (178). In contrast to Bcl-2, Bax is an inducer of apoptosis. The bax gene contains several p53 binding sites, and recently, p53 was confirmed to be a transcriptional activator of bax (157). Irradiation of the whole mouse results in rapid expression of Bax in radiosensitive organs, followed by massive apoptosis at these sites (220). Cells with mutant p53 fail to increase bax mRNA expression upon γ–irradiation (221). Although the above results suggest a direct role for Bax in apoptosis, p53 mediated apoptosis can occur without changes in Bax levels (207, 209) and cells from bax$^{-/-}$ mice show a normal apoptotic response following exposure to ionising radiation (51).

Fas, the cell surface receptor of Fas ligand, triggers apoptosis upon binding to Fas ligand. Recently it was shown that p53 can induce expression of the Fas gene (159). This result suggests that p53 may play a role in Fas-Fas ligand dependent apoptosis.

Retinoblastoma protein: Regulator of p53 mediated apoptosis?

It has been proposed that the ability of wild type p53 to inhibit the cell cycle or to induce apoptosis is determined by Rb. In the absence of Rb, p53 induces apoptosis. Oncogenic viral proteins, including adenovirus E1A, and human papillomavirus E7, bind Rb directly and reverse its growth suppressive effects (222, 223). An indication that Rb may inhibit apoptosis came from the observations that these proteins can induce apoptosis (224, 225). Overexpression of HPV E7 in human fibroblasts leads to apoptosis upon irradiation, whereas a similar radiation exposure leads to growth arrest in the absence of E7 (225). The above is suggestive evidence but the breakthrough in making a definitive link came from two approaches: inactivation of Rb function by expression of E7 in lens fibre cells (226) and examination of the failure of lens differentiation in Rb$^{-/-}$ mouse embryos (227). Rb$^{-/-}$ mice die in mid-gestation and there is widespread cell death in the nervous system (228-230). In Rb$^{-/-}$ mice, lens fibres which normally exit the cell cycle and differentiate, continue to duplicate DNA and die by apoptosis. Apoptosis in lens is p53 dependent since embryos that are negative for both p53 and Rb undergo significantly less death (227). Further, embryos doubly transgenic for E7 and E6 expression (which target p53 for degradation) display reduced apoptosis in the lens and adult mice go on to develop lens tumours (226).

How does Rb protect cells from apoptosis? Rb exerts its growth inhibitory effect by binding to transcription factors of the E2F family. Control of E2F by Rb can regulate progression from G0/G1 into S phase. There is a link between p53 and Rb action in the G1 arrest. Cyclin dependent kinases are responsible for phosphorylation and removal of Rb from E2F-1. p53 induces the expression of p21, an inhibitor of cyclin dependent kinases (Fig. 3). Increased levels of p21 would prevent the release of Rb, thus arresting cells in G1. While p53 can also induce growth arrest, this function may be redundant in the presence of intact Rb. In the absence of Rb, E2F stimulates inappropriate entry into S phase. Possibly, in the absence of Rb, p53 attempts to inhibit cell cycle progression through inhibition of cdk-cyclin kinases. The conflicting signals, stimulation of entry into S phase by E2F and inhibition cell cycle progression by p53, results in apoptosis. Furthermore, p53 has been shown to cooperate with E2F to induce apoptosis (202). In the absence of both p53 and Rb, E2F stimulates cell proliferation and, unhindered by p53, permits tumour formation. p53 and Rb may interact in yet another way. p53 may regulate the expression of Rb, as a p53 binding site is present in the Rb promoter (231).

Ataxia Telangiectasia Mutated (ATM): An upstream regulator of p53?

The integrity of DNA is monitored at several stages during the cell cycle. Radiation damage to DNA delays the transition from G1 to S phase (G1/S checkpoint), halts DNA synthesis (S phase progression checkpoint), or delays entry into mitosis (G2/M checkpoint). Incomplete DNA synthesis due to errors delays entry into mitosis (S-G2/M checkpoint). In its simplest form, the checkpoint controls require a system to detect DNA damage, to transmit this information, and an effector mechanism that interacts with the cell cycle machinery (reviewed in 232). The ATM gene is mutated in patients with the genetic disorder Ataxia Telangiectasia. The disease is characterised by immunodeficiency, neurological abnormalities, predisposition to a number of tumours, and failure to arrest properly at radiation induced cell cycle checkpoints. Radiation sensitivity of A-T cells has long been recognised (233, 234). Unlike normal cells which exhibit a marked inhibition of DNA synthesis after ionising radiation and delay for repair before resuming cell cycle progression, A-T cells fail to arrest in G1 or in S phase after receiving ionising radiation, but continue DNA synthesis and accumulate in G2 (235). A-T cells exhibit normal cell cycle arrest following UV irradiation or upon treatment with inhibitors of DNA synthesis. Ionising radiation and not UV induced DNA damage has been reported to lead to apoptosis in A-T lines (236).

The ATM gene belongs to a gene family called the phosphatidylinositol (PI)3-kinase family (reviewed in 237). Other members of this family that are strikingly similar in function to the ATM gene are the *rad3* gene in *Schizosaccharomyces pombe* and its homolog *MEC1* in *Saccharomyces cerevisiae*. These genes are involved in checkpoint response to DNA damage and in the dependence of mitotic entry on completion of DNA replication (reviewed in 238). The *Drosophila melanogaster mei-4* gene product which shares homology with these proteins similarly regulates checkpoint control to DNA damage. The homology of the ATM gene product with other proteins that respond to DNA damage is consistent with the function of the ATM protein. Failure of checkpoint controls results in entry of cells with damaged DNA into mitosis with consequent increase in chromosomal aberrations.

In response to ionising irradiation, p53 was shown to have reduced induction in A-T cells and to be produced with delayed kinetics (239-243). In contrast, induction of p53 appears relatively normal following UV irradiation or upon treatment with inhibitors of topoisomerase (241). Consistent with lower p53 levels, there is a suppression in the normal mRNA for p21 and Gadd45, genes which are transactivated by p53 (241-243). The failure to arrest in G1 correlates with the lack of inhibition of cyclin E dependent kinase by p21 and the presence of de-phosphorylated Rb in A-T cells after radiation (124, 243). These studies suggest that the ATM gene product may lie upstream of p53, and that the checkpoint signal pathway that operates through downstream effectors such as p53, p21, cyclin dependent kinases and Rb is defective in A-T cells. Lack of a functional ATM gene results in an inability to respond to multiple checkpoints upon DNA damage, to activate damage inducible genes to permit DNA repair, and to prevent apoptosis due to failure to arrest the cell cycle following DNA damage (244).

CONCLUSIONS

It is now clear that apoptosis is regulated by a complex circuitry of inhibitors and activators (Fig. 4), but the functional relationship between the different regulators of apoptosis still needs to be defined. It is of primary importance to understand how tumour progression undermines the normal cell cycle linkage to apoptosis. In this respect, it is evident that the tumour suppressor p53 plays a central role in regulating the cell cycle following DNA damage, and can induce apoptosis if this checkpoint control is overridden. This phenomenon has established a link between cell cycle control and apoptosis. p53 mediated death is one of several possible pathways that cooperate to induce cell

CHAPTER 15 / APOPTOSIS AND CELL CYCLE

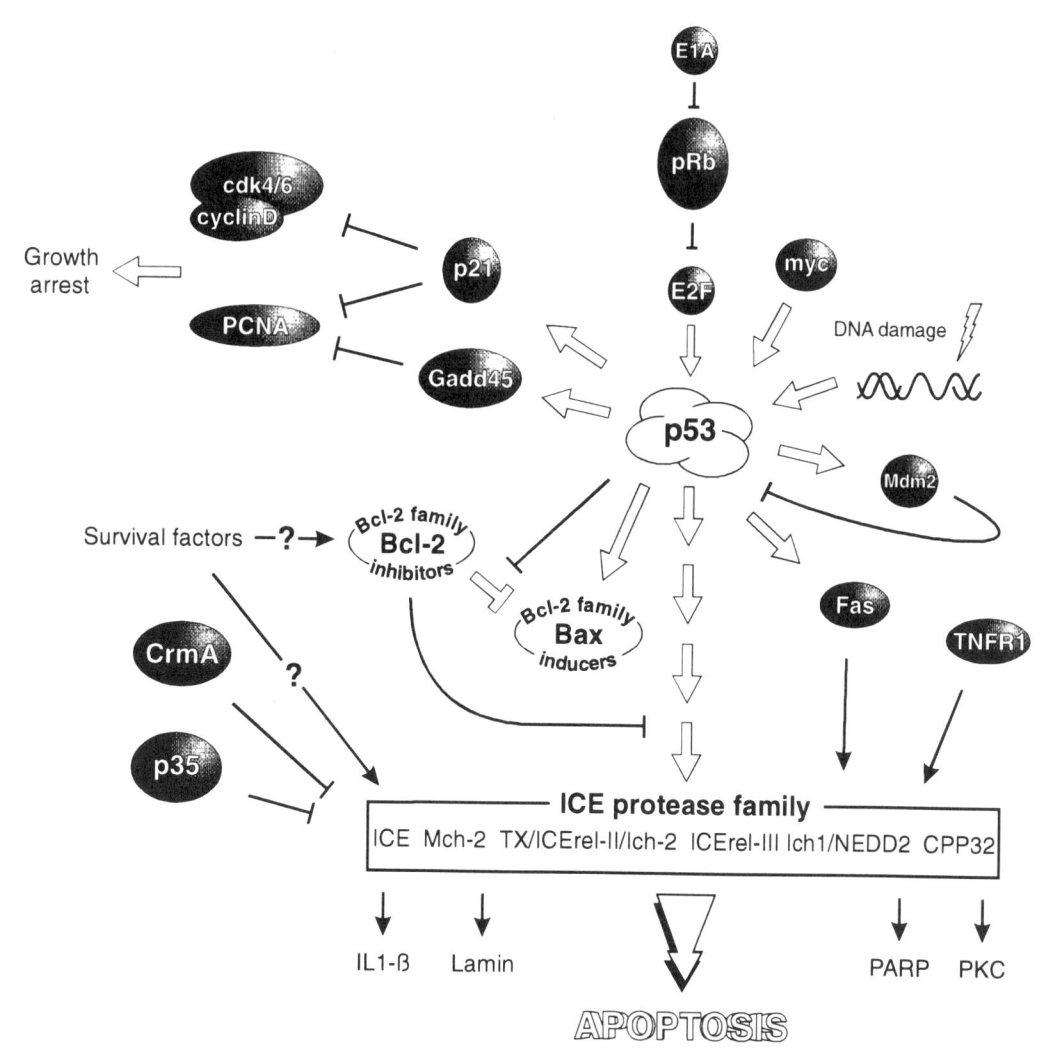

Figure 4. Regulators and executioners of apoptosis.

death. It remains for future studies to provide insight into the mechanism that links the activity of the various cell cycle components to activation of apoptosis.

ACKNOWLEDGEMENTS

We are grateful to Robert L. Margolis for critical reading of this review. The work in the authors laboratory was supported by grants from Programme région Rhone Alpes, INSERM and ARC to RF and by grants from American Cancer Society and NIH to AF.

REFERENCES

1. Ellis, H. M. and Horvitz, H. R. (1986) *Cell*. **44**, 817-829.
2. Wyllie, A. H., Kerr, J. F., and Currie, A. R. (1980) *Int Rev Cytol*. **68**, 251-306.
3. Peitsch, M. C., Mannherz, G. H., and Tschopp, J. (1994) *Trends Cell Biol*. **4**, 37-41.
4. Schwartz, L. M. and Osborne, B. A. (1993) *Immunol Today*. **14**, 582-590.
5. Krammer, P. H., Behrmann, I., Dhein, J., and Debatin, K. M. (1994) *Curr. Biol*. **6**, 279-289.
6. Nagata, S. and Golstein, P. (1995) *Science*. **267**, 1449-1456.
7. Trauth, B. C., Klas, C., Peters, A. M., Matzku, S., Moller, P., Falk, W., Debatin, K. M., and Krammer, P. H. (1989) *Science*. **245**, 301-305.

8. Itoh, N., Yonehara, S., Ishii, A., Yonehara, M., Mizushima, S., Sameshima, M., Hase, A., Seto, Y., and Nagata, S. (1991) *Cell.* **66,** 233-243.
9. Owen-Schaub, L. B., Yonehara, S., Crump, W. L. d., and Grimm, E. A. (1992) *Cell Immunol.* **140,** 197-205.
10. Suda, T., Takahashi, T., Golstein, P., and Nagata, S. (1993) *Cell.* **75,** 1169-1178.
11. Schulze-Osthoff, K. (1994) *Trends Cell Biol.* **4,** 421-426.
12. Oltvai, Z. N. and Korsmeyer, S. J. (1994) *Cell.* **79,** 189-192.
13. Itoh, N., Tsujimoto, Y., and Nagata, S. (1993) *J Immunol.* **151,** 621-627.
14. Takayama, S., Sato, T., Krajewski, S., Kochel, K., Irie, S., Millan, J. A., and Reed, J. C. (1995) *Cell.* **80,** 279-284.
15. Los, M., Van de Craen, M., Penning, L. C., Schenk, H., Westendorp, M., Baeuerle, P. A., Droge, W., Krammer, P. H., Fiers, W., and Schulze-Osthoff, K. (1995) *Nature.* **375,** 81-83.
16. Enari, M., Hug, H., and Nagata, S. (1995) *Nature.* **375,** 78-81.
17. Schlegel, J., Peters, I., Orrenius, S., Miller, D. K., Thornberry, N. A., Yamin, T. T., and Nicholson, D. W. (1996) *J. Biol. Chem.* **271,** 1841-1844.
18. Beidler, D. R., Tewari, M., Friesen, P. D., Poirier, G., and Dixit, V. M. (1995) *J Biol Chem.* **270,** 16526-16528.
19. Tewari, M. and Dixit, V. M. (1995) *J Biol Chem.* **270,** 3255-3260.
20. Yuan, J., Shaham, S., Ledoux, S., Ellis, H. M., and Horvitz, H. R. (1993) *Cell.* **75,** 641-652.
21. Kuida, K., Lippke, J. A., Ku, G., Harding, M. W., Livingston, D. J., Su, M. S., and Flavell, R. A. (1995) *Science.* **267,** 2000-2003.
22. Li, P., Allen, H., Banerjee, S., Franklin, S., Herzog, L., Johnston, C., McDowell, J., Paskind, M., Rodman, L., Salfeld, J., Towne, E., Tracey, D., Wardwell, S., Wei, F.-Y., Wong, W., Kamen, R., and Seshadri, T. (1995) *Cell.* **80,** 401-411.
23. Rathmell, J. C. and Goodnow, C. C. (1995) *Current Biology.* **5,** 1218-1221.
24. Adachi, M., Suematsu, S., Kondo, T., Ogasawara, J., Tanaka, T., Yoshida, N., and Nagata, S. (1995) *Nature genetics.* **11,** 294-299.
25. Amakawa, R., Hakem, A., Kundig, T. M., Matsuyama, T., Simard, J. J. L., Timms, E., Wakeham, A., Mittruecker, H.-W., Griesser, H., Takimoto, H., Schmits, R., Shahinian, A., Ohashi, P. S., Penninger, J. M., and Mak, T. W. (1995) *Cell.* **84,** 551-562.
26. White, K., Grether, M. E., Abrams, J. M., Young, L., Farrell, K., and Steller, H. (1994) *Science.* **264,** 677-683.
27. Steller, H. (1995) *Science.* **267,** 1445-1449.
28. White, K., Tahaoglu, E, and Steller, H. (1996) *Science.* **271,** 805-807.
29. Pronk, G. J., Ramer, K., Amiri, P., and Williams, L.T. (1996) *Science.* **271,** 808-810.
30. Cleveland, J. L. and Ihle, J. N. (1995) *Cell.* **81,** 479-482.
31. Baker, S. J. and Reddy, P. (1996) *Oncogene.* **12,** 1-12.
32. Chu, K., Niu, X., and Williams, L. T. (1995) *Proc Natl Acad Sci USA.* **92,** 11894-11898.
33. Kischkel, F. C., Hellbardt, S., Behrmann, I., Germer, M., Pawlita, M., Krammer, P. H., and Peter, M. E. (1995) *EMBO J.* **14,** 5579-5588.
34. Vaux, D. L., Cory, S., and Adams, J. M. (1988) *Nature.* **335,** 440-442.
35. Reed, J. C. (1994) *J Cell Biol.* **124,** 1-6.
36. White, E. (1996) *Genes Dev.* **10,** 1-15.
37. Hengartner, M. O. and Horvitz, H. R. (1994) *Cell.* **76,** 665-676.
38. Korsmeyer, S. J. (1995) *Trends Genet.* **11,** 101-105.
39. Merry, D. E., Veis, D. J., Hickey, W. F., and Korsmeyer, S. J. (1994) *Development.* **120,** 301-311.
40. Hockenbery, D. M., Zutter, M., Hickey, W., Nahm, M., and Korsmeyer, S. J. (1991) *Proc Natl Acad Sci USA.* **88,** 6961-6965.
41. Veis, D. J., Sorenson, C. M., Shutter, J. R., and Korsmeyer, S. J. (1993) *Cell.* **75,** 229-240.
42. Sentman, C. L., Shutter, J. R., Hockenbery, D., Kanagawa, O., and Korsmeyer, S. J. (1991) *Cell.* **67,** 879-888.
43. Siegel, R. M., Katsumata, M., Miyashita, T., Louie, D. C., Greene, M. I., and Reed, J. C. (1992) *Proc Natl Acad Sci USA.* **89,** 7003-7007.
44. Strasser, A., Harris, A. W., von Boehmer, H., and Cory, S. (1994) *Proc Natl Acad Sci USA.* **91,** 1376-1380.
45. Linette, G. P. and Korsmeyer, S. J. (1994) *Curr Opin Cell Biol.* **6,** 809-815.
46. Cory, S. (1995) *Annu Rev Immunol.* **13,** 513-543.
47. Farrow, S. N. and Brown, R. (1996) *Curr. Biol.* **6,** 45-49.
48. Chiou, S. K., Rao, L., and White, E. (1994) *Mol Cell Biol.* **14,** 2556-2563.
49. Schott, A. F., Apel, I. J., Nunez, G., and Clarke, M. F. (1995) *Oncogene.* **11,** 1389-1394.
50. Kiefer, M. C., Brauer, M. J., Powers, V. C., Wu, J. J., Umansky, S. R., Tomei, L. D., and Barr, P. J. (1995) *Nature.* **374,** 736-739.
51. Knudson, C. M., Tung, K. S., Tourtellotte, W. G., Brown, G. A., and Korsmeyer, S. J. (1995) *Science.* **270,** 96-99.
52. Motoyama, N., Wang, F., Roth, K. A., Sawa, H., Nakayama, K., Nakayama, K., Negishi, I., Senju, S., Zhang, Q., Fujii, S., and et al. (1995) *Science.* **267,** 1506-1510.

53. Yuan, J., Shaham, S., Ledoux, S., Ellis, H. M., and Horvitz, H. R. (1993) *Cell*. **75**, 641-652.
54. Thornberry, N. A., Bull, H. G., Calaycay, J. R., Chapman, K. T., Howard, A. D., Kostura, M. J., Miller, D. K., Molineaux, S. M., Weidner, J. R., Aunins, J., and et al. (1992) *Nature*. **356**, 768-774.
55. Walker, N. P., Talanian, R. V., Brady, K. D., Dang, L. C., Bump, N. J., Ferenz, C. R., Franklin, S., Ghayur, T., Hackett, M. C., Hammill, L. D., and et al. (1994) *Cell*. **78**, 343-352.
56. Wilson, K. P., Black, J. A., Thomson, J. A., Kim, E. E., Griffith, J. P., Navia, M. A., Murcko, M. A., Chambers, S. P., Aldape, R. A., Raybuck, S. A., and et al. (1994) *Nature*. **370**, 270-275.
57. Miura, M., Zhu, H., Rotello, R., Hartwieg, E. A., and Yuan, J. (1993) *Cell*. **75**, 653-660.
58. Ray, C. A., Black, R. A., Kronheim, S. R., Greenstreet, T. A., Sleath, P. R., Salvesen, G. S., and Pickup, D. J. (1992) *Cell*. **69**, 597-604.
59. Gagliardini, V., Fernandez, P. A., Lee, R. K., Drexler, H. C., Rotello, R. J., Fishman, M. C., and Yuan, J. (1994) *Science*. **263**, 826-828.
60. Komiyama, T., Ray, C. A., Pickup, D. J., Howard, A. D., Thornberry, N. A., Peterson, E. P., and Salvesen, G. (1994) *J Biol Chem*. **269**, 19331-19337.
61. Tewari, M., Telford, W. G., Miller, R. A., and Dixit, V. M. (1995) *J Biol Chem*. **270**, 22705-22708.
62. Sugimoto, A., Hozak, R. R., Nakashima, T., Nishimoto, T., and Rothman, J. H. (1995) *EMBO J*. **14**, 4434-4441.
63. Hay, B. A., Wolff, T., and Rubin, G. M. (1994) *Development*. **120**, 2121-2129.
64. Bump, N. J., Hackett, M., Hugunin, M., Seshagiri, S., Brady, K., Chen, P., Ferenz, C., Franklin, S., Ghayur, T., Li, P., and et al. (1995) *Science*. **269**, 1885-1888.
65. Xue, D. and Horvitz, H. R. (1995) *Nature*. **377**, 248-251.
66. Takahashi, A. and Earnshaw, W. C. (1996) *Curr Opin Genet Dev*. **6**, 50-55.
67. Tewari, M., Quan, L. T., O'Rourke, K., Desnoyers, S., Zeng, Z., Beidler, D. R., Poirier, G. G., Salvesen, G. S., and Dixit, V. M. (1995) *Cell*. **81**, 801-809.
68. Faucheu, C., Diu, A., Chan, A. W., Blanchet, A. M., Miossec, C., Herve, F., Collard-Dutilleul, V., Gu, Y., Aldape, R. A., Lippke, J. A., and et al. (1995) *EMBO J*. **14**, 1914-1922.
69. Kaufmann, S. H., Desnoyers, S., Ottaviano, Y., Davidson, N. E., and Poirier, G. G. (1993) *Cancer Res*. **53**, 3976-3985.
70. Lazebnik, Y. A., Kaufman, S.H., Desnoyers, S., Poirier, G.G. and Earnshaw, W.C. (1994) *Nature*. **371**, 346-347.
71. Gu, Y., Sarnecki, C., Aldape, R. A., Livingston, D. J., and Su, M. S. (1995) *J Biol Chem*. **270**, 18715-18718.
72. Nicholson, D. W., Ali, A., Thornberry, N. A., Vaillancourt, J. P., Ding, C. K., Gallant, M., Gareau, Y., Griffin, P. R., Labelle, M., Lazebnik, Y. A., and et al. (1995) *Nature*. **376**, 37-43.
73. Lazebnik, Y. A., Takahashi, A., Moir, R. D., Goldman, R. D., Poirier, G. G., Kaufmann, S. H., and Earnshaw, W. C. (1995) *Proc Natl Acad Sci USA*. **92**, 9042-9046.
74. Casciola-Rosen, L. A., Miller, D. K., Anhalt, G. J., and Rosen, A. (1994) *J. Biol. Chem*. **269**, 30757-30760.
75. Tewari, M., Beidler, D. R., and Dixit, V. M. (1995) *J Biol Chem*. **270**, 18738-18741.
76. Brancolini, C., Benedetti, M., and Schneider, C. (1995) *EMBO J*. **14**, 5179-5190.
77. Casciola-Rosen, L. A., Anhalt, G. J., and Rosen, A. (1995) *J Exp Med*. **182**, 1625-1634.
78. Emoto, Y., Manome, Y., Meinhardt, G., Kisaki, H., Kharbanda, S., Robertson, M., Ghayur, T., Wong, W. W., Kamen, R., and Weichselbaum, R. (1995) *EMBO J*. **14**, 6148-6156.
79. Pines, J. (1993) *Trends Biochem Sci*. **18**, 195-197.
80. Pagano, M., Pepperkok, R., Lukas, J., Baldin, V., Ansorge, W., Bartek, J., and Draetta, G. (1993) *J Cell Biol*. **121**, 101-111.
81. Tsai, L. H., Lees, E., Faha, B., Harlow, E., and Riabowol, K. (1993) *Oncogene*. **8**, 1593-1602.
82. Reed, S. I. (1992) *Annu Rev Cell Biol*. **8**, 529-561.
83. van den Heuvel, S. and Harlow, E. (1993) *Science*. **262**, 2050-2054.
84. Fotedar, R. and Fotedar, A. (1995) in *Progress in Cell Cycle Research* (L. Meijer, Guidet, S and Tung, H.Y.L. eds.) Vol. 1, pp. 73-89, Plenum Press, New York.
85. Sherr, C. J. (1994) *Cell*. **79**, 551-555.
86. Quelle, D. E., Ashmun, R. A., Shurtleff, S. A., Kato, J. Y., Bar-Sagi, D., Roussel, M. F., and Sherr, C. J. (1993) *Genes Dev*. **7**, 1559-1571.
87. Resnitzky, D., Gossen, M., Bujard, H., and Reed, S. I. (1994) *Mol Cell Biol*. **14**, 1669-1679.
88. Ohtsubo, M. and Roberts, J. M. (1993) *Science*. **259**, 1908-1912.
89. Nigg, A. E. (1995) *BioEssays*. **17**, 471-480.
90. Elledge, S. J. and Harper, J. W. (1994) *Curr Opin Cell Biol*. **6**, 847-852.
91. Hunter, T. and Pines, J. (1994) *Cell*. **79**, 573-582.
92. Sherr, C. J. and Roberts, J. M. (1995) *Genes Dev*. **9275**, 1149-1163.
93. el-Deiry, W. S., Tokino, T., Velculescu, V. E., Levy, D. B., Parsons, R., Trent, J. M., Lin, D.,

Mercer, W. E., Kinzler, K. W., and Vogelstein, B. (1993) *Cell*. **75**, 817-825.
94. Gu, Y., Turck, C. W., and Morgan, D. O. (1993) *Nature*. **366**, 707-710.
95. Harper, J. W., Adami, G. R., Wei, N., Keyomarsi, K., and Elledge, S. J. (1993) *Cell*. **75**, 805-816.
96. Xiong, Y., Hannon, G. J., Zhang, H., Casso, D., Kobayashi, R., and Beach, D. (1993) *Nature*. **366**, 701-704.
97. Noda, A., Ning, Y., Venable, S. F., Pereira-Smith, O. M., and Smith, J. R. (1994) *Exp Cell Res*. **211**, 90-98.
98. Polyak, K., Lee, M. H., Erdjument-Bromage, H., Koff, A., Roberts, J. M., Tempst, P., and Massague, J. (1994) *Cell*. **78**, 59-66.
99. Toyoshima, H. and Hunter, T. (1994) *Cell*. **78**, 67-74.
100. Lee, M. H., Reynisdottir, I., and Massague, J. (1995) *Genes Dev*. **9**, 639-649.
101. Matsuoka, S., Edwards, M. C., Bai, C., Parker, S., Zhang, P., Baldini, A., Harper, J. W., and Elledge, S. J. (1995) *Genes Dev*. **9**, 650-662.
102. Chen, J., Jackson, P. K., Kirschner, M. W., and Dutta, A. (1995) *Nature*. **374**, 386-388.
103. Goubin, F. and Ducommun, B. (1995) *Oncogene*. **10**, 2281-2287.
104. Luo, Y., Hurwitz, J., and Massague, J. (1995) *Nature*. **375**, 159-161.
105. Nakanishi, M., Robetorye, R. S., Adami, G. R., Pereira-Smith, O. M., and Smith, J. R. (1995) *EMBO J*. **14**, 555-563.
106. Warbrick, E., Lane, D. P., Glover, D. M., and Cox, L. S. (1995) *Curr Biol*. **5**, 275-282.
107. Fotedar, R., Fitzgerald, P., Rousselle, T., Cannella, D., Doree, M., Messier, H., and Fotedar, A. (1996) *Oncogene*. **12**, In Press.
108. Michieli, P., Chedid, M., Lin, D., Pierce, J. H., Mercer, W. E., and Givol, D. (1994) *Cancer Res*. **54**, 3391-3395.
109. Sheikh, M. S., Li, X. S., Chen, J. C., Shao, Z. M., Ordonez, J. V., and Fontana, J. A. (1994) *Oncogene*. **9**, 3407-3415.
110. Guan, K. L., Jenkins, C. W., Li, Y., Nichols, M. A., Wu, X., O'Keefe, C. L., Matera, A. G., and Xiong, Y. (1994) *Genes Dev*. **8**, 2939-2952.
111. Hannon, G. J. and Beach, D. (1994) *Nature*. **371**, 257-261.
112. Serrano, M., Hannon, G. J., and Beach, D. (1993) *Nature*. **366**, 704-707.
113. Hirai, H., Roussel, M. F., Kato, J. Y., Ashmun, R. A., and Sherr, C. J. (1995) *Mol Cell Biol*. **15**, 2672-2681.
114. Chan, F. K., Zhang, J., Cheng, L., Shapiro, D. N., and Winoto, A. (1995) *Mol Cell Biol*. **15**, 2682-2688.
115. Weinberg, R. A. (1995) *Cell*. **81**, 323-330.
116. Ohtani, K., DeGregori, J. and Nevins, J.R. (1995) *Proc. Natl. Acd. Sci. USA*. **92**, 12146-12150.
117. Schulze, A., Zerfass, K., Spitkovsky, D., Middendorp, S., Berges, J., Helin, K., Jansen-Durr, P., and Henglein, B. (1995) *Proc Natl Acad Sci USA*. **92**, 11264-11268.
118. Laiho, M., DeCaprio, J. A., Ludlow, J. W., Livingston, D. M., and Massague, J. (1990) *Cell*. **62**, 175-185.
119. Kato, J. Y., Matsuoka, M., Polyak, K., Massague, J., and Sherr, C. J. (1994) *Cell*. **79**, 487-496.
120. Polyak, K., Kato, J. Y., Solomon, M. J., Sherr, C. J., Massague, J., Roberts, J. M., and Koff, A. (1994) *Genes Dev*. **8**, 9-22.
121. Slingerland, J. M., Hengst, L., Pan, C. H., Alexander, D., Stampfer, M. R., and Reed, S. I. (1994) *Mol Cell Biol*. **14**, 3683-3694.
122. Ewen, M. E., Sluss, H. K., Whitehouse, L. L., and Livingston, D. M. (1993) *Cell*. **74**, 1009-1020.
123. Fornace, A. J., Jr. (1992) *Annu Rev Genet*. **26**, 507-526.
124. Dulic, V., Kaufmann, W. K., Wilson, S. J., Tlsty, T. D., Lees, E., Harper, J. W., Elledge, S. J., and Reed, S. I. (1994) *Cell*. **76**, 1013-1023.
125. el-Deiry, W. S., Harper, J. W., O'Connor, P. M., Velculescu, V. E., Canman, C. E., Jackman, J., Pietenpol, J. A., Burrell, M., Hill, D. E., Wang, Y., and et al. (1994) *Cancer Res*. **54**, 1169-1174.
126. Fotedar, R., Flatt, J., Gupta, S., Margolis, R. L., Fitzgerald, P., Messier, H., and Fotedar, A. (1995) *Mol Cell Biol*. **15**, 932-942.
127. Shi, L., Nishioka, W. K., Th'ng, J., Bradbury, E. M., Litchfield, D. W., and Greenberg, A. H. (1994) *Science*. **263**, 1143-1145.
128. Lahti, J. M., Xiang, J., Heath, L. S., Campana, D., and Kidd, V. J. (1995) *Mol Cell Biol*. **15**, 1-11.
129. Meikrantz, W., Gisselbrecht, S., Tam, S. W., and Schlegel, R. (1994) *Proc Natl Acad Sci USA*. **91**, 3754-3758.
130. Hoang, A. T., Cohen, K. J., Barrett, J. F., Bergstrom, D. A., and Dang, C. V. (1994) *Proc Natl Acad Sci USA*. **91**, 6875-6879.
131. Li, C. J., Wang, C., and Pardee, A. B. (1995) *Cancer Res*. **55**, 3712-3715.
132. Oberhammer, F. A., Hochegger, K., Froschl, G., Tiefenbacher, R., and Pavelka, M. (1994) *J Cell Biol*. **126**, 827-837.
133. Norbury, C., MacFarlane, M., Fearnhead, H., and Cohen, G. M. (1994) *Biochem Biophys Res Commun*. **202**, 1400-1406.
134. Massague, J. and Polyak, K. (1995) *Curr Opin Genet Dev*. **5**, 91-96.
135. Markowitz, S., Wang, J., Myeroff, L., Parsons, R., Sun, L., Lutterbaugh, J., Fan, R. S.,

Zborowska, E., Kinzler, K. W., Vogelstein, B., and et al. (1995) *Science.* **2685,** 1336-1338.
136. Geng, Y. and Weinberg, R. A. (1993) *Proc Natl Acad Sci USA.* **90,** 10315-10319.
137. Reynisdottir, I., Polyak, K., Iavarone, A., and Massague, J. (1995) *Genes Dev.* **9,** 1831-1845.
138. Li, J. M., Nichols, M. A., Chandrasekharan, S., Xiong, Y., and Wang, X. F. (1995) *J Biol Chem.* **270,** 26750-26753.
139. Datto, M. B., Yu, Y., and Wang, X. F. (1995) *J Biol Chem.* **270,** 28623-28628.
140. Oberhammer, F. A., Pavelka, M., Sharma, S., Tiefenbacher, R., Purchio, A. F., Bursch, W., and Schulte-Hermann, R. (1992) *Proc Natl Acad Sci USA.* **89,** 5408-5412.
141. Alam, R., Forsythe, P., Stafford, S., and Fukuda, Y. (1994) *J. Exp. Med.* **179,** 1041-1045.
142. Barlat, I., Henglein, B., Plet, A., Lamb, N., Fernandez, A., McKenzie, F., Pouyssegur, J., Vie, A., and Blanchard, J. M. (1995) *Oncogene.* **11,** 1309-1318.
143. Lomo, J., Blomhoff, H. K., Beiske, K., Stokke, T., and Smeland, E. B. (1995) *J Immunol.* **154,** 1634-1643.
144. Cerwenka, A., Kovar, H., Majdic, O., and Holter, W. (1996) *J Immunol.* **156,** 459-464.
145. Fotedar, R. and Fotedar, A. *Unpublished observations.*
146. Hollstein, M., Sidransky, D., Vogelstein, B., and Harris, C. C. (1991) *Science.* **253,** 49-53.
147. Bakalkin, G., Yakovleva, T., Selivanova, G., Magnusson, K. P., Szekely, L., Kiseleva, E., Klein, G., Terenius, L., and Wiman, K. G. (1994) *Proc Natl Acad Sci USA.* **91,** 413-417.
148. Oberosler, P., Hloch, P., Ramsperger, U., and Stahl, H. (1993) *EMBO J.* **12,** 2389-2396.
149. Brain, R. and Jenkins, J. R. (1994) *Oncogene.* **9,** 1775-1780.
150. Prives, C. (1994) *Cell.* **78,** 543-546.
151. Crook, T., Marston, N. J., Sara, E. A., and Vousden, K. H. (1994) *Cell.* **79,** 817-827.
152. Pietenpol, J. A., Tokino, T., Thiagalingam, S., el-Deiry, W. S., Kinzler, K. W., and Vogelstein, B. (1994) *Proc Natl Acad Sci USA.* **91,** 1998-2002.
153. Zhan, Q., Bae, I., Kastan, M. B., and Fornace, A. J., Jr. (1994) *Cancer Res.* **54,** 2755-2760.
154. Okamoto, K. and Beach, D. (1994) *EMBO J.* **13,** 4816-4822.
155. Zauberman, A., Lupo, A., and Oren, M. (1995) *Oncogene.* **10,** 2361-2366.
156. Barak, Y., Juven, T., Haffner, R., and Oren, M. (1993) *EMBO J.* **12,** 461-468.
157. Miyashita, T. and Reed, J. C. (1995) *Cell.* **80,** 293-299.
158. Buckbinder, L., Talbott, R., Velasco-Miguel, S., Takenaka, I., Faha, B., Seizinger, B. R., and Kley, N. (1995) *Nature.* **377,** 646-649.
159. Owen-Schaub, L. B., Zhang, W., Cusack, J. C., Angelo, L. S., Santee, S. M., Fujiwara, T., Roth, J. A., Deisseroth, A. B., Zhang, W. W., Kruzel, E., and et al. (1995) *Mol Cell Biol.* **1555,** 3032-3040.
160. Brugarolas, J., Chandrasekaran, C., Gordon, J. I., Beach, D., Jacks, T., and Hannon, G. J. (1995) *Nature.* **377,** 552-557.
161. Deng, C., Zhang, P., Harper, J. W., Elledge, S. J., and Leder, P. (1995) *Cell.* **82,** 675-684.
162. Smith, M. L., Chen, I. T., Zhan, Q., Bae, I., Chen, C. Y., Gilmer, T. M., Kastan, M. B., O'Connor, P. M., and Fornace, A. J., Jr. (1994) *Science.* **266,** 1376-1380.
163. Hall, P. A., Kearsey, J. M., Coates, P. J., Norman, D. G., Warbrick, E., and Cox, L. S. (1995) *Oncogene.* **10,** 2427-2433.
164. Kearsey, J. M., Coates, P. J., Prescott, A. R., Warbrick, E., and Hall, P. A. (1995) *Oncogene.* **11,** 1675-1683.
165. Zhan, Q., Carrier, F., and Fornace, A. J., Jr. (1993) *Mol Cell Biol.* **13,** 4242-4250.
166. Momand, J., Zambetti, G. P., Olson, D. C., George, D., and Levine, A. J. (1992) *Cell.* **69,** 1237-1245.
167. Oliner, J. D., Pietenpol, J. A., Thiagalingam, S., Gyuris, J., Kinzler, K. W., and Vogelstein, B. (1993) *Nature.* **362,** 857-860.
168. Chen, J., Marechal, V., and Levine, A. J. (1993) *Mol Cell Biol.* **13,** 4107-4114.
169. Jones, S. N., Roe, A. E., Donehower, L. A., and Bradley, A. (1995) *Nature.* **378,** 206-208.
170. Montes de Oca Luna, R., Wagner, D. S., and Lozano, G. (1995) *Nature.* **378,** 203-206.
171. Tamura, K., Kanaoka, Y., Jinno, S., Nagata, A., Ogiso, Y., Shimizu, K., Hayakawa, T., Nojima, H., and Okayama, H. (1993) *Oncogene.* **8,** 2113-2118.
172. Ginsberg, D., Mechta, F., Yaniv, M., and Oren, M. (1991) *Proc Natl Acad Sci USA.* **88,** 9979-9983.
173. Seto, E., Usheva, A., Zambetti, G. P., Momand, J., Horikoshi, N., Weinmann, R., Levine, A. J., and Shenk, T. (1992) *Proc Natl Acad Sci USA.* **89,** 12028-12032.
174. Mack, D. H., Vartikar, J., Pipas, J. M., and Laimins, L. A. (1993) *Nature.* **363,** 281-283.
175. Kley, N., Chung, R. Y., Fay, S., Loeffler, J. P., and Seizinger, B. R. (1992) *Nucl Acids Res.* **20,** 4083-4087.
176. Jackson, P., Ridgway, P., Rayner, J., Noble, J., and Braithwaite, A. (1994) *Biochem Biophys Res Commun.* **203,** 133-140.
177. Shivakumar, C. V., Brown, D. R., Deb, S., and Deb, S. P. (1995) *Mol Cell Biol.* **15,** 6785-6793.
178. Miyashita, T., Harigai, M., Hanada, M., and Reed, J. C. (1994) *Cancer Res.* **54,** 3131-3135.
179. Symonds, H., Krall, L., Remington, L., Saenz-Robles, M., Lowe, S., Jacks, T., and Van Dyke, T. (1994) *Cell.* **78,** 703-711.
180. Bardeesy, N., Beckwith, J. B., and Pelletier, J. (1995) *Cancer Res.* **55,** 215-219.

181. DeCaprio, J. A., Ludlow, J. W., Figge, J., Shew, J. Y., Huang, C. M., Lee, W. H., Marsilio, E., Paucha, E., and Livingston, D. M. (1988) *Cell.* **54**, 275-283.
182. Lowe, S. W., Jacks, T., Housman, D. E., and Ruley, H. E. (1994) *Proc Natl Acad Sci USA.* **91**, 2026-2030.
183. Ziegler, A., Jonason, A. S., Leffell, D. J., Simon, J. A., Sharma, H. W., Kimmelman, J., Remington, L., Jacks, T., and Brash, D. E. (1994) *Nature.* **372**, 773-776.
184. Hall, P. A., McKee, P. H., Menage, H. D., Dover, R., and Lane, D. P. (1993) *Oncogene.* **8**, 203-207.
185. Clarke, A. R., Purdie, C. A., Harrison, D. J., Morris, R. G., Bird, C. C., Hooper, M. L., and Wyllie, A. H. (1993) *Nature.* **362**, 849-852.
186. Lowe, S. W., Schmitt, E. M., Smith, S. W., Osborne, B. A., and Jacks, T. (1993) *Nature.* **362**, 847-849.
187. Clarke, A. R., Gledhill, S., Hooper, M. L., Bird, C. C., and Wyllie, A. H. (1994) *Oncogene.* **9**, 1767-1773.
188. Merritt, A. J., Potten, C. S., Kemp, C. J., Hickman, J. A., Balmain, A., Lane, D. P., and Hall, P. A. (1994) *Cancer Res.* **54**, 614-617.
189. Midgley, C. A., Owens, B., Briscoe, C. V., Thomas, D. B., Lane, D. P., and Hall, P. A. (1995) *J Cell Sci.* **108**, 1843-1848.
190. Strasser, A., Harris, A. W., Jacks, T., and Cory, S. (1994) *Cell.* **79**, 329-339.
191. Tamura, T., Ishihara, M., Lamphier, M. S., Tanaka, N., Oishi, I., Aizawa, S., Matsuyama, T., Mak, T. W., Taki, S., and Taniguchi, T. (1995) *Nature.* **376**, 596-599.
192. Lotem, J. and Sachs, L. (1993) *Blood.* **82**, 1092-1096.
193. Gottlieb, E., Haffner, R., von Ruden, T., Wagner, E. F., and Oren, M. (1994) *EMBO J.* **13**, 1368-1374.
194. Zhu, Y. M., Bradbury, D. A., and Russell, N. H. (1994) *Br J Cancer.* **69**, 468-472.
195. Yin, Y., Tainsky, M. A., Bischoff, F. Z., Strong, L. C., and Wahl, G. M. (1992) *Cell.* **70**, 937-948.
196. Almasan, A., Linke, S. P., Paulson, T. G., Huang, L. C., and Wahl, G. M. (1995) *Cancer Metastasis Rev.* **14**, 59-73.
197. Debbas, M. and White, E. (1993) *Genes Dev.* **7**, 546-554.
198. Lowe, S. W. and Ruley, H. E. (1993) *Genes Dev.* **7**, 535-545.
199. Hermeking, H. and Eick, D. (1994) *Science.* **265**, 2091-2093.
200. Wagner, A. J., Kokontis, J. M., and Hay, N. (1994) *Genes Dev.* **8**, 2817-2830.
201. Qin, X. Q., Livingston, D. M., Kaelin, W. G., Jr., and Adams, P. D. (1994) *Proc Natl Acad Sci USA.* **91**, 10918-10922.
202. Wu, X. and Levine, A. J. (1994) *Proc Natl Acad Sci USA.* **91**, 3602-3606.
203. Haupt, Y., Rowan, S., Shaulian, E., Vousden, K. H., and Oren, M. (1995) *Genes Dev.* **9**, 2170-2183.
204. Caelles, C., Helmberg, A., and Karin, M. (1994) *Nature.* **370**, 220-223.
205. Ryan, J. J., Danish, R., Gottlieb, C. A., and Clarke, M. F. (1993) *Mol Cell Biol.* **13**, 711-719.
206. Guillouf, C., Grana, X., Selvakumaran, M., De Luca, A., Giordano, A., Hoffman, B., and Liebermann, D. A. (1995) *Blood.* **85**, 2691-2698.
207. Allday, M. J., Inman, G. J., Crawford, D. H., and Farrell, P. J. (1995) *EMBO J.* **14**, 4994-5005.
208. Han, Z., Chatterjee, D., He, D. M., Early, J., Pantazis, P., Wyche, J. H., and Hendrickson, E. A. (1995) *Mol Cell Biol.* **15**, 5849-5857.
209. Canman, C. E., Gilmer, T. M., Coutts, S. B., and Kastan, M. B. (1995) *Genes Dev.* **9**, 600-611.
210. Wang, Y., Okan, I., Szekely, L., Klein, G., and Wiman, K. G. (1995) *Cell Growth Differ.* **6**, 1071-1075.
211. Cross, S. M., Sanchez, C. A., Morgan, C. A., Schimke, M. K., Ramel, S., Idzerda, R. L., Raskind, W. H., and Reid, B. J. (1995) *Science.* **267**, 1353-1356.
212. Di Leonardo, A., Linke, S. P., Clarkin, K., and Wahl, G. M. (1994) *Genes Dev.* **8**, 2540-2551.
213. Rowan, S., Ludwig, R. L., Haupt, Y., Bates, S., Lu, X., Oren, M., and Vousden, K. H. (1996) *EMBO J.* **15**, 827-838.
214. Lin, J., Chen, J., Elenbaas, B., and Levine, A. J. (1994) *Genes Dev.* **8**, 1235-1246.
215. Sabbatini, P., Lin, J., Levine, A. J., and White, E. (1995) *Genes Dev.* **9**, 2184-2192.
216. Yonish-Rouach, E., Deguin, V., Zaitchouk, T., Breugnot, C., Mishal, Z., Jenkins, J. R., and May, E. (1996) *Oncogene.* **12**, 2197-2205.
217. Shen, Y. and Shenk, T. (1994) *Proc Natl Acad Sci USA.* **91**, 8940-8944.
218. Miyashita, T., Krajewski, S., Krajewska, M., Wang, H. G., Lin, H. K., Liebermann, D. A., Hoffman, B., and Reed, J. C. (1994) *Oncogene.* **9**, 1799-1805.
219. Wang, Y., Szekely, L., Okan, I., Klein, G., and Wiman, K. G. (1993) *Oncogene.* **8**, 3427-3431.
220. Kitada, S., Krajewski, S., Miyashita, T., Krajewska M and Reed, J.C. (1996) *Oncogene.* **12**, 187-192.
221. Zhan, Q., Fan, S., Bae, I., Guillouf, C., Liebermann, D. A., O'Connor, P. M., and Fornace, A. J., Jr. (1994) *Oncogene.* **9**, 3743-3751.
222. Dyson, N., Howley, P. M., Munger, K., and Harlow, E. (1989) *Science.* **243**, 934-937.
223. Whyte, P., Williamson, N. M., and Harlow, E. (1989) *Cell.* **56**, 67-75.

224. Rao, L., Debbas, M., Sabbatini, P., Hockenbery, D., Korsmeyer, S., and White, E. (1992) *Proc Natl Acad Sci USA.* **89,** 7742-7746.
225. White, A. E., Livanos, E. M., and Tlsty, T. D. (1994) *Genes Dev.* **8,** 666-677.
226. Pan, H. and Griep, A. E. (1995) *Genes Dev.* **9,** 2157-2169.
227. Morgenbesser, S. D., Williams, B. O., Jacks, T., and DePinho, R. A. (1994) *Nature.* **371,** 72-74.
228. Lee, E. Y., Chang, C. Y., Hu, N., Wang, Y. C., Lai, C. C., Herrup, K., Lee, W. H., and Bradley, A. (1992) *Nature.* **359,** 288-294.
229. Jacks, T., Fazeli, A., Schmitt, E. M., Bronson, R. T., Goodell, M. A., and Weinberg, R. A. (1992) *Nature.* **359,** 295-300.
230. Clarke, A. R., Maandag, E. R., van Roon, M., van der Lugt, N. M., van der Valk, M., Hooper, M. L., Berns, A., and te Riele, H. (1992) *Nature.* **359,** 328-330.
231. Osifchin, N. E., Jiang, D., Ohtani-Fujita, N., Fujita, T., Carroza, M., Kim, S. J., Sakai, T., and Robbins, P. D. (1994) *J Biol Chem.* **269,** 6383-6389.
232. Carr, A. M. (1996) *Science.* **271,** 314-315.
233. Painter, R. B. and Young, B. R. (1980) *Proc Natl Acad Sci USA.* **77,** 7315-7317.
234. Houldsworth, J. and Lavin, M. F. (1980) *Nucl Acids Res.* **8,** 3709-3720.
235. Beamish, H. and Lavin, M. F. (1994) *Int J Radiat Biol.* **65,** 175-184.
236. Meyn, M. S., Strasfeld, L., and Allen, C. (1994) *Int J Radiat Biol.* **66,** S141-149.
237. Lavin, M. F., Khanna, K. K., Beamish, H., Spring, K., Watters, D., and Shiloh, Y. (1995) *Trends Biochem Sci.* **20,** 382-383.
238. Lehmann, A. R. and Carr, A. M. (1995) *Trends Genet.* **11,** 375-377.
239. Kastan, M. B., Zhan, Q., el-Deiry, W. S., Carrier, F., Jacks, T., Walsh, W. V., Plunkett, B. S., Vogelstein, B., and Fornace, A. J., Jr. (1992) *Cell.* **71,** 587-597.
240. Lu, X. and Lane, D. P. (1993) *Cell.* **75,** 765-778.
241. Canman, C. E., Wolff, A. C., Chen, C. Y., Fornace, A. J., Jr., and Kastan, M. B. (1994) *Cancer Res.* **54,** 5054-5058.
242. Artuso, M., Esteve, A., Bresil, H., Vuillaume, M., and Hall, J. (1995) *Oncogene.* **11,** 1427-1435.
243. Khanna, K. K., Beamish, H., Yan, J., Hobson, K., Williams, R., Dunn, I., and Lavin, M. F. (1995) *Oncogene.* **11,** 609-618.
244. Meyn, M. S. (1995) *Cancer Res.* **55,** 5991-6001.
245. Seto, E., Usheva, A., Zambetti, G. P., Momand, J., Horikoshi, N., Weinmann, R., Levine, A. J., and Shenk, T. (1992) *Proc Natl Acad Sci USA.* **89,** 12028-12032.
246. Thut, C. J., Chen, J. L., Klemm, R., and Tjian, R. (1995) *Science.* **267,** 100-104.

chapter 16
DNA damage checkpoints: Implications for cancer therapy

Patrick M. O'Connor[1] and Saijun Fan

Laboratory of Molecular Pharmacology, Division of Basic Science,
Room 5C-25, Bldg 37, National Cancer Institute, National Institutes of Health, Bethesda, MD 20892 USA
[1]To whom correspondence should be addressed

DNA damage evokes a complex array of cellular responses, including cycle arrest in late G1 and/or G2 phases, and delayed progression through S phase. Arrest at these points in the cell cycle is governed, in large part, by a series of control systems, commonly termed "checkpoints". Activation of these checkpoints tends to protect cells from DNA damage by providing cells additional time to complete DNA repair. We discuss the impact of these DNA damage checkpoints on the chemosensitivity of human cancer cells. We focus on some of the complexities of the p53-dependent G1 checkpoint and review some recently discovered vulnerabilities in p53 disrupted cells that might be pharmacologically exploited for cancer treatment.

INTRODUCTION

Mammalian cells exhibit a multitude of cellular responses to DNA damage including cell cycle arrest, the induction of DNA repair processes and in some cases apoptosis (1, 2). Cell cycle arrest following DNA damage is mediated through a series of negative-feedback control systems or "checkpoints" (1-3). In normal human cells, DNA damage arrests cell cycle progression in late G1 and G2 phases (1, 2). There also appears to be a DNA damage checkpoint operational during, S phase that suppresses both the initiation and elongation phases of DNA replication (4, 5) (Figure 1).

DNA damage-checkpoints were originally thought to protect cells from the deleterious consequences of genotoxic stress by allowing sufficient time for DNA repair to correct the damage before cell division (3). An inability to arrest cell cycle progression in the presence of DNA damage has been shown, in yeast, to result in genomic instability and decreased cell survival (3, 6). Disruption of the G1 and/or G2 checkpoints in mammalian cells has also been shown to increase the likelihood of genetic mutations (2, 7). However, in contrast to yeast, the effect of checkpoint disruption on the cytotoxicity of DNA damaging agents in mammalian cells depends upon [1] the particular checkpoint disrupted, [2] the status of the p53 tumor suppressor gene, [3] the cell type, and [4] the type of DNA damaging agent used to treat the cells. We will review each of these topics in light of emerging knowledge on the roles of the p53 tumor suppressor pathway in cell cycle checkpoint regulation, DNA repair and apoptosis. We also describe a vulnerability in p53-disrupted cells that might be exploitable for cancer treatment.

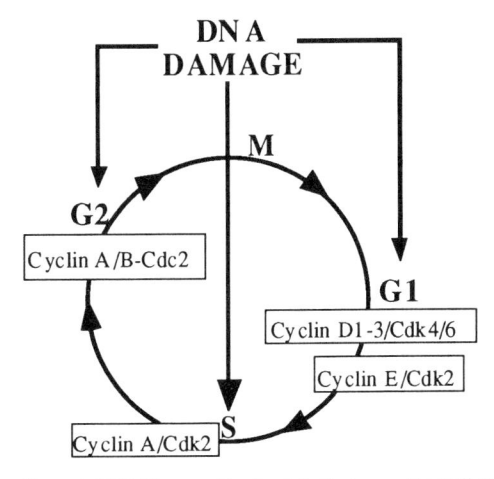

Figure 1. DNA Damage Checkpoints: Brakes on the Cell Cycle Engine.
DNA damage induces the arrest of cell cycle progression in late G1 and/or G2 phases, and retards progression through S phase. These responses are regulated through a series of negative-feedback systems commonly termed checkpoints (1, 2). In order for these checkpoint systems to arrest progression they must interact with the cyclin-dependent kinases that regulate all major cell cycle transitions. Evidence has been presented for the interaction of the G1 and G2 checkpoint control systems with the cyclin-dependent kinases that regulate the G1/S and G2/M phase transitions (30, 31, 82-90). The interaction of the mammalian S phase DNA damage checkpoint with cyclin-dependent kinases has yet to be reported.

THE G1 CHECKPOINT AND THE P53 TUMOR SUPPRESSOR

In order for mammalian cells to arrest in G1 phase following DNA damage the cells must contain wild-type p53 (8) and the pathway in which p53 operates must be intact (9, 10). However, since the p53 gene is also the most commonly mutated gene in human cancer (11), many cancer cells lack the ability to G1 arrest in the presence of

DNA damage. P53 function can also be inactivated through a number of other routes including the expression of certain viral proteins that inactivate p53 function. These include the human papillomavirus E6 protein, the SV40 T antigen, and the adenovirus E1b 55 kDa protein (9, 10, 12, 13). As described below, the p53 pathway can also be inactivated by overexpression of certain cellular oncogenes as well as mutations in checkpoint components required to activate p53 following DNA damage. These disruptions alter a wide-variety of cellular processes and have dramatic effects on genomic stability, cell survival and the success of anticancer therapy.

An evolving outline of the p53 pathway is illustrated in Figure 2. p53 encodes a 393 amino acid nuclear phosphoprotein that regulates the transcriptional activity of a diverse number of genes. p53 can directly induce transcription through binding of a cognate DNA binding sequence located either in the promoter or within an intron of a p53-regulated gene (13). Three examples of p53-regulated genes: MDM2, GADD45 and CIP1/WAF1 are discussed below. In some cases, such as the p53-regulated gene, BAX, γ-ray inducibility is only seen in those cell types that are wild-type for p53 and which subsequently undergo apoptosis following γ-irradiation (14). These findings suggest other factors can modulate the ability of p53 to transactivate downstream effector genes. Furthermore, it has recently been shown that p53 can also indirectly induce the transcription of some genes such as cyclin D1 (15). These finding highlight the need to characterise whether a p53 regulated gene is either directly or indirectly activated by p53, and whether other cellular factors can modulate p53-dependent transactivation. p53 can also suppress gene transcription, of at least genes that contain a TATA element in their promoter (13). The biological significance of this aspect of p53 function has, however, not been extensively evaluated.

MDM2 was the first identified cellular oncogene product that bound p53 (16-18). MDM2 derives its name through its location on a murine double minute chromosome present in a cancer cell line (16). p53 transactivates MDM2 transcription through binding of a p53 response element in the first intron of the gene. Mdm2 then binds to the amino-terminal transactivating domain of p53 and in doing so, blocks p53 mediated transcription (17-19). The p53-Mdm2 negative-feedback loop thus limits the duration of p53-mediated transactivation and sets the duration of G1 arrest to the stability of checkpoint proteins that interact with the cell cycle engine to cause arrest. Constitutive overexpression of Mdm2 has been observed in a number of cancers, most notably soft-tissue sarcoma's

(20) and such cells have markedly reduced p53 transactivating activity, despite the presence of wild-type p53.

The GADD45 gene was discovered as one of a number of UV-inducible transcripts isolated from Chinese hamster ovary cells using a subtraction hybridisation procedure (21). GADD45 is also inducible by γ-irradiation, methylmethane sulfonate and growth arrest (21, 22) and the γ-ray inducibility of GADD45 is strictly dependent upon wild-type p53 function (23). GADD45 derives its name from being a growth arrest and DNA damage-inducible gene. p53 transactivates GADD45

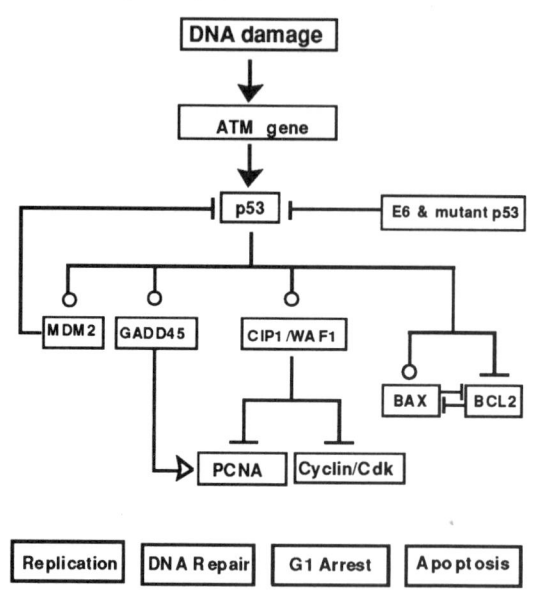

Figure 2. Schematic view of part of the p53 pathway and its response to DNA damage. The p53 control system has been implicated in a number of cellular responses to DNA damage, including G1 arrest, apoptosis, direct inhibition of DNA synthesis and the induction of DNA repair (1, 2). DNA damage activates p53 and requires a functional ATM gene product to do so (23, 40, 41). Activation of p53 is associated with accumulation of the normally short-lived protein and transcriptional activation of a number of downstream effector genes which contain a p53 binding element. Shown are four genes transcriptionally induced by p53: MDM2, GADD45, CIP1/WAF1, BAX. Also shown is the p53-dependent transcriptional repression of one gene: BCL2. The gene product of the CIP1/WAF1 gene can bind to and inhibit cyclin-dependent kinases (29) and this action has been implicated in the mechanism by which Cip1/Waf1 induces G1 arrest (30, 31). Cip1/Waf1 can also bind to PCNA, suggesting that Cip1/Waf1 might also directly regulate DNA replication and/or DNA repair following DNA damage. The p53-regulated gene product Mdm2 can feedback to p53 and inhibit p53 transcriptional activity. This has suggested that Mdm2 provides the means to limit the duration of p53 transcriptional activity (17-19). Gadd45, like p21, can bind to PCNA and this interaction has been implicated as a means by which Gadd45 induces DNA repair and/or inhibits cell cycle progression (24, 25). P53 can also transcriptionally induce the apoptosis-inducing gene, BAX (14). The BAX gene product binds to and inhibits the anti-apoptotic effect afforded by Bcl2. Open circles represent transcriptional activation. Open arrows represent interaction between two components. Closed arrows represent the relay of a signal generated by DNA damage. Blunt-ended lines represent inhibition.

transcription through binding of a p53 response element in the third intron of the gene. Gadd45 binds to PCNA, a protein involved in DNA replication and DNA repair (24, 25). Although, Gadd45 has been implicated in nucleotide excision repair processes (24), this function is obviously dependent upon other factors, since enhanced nucleotide excision repair has not been observed in all studies (26, 27). Overexpression of Gadd45 in cells can also lead to growth suppression (24), indicating that this gene product might also participate in cell cycle arrest following DNA damage.

WAF1 was discovered as a p53-inducible transcript that inhibited cell division (*WAF1*: wild-type p53 activated factor-1, 28). *WAF1* contains p53 binding sites in the promoter of the gene (28). The same gene was simultaneously discovered using a yeast 2-hybrid system to identify proteins that interacted with the Cdk2 protein (*CIP1*: Cdk2 interacting protein, 29). Cip1 turned out to be a potent inhibitor of cyclin-dependent kinase activity and taken together, these two studies linked, for the first time, the p53 pathway to the cell cycle engine (28, 29). DNA damage-induction of *CIP1/WAF1* was linked to G1 arrest through the discovery that the G1/S phase cyclin-dependent kinase, cyclin E/Cdk2, became inhibited as these complexes accumulated Cip1/Waf1 (30, 31). Inhibition of the G1 cyclin-dependent kinases blocks their ability to phosphorylate/inactivate the retinoblastoma (Rb) protein that normally inhibits the G1/S phase transition (32). The Rb protein blocks cells in G1 phase by sequestering the E2F class of transcription factors and repressing genes regulated by E2F (32). Such genes include dihydrofolate reductase, thymidylate synthetase, DNA polymerase alpha, c-myc, cyclin A and B-myb: gene products essential for G1/S phase progression (32).

Recent studies in cells from mice lacking *CIP1/WAF1* genes, confirm the important contribution of this gene product to DNA damage-induced G1 arrest (33). Similar studies performed in the colon cancer cell line, HCT-116, suggest that *CIP1/WAF1* is essential for G1 arrest following DNA damage (34). Cip1/Waf1 can, like Gadd45, bind to PCNA and studies conducted in *in vitro* replication assays suggest that Cip1/Waf1 can inhibit DNA replication directly (35). The effects of Cip1/Waf1 on DNA repair are still open to interpretation with one study reporting no effect on DNA repair (35), while another reporting inhibition of DNA repair (36). This is a complex issue because Cip1/Waf1 also mediates cell cycle arrest following damage, and as such might indirectly aid DNA repair by preventing replication through damaged replicons. Studies in mice lacking the *CIP1/WAF1* gene have shown that this gene product is not required for p53-mediated apoptosis (33) Whether Cip1/Waf1 can protect against apoptosis remains to be determined.

Activation of p53 (37) and subsequent G1 arrest (38) have most notably been seen following treatment with agents that induce DNA strand breaks (Table 1). Such agents include ionising radiation, topoisomerase II inhibitors (etoposide, adriamycin) and bleomycin. DNA base damaging agents, such as the DNA crosslinking agents, cisplatin and nitrogen mustard, produce less G1 arrest than that seen with an equitoxic dose of agents inducing primarily DNA strand breaks (38). Indeed, it has been suggested that DNA strand breaks are sufficient and probably necessary for DNA damaging agents to activate p53 (37). Activation of p53 following DNA damage is associated with accumulation of the normally short-lived protein (half-life approximately 30 minutes) and this occurs without marked changes in p53 mRNA levels (8, 39). P53 accumulation and activation of G1 arrest following DNA damage requires the product of the recently cloned ataxia telangectasia gene (*ATM*) (23, 40, 41). The sequence of the *ATM* gene bears similarities to that of the MEC1 and RAD3 genes in yeast, which are important for cell cycle arrest following DNA damage (40). The *ATM* gene encodes a putative PI-3 kinase and curiously the PI-3 kinase inhibitor, wortmanin, can inhibit p53 activation following DNA damage (42).

APOPTOSIS AND THE P53 TUMOR SUPPRESSOR

Overexpression of wild-type p53 has been shown to induce apoptosis in some cell types (43), while in others, p53 induces a stable G1 arrest (44). The molecular basis behind these different responses has yet to be elucidated, however, some of the most telling cases are from cells overexpressing transcription factors, such as c-myc or E2F-1 (45, 46, 47). Both of these transcription factors along with the transcription factor, B-myb (48), can drive cells into S phase, despite active p53, and it is within S phase that such cells tend to die by apoptosis. Progression into S phase in the presence of a negative growth signal (p53) presumably "confuses" a cell by delivering two conflicting signals: *Arrest* (p53) versus *Progress* (E2F-1, c-myc). Presumably this clash of signals is responsible for the induction of apoptosis (2, and references therein). Whether this occurs directly or indirectly due to DNA damage remains to be determined. Cancer cell types that do not normally undergo apoptosis following wild-type p53-overexpression may not harbor deregulated E2F or c-myc, or alternatively may combat apoptosis by alternative mechanisms such as overexpression of Bcl2 (49).

Table 1. Anticancer agents, targets and cell cycle perturbations.

Anticancer Agent	Target	Mode of Action	Cell Cycle Arrest
Cisplatin Carboplatin Nitrogen Mustard Melphalan Chlorambucil Mitomycin C	DNA	DNA crosslinking	G2, S > G1 phase
Gamma Rays Bleomycin	DNA	DNA strand breaks	G1, G2 > S phase
Camptothecin Topotecan CPT-11	Topoisomerase I	Stabilization of Topo I-DNA cleavable complexes	G2, S > G1 phase
Etoposide Adriamycin mAMSA	Topoisomerase II	Stabilization of Topo II-DNA cleavable complexes	G1, G2 > S phase
Methotrexate 5-Fluorouracil 6-Mercaptopurine	Enzymes	DNA synthesis inhibition	S phase
Taxol Vincristine Vinblastine	Microtubules	Microtubule Stabilizer Microtubule Destabilizer Microtubule Destabilizer	G2/M phase

It appears from several studies that p53-mediated apoptosis can be suppressed by increasing the concentration of serum in the medium or by addition of growth factors (50 and references therein). Furthermore, suppression of apoptosis is associated with a more stable G1 arrest (50). These results suggest that growth factor stimulation relays signals to the decision making processes which determines G1 arrest or apoptosis. In doing so, these signals sway judgement towards G1 arrest and survival. The latter is presumably dependent on the efficiency of DNA repair. These findings are important because they illustrate the contribution of extracellular signals to the final outcome of p53 activation.

One p53-regulated gene, *BAX*, encodes a gene product that counteracts the ability of Bcl2 to protect cells from apoptosis, and overexpression of the *BAX* gene in cells can induce apoptosis (51). Interestingly, *BAX* is only up-regulated in those wild-type p53 cells that will subsequently die by apoptosis (14), suggesting that some cell types can modulate p53's ability to activate this and possibly other downstream effector genes. Such modulation could have profound influences upon the outcome of p53 activation. Despite our knowledge of the relationship between *BAX* induction and p53-mediated apoptosis, it is still unclear how the apoptotic machinery becomes activated and whether the transactivating function of p53 is even required for apoptosis in all cell types (52). The ability of p53 to induce apoptosis in cells with deregulated cell cycle control, highlights one possible mechanism by which p53 could limit tumorigenesis. Other possibilities also exist and will be discussed below.

A ROLE FOR P53-INDUCED APOPTOSIS IN CHEMOTHERAPY

Viral proteins that disrupt the ability of the Rb to repress E2F-1 mediated transcription have also been shown to promote apoptosis in wild-type p53 cells. An example is the adenoviral E1A gene product, which binds to Rb and displaces E2F-1 (53). Furthermore, the E1A gene sensitises murine embryonic fibroblasts to DNA damaging agents (ionising radiation, adriamycin and etoposide) and some DNA synthesis inhibitors (5-fluorouracil), and this increased sensitivity is associated with enhanced susceptibility to apoptosis (54). Wild-type p53 function is important in this sensitisation because E1A-transformed embryonic fibroblasts from transgenic mice that lack p53 are much less susceptible to apoptosis than cells from normal littermates (54). These studies support the notion that activation of p53, in this case, by the use of DNA damaging agents, results in apoptosis in transformed cells that have deregulated E2F-1 function. Apoptosis presumably results from the inability of cells to G1 arrest.

A tumor type with deregulated *c-myc* and in which wild-type p53 cells tend to be more sensitive to DNA damaging agents than cells with mutant-p53, is Burkitt's lymphoma (55, 56). Burkitt's lymphoma is characterised by a chromosome translocation, which in most cases, involves transfer of the *c-myc* gene, normally located on chromosome 8, to the immunoglobulin Heavy chain locus, located on chromosome 14 (57). Studies on a series of human Burkitt's lymphoma cell lines revealed that those cell lines with a functionally intact p53 pathway were, on average, more sensitive to ionising radiation, etoposide, nitrogen mustard and cisplatin than counterpart cell lines with mutant p53 (55, 56). The decreased sensitivity of the mutant p53 cell lines correlated with an evasion of p53-mediated apoptosis, illustrating that the mutant p53 lines had a survival advantage over the lines with wild-type p53. However, other factors contributed to the final outcome of cytotoxic therapy (55, 56). Ionising radiation-induced G1 arrest was clearly evident in the wild-type p53 cell lines prior to apoptosis, suggesting that G1 arrest can occur in cells that will ultimately die of apoptosis. Whether, the cells undergoing apoptosis were those that prematurely escaped G1 arrest despite persistent p53 activation was not determined.

Studies in a variety of cell line systems and mice have created a single impression of the role of p53 in radiosensitivity and chemosensitivity (54-56, 58-61). This paradigm sets forth that p53 disruption, either by mutation or by other means, results in resistance to a wide-variety of DNA damaging agents. Inherent within this proposal is that resistance arises from a lack of p53-induced apoptosis (54). Such a paradigm has gained widespread acceptance and has been supported by findings on the poor prognosis of patients with mutant p53 tumors (62, 63). However, discussed below is an increasing number of studies that have brought to the fore "cellular context" as an important determinant of both survival and possibly the route to resistance.

CELL TYPES NOT INHERENTLY PRONE TO P53-MEDIATED APOPTOSIS: ALTERED SURVIVAL DEPENDS ON THE DNA DAMAGING AGENT!

The effect of p53 disruption on the sensitivity of cells not inherently prone to p53-mediated apoptosis is complex and depends on several factors including the type of DNA damaging agent used and the cell cycle phase in which treatment is administered. Studies performed in the human colon carcinoma RKO cell line showed that p53 disruption did not affect radiosensitivity (64). Also, untransformed murine embryonic fibroblasts from p53 deficient mice exhibited similar radiosensitivity as embryonic fibroblasts from normal littermates (65). Furthermore, disruption of p53 function in the breast carcinoma MCF-7 cell line (66) or studies in a series of head and neck tumor cell lines (67) did not reveal a relationship between p53 status and radiosensitivity. These results suggest that p53 is not a determinant of radiosensitivity in these cell types. In contrast, two studies performed on cells which failed to undergo p53-mediated apoptosis, revealed that the wild-type p53 cells were more sensitive to ionising radiation than the p53 disrupted cells (68, 69). In these latter cases a prolonged G1 arrest without obvious apoptosis was observed, and in one study (68) this G1 arrest was related to a premature senescent-like state driven by prolonged p53 and Cip1/Waf1 activation. Further complicating this issue are the findings of Biard et al., (70) who recently observed a positive relationship between p53 mutation and increased radiosensitivity. This may have been related to the marked genetic instability of the cells in the study of Biard et al., (70), possibly implicating disruption of other DNA damage checkpoints: G2? (see below).

Studies performed in our laboratory using the breast cancer MCF-7 and colon cancer RKO cells revealed that p53 disruption sensitised these cells to cisplatin (66). These findings agreed with an earlier report (59) that showed mutant p53 transfection into a cisplatin-resistant ovarian cancer cell line (A2780/cp70) sensitised the cells to cisplatin. Similar cisplatin-sensitization has recently been reported by Hawkins et al., (71) using the untransformed murine embryonic fibroblasts from p53 "knockout" mice and normal diploid human fibroblasts. We have added another cell line to this group by the finding that human colon cancer HCT-116 cells become more sensitive to cisplatin and nitrogen mustard following inactivation of p53 function (Fan, S. and O'Connor, PM. unpublished observations). Thus, at least 7 cell line systems have been reported to exhibit enhanced cisplatin-sensitivity following p53 disruption. The precise mechanism behind this increased sensitivity still remains to be determined, however, initial reports suggest that p53 disrupted cells have a reduced ability to remove DNA lesions repaired through nucleotide excision (66, 72). This could result from the lack of G1 arrest and/or reduced DNA repair capacity.

Evidence supporting a role for the p53 pathway in DNA repair has come from several quarters including [1] the involvement of Gadd45 in nucleotide excision repair (24), [2] the finding that p53 can bind to the nucleotide excision repair gene product, ERCC3 (73) and [3] the finding that p53 can bind transcription-replication repair factors (74). The dual role of p53 in cell cycle arrest and DNA repair raises the possibility that other checkpoint components might be involved in both processes. Support for this notion is beginning to emerge, most notably from studies in yeast, which have suggested components of the G2 checkpoint are involved in both cell cycle arrest and DNA damaging processing (75). The suggestion that p53 and/or components of the p53 pathway are involved in aspects of DNA repair supports the "Guardian of the Genome" function of p53, which suggests that wild-type p53 suppresses tumorigenesis by maintaining genome stability through cell cycle arrest and enhanced DNA repair (76). Furthermore, the genomic instability of p53 disrupted cells could provide an alternative route to chemotherapy resistance independent of the evasion of p53-mediated apoptosis.

THE G2 CHECKPOINT PROTECTS CELLS AGAINST DNA DAMAGE AND THIS CHECKPOINT CAN BE SELECTIVELY ABROGATED IN CELLS WITH DISRUPTED P53 FUNCTION.

Despite the fact that p53 disrupted cells fail to G1 arrest following DNA damage these cells still go on to arrest in G2 phase (8, 55). Arrest at the G2 checkpoint could partially compensate for lack of p53 function by allowing cells an alternative point in the cell cycle to arrest and conduct DNA repair. If this was the case then disruption of the G2 checkpoint should sensitise cells to DNA damaging agents. In agreement with this possibility chemical abrogators of the G2 checkpoint: caffeine and pentoxifylline, have been shown to increase the cytotoxicity of DNA damaging agents (77, 78). Strikingly, studies performed in MCF-7 cells with intact versus disrupted p53 revealed that a combination of cisplatin and pentoxifylline was more cytotoxic to MCF-7 derivatives with disrupted p53 function (66). Flow cytometric studies on these cells revealed that pentoxifylline preferentially abrogated the G2 checkpoint in cells with disrupted p53 and that enhanced G2 checkpoint abrogation correlated with the enhanced cell killing observed (66). These studies were among the first to reveal vulnerabilities in p53 disrupted cells that could be exploited pharmacologically. Similar studies performed with caffeine also revealed the enhanced sensitivity of p53 disrupted cells (65, 79).

Although, pentoxifylline is presently undergoing clinical trials in the treatment of cancer, tolerable plasma levels of this agent are limited to 30-50 uM (80), well below that needed for G2 checkpoint abrogation *in vitro* (500 uM to 2 mM). We have recently found that UCN-01 (7-hydroxystaurosporine) is also capable of preferentially abrogating the G2 checkpoint in cells with disrupted p53 (81). Furthermore, like pentoxifylline, UCN-01 synergizes with cisplatin to preferentially kill MCF-7 cells with disrupted p53 function (81). UCN-01 has certain advantages over pentoxifylline including a 10,000-fold greater potency than pentoxifylline *in vitro*, and that plasma levels of UCN-01 needed to abrogate G2 checkpoint function *in vitro* can be achieved in rodents without marked toxicity. UCN-01 has just commenced phase I trials, and encouragement is given to investigating the activity of combinations of UCN-01 and cisplatin in patients with wild-type versus mutant p53 gene status.

The molecular basis for the difference in sensitivity of p53 intact versus p53 disrupted cells to these G2 checkpoint abrogators remain to be determined. Focus has been placed on determining whether there are differences in either the uptake of these agents into the cells or alternatively the mechanism of G2 arrest following DNA damage. Most of the work on the mechanism of DNA damage-induced G2 arrest in mammalian cells has concentrated on the status of the Cdc2 kinase (82-90). In human lymphoma CA46 cells, which have mutant p53, DNA damage-induced G2 arrest has been associated with accumulation of hyperphosphorylated and relatively inactive cyclin B1/Cdc2 and cyclin A/Cdc2 complexes, while at the same time cyclin A/Cdk2 complexes remain activated in the G2 arrested cells (83). Failure to activate Cdc2 in these G2 arrested cells appears to stem from a failure to remove inhibitory phosphorylations from the ATP-binding pocket of Cdc2 (threonine-14 and tyrosine-15 phosphorylations). The phosphatase that activates Cdc2 is the Cdc25C phosphatase, and we have found that G2 arrest suppresses the activation of Cdc25C, possibly by blocking its interaction with Cdc2 (84). Such an inhibitor would block the formation of the Cdc2-Cdc25C-autocatalytic-feedback loop that normally brings about entry into mitosis (84).

The fact that p53 disrupted cells still arrest in G2 phase following DNA damage has suggested that p53 is not involved in G2 checkpoint regulation. However, the only conclusion that can be drawn from such studies is that p53 is not essential for G2 arrest following DNA damage. This leaves open the possibility that p53 could contribute an extra layer of protection to the G2 checkpoint and that when p53 function is disrupted, this layer of protection is peeled away to leave a cell still capable of G2 arrest but now more vulnerable to manipulation. Our selective manipulation of the G2 checkpoint in p53 disrupted cells with pentoxifylline and UCN-01 (66, 81) support this possibility. A molecular dissection of this vulnerability is still awaited.

NEW TERRITORY: THE S PHASE DNA DAMAGE CHECKPOINT

S phase progression is also delayed following DNA damage and this delay is associated with a block to new replicon synthesis as well as suppression of DNA elongation (4, 5, 91). Evidence that this delay is mediated through a DNA damage checkpoint comes from the observations that the *ATM* gene product is required to induce S phase arrest (4, 5) and that chemical agents can abrogate S phase arrest following DNA damage (83). Also, recent studies have shown that the homolog of the *ATM* gene: *MEC1* as well as the *RAD53* gene are required for S phase arrest following DNA damage in yeast (5). Both *MEC1* and *RAD53* are also required to halt cells in G2 phase following DNA damage as well as halt entry into mitosis when DNA synthesis is inhibited by hydroxyurea (92). Taken together these findings indicate overlap in the operation of the S and G2 checkpoints. S phase delay is induced by a wide variety of DNA damaging agents including alkylating (5, 83) and platinating (93) agents. The S phase delay induced by the DNA crosslinking agent, nitrogen mustard, can be abrogated by pentoxifylline (83), suggesting that abrogation of the S phase checkpoint could contribute to the synergy that has been seen between DNA crosslinking agents and pentoxifylline (66). However, given the overlap between the S and G2 checkpoints, it will be necessary to develop novel approaches to distinguish the contribution of each checkpoint to survival.

SUMMARY

A number of lines of evidence from organisms as diverse as yeast and human point to the importance of cell cycle checkpoints in determining both genome stability as well as the survival of cells treated with DNA damaging agents. The most widely studied checkpoint component in human cells, the p53 tumor suppressor, maintains genome stability and suppresses tumorigenesis. Disruption of the p53 pathway not only accelerates tumorigenesis but also imparts altered sensitivity to currently available chemotherapeutic agents. Furthermore, cellular context is an important consideration in predicting the survival of p53 disrupted cancer cells. The G1, S and G2 phase DNA damage checkpoints are complex and involve considerable overlap in components that not only participate in cell cycle arrest but probably also in DNA damage processing. Pharmacological exploitation of checkpoint control defects in cancer cells is possible with agents like pentoxifylline and UCN-01, and these agents are presently undergoing clinical trials. Continued exploration of these checkpoint systems could uncover further vulnerabilities to focus new drug discovery efforts towards.

REFERENCES

1. O'Connor, P.M. and Kohn, K.W. (1992) *Semin. Cancer Biol.*, **3**, 409-416.
2. Hartwell, L.H. and Kastan, M.B. (1994) *Science*, **266**, 1821-1825.
3. Weinert, T.A. and Hartwell, L.H. (1988) *Science*, **241**, 317-322.
4. Painter, R.B. and Young, B.R. (1980) *Proc. Natl. Acad. Sci. (USA)*, **77**, 7315-7317.
5. Paulovich, A.G. and Hartwell, L.H. (1995) *Cell*, **82**, 841-847.
6. Weinert, T.A. and Hartwell, L.H. (1990) *Mol. Cell. Biol.*, **10**, 6554-6564.
7. Livingstone, L. R., White, A., Sprouse, J., Livanos, E., Jacks, T. and Tlsty, T. D. (1992) *Cell* **70**, 923-935.
8. Kastan, M.B., Onyekwere, O., Sidransky, D., Vogelstein, B. and Craig, R.W. (1991) *Cancer Res.* **51**, 6304-6311.
9. Kuebitz, S.J., Plunkett, B.S., Walsh, W.V. and Kastan, M.B. (1992) *Proc. Natl. Acad. Sci. (USA)*, **89**, 7491-7495.
10. Kessis, T.D., Slebos, R.J., Nelson, W.G., Kastan, M.B., Plunkett, B.S., Han, S.M., Lorincz, A.T. Hedrick, L. and Cho, K.R. (1993) *Proc. Natl. Acad. Sci. (USA)*, **90**, 3988-3992.
11. Hollstein, M., Sidransky, D., Vogelstein, B and Harris, C.C. (1991) *Science*, **253**, 49 53.
12. Slebos, R.J.C., Lee, M.H., Plunkett, B.S., Kessis, T.D., Jacks, T., Hedrick, L.,Kastan, M.B., and Cho, K.R. (1994) *Proc. Natl. Acad. Sci. USA* **91**, 5320-5324.
13. Zambetti, G.P. and Levine, A.J. (1993) *FASEB J.* **7**, 855-865.
14. Zhan, Q., Fan, S., Bae, I., Guillouf, C., Liebermann, D.A., O'Connor, P.M. and Fornace, A. J. Jr. (1994) *Oncogene*, **9**, 3743-3751.
15. Chen, X., Bargonetti, J. and Prives, C. (1995) *Cancer Res.*, **55**, 4257-63.
16. Fakharzadeh, S.S., Trusko, S.P. and George, D.L. (1991) *EMBO J.* **10**, 1565-1569.

17. Momand, J., Zambetti, G.P., Olson, D.C., George, D. and Levine, A.J. (1992) *Cell* **69**, 1237-1245.
18. Oliner, J.D., Pietenpol, J.A., Thiagalingam, S., Gyuris, J., Kinzler, K. W. and Vogelstein, B. (1993) *Nature*, **362**, 857-860.
19. Wu, X., Bayle, J.H., Olson, D. and Levine, A. J. (1993) *Genes & Dev.* **7**, 1126-1132.
20. Leach, F.S., Tokino, T., Meltzer, P., Burrell, M., Oliner, J.D., Smith, S., Hill, D.E., Sidransky, D. Kinzler, K.W., and Vogelstein, B. (1993) *Cancer Res.*, **53**, 2231-4.
21. Fornace, A.J.Jr., Nebert, D.W., Hollander, M.C., Luethy, J.D., Papathanasiou, M., Fargnoli, J. and Holbrook, (1989) *N.J. Mol. Cell. Biol.*, **9**, 4196-4203.
22. Papathanasiou, M., Kerr, N., Robbins, J.H., McBride, O.W., Alamo, I.J., Barrett, S.F., Hickson, I.D. and Fornace, A.J.Jr. (1991) *Molec. Cell. Biol.*, **11**, 1009-1016.
23. Kastan, M.B., Zhan, Q., El-Deiry, W.S., Carrier, F. Jacks, T., Walsh, W.V., Plunkett, B.S., Vogelstein, B. and Fornace, A.J.Jr. (1992) *Cell*, **71**, 587-597.
24. Smith, M.L., Chen, I., Zhan, Q. Bae, I., Chen, C., Gilmer, T., Kastan, M.B., O'Connor, P.M. and Fornace, A.J.Jr. (1994) *Science* **266**, 1376-1380.
25. Chen, I.T., Smith, M.L., O'Connor, P.M and Fornace, A.J. Jr. (1995) *Oncogene*, **11**, 1931-7.
26. Kearsey, J.M., Shivji, M.K., Hall, P.A. and Wood, R.D. (1995) *Science*, **270**, 1004-5.
27. Kazantsev, A. and Sancar, A. (1995) *Science*, **270**, 1003-4.
28. El-Deiry, W.S., Tokino, T., Velculescu, V.E., Levy, D.B., Parsons, R., Trent, J.M., Lin, D., Mercer, W.E., Kinzler, K.W. and Vogelstein, B. (1993) *Cell* **75**, 817-825.
29. Harper, J.W., Adami, G.R., Wei, N., Keyomarsi, K. and Elledge, S.J. (1993) *Cell* **75**, 805-816.
30. El-Deiry, W.S., Harper, J.W., O'Connor, P.M., et al., (1994) *Cancer Research*, **54**, 1169-1174.
31. Dulic, V., Kaufmann, W.K., Wilson, S.J., Tlsty, T.D., Lees, E., Harper, J.W., Elledge, S.J. and Reed, S.I. (1994) *Cell* **76**, 1013-1023.
32. Adams, P.D. and Kaelin, W.G.Jr. (1995) *Semin. Cancer Biol.*, **6**, 99-108.
33. Deng, C., Zhang, P., Harper, J.W., Elledge, S.J. and Leder, P. (1995) *Cell*, **82**, 675-684.
34. Waldman, T., Kinzler, K. and Vogelstein, B. (1995) *Cancer Res.*, **55**, 5187-5190.
35. Waga, S., Hannon, G.J., Beach, D., and Stillman, B. (1994) *Nature*, **369**, 574-578.
36. Pan, Z.Q., Reardon, J.T., Li, L., Flores-Rozas, H., Legerski, R., Sancar, A. and Hurwitz, J. (1995) *J. Biol. Chem.*, **270**, 22008-16.
37. Nelson, W.G. and Kastan, M.B. (1994) *Mol. Cell. Biol.*, **14**, 1815-1823.
38. Fan, S., El-Deiry, W. S., Bae, I., Freeman, J., Jondle, D., Bhatia, K., Fornace, Jr, A. J., Magrath, I., Kohn, K. W., and O'Connor, P. M. (1994) *Cancer Res.*, **54**, 5824-5830.
39. Maltzman, W. and Czyzyk, L. (1984) *Molec. Cell. Biol.*, **4**, 1689-1694.
40. Savitsky, K., Bar-Shira, A., Gilad, S., et al., (1995) *Science*, **268**, 1169-1173.
41. Canman, C. E., Wolff, A. C., Chen, C. Y., Fornace, Jr, A. J., and Kastan, M. B. (1994) *Cancer Res.*, **54**, 5054-5058.
42. Price, B.D. and Youmell, M.B. (1996) *Cancer Res.*, **56**, 246-250.
43. Wang, Y., Ramqvist, T., Szekely, L., Axelson, H., Klein, G. and Wiman, K. G. (1993) *Cell Growth & Diff.* **4**, 467-473.
44. Lin, J., Wu, X., Chen, J., Chang, A. and Levine A.J. (1994) *Cold Spring Harb. Symp. Quant. Biol.*, **59**, 215-23.
45. Wu, X. and Levine, A.J. (1994) *Proc. Natl. Acad. Sci., USA* , **91**, 3602-3606.
46. Qin, X., Livingston, D. M., Kaelin, Jr, W. G., and Adams, P. D. (1994) *Proc. Natl. Acad. Sci. USA*, **91**, 10918-10922.
47. Hermeking, H., and Eick, D. (1994) *Science*, **265**, 2091-2093.
48. Lin, D., Fiscella, M., O'Connor, P.M., Jackman, J., Chen, M., Luo, L.L., Sala, A., Travali, S., Apella, E. and Mercer, E. (1994) *Proc. Natl. Acad. Sci. (USA)* **91**, 10079-10083.
49. Walton, M.I., Whysong, D., O'Connor, P.M., Korsmeyer, S.J. and Kohn, K.W. (1993) *Cancer Res.*, **53**, 1853-1861.
50. Canman, C.E., Gilmer, T.M., Coutts, S.B. and Kastan, M.B. (1995) *Genes Dev.*, **9**, 600-11.
51. Oltvai, Z.N., Milliman, C.L. and Korsmeyer, S.J. (1993) *Cell*, **74**, 609-19.
52. Caelles, C., Helmberg, A. and Karin, M. (1994) *Nature*, **370**, 220-223.
53. White, E. and Gooding, L.R. (1994) *Apoptosis II* (Tomei, L.D. and Cope, F.O., eds.) pp. 111-142, Cold Spring Harbor Laboratory Press, NY.
54. Lowe, S.W., Ruley, H.E., Jacks, T., and Houseman, D.E. (1993) *Cell*, **74**, 957-967.
55. O'Connor, P., Jackman , J., Jondle, D., Bhatia K., Magrath, I., and Kohn, K.W. (1993) *Cancer Res.*, **53**, 4776-4780.
56. Fan, S., El-Deiry, W. S., Bae, I., Freeman, J., Jondle, D., Bhatia, K., Fornace, Jr, A. J., Magrath, I., Kohn, K. W., and O'Connor, P. M. (1994) *Cancer Res.*, **54**, 5824-5830.
57. Magrath, I. (1990) *Adv. Cancer Res.*, **5**, 133-270.
58. McIlwrath, A. J., Vasey, P. A., Ross, G. M., and Brown, R. (1994) *Cancer Res.*, **54**, 3718-3722.
59. Brown, R., Clugston, C., Burns, P., Edlin, A., Vasey, P., Vojtesek, B. and Kaye, S. (1993) *Int. J. Cancer*, **55**, 678-683.

60. Lee, J.M., and Bernstein, A. (1993) *Proc. Natl. Acad. Sci. USA*, **90**, 5742-5746.
61. Lowe, S.W., Bodis, S., McClatchey, A., Remington, L., Ruley, H.E., Fisher, D.E., Housman, D.E. and Jacks, T. (1994) *Science*, **266**, 807-10.
62. Righetti, S.C., Torre, G.D., Pilotti, S. et al., (1996) *Cancer Res.*, **56**, 689-693.
63. Nakai, H. and Misawa, S. (1995) *Leuk. Lymphoma*, **19**, 213-21.
64. Slichenmyer, W.J., Nelson, W.G., Slebos, R.J., and Kastan, M.B. (1993) *Cancer Res.*, **15**, 4164-4168.
65. Powell, S.N., DeFrank, J.S., Connell, P., Preffer, D., Dombkowski, D., Tang, W. and Friend, S. (1995) *Cancer Res.*, **55**, 1643-1648.
66. Fan, S., Smith, M.L., Rivet, D., Duba, D., Zhan, Q., Kohn, K.W., Fornace, A.J.Jr. and O'Connor, P.M. (1995) *Cancer Res.*, **55**, 1649-1657.
67. Brachman, D.G., Beckett, M., Graves, D., Haraf, D., Vokes, E. and Weichselbaum, R. (1993) *Cancer Res.*, **53**, 3666-3670.
68. Di Leonardo, A., Linke, S.P., Clarkin, K., and Wahl, G. M. (1994) *Genes & Develop.* **8**, 2540-2551.
69. Yount, G.L., Haas-Kogan, D.A., Viadair, C.A., Haas, M., Dewey, W.C. and Israel, M.A. (1996) *Cancer Res.*, **56**, 500-506.
70. Biard, D.S., Martin, M., Rhun, Y.L., Duthu, A., Lefaix, J.L., May, E. and May, P. (1994) *Cancer Res.*, **54**, 3361-3364.
71. Hawkins, D.S., Demers, G.W. and Galloway, D.A. (1996) *Cancer Res.*, **56**, 892-898.
72. Smith, M.L., Chen, I.T., Zhan, Q., O'Connor, P.M. and Fornace, A.J.Jr. (1995) *Oncogene*, **10**, 1053-1057.
73. Wang, X. W., Forrester, K., Yeh, H., Feitelson, M. A., Gu, J. R., and Harris, C. C. (1994) *Proc. Natl. Acad. Sci. USA*, **91**, 2230-2234.
74. Wang, X.W., Yeh, H., Schaeffer, L., et al., (1995) *Nature Genetics*, **10**, 188-195.
75. Lydall, D. and Weinert, T. (1995) *Science*, **270**, 1488-91.
76. Lane, D. P. (1992) *Nature*, **358**, 15-16.
77. Fingert H.J., Chang J.D., Pardee A.B. (1986) *Cancer Res.*, **46**, 2463-2467.
78. Fingert H.J., Pu A.T., Chen Z., Googe P.B., Alley M.C., Pardee A.B. (1988) *Cancer Res.*, **48**, 4375-4381.
79. Russell, K., Weins, L.W., Galloway, D.A. and M. Groudine. (1995) *Cancer Res.*, **55**, 1639-1642.
80. Shapiro, C.L., Dezube, B.J., Wright, J., Teicher, B.A., Pardee, A.B., Rei, E, III. and Henderson, I.C. (1992) *Pentoxifylline, Leukocytes and Cytokines* (Mandell, G.L and Novick, W.J.Jr. eds), pp. 34-39, Hoechst-Roussel Pharmaceuticals, NJ.
81. Wang, Q., Fan, S., Eastman, A., Worland, P.J., Sausville, E.A. and O'Connor, P.M. (1996) *J. Natl. Cancer Inst.*, in press.
82. O'Connor P.M., Ferris D.K., White G.A., Pines J., Hunter T., Longo D.L., Kohn K.W. (1992) *Cell Growth & Diff.*, **3**, 43-52.
83. O'Connor, P.M., Ferris, D. K., Pagano, M., Draetta, G., Pines, J., Hunter, T., Longo, D.L. and Kohn, K.W. (1993) *J. Biol. Chem.*, **268**, 8298-8308.
84. O'Connor, P. M., Ferris, D. K., Hoffmann, I., Jackman, J., Draetta, G. and Kohn, K.W. (1994) *Proc. Natl. Acad. Sci. USA*, **91**, 9480-9484.
85. Lock R.B., Ross W.E. (1990) *Cancer Res.*, **50**, 3761-376.
86. Lock R.B., Ross W.E. (1990) *Cancer Res.*, **50**, 3767-3771.
87. Lock R.B. (1992) *Cancer Res.*, **52**, 1817-1822.
88. Tsao, Y-P., D'Arpa, P., Liu, L.F. (1992) *Cancer Res.*, **52**, 1823-1829.
89. Muschel, R.J., Zhang, H.B., Iliakis, G., Mckenna, W.G. (1991) *Cancer Res.*, **51**, 5113-5117.
90. Maity, A., McKenna, W.G. and Muschel, R.J. (1995) *EMBO J.*, **14**, 603-9.
91. Larner, J.M., Lee, H. and Hamlin, J.L. (1994) *Molec. Cell. Biol.*, **14**, 1901-1908.
92. Weinert, T.A., Kiser, G.L. and Hartwell, L.H. (1994) *Genes & Develop.* **8**, 652-655.
93. Sorenson, C.M., Barry, M.A., Eastman, A. (1990) *J. Natl. Cancer Inst.*, **82**, 749-755.

chapter 17
Cellular responses to antimetabolite anticancer agents: cytostasis versus cytotoxicity

Janet A. Houghton and Peter J. Houghton[1]

Department of Molecular Pharmacology St. Jude Children's Research Hospital,
332 N. Lauderdale, Memphis, TN 38105-2794, USA
[1]To whom correspondence should be addressed

Thymineless death is an important cytotoxic response to several classes of antimetabolite agents used in the treatment of patients with carcinomas and hematopoeitic malignancies. Cell death induced by lack of dThd results in the formation of DNA nucleosomal ladders, and hence would be defined as a form of apoptosis. Although drug resistance to these agents has been extensively studied, relatively little attention has been focused on events downstream of dTTP depletion that determine the ultimate fate of the cancer cell. In this article we review some of the emerging data that suggests the role of p53 in determining whether the cellular response to dThd deprivation is cytostasis or cytotoxicity (apoptosis).

INTRODUCTION

Cellular responses to cytotoxic agents involve complex pathways that ultimately determine whether a cell survives or dies. Death may be through a process referred to as programmed cell death, or apoptosis, that occurs as a natural physiological process with well recognised morphologic and biochemical markers (chromatin condensation, internucleosomal DNA cleavage), observed, for example, during normal development and maintenance of tissue homeostasis (1,2). Alternatively, cells may lose viability and die through necrosis or other through pathways that do not result in internucleosomal cleavage of DNA, characteristic of classic apoptosis. In measurement of response to cytotoxic agents used to treat cancer cells, death may be defined as reproductive death. Measurement of apoptosis *per se* may not give a true indication of whether a drug exerts a cytotoxic effect, as the rate of apoptosis is a dynamic variable, changing with time; hence failure to see early apoptosis may not equate with lack of drug effect as onset may be delayed, and hence missed. Conversely, colony forming assays that measure putative 'stem cells' in which cells are exposed to drugs for prolonged periods may overestimate cytotoxic effects, and may reflect purely cytostatic events if cytostasis is prolonged. Conceptually, at least, certain classes of anticancer agents that inhibit DNA replication would be anticipated to cause cytostasis rather than cytotoxicity. These include antimetabolite drugs including 5-fluoropyrimidines, antifolates, and inhibitors of pyrimidine biosynthesis such as PALA, each of which either directly or indirectly inhibits formation of dTMP. Of interest mechanistically, is that agents which induce 'thymineless death' represent major chemotherapeutic classes used for palliative treatment of carcinomas of colon, breast, head and neck and certain lung tumours, as well as certain hematopoietic malignancies. However, molecular events controlling the cell cycle that influence or control the commitment process in 'thymineless death' are poorly understood.

Understanding genetic events that dictate whether the cellular response to a potentially toxic insult is growth arrest (cytostasis) or death (cytotoxicity), would appear important, as potentially tumour cells may enter a period of reversible cytostasis during the period of treatment thereby causing resistance to the cytotoxic actions of these drugs. Also, more comprehensive understanding of how genetic events involved in transformation and progression of tumour cells impact on cellular responses to these antitumour agents may lead to approaches to selectively modulate cytotoxicity in neoplastic cells.

THYMIDYLATE SYNTHASE AS A TARGET IN CANCER CHEMOTHERAPY

Thymidylate synthase is the final, and rate limiting, enzyme involved in *de novo* synthesis of dTMP. Thymidylate synthase is a two-substrate enzyme forming a ternary complex between dUMP and CH_2-H_4PteGlu, resulting in the reductive methylation of dUMP, and in the oxidation of CH_2-H_4PteGlu, which acts as a one carbon donor, to H_2PteGlu (3). Inhibition of thymidylate synthase leads to depletion of dTTP, and 'thymineless

Abbreviations: FUra, 5-fluorouracil; FdUrd, 5-fluoro-2'-deoxyuridine; dThd, thymidine; dTMP, thymidine monophosphate; H_4PteGlu, tetrahydrofolate; 10-CHO-H_4PteGlu, 10-formyltetrahydrofolate; CH_2-H_4PteGlu, 5,10-methylenetetrahydrofolate; H_2PteGlu, dihydrofolate; PALA, N-(phosphonacetyl)-L-aspartate; dNTP, deoxyribonucleotide triphosphates.

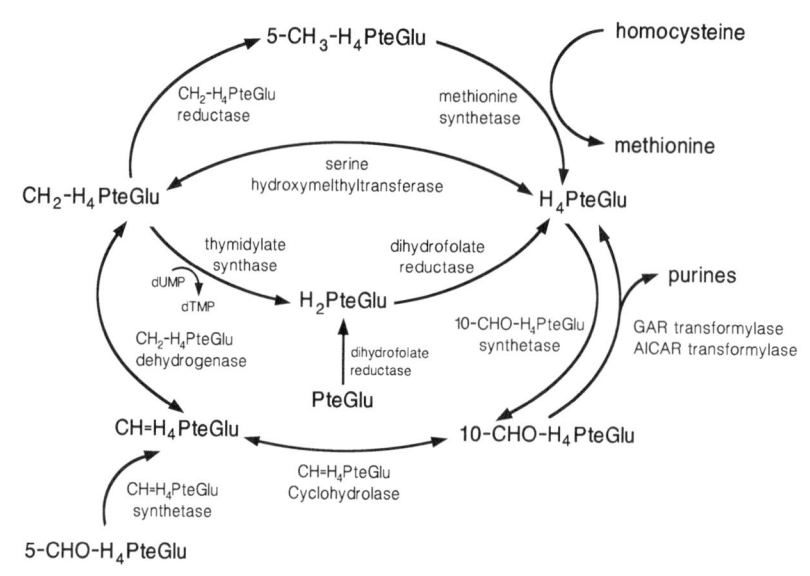

Figure 1. Metabolic pathways for natural folates. (Reprinted from Tew K.D., Houghton, P.J., and Houghton, J.A. (1993) In: *Preclinical and Clinical Modulation of Anticancer Drugs*. 197-321. Boca Raton. CRC Press with permission).

death'. Classically, the process of thymineless death has been described by the inhibition of DNA synthesis followed by a period of unbalanced growth (i.e. continued protein and RNA synthesis) before loss of viability (4,5).

Metabolic pathways for natural folates are presented in Figure 1, and show the interconversion of reduced folates that serve as one carbon unit donors in synthesis of purines, pyrimidines and methionine. Of the interconversions, that involving the formation of dTMP is unique in that CH_2-H_4PteGlu is converted to H_2PteGlu, whereas other interconversions result in tetrahydrofolate products. Thus, in cells in which dihydrofolate reductase is inhibited by antifolate agents such as methotrexate, continued activity of thymidylate synthase ultimately results in depletion of tetrahydrofolate pools (6-8), leading to accumulation of H_2PteGlu. The higher polyglutamate forms of the accumulated H_2PteGlu may act to inhibit other enzymes involved in purine biosynthesis (9). Thus, agents that inhibit dihydrofolate reductase indirectly inhibit thymidylate synthase, through depletion of CH_2-H_4PteGlu. In addition, higher polyglutamate forms of methotrexate are potent direct inhibitors of this enzyme (10). More recently, folate-based specific inhibitors of thymidylate synthase have entered clinical trial. The prototype agent, CB3717, a 2-amino-4-hydroxy quinazoline with a propargyl group at the N^{10} position, inhibited thymidylate synthase with a Ki of 2.7 nM, and was competitive with CH_2-H_4PteGlu (11). Polyglutamylation to the triglutamate species increased its potency as an inhibitor >100-fold. Further development led to clinical evaluation of ZD1694 (tomudex), N-(5-[N-(3,4-dihydro-2 methyl-4-oxoquinazolin-6-ylmethyl) -N-methyl-1 amino]-2-thenoyl)-L-glutamic acid (12). Four other inhibitors are in currently in clinical trial (13), ZD9331, 1843U89, and two novel structures (AG-331 and AG-337) which were designed based on the high resolution crystal structure of E. coli thymidylate synthase (14). Development of these newer inhibitors has been reviewed recently (13). The other major therapeutic class of agent that targets thymidylate synthase are the 5-fluoropyrimidines. Thymidylate synthase constitutes the target for the pyrimidine analogues 5-fluorouracil (FUra), and 5-fluoro 2'-deoxyuridine (FdUrd). After intracellular conversion of FUra or FdUrd to FdUMP, this metabolite forms a tight binding quasi-irreversible complex with the enzyme and CH_2-H_4PteGlu, the cofactor used in the conversion of dUMP to dTMP. This interaction and the consequences of inhibiting dTMP synthesis de novo have been reviewed extensively (15,16). Although there is a vast literature concerning the mechanism(s) of action of each of these classes of anticancer drugs, each is complicated by potentially multiple mechanisms of action.

In addition to the lack of availability of dTTP to maintain semi conservative DNA synthesis, or DNA repair, elevated dUTP as a consequence of inhibition of dihydrofolate reductase by methotrexate, or thymidylate synthase inhibition or FdUTP (after conversion of FUra or FdUrd) synthesised from dUMP or FdUMP, respectively, may be incorporated into DNA and subsequently excised to produce a futile round of incorporation/excision that may contribute to the process of thymineless death (17-20). The action of

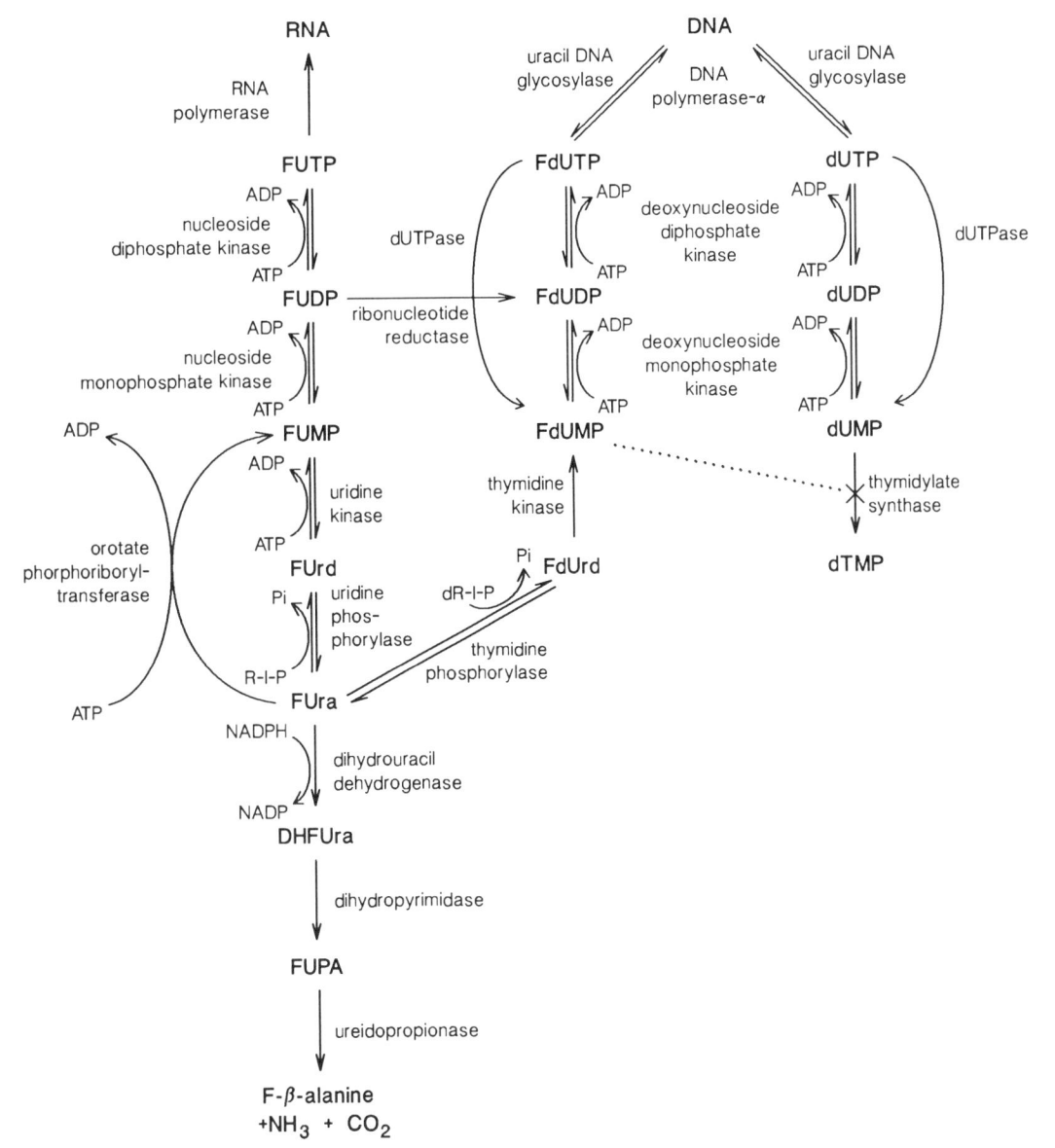

Figure 2. Pathways for metabolism of Fura and Ura deoxyribonucleotides. Reprinted from Tew K.D., Houghton, P.J., and Houghton, J.A. (1993) In: *Preclinical and Clinical Modulation of Anticancer Drugs*. 197-321. Boca Raton. CRC Press with permission).

methotrexate is further complicated due to accumulation of H$_2$PteGlu polyglutamates, which may have inhibitory activity against enzymes involved in purine synthesis *de novo* (9). Similarly, the action of 5-fluoropyrimidines is very complex, in part due to complex metabolic interconversions (Figure 2), where the analogue utilises pathways of normal pyrimidine metabolism. In addition to incorporation into DNA, FUTP may be incorporated into all RNA species, resulting in altered processing (reviewed in 16,21), or converted to FUDP-sugars by glycosyltransferases, although the significance to cellular response is unknown (22,23). *In vivo*, catabolism of 5-fluoropyrimidines results in production of F-β-alanine, which may be incorporated into peptides, or may be converted to fluoroacetate, a compound that may be responsible for neurotoxicity of this class of agent (24). Consequently, determination of the mechanism by which these agents cause cytotoxicity or cytostasis is very complex. One may anticipate that a 'pure' inhibitor of thymidylate synthase would be selectively toxic to cells undergoing DNA replication (S-phase specific), and hence self-limiting in its action, giving a biphasic survival curve in a typical colony forming assay, when cells are exposed for brief periods to the agent. Most evidence suggests that cells lose clonogenic potential when exposed to a thymidylate synthase inhibitor whilst in S-phase. However, of particular interest is the response of cells exposed to these agents while traversing G1 phase of the cell cycle. Two approaches to understanding these responses have been undertaken. The first has been to transfect specific genes, and determine whether the cellular response is altered. The other is to create the biochemical lesion through deletion of thymidylate synthase, and to withdraw exogenous dThd. The latter approach has allowed a more specific understanding of the molecular process of thymineless death, and the sequence of events that are associated with irreversible commitment.

DEOXYRIBONUCLEOSIDE TRIPHOSPHATE POOL PERTURBATIONS DURING THYMINELESS DEATH

The response to agents that deplete cellular pools of dTTP can be specifically studied in cells deficient in thymidylate synthase (25-27). Such cells are dThd auxotrophs, requiring exogenous nucleoside to maintain DNA synthesis. Consequently, withdrawal of dThd recapitulates the effect of pharmacologic inhibition of thymidylate synthase without potential confounding effects such as incorporation of fraudulent nucleotides into nucleic acids, or effects in the purine biosynthetic pathway. The system has advantage, also, in that irreversible commitment to reproductive cell death may be determined by an inability to rescue cells by re-addition of dThd. Several groups have reported derivation of thymidylate synthase deficient mammalian cells, and these have been used to understand alterations in deoxyribonucleoside triphosphate pools (dNTP) and their biological significance, following dThd withdrawal. In TS⁻ murine FM3A cells, withdrawal of dThd was associated with rapid loss of clonogenic potential and DNA fragmentation ($t_{1/2}$ ~6 hours), associated with a rapid loss of dTTP, dGTP and a modest elevation of dATP. It was concluded that the imbalance between dGTP and dATP may be the signal for cell death (28). The kinetics with which GC3/c1 human colon adenocarcinoma cells lacking thymidylate synthase lost colony formation after dThd withdrawal differs markedly from that described for the murine cell line (29). In asynchronous populations of cells 50% could not be rescued (i.e. had lost clonogenic potential) when dThd was added after 65 hours of dThd starvation. Studies of dNTP pool perturbations demonstrated characteristics different from those in murine models following induction of the thymineless state. Specifically, the dGTP pool was not depleted, and after a brief decline dCTP pools recovered to control levels within 24 hours, Figure 3. Thymineless death, as determined by nucleosomal

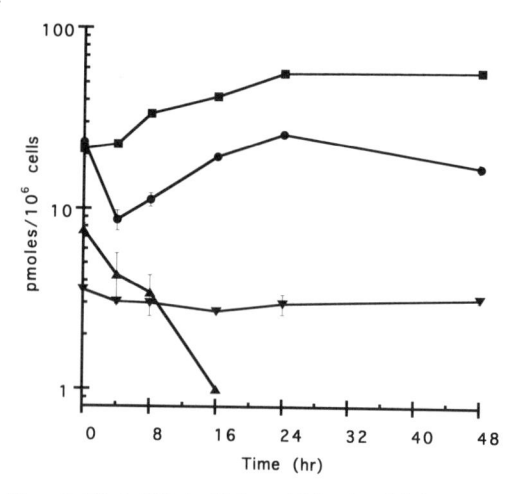

Figure 3. Effect of dThd withdrawal following dThd starvation in human colon cancer cells deficient in thymidylate synthase. Following dThd deprivation, changes in dNTP pools were determined in asynchronously growing TS⁻ cells for periods up to 48 hours. Data represent mean ± SD of two to four determinations per point ■,dATP; ▲,dTTP; ●, dCTP; ▼, dGTP. Basal levels of dNTP pools were 21 ± 0.7, 22 ± 0.2, and 3 ± 1 pmol/10⁶ cells respectively. (From Houghton, J.A. et al, *Clin. Cancer Res.* 1 723-730 with permission).

DNA ladder formation and loss in clonogenic survival, correlated with a temporally associated decrease in dTTP and rise in the dATP pool. In cell lines where dUTP (or FdUTP following treatment with FUra or FdUrd) has been considered to play a significant role in the mechanism of induction of thymineless death dUTP has been detected at relatively high levels within cells (90-338 pmol/10⁶ cells). However, in thymidylate synthase negative GC3/c1 cells dUTP was not accumulated to detectable levels following dThd withdrawal (29)

THE ROLE OF THE TUMOUR SUPPRESSOR GENE P53 IN ANTICANCER DRUG RESPONSE

There is considerable evidence that the p53 tumour suppressor gene is involved in the control of cell growth and increases the susceptibility of cells to drug-induced toxicity (30-32). The encoded protein functions as a G1 cell cycle checkpoint, and becomes elevated in response to DNA-damaging agents (33,34). Inactivation of p53 by mutation or deletion results in genomic instability (35,36), and in p53-deficient transgenic mice, allows normal foetal development, but predisposes the animals to the development of several different types of neoplasm (37,38). The loss of G1 checkpoint control in movement of cells from G1 to S phase has also reduced cellular sensitivity to cytotoxic agents (30) and ionising radiation (39-41).

The functional activity of wild-type p53 (wtp53), which is frequently lost following mutation, is mediated via transcriptional activation or down-regulated expression of genes involved in growth arrest, survival or cell death responses, through sequence-specific interactions with DNA (42-44). These genes include 1) WAF1 (45), which encodes a 21kD protein (p21$^{Waf1/Cip1}$) that binds to and inhibits G1 cyclin-dependent kinases, thereby preventing entry of cells from G1 to S phase, 2) the Bax gene, which is also directly transcriptionally activated by wtp53 (46), and is a regulator of the induction of apoptosis in response to toxic stress (47-51), and 3) the homologous protein, Bcl-2, which opposes the function of Bax, and can prolong cell survival (47,52). Expression of the Bcl-2 gene, which contains a p53-dependent negative response element (44), may be down-regulated by wtp53, and the coordinate regulation of Bax and Bcl-2 can occur simultaneously (48). Both Bax and Bcl-2 are expressed in the normal colon, Bax immunoreactivity being stronger in the epithelial cells of the upper portions of the crypts (53), and Bcl-2, in the lower crypt cells (54,55). During neoplastic progression, deregulated expression of Bcl-2 has been reported (55). Constitutive expression of Bcl-2 has resulted in protection from p53-induced apoptosis (56,57), and the cytotoxicity of anticancer agents (58-61), including drugs that inhibit thymidylate synthase (62,63).

P53 AND THE CELLULAR RESPONSE TO ANTIMETABOLITE AGENTS

The initial work by Lowe (30) demonstrated p53-dependent cytotoxicity for agents frequently used in the treatment of cancer in murine embryo fibroblasts (MEFs) that expressed the E1A or were transformed with E1A plus T24-*ras* oncogenes. This study showed that MEFs derived from p53$^{+/+}$, p53$^{+/-}$ or p53$^{-/-}$ animals were quite resistant to ionising radiation, FUra and inhibitors of topoisomerase II inhibitors. However, when cells were transfected with the viral E1A oncogene cells from p53$^{+/+}$ or p53$^{+/-}$ embryos became sensitive to these agents, whereas MEFs from p53$^{-/-}$ embryos were resistant, Table 1. In cells expressing E1A and E1B genes, the latter of which counteracts wtp53 (64), p53$^{+/+}$ cells were also resistant. Further, studies where MEFs were transformed with E1A and T24-*ras* also demonstrated p53-dependent sensitivity to radiation, FUra and topoisomerase II inhibitors. In contrast, cell killing by sodium azide, which uncouples mitochondrial oxidative phosphorylation and depletes ATP was p53-independent. While these studies are seminal in defining p53-dependent sensitivity to various cytotoxic agents, including FUra, the role of p53 is complicated by an enhanced rate of spontaneous apoptosis in E1A-expressing p53$^{+/+}$ cells, particularly under conditions of stress, such as serum depletion. In addition, it is unclear from this study whether the mechanism of FUra-induced cytotoxicity was directed at thymidylate synthase, as the ability of dThd to protect cells was not examined. Thus, the cytotoxicity observed could have been a consequence of any of the several mechanisms of action discussed previously. However, that p53 may be implicated in determining the cytotoxic response to antimetabolites is becoming more evident. Using PALA an inhibitor of carbamoylphosphate synthetase, it was found that Li-Fraumeni fibroblasts (p53$^{+/-}$), and murine fibroblasts that expressed mutant alleles, but transfected with a wt p53 allele, growth arrested in presence of drug. In contrast, cells lacking wtp53 proceeded from G1 to S-phase, resulting in amplification of the CAD gene (encoding the trifunctional enzyme carbamoyl-phosphate synthetase, aspartate transcarbamylase, dihydroorotase) and drug resistance (35). Similar results were reported by Yin et al (36), where growth arrest in cells exposed to PALA occurred only in cells expressing a wtp53 allele. Interestingly, MDAH087 cells expressing wild type and a mutant allele, or where wild type p53 was expressed in clones expressing two mutant alleles, the 'normal' phenotype was observed. i.e. cells growth arrested without amplification of the gene CAD), Table 2. These data suggested that wtp53 was involved in establishing a G1 checkpoint that arrests cells in response to PALA treatment. Results obtained in clones expressing mutant and wtp53 alleles where entry into S-phase was partially inhibited, suggests that the ratio of mutant to wt protein, rather than the absolute level of wtp53, is the critical determinant of PALA-induced G1 arrest (36). Since PALA depletes both pyrimidines and purines, including UTP, CTP, dCTP, dGTP and dTTP (65), it is not possible to attribute the cytostatic effect to depletion of any particular nucleotide pool, or to infer that similar events would occur with specific inhibitors of thymidylate synthase. Growth arrest in cells exposed to the dihydrofolate reductase inhibitor methotrexate is less clear, as cells with either wild-type (35,66) or mutant (36) p53 alleles arrested in G1.

P53 AND THE CELLULAR RESPONSE TO THYMIDYLATE SYNTHASE INHIBITORS

Preliminary clinical studies implicate p53 in influencing the responses of patients with advanced colon cancer to treatment with FUra administered with leucovorin (67). Since the p53 gene is mutated in high frequency (>75%) in colon carcinomas, occurring during progression from adenoma to carcinoma (68), this observation is of considerable importance. The basis for modulation of FUra with a reduced folate is to increase the level of thymidylate synthase inhibition by promoting efficient and tight binding of the proximal metabolite FdUMP to the enzyme by increasing the concentration of higher polyglutamate forms of

Table 1. Colony regression following irradiation or treatment with chemotherapeutic agents.

Treatment	Exogenous Genes	p53 Genotype	Colony Viability (72 hr)		
			Regressing	Resistant	Resistant (%)
Radiation 5 Gy	E1A	(+/+)	5	1	17
	E1A	(+/-)	25	0	0
	E1A	(-/-)	1	24	96
	E1A + E1B	(+/+)	4	9	69
	E1A + E1B	(+/-)	9	16	64
	E1A + E1B	(-/-)	ND	ND	ND
5-Fluorouracil 1 µM	E1A	(+/+)	5	0	0
	E1A	(+/-)	23	2	8
	E1A	(-/-)	1	26	96
Etoposide 0.2 µM	E1A	(+/+)	5	1	17
	E1A	(+/-)	20	5	20
	E1A	(-/-)	0	25	100
Adriamycin 0.2 µg/ml	E1A	(+/+)	3	0	0
	E1A	(+/-)	22	3	12
	E1A	(-/-)	0	25	100

Reproduced from Lowe, S.W. et al. 1993 Cell 74, 957-967 with permission from Cell Press.

Table 2. Gene Amplification in Normal Cells and Precrisis and Postcrisis Li-Fraumeni Fibroblasts.

Cell Source	Passages in Culture	p53	PALA Colonies		CAD Gene Amplification	Growth Arrest in PALA
			Number of Cells Selected	PALA Frequency		
MDAH170		WT/WT	$0/1 \times 10^8$	$<1 \times 10^{-8}$	-	+
NHF-3		WT/WT	$0/2 \times 10^8$	$<5 \times 10^{-9}$	-	+
IMR-90		WT/WT	$0/1 \times 10^8$	$<1 \times 10^{-8}$	-	+
MDAHO41	5-10	WT/FS/184	$0/2 \times 10^7$	$<5 \times 10^{-8}$	-	+
MDAHO41	>100	FS184/FS184	$28/5 \times 10^6$	5.6×10^{-6}	+	-
MDAHO87	10-15	WT/248	$0/2 \times 10^7$	$<5 \times 10^{-8}$	-	+
MDAHO87	>100	248/248	$22/5 \times 10^6$	4.4×10^{-6}	+	-
MDAHO87 C4		248/248/WT	$0/3 \times 10^6$	$<3.3 \times 10^{-7}$	-	+
MDAHO87 C9E		248/248/WT	$0/1 \times 10^7$	$<1 \times 10^{-7}$	-	+
MDAHO87 C10		248/248/WT	$0/1 \times 10^7$	$<1 \times 10^{-7}$	-	+
MDAHO87 C5, C6, C7, C8, C9L		248/248/WT	$3-6/1 \times 10^6$	4×10^{-6} (average)	+	-

Reproduced from Yin, Y. et al. 1992 Cell 70, 937-948 with permission from Cell Press.

CH_2-H_4PteGlu (reviewed in 16). In the study reported, 6 of 14 patients had mutated p53 alleles, and of these 4 of 5 did not respond to treatment. In contrast, of 9 patients with tumours that expressed wt p53, 7 demonstrated response to therapy. Although used less frequently in patients, FdUrd is converted to FdUMP by thymidine kinase, and is a potent inhibitor of thymidylate synthase (15,21). Transfection of wtp53 into HL60 leukemic cells that lack p53 expression was found to sensitise cells ~ 10-fold to this inhibitor. Additional studies, in which cells were synchronised in G1 phase using mimosine showed a further 10-fold sensitisation, as cells appeared to remain blocked in G1, while p53- cells proceeded into S-phase (69). It remains unclear whether such arrested cells maintain proliferative potential, or whether they are reproductively dead. In the model proposed by Lane (70), the p53 tumour suppressor gene is not required during a normal cell cycle; this is supported by the relative lack of a phenotype in transgenic mice where p53 has been disrupted (71). Where the cell has been stressed by exposure to radiation or other genotoxic agents, p53 is considered to cause G1 arrest, followed by repair, or apoptosis. According to this model, cells lacking p53 would not arrest in G1, but would progress through DNA replication which would lead to additional mutations, and mitotic failure leading to death. In this context it is appropriate to question whether G1 arrest after ionising radiation does protect cells from reproductive death (i.e. failure to form colonies). Available data suggest not. Cells lacking p53 may be more resistant to ionising radiation than cells with functional p53 when exposed in G1 phase (41). Thus, synchronised populations of U-87 MG human glioblastoma cells in which p53 had been inactivated by expression of a dominant-negative allele were more resistant to ionising radiation when exposed in G1 than were cells expressing wild-type alleles. Cells expressing functional p53 appeared to arrest for prolonged periods in G1 phase whereas cells with functionally attenuated p53 continued to replicate and form colonies. However, cells expressing wtp53 did not undergo apoptosis. These results are consistent with those reported for PALA-induced cytostasis being p53-dependent, although different from results obtained with transformed MEF cells where expression of p53 was associated with apoptosis in drug- and radiation-treated cells.

P53 AND CELLULAR RESPONSE TO dTTP DEPLETION

While the biological significance of G1 arrest of irradiated cells is not apparent as a mechanism to promote reproductive cell survival, there is accumulating evidence that p53 may determine the fate of cells exposed to antimetabolites that target thymidylate synthase (67,69). However, there are several concerns over these studies; in the clinical studies reported it is assumed that responses to the combination of FUra and leucovorin is mediated by inhibition of thymidylate synthase, and although conceptually appealing, and probable, is not possible to confirm. In studies where either wt or dominant-negative p53 alleles are transfected, the transgene is not under control of the endogenous promoter, hence regulation may not recapitulate the endogenous gene. Consequently, it can only be assumed that the product of a gene under these conditions mimics that of the endogenous protein. This may not be correct, as often exogenous genes are expressed from viral promoters, which may result in very high levels of expression, and because of constitutive expression are not regulated in a cell cycle phase-specific manner.

REGULATION OF P53 IN COMMITMENT TO THYMINELESS DEATH

Despite the concerns raised, more direct evidence suggests that thymineless death, the end product involved in targeting thymidylate synthase with inhibitors, is a p53-dependent process (72,73). As discussed above, cells deficient in thymidylate synthase (TS-) commit to reproductive death upon withdrawal of exogenous dThd. In the human colon cancer cell line (GC_3/c1), the TS- variant loses the

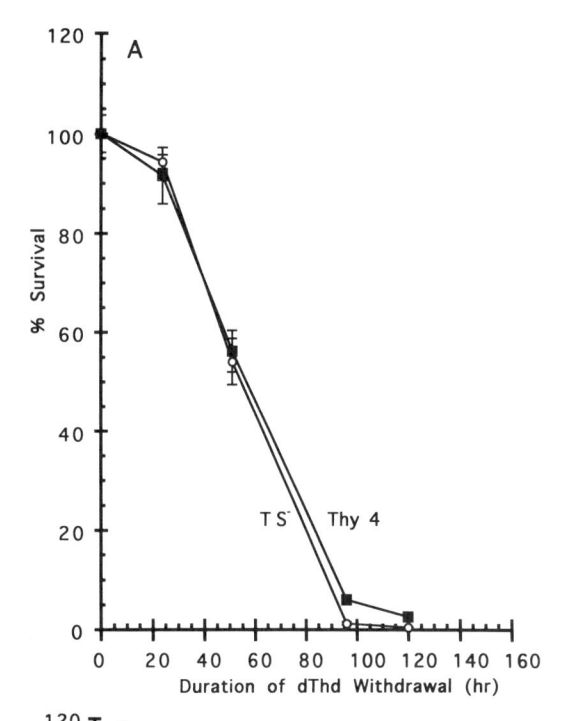

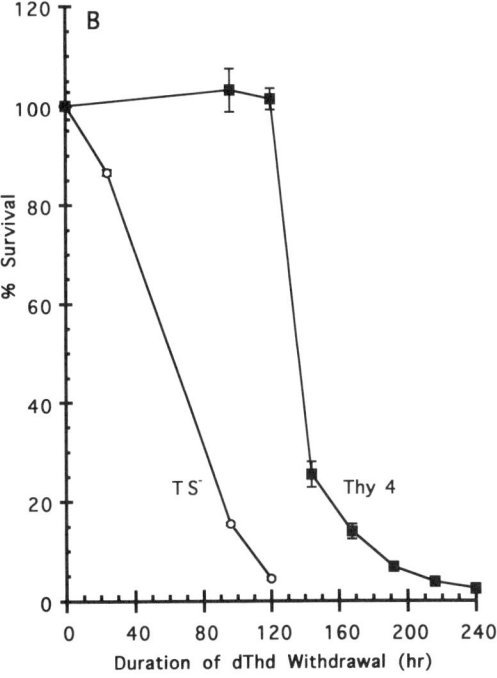

Figure 4. A, Clonogenic survival of asynchronous populations of TS- (●) and Thy4 (○) cells following dThd withdrawal. B, clonogenic survival of G0 synchronized TS- (○) and Thy4 (□) cells following leucine restoration and dThd withdrawal. Cells were rescued with dThd (20μM) for intervals up to 10 days, and clonogenic survival was determined 11 days after rescue. (From Houghton, J.A. et al. 1994 *Cancer Res.* **54** 4967-4973 with permission).

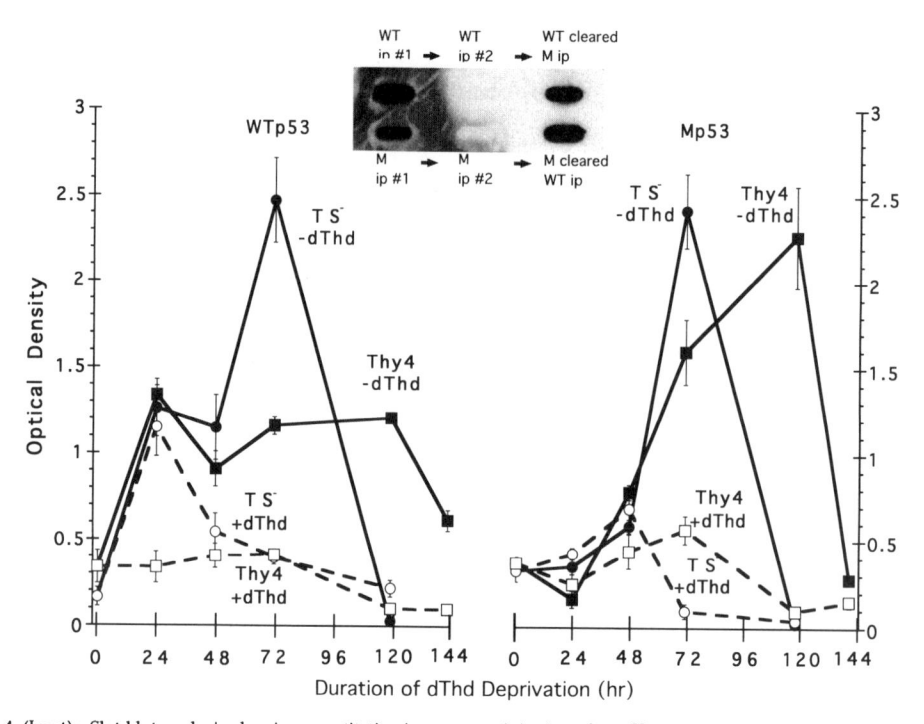

Figure 5. *A* (Inset): Slot-blot analysis showing quantitative immunoprecipitation of wtp53 protein by the PAb1620 antibody, and of mp53 protein by the PAb240 antibody in extracts derived from TS⁻cells. Lysates were immunoprecipitated twice using PAb1620 then further immunoprecipitated with PAb240 (upper panel), or alternatively, precipitated twice to remove mutant p53 (PAb240) and then with PAb1620 specific for wtp53 (lower panel). Immunoprecipitated proteins were detected following slot-blot analysis on nitrocellulose membranes, using the pan p53 antibody DO-I (which recognizes both wtp53 and mp53) conjugated to horseradish peroxidase, and ECL reagents. No cross reactivity between antibodies and proteins was evident.
B: Quantitative immunoprecipitation of wtp53 and mp53 proteins from TS⁻ and Thy4 cells following release from G0 synchrony in the absence or presence of dThd. After immunoprecipitation, proteins were quantitated following slot blot analysis. Data represent the mean ± SD of 2-4 determinations from 2 independently derived sets of samples for each of the p53 proteins. Extracts derived from equivalent cell numbers were analyzed. (Reproduced from Harwood et al. (1996) *Oncogene* **12**, 2057-2067 with permission).

ability to be rescued by dThd with a $t_{1/2} \sim 65$ hours. We initially speculated that it should be possible to select a subpopulation of TS⁻ cells that may be resistant to thymineless death, and that comparison between these variants and the TS⁻ cells would be valuable in defining critical regulatory steps of the thymineless death process. To examine whether a subpopulation of TS⁻ cells were resistant to dThd starvation, cells were initially arrested in G1 by leucine deprivation, thus avoiding dTTP depletion during S-phase which is lethal. Cells were then released from G1 block in the absence of dThd. Under this condition loss of clonogenic potential was similar to that in asynchronous populations ($t_{1/2} \sim 65$ hours), but a low frequency of cells were recovered by addition of dThd after periods of up to 28 days of dThd starvation (74). A clonal population (Thy4) was derived from these surviving cells and compared to parental TS⁻ cells. As shown in Figure 4A, in asynchronous populations the loss of clonogenic potential was similar in TS⁻ and Thy4 populations. In contrast, in G1 synchronised cells Thy4 cells maintained clonogenic potential for up to 5 days in the absence of dThd, whereas over the same period there was >90% loss of clonogenic potential in TS⁻ cells, Figure 4B. Loss of clonogenicity in both TS⁻ and Thy4 cells was associated with intranucleosomal cleavage of DNA, characteristic of apoptosis, and uptake of the vital dye trypan blue. These data suggest that Thy4 cells are able to arrest in G1 phase, and maintain viability for at least 5 days, after which time there is a precipitous loss of viability, indicating delayed apoptosis. Subsequent studies with these cells indicated that acute and delayed apoptosis induced by dThd deprivation correlated with expression of p53 and p53-regulated genes (73). Similar to the parental GC₃/c1 cells, both TS⁻ and Thy4 cells were found heterozygous for p53, expressing both wt and mutant (240 Ser to Arg) alleles. Examination of p53 proteins in TS⁻ and Thy4 cells released from G1 block in the absence of dThd revealed interesting differences between these clones (73). In TS⁻ cells, which commit to acute apoptosis, wt p53 predominated, whereas in Thy4 levels of mutant protein exceeded those of wt for up to 5 days, Figure 5. Functional activity of p53 was confirmed in TS⁻ cells and increased in cells released from G1 block under conditions of dThd starvation, whereas p53 function was not detected in Thy4 cells, after transient transfection of the p53-reporter plasmid p50-2-CAT (75). In agreement with these data, levels of MDM2 expression increased in TS- cells within 24 hours, and continued to increase over the time course of the experiment in cells released from G1 block in the presence or absence of dThd. Of interest was the change in the ratio of wt to mutant p53 protein in Thy4 cells as they committed to delayed apoptosis, where the level of mutant protein, determined in immunoprecipitates, decreased whereas the level of wt protein demonstrated a lesser fall resulting in wt:mutant > 1. Upon onset of delayed apoptosis, levels of MDM2 increased ~ 4-fold, consistent with functional p53. These results strengthen the contention that the effect of p53 on cellular response may be defined by the ratio of wt to mutant protein, as proposed by Yin et al (36).

CELLULAR RESPONSES: GROWTH ARREST OR DEATH?

The studies discussed have focused predominantly on the cellular responses to antimetabolite agents, particularly those that ultimately deplete dTTP pools. However, expression of a functional p53 reduces clonogenic potential in most studies irrespective of the DNA damaging agent, either chemical or ionising radiation, giving the impression of p53 sensitising cells to these agents. However, expression of functional p53 may result in prolonged cytostasis (Li Fraumeni fibroblasts, HL-60 leukemic cells, and U-87 glioblastoma cells) or alternatively trigger acute cell death (E1A expressing or transformed MEF cells, and TS- colon adenocarcinoma cells). Some insight as to what may trigger apoptosis may be derived from these studies. In progression of colon carcinoma, mutation in one p53 allele is frequently followed by deletion of the second, demonstrating inactivation of function during oncogenesis (68). Several studies have demonstrated spontaneous apoptosis in human colon carcinoma cells into which wt p53 is transfected, suggesting that co-expression of wt p53 in cells transformed with specific oncogenes (e.g. activated Ki-*ras*) may be incompatible with viability. Thus the cell cycle regulatory signal is in conflict with the oncogene driving the cell through the cell cycle, leading to the 'clash' of conflicting signals that activates the apoptotic cascade (76). Similarly, E1A expressing p53$^{+/+}$ MEFs have a higher rate of spontaneous apoptosis when grown under conditions of stress (low serum), suggesting that these cells are now susceptible to many forms of damage interpreted as stress. The implication is that such tumour cells would be susceptible to many classes of chemotherapeutic agent, as the cell is already 'primed' to undergo apoptosis under such stress (i.e. DNA damage). Whether this applies in general to clinical cancer, or is restricted to certain malignancies associated with specific transforming events remains to be resolved. However, in addition to p53, an understanding of the interaction of effector molecules downstream of p53 that regulate cell death or survival responses, including Bax and Bcl-2, respectively (49,52,77), will be essential.

ACKNOWLEDGEMENTS

This work was supported in part by PHS awards R37CA32613, CA21765, and CA23099 from the National Cancer Institute, and by American, Lebanese, Syrian Associated Charities (ALSAC)

REFERENCES

1. Raff, M.C. (1992) *Nature* (London) **356**: 397-400.
2. Wyllie, A.H. (1980) *Nature* (London) **284**: 555-556.
3. Friedkin, M. (1973) *Thymidylate Synthetase, Adv. Enzymol.* **38**: 235-292.
4. Cohen, S.S. (1971) *Ann N.Y. Acad. Sci.* **106**: 292-301.
5. Goulian, M., Bleile, B.M., Dickey, L.M., Grafstrom, R.H., Ingraham, H.A., Neynaber, S.A., Peterson, M.S., and Tseng, B.Y. (1986) *Adv. Exp. Med. Biol.* **195**: 89-95.
6. Harrap, K.R., Jackman, A.L., Newell, D.R., Taylor, G.A., Hughes L.R., and Calvert, A.H. (1989) *Adv. Enzyme Regul.* **29**:161-179.
7. Roth, B. and Cheng, C.C. (1982) *Prog .Med. Chem.* **19**: 269-331.
8. Sirotnak, F.M., Burchall, J.J., Ensminger, W.D., and Montgomery, J.A. (1984). *Folate Antagonists as therapeutic Agents*. Vol 1, Cambridge. Academic press.
9. Allegra, C.J., Drake, J.C., Jolivet, J., and Chabner, B.A. (1985) *Proc. Natl .Acad .Sci. (USA)* **82**: 4881-4885.
10. Allegra, C.J., Chabner, B.A., Drake, J.C., Lutz, R., Rodbard, D., and Jolivet J. (1985) *J. Biol. Chem.* **260**: 9720-9726.
11. Jones, T.R., Calvert, A.H., Jackman, A.L., Brown, S.J., Jones, M., and Harrap, K.R. (1981) *Eur J. Cancer*, **17**: 11-19.
12. Jackman, A.L., Jodrell, D.I., Gibson, W., and Stephens T.C. (1991) In: R.A. Harkness, G.B. Elion and N. Zollner (eds), *Purine and Pyrimidine Metabolism in Man*. Vol 7, Part A pp 19-23. New York, Plenum Press.
13. Touroutoglou, N., and Pazdur, R. (1996) *Clin. Cancer Res.* **2**: 227-243.
14. Matthews, D.A., Appelt, K., Oatley, S.J., and Xuong, N.H. (1990) *J. Mol. Biol.* **214**: 923-936.
15. Danenberg, p.V. (1977) *Biochim. Biophys. Acta.* **473**: 73-92.

16. Tew K.D., Houghton, P.J., and Houghton, J.A. (1993) In: *Preclinical and Clinical Modulation of Anticancer Drugs.* 197-321. Boca Raton. CRC Press.
17. Goulian, M., Bleile, B., and Tseng, B.Y. (1980) *J .Biol. Chem.* 255: 10630-10637.
18. Jackson, R.C., Jackman, A.L., and calvert, A.H. (1983) *Biochem. Pharmacol.* 32: 3783-3790.
19. Danenberg, P.V., and Lockshin, A. (1981) *Pharmacol. Ther.* 13: 69-90.
20. Lonn, U., and Lonn, S. (1984) *Cancer Res.* 44: 3414-3418.
21. Houghton, J.A., and Houghton, P.J. (1984) In B.W. Fox (ed). *Handbook of Experimental Pharmacology* 72: 515-549.
22. Pogolotti, A.L., Nolan, P.A., and santi, D.V. (1981) *Anal. Biochem.* 117: 178-186.
23. Peters, G.J., Laurensse, E., Lankelma, J., Leyva, A., and Pinedo, H.M. (1984) *J Cancer Clin. Oncol.* 20: 1425-1431.
24. Koenig, H., and patel, A. (1970) *Arch. Neurol.* 1: 181-183.
25. Li, I.C., and Chu, E.H.Y. (1984) *J. Cell Physiol.* 120: 109-116.
26. Ayusawa, D., Shimizu, K., Koyama, H., Takeishi, K., and seno, T. (1983) *J. Biol. Chem.* 258: 12448-12454.
27. Houghton, P.J., Germain, G.S., Hazelton, B.J., Pennington, J.W., and Houghton, J.A. (1989) *Proc. Natl. Acad. Sci. (USA)* 86: 1377-1381.
28. Yoshioka, A., tanaka, S., Hiraoka, O., Koyama, Y., Hirota, Y., Ayusawa, D., Seno, T., Garrett, C., and Wataya, Y. (1987) *J . Biol. Chem.* 282: 8235-8241.
29. Houghton, J.A., Tillman, D.M., and Harwood, F.G. (1995) *Clin. Cancer Res.* 1: 723-730.
30. Lowe, S.W., Ruley, H.E., Jacks, T., and Housman, D.E. (1993) *Cell* 74: 957-967.
31. Fan, S., El-Deiry, W.S., Bae, I., Freeman, J., Jondle, D., Bhatia, K., Fornace, A.J., Magrath, I., Kohn, K.W., and O'Connor, P.M. ((1994) *Cancer Res.* 54: 5824-5830.
32. Lowe, S.W., Bodis, B., McCarthy, A., Remington, L.H., Ruley, H.E., Fisher, D., Housman, D.E. and Jacks, T. (1994) *Science* (Washington DC) 266: 807-810.
33. Kastan, M.B., Onyekwere, O., Sidransky, D., Vogelstein, B. and Craig, R.W. (1991). *Cancer Res.* 51, 6304-6311.
34. Kuerbitz, S.J., Plunkett, B.S., Walsh, W.V. and Kastan, M.B. (1992). *Proc. Natl. Acad. Sci.* 89, 7491 -7495.
35. Livingstone, L.R., White, A., Spouse, J., Livanos, E., Jacks, T. and Tlsty, T.D. (1992). *Cell* 70, 923-935.
36. Yin, Y., Tainsky, M.A., Bischoff, F.Z., Strong, L. C. and Wahl, G.M. (1992). *Cell* 70:937-948.
37. Donehower, L.A., Harvey, M., Slagle, B.L., McArthur, M.J., Montgomery, C.A., Butel, J.S. and Bradley, A. (1992). *Nature* 356, 215-221.
38. Jacks, T., Remington, L., Williams, B.O., Schmitt, E.M., Halachmi, S., Bronson, R.T. and Weinberg, R.A. (1994). *Curr. Biol.* 4, 1-7.
39. Lee, J.L. and Bernstein, A. (1993). *Proc. Natl. Acad. Sci.* 90, 5742-5746.
40. McIlwrath, A.J., Vasey, P.A., Ross, G.M. and Brown, R. (1994). *Cancer Res.* 54, 3718-3722.
41. Yount, G.L., Haas-Kogan, D.A., Vidair, C.A., Haas, M., Dewey, W.C., and Israel, M.A. (1996) *Cancer Res.* 56:500-506.
42. El-Deiry, W.S., Kern, S.E., Pietenpol, J.A., Kinzler, K.W. and Vogelstein, B. (1992). *Nature Genet.* 1, 45-49.
43. Pietenpol, J.A., Papadopoulos, N., Markowitz, S., Willson, J.K.V., Kinzler, K.W., and Vogelstein, B. (1994). *Cancer Res.* 54, 3714-3717.
44. Miyashita, T., Harigai, M., Hanada, M. and Reed, J.C. (1994a). *Cancer Res.* 54: 3131-3135.
45. El-Deiry, W.S., Tokino, T., Velculescu, V.E., Levy, D.B., Parsons, R., Trent, J.M., Lin, D., Mercer, W.E., Kinzler, K.W. and Vogelstein, B. (1993). *Cell* 75: 817-825.
46. Miyashita, T. and Reed, J.C. (1995). *Cell* 80:293-299.
47. Oliver, F.J., Marvel, J., Collins, M.K.L. and Lopez-Rivas, A. (1993). *Biochem. Biophys. Res. Commun.* 194: 126-132.
48. Oltvai, Z.N., Milliman, C.L., and Korsmeyer, S.J. (1993). *Cell* 74: 609-619.
49. Oltvai, Z.N. and Korsmeyer, S.J. (1994). *Cell* 79: 189-192.
50. Selvakumaran, M., Lin, H-K., Miyashita, T., Wang, H.G., Krajewski, S., Reed, C., Hoffman, B. and Liebermann, D. (1994). *Oncogene* 9: 1791-1798.
51. Zhan, Q., Fan, S., Bae, I., Guillouf, C., Liebermann, D.A., O'Connor, P.M. and Fornace, A.J. (1994). *Oncogene* 9: 3743-3751.
52. Reed, J.C. (1994). *J. Cell Biol.* 124: 1-6.
53. Krajewski, S., Krajewski, M., Shabaik, A., Wang, H-G, Irie, S., Fong, L. and Reed, J.C. (1994). *Cancer Res.* 54: 5501-5507.
54. Lu, Q-L., Poulsom, R., Wong, L. and Hanby, A.M. (1993). *J. Pathol.* 169: 431-437.
55. Bronner, M.P., Culin, C., Reed, J.C. and Furth, E.E. (1995). *Am. J. Pathol.* 146: 20-26.
56. Chiou, S-K., Rao, L., and White, E. (1994). *Mol. Cell Biol.* 14: 2556-2563.
57. Wang, Y., Szekely, L., Okan, I., Klein, G. and Wiman, K.G. (1993). *Oncogene* 8: 3427-3431.
58. Dole, M., Nunez, G., Merchant, A.K., Maybaum, J., Rode, C.K., Bloch, C.A. and Castle, V.P. (1994). *Cancer Res.* 54: 3253-3259.
59. Miyashita, T. and Reed, J.C. (1992). *Cancer Res.* 52: 5407-5411.

60. Walton, M.I., Whysong, D., O'Connor, P.M., Hockenberry, D., Korsmeyer, S.J. and Kohn, K.W. (1993). *Cancer Res.* **53**: 1853-1861.
61. Kamesaki, S., Kamesaki, H., Jorgensen, T.J., Tanizawa, A., Pommier, Y. and Cossman, J. (1993). *Cancer Res.* **53**: 4251-4256.
62. Fisher, T.C., Milner, A.E., Gregory, C.D., Jackman, A.L., Aherne, G.W., Hartley, J.A., Dive, C. and Hickman, (1993) J.A. *Cancer Res.* **53**: 3321-3326.
63. Oliver, F.J., Marvel, J., Collins, M.K.L. and Lopez-Rivas, A. (1993). *Biochem. Biophys. Res. Commun.* **194**: 126-132.
64. Debbas, M., and White, E. (1993) *Genes Dev.* **7**: 546-554.
65. Moyer, J.D., Smith, P.A., Levy, E.J., and Handschumacher, R.E. (1992) *Cancer Res.* **42**: 4525-4531.
66. Tltsy, T. (1992) *Proc. Natl. Acad. Sci.(USA)* **87**:3132-3136.
67. Lenz H-J, Danenberg, K.D. Leichman, L., Leichman, C., and Danenberg, P.V. (1995) *Proc Am. Assoc. Cancer .Res.* **36**: 3332A.
68. Fearon, E.R., and Vogelstein, B. (1990) *Cell* **61**: 759-767.
69. Lenz, H-J., Ju, J-F., Danenberg, K.D., Banergee, D., Bertino, J.R., and Danenberg, P.V. (1995) *Proc. Am .Assoc .Cancer Res.* **36**: 121A
70. Lane, D.P. (1992) *Nature* (London) **358**: 15-16.
71. Jacks, T., Remington, L., Williams, B., Scmitt, E., Halachmi, S., Bronson, R., and Weinberg, R. (1994) *Curr .Biol.* **4**: 1-7.
72. Houghton, J.A., Harwood, F.G., Tillman, D.M. (1995) *Proc. American Assoc. Cancer Res.* **36**: 18A
73. Harwood, F.G., Frazier, M.W., Krajewski, S., Reed, J.C., and Houghton, J.A. (1996) *Oncogene* **12**, 2057-2067.
74. Houghton, J.A., Harwood, F.G., and Houghton, P.J. (1994) *Cancer Res.* **54**: 4967-4973.
75. Zambetti, G.P., Bargonetti, J., Walker, K., Prives, C., and Levine, A.J. (1992) *Genes Dev.* **6**: 1143-1152.
76. Fisher, D.E. (1994) *Cell* **78**: 539-542.
77. Miyashita, T., Krajewski, S., Krajewska, M., Wang, H.G., Lin, H.K., Liebermann, D.A., Hoffman, P., and Reed, J.C. (1994). *Oncogene* **9**: 1799-1805.

chapter 18
Telomeres, telomerase, and the cell cycle

Karen J. Buchkovich

University of Illinois at Chicago, Department of Pharmacology
835 S. Wolcott Avenue, Chicago, IL 60612, USA

Telomeres protect the ends of chromosomes from degradation and fusion. In most eukaryotes telomeres are replicated by a specialised polymerase, telomerase. Telomerase synthesises one strand of the telomere; while conventional DNA polymerases synthesise the complementary strand. Additional processing of telomeres occurs in ciliates and yeast during each cell cycle. Telomerase activity and RNA levels change as cells enter and exit the cell cycle. Gradual telomere shortening in the absence of telomerase does not immediately affect cell cycling; however, "critically" short telomeres are hypothesised to play a role in senescence and the triggering of DNA damage checkpoints.

Telomeres are the specialised nucleoprotein structures located at the ends of linear chromosomes. Telomeres play an essential role in maintaining chromosome stability. The absence of telomeres leads to chromosome fusion and breakage cycles (1-3).

Most eukaryote chromosomes end in a simple, tandemly repeated DNA sequence (reviewed in 4, 5). In humans, the sequence TTAGGG is repeated hundreds to thousands of times (6-10). The number of telomeric repeats varies from organism to organism with some ciliates having fewer than 50 nucleotides of repeated DNA and some strains of mice having greater than 100 kilobase of repeats (reviewed in 4, 5). The mean length of telomeres differs from strain to strain in yeast, but is maintained around an average "set point" for each individual strain (11). Mammals show tissue-to-tissue variation in average telomere length (9, 10, 12-14). Within a single mammalian cell, telomere length varies from chromosome to chromosome (15, 16)

Several physical characteristics of telomeres are conserved between distantly related organisms. One DNA strand is usually rich in guanine and thymidine residues and forms the 3' end of the telomere; the complementary cytosine- and adenine-rich strand forms the 5' end (4, 5, 17). The GT-rich strand in several species of ciliates and slime mold extends beyond the other strand, resulting in a 3' overhang (18-20). This specialised DNA structure of telomeres binds telomere-specific proteins and serves as a substrate for the enzymes that maintain telomeres (reviewed in 4, 17, 21-24).

TELOMERE REPLICATION

The replication of linear DNA molecules by conventional DNA polymerases is predicted to result in the gradual shortening of telomeres (reviewed in 4, 25, 26). Because DNA polymerases work only in the 5' to 3' direction and require short oligonucleotides as primers, shortening of telomeric DNA by the length of a terminal primer is expected in each cell cycle. Consistent with this prediction, the average length of telomeres in some mammalian somatic cells shorten as the cells proliferate *in vitro* and *in vivo* (10, 27-32). Single-cell eukaryotes and mammalian germ cells maintain telomeres at relatively constant length (11, 31; reviewed in 33, 34). Thus, mechanisms must exist in these cells to compensate for the telomere shortening that results from conventional DNA replication. Although recombination may be a secondary mechanism (35-38), telomere synthesis *de novo* is the primary mechanism for maintaining telomere length (reviewed in 34).

Telomerase is a specialised DNA polymerase that synthesises telomeric repeats *de novo*. The enzyme does not require a DNA template, because it uses an integral RNA molecule as a template (reviewed in 21, 39). Telomerase activity was identified using biochemical experiments to test for *de novo* addition of DNA repeats to telomere-like oligonucleotides (40). The biochemistry of telomerase has been best studied in ciliates, particularly in *Tetrahymena* (reviewed in 21, 22; see also 41-43). This organism was chosen for early studies because genome fragmentation during macronuclear development generates tens of thousands of telomeres (reviewed in 44), and thus the enzyme(s) that synthesises telomere repeats was expected to be abundant in ciliates. Telomerase activity has been identified in cellular extracts of ciliates (*Tetrahymena, Euplotes, Oxytricha*), yeasts (*Saccharomyces cerevisiae* and *S. casstellii*), frog (*Xenopus*), mouse and human. The sequence synthesised by each telomerase *in vitro* corresponds to the telomere sequence of the organism from which it was cloned (reviewed in 21). Telomerase preferentially binds and elongates telomeric oligonucleotides over nontelomeric

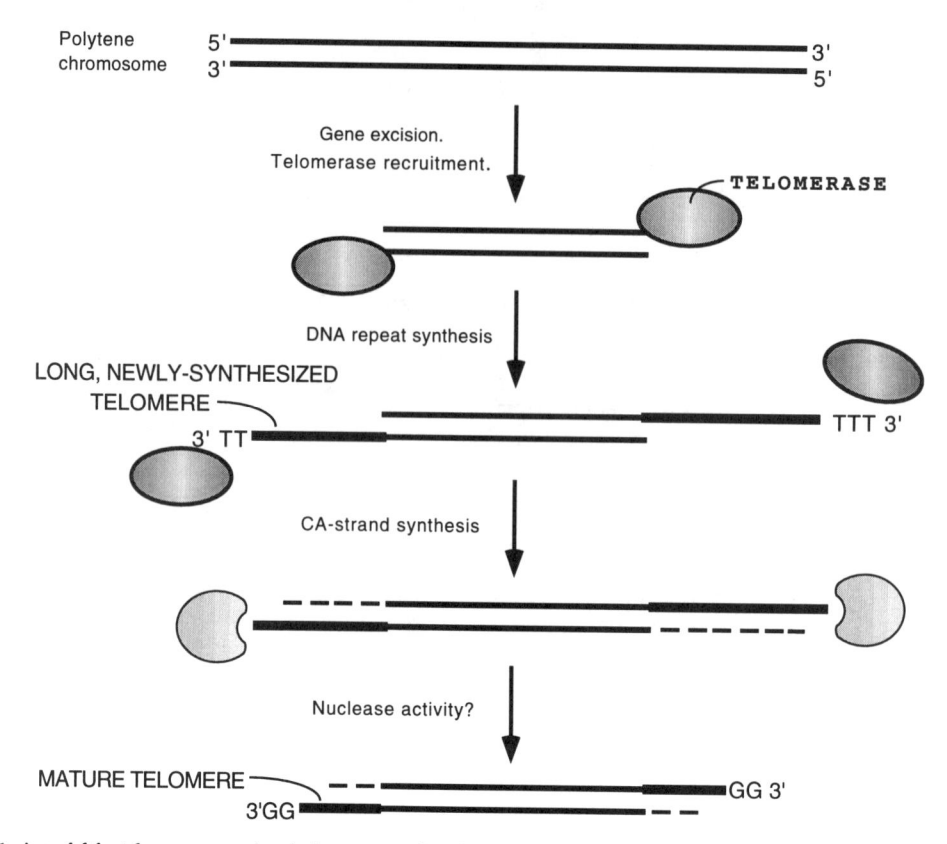

Figure 1. A model for telomere processing during macronucleus formation in ciliates. In *Euplotes crassus* telomeres are added to gene-size DNA fragments following genome fragmentation. Telomeres synthesized by telomerase are long and heterogeneous in length. CA-strand synthesis and processing by a putative nuclease results in a mature telomere of precise length. (Adapted from Roth and Prescott 1985; Vermeesch and Price 1994).

oligonucleotides *in vitro* (45-47). When telomerase is incubated with telomeric oligonucleotides, the terminal DNA repeat is base-paired with the enzyme's RNA template and a portion of the RNA template is copied. The enzyme then translocates and begins another round of template copying (reviewed in 21, 22). An interaction between a second, non-template site of the enzyme ("anchor site") and the oligonucleotide substrate tethers the substrate to the enzyme during processive elongation (reviewed in 21, 22).

The RNA components of telomerase have been cloned from ciliates (*Tetrahymena, Euplotes, Oxytricha*), yeasts (*Saccharomyces cerevisiae* and *K. lactis*), mouse and human (reviewed in 21, 22). Protein components have been identified and cloned from *Tetrahymena*. The binding specificity of the protein components to nucleic acids is consistent with the function of telomerase: one binds specifically to telomeric primer DNA, while the other binds to the telomerase RNA (48).

While telomerase uses a novel mechanism to synthesise the GT-rich strand of telomeres, the CA-rich strand is likely to be replicated by conventional DNA polymerases, according to current models (20, 49, 50). The polymerase α-primase complex is a good candidate for initiation of CA-strand synthesis, since it synthesises short RNA-DNA primers *de novo*.

A mutation in polymerase α (*cdc17*) causes a striking increase in telomere length (51). Recently additional mutations that affect DNA synthesis were screened for their effects on telomere length. (50). The only other replication component tested in which mutation affects telomere length is replication factor C (*rfc1* or *cdc44*; 50). Mutations in RFC, like polymerase α, cause telomere elongation in an allele-specific manner. In the SV40 *in vitro* replication system, the binding of RFC to DNA displaces polymerase α-primase. RFC subsequently recruits polymerase ∂ via PCNA, and elongation of the DNA proceeds (52-54). The functional proximity of polymerase α and RFC during the polymerase exchange step of DNA replication may suggest that this step is defective in the telomere-elongating mutants (50). Interestingly the *cdc17* effect on telomeres required the presence of the telomerase RNA; therefore, the elongation of telomeres in the mutant is likely due to telomerase. The mutations may result in changes in the telomeric DNA, such as gaps or nicks, that directly or indirectly regulate telomerase activity (50). Alternatively, the phenotype may result from interactions between *cdc17*, RFC, and telomerase that do not involve underreplicated DNA.

TELOMERE PROCESSING DURING CELL CYCLING

Telomeres are dynamic structures that in some cases respond to cell growth conditions and developmental cues. Mature telomeres result from a combination of synthetic and processing activities. Some of the processing activities, in addition to the synthetic activities, are regulated during the cell cycle.

In the ciliate *Euplotes*, telomere processing is best characterised during the developmental step of macronucleus formation (Figure 1), but a similar mechanism of telomere processing is likely to take place during vegetative cell cycling. The newly synthesised telomeres that are added following genome fragmentation are longer than mature telomeres (55). Whereas mature telomeres are precisely 42 nucleotides long (28 double-stranded and 14 single-stranded), newly synthesised telomeres are longer by an average of 50 nucleotides and are more heterogeneous in length (55, 56). The shortening of telomeres to the mature size does not result from incomplete DNA replication, because it occurs in cells where DNA synthesis is blocked by aphidicolin. This suggested that an active process may be responsible for the trimming of the telomeres (56). Newly synthesised and mature telomeres also differ in the terminal nucleotides: newly synthesised telomeres are heterogeneous ending in -TT, -TTT, or -GG, while mature telomeres terminate in -GG 3'. The terminal nucleotides of mature telomeres do not correspond with the position at which *Euplotes* telomerase preferentially dissociates during telomere elongation, supporting the model that another enzyme is responsible for processing the telomeres after telomerase synthesises the GT-rich strand (20 and references therein). This enzyme may be an endo- or exonuclease (20). *Euplotes* telomere processing is likely to occur during each cell cycle, and not just during conjugation and macronuclear development, since the unique -GG 3' end is maintained at telomeres during vegetative growth (reviewed in 34).

In *S. cerevisiae* telomeres are processed during S phase. At the end of S phase a >30 base long overhang of the GT-rich strand is detected on yeast telomeres ("TG$_{1-3}$ tails"; 57, 58). At other points in the cell cycle, TG$_{1-3}$ tails are not detected (58). Using synthetic molecules that mimic putative telomere replication intermediates, Wellinger *et al.* (59) showed that TG$_{1-3}$ tails were generated on both the lagging and leading strands, the tails could interact with each other, and most importantly, could be generated in a strain of yeast lacking active telomerase. This led the authors to propose a model in which a strand-specific exonuclease generates a TG$_{1-3}$ tail at each telomere during S phase.

Several telomere binding proteins have been identified (reviewed in 23, 24), and their role in telomere structure and length regulation is an active area of research. The telomeres of the ciliate *Oxytricha nova* are protected from nuclease digestion and chemical modification by a heterodimeric telomere binding protein complex (60-62). Both subunits of this protein are phosphorylated *in vivo*. The β subunit is phosphorylated by M-phase *Xenopus* egg extracts and multiple recombinant cyclin-dependent kinases *in vitro* (63), raising the possibility that the function of this complex is regulated during cell cycling *in vivo*.

TELOMERE SHORTENING AND CELL CYCLING

Gradual telomere shortening does not interfere with cell cycling

In multiple types of postnatal human somatic cells, telomeres shorten over time both *in vitro* and *in vivo*. When primary human fibroblasts are grown in culture, mean telomere length shortens with each cell division (64). When the average telomere length of cells from individuals of different ages are compared, telomere length is longer in young individuals than in old individuals (10, 28, 32), suggesting that telomeres shorten with ageing *in vivo*. Telomere shortening in some cells can be explained by the absence of telomerase and incomplete replication of DNA ends. Other cells including peripheral blood leukocytes exhibit shortening in the presence of telomerase; presumably the telomerase levels are inadequate to prevent telomere attrition, or other components of telomere length maintenance are limiting.

Whether due to inadequate telomerase or other telomere components, gradual telomere shortening does not interfere with cell cycling, at least not until after many population doublings. Yeast lacking the telomerase RNA component exhibit gradual telomere shortening and continue to divide for at least 40 generations without arresting or losing viability (65, 66). Telomere shortening in ciliates resulting from the expression of mutant *Tetrahymena* telomerase RNA (67) or in human cells expressing antisense human telomerase RNA does not result in immediate cell cycle arrest (68). Thus, the incomplete replication or gradual shortening of telomeres does not interfere with cell cycling. No checkpoint alarms are sounded. No arrest occurs. Gradual shortening apparently does not interfere with normal structure and function of telomeres, since chromosomes are replicated and segregated in an orderly fashion.

Critically short telomeres may trigger a DNA damage checkpoint and lead to withdrawal from the cell cycle

Telomere shortening in several types of human cells correlates with senescence after 20-50 population doublings (29, 30, 69, 70). Several lines of evidence suggest that telomere loss is biologically relevant to the senescence of human cells (reviewed in 71): 1) there appears to be a common mean telomere length at senescence (29, 72, 73); 2) quantitative estimates of the number of telomere repeats in senescent cells suggest that some chromosomes have lost most, if not all, of their repeats (74, 75); 3) chromosomal abnormalities associated with telomeric loss accumulate in senescent cells (reviewed in 76); and 4) experimental elongation of telomeres extends the life span of a cell hybrid (77).

Two models have been proposed to explain how and why critically short telomeres lead to senescence (reviewed in 78). One model postulates that critically short telomeres are recognised as damaged DNA, triggering cell cycle checkpoints and proliferative arrest. A complementary model suggests that critically short telomeres result in the derepression of one or more subtelomeric genes that direct the senescence program. The two models are not mutually exclusive: DNA damage, transcriptional derepression, as well as additional molecular events, may contribute to the onset of senescence. There is no direct evidence for these models.

In mammalian cells the p53 protein is involved in DNA damage-cell cycle checkpoint activities. Oncoproteins that inactivate p53 have been shown to overcome senescence, allowing cells to continue proliferating in culture beyond their usual life span (29, 30, 69, 70, 79, 80). This raises the possibility that p53 is involved in the establishment of senescence. In mammary epithelial cells, the expression of an oncoprotein that inactivates p53 is sufficient to overcome senescence; whereas, oncoproteins that inactivate both pRB and p53 are required in fibroblasts (69). If telomere shortening plays a causative role in senescence, p53 may be involved in sensing critically short telomeres.

In *S. cerevisiae RAD 9* is critical for cell cycle arrest in response to DNA damage (81). RAD9 may be involved in sensing telomere abnormalities. A *cdc13* temperature sensitive mutation at high temperature results in the appearance of single-stranded DNA at telomeres corresponding to the GT-rich strand. Another allele of *cdc13* (*est4*) was identified during a screen for mutants with a telomerase-like defect (V. Lundblad, personal communication). This mutation results in gradual telomere shortening and eventual loss of cell viability. Cdc13p/Est4p binds single-stranded yeast telomeres and is likely to be involved in telomere replication (V. Lundblad, personal communication). *rad9 cdc13* cells have a lower maximum temperature at which the cells can grow than *cdc13* cells (82). This suggests that *rad9cdc13* cells are more sensitive to sub-lethal doses of *cdc13*-associated DNA damage. Thus, Rad9p may be sensing single strand DNA abnormalities at telomeres (82). A model has been proposed for the role of Rad9p and several other checkpoint genes in the processing of *cdc13*-induced telomere damage (83).

Whether short telomeres that result from many rounds of proliferation in the absence of *de novo* telomere synthesis are indeed recognised as damaged DNA remains to be determined. Disruption of the telomerase RNA component in yeast does not lead to immediate cell cycle arrest; decreases in population doubling rates and viability become evident only after > 40 generations (65, 66). Determining if RAD9 or other DNA damage checkpoints play a role in the delayed phenotype of yeast carrying a telomerase RNA disruption will reveal if critically short telomeres trigger DNA damage checkpoints. This may shed light on the role of telomere shortening in cellular senescence.

TELOMERASE REGULATION AND MAMMALIAN DEVELOPMENT

Telomerase activity and expression of the RNA component are regulated in a tissue-specific manner during mammalian development. Many somatic tissues of adult humans do not express telomerase activity, at least not at detectable levels when bulk samples of the tissue are tested in a sensitive PCR-based assay. In a survey of six normal tissues and approximately twenty cell lines derived from non-tumorigenic adult tissues, telomerase activity was not detected in most of the samples (84). Embryos at the blastocyst stage and many fetal tissues at 16-20 weeks gestation do express telomerase activity when using a similar assay (85). The telomerase activity present in fetal tissues is no longer detected from the neonatal period onward (85). These data suggest that telomerase activity is down-regulated during prenatal development in multiple human tissues. Developmental down-regulation of the human telomerase RNA (hTR) component is suggested by differences in the steady state level of hTR in fetal and adult livers (68). While developmental regulation of telomerase in somatic tissues is observed in mice as well as humans, the timing and the degree of the decrease in activity may differ between the two genuses. Unlike many adult human tissues, adult mouse tissues express telomerase

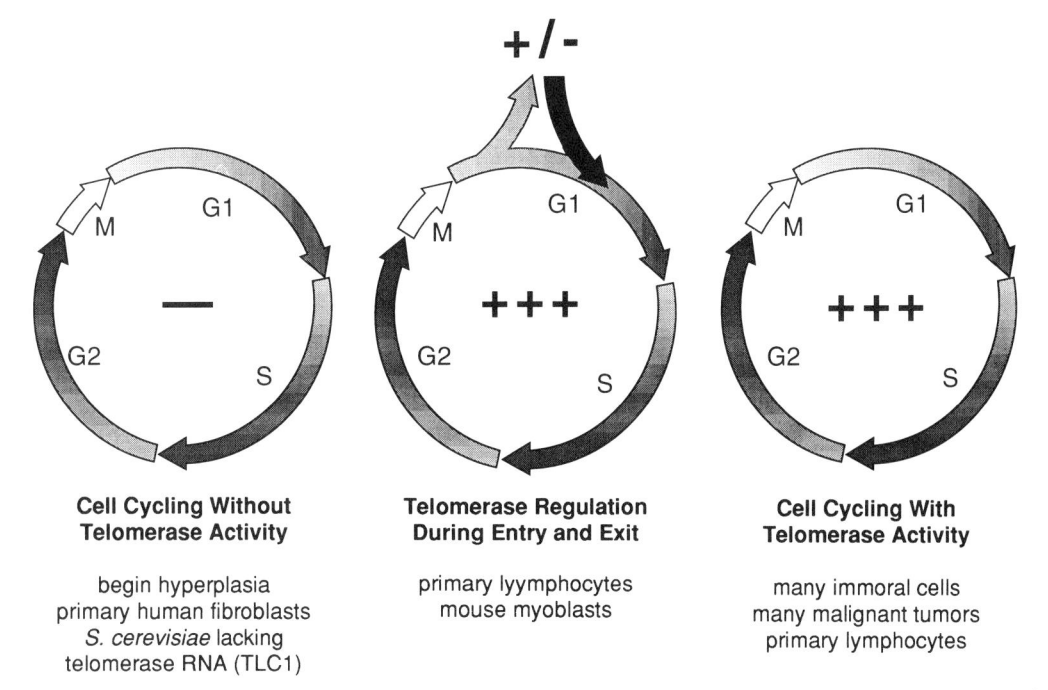

Figure 2. Telomerase activity during cell cycling. Many types of cells progress through the cell cycle in the absence of telomerase (left diagram; 84, 88, 89, 91, 92, 115). A few examples are listed. Most immortal and tumor-derived cells, and some types of normal mammalian cells, express telomerase (right diagram; 29, 84, 99, 100, 116, 117). Telomerase activity has been detected in G1, S, and G2/M phase in these cells (103, 104). Telomerase activity increases as T lymphocytes enter the cell cycle and decreases as mouse myoblasts arrest proliferation (middle diagram; 101, 103, 104, 106-108).

activity that is detected in the PCR-based telomerase assay (14, 86-89). Post-natal decreases in mouse telomerase RNA (mTR) and activity have been observed in multiple tissues (87; A. Kass-Eisler, personal communication); prenatal studies have not been carried out to date.

Recent data (90) suggest that results obtained from assaying bulk tissue samples may be misleading regarding the absence of telomerase activity in postnatal human tissues. In this study, greater than fifty percent of primary total skin samples were deficient for telomerase activity. However, when primary skin samples were enzymatically dissociated into epidermal and dermal layers, the small basal cells of the epidermis exhibited telomerase activity, while the dermal cells were devoid of activity. The absence of activity in some bulk skin samples most likely resulted from a high proportion and/or an inhibitory effect of telomerase-negative dermal fibroblasts. The dissection of other complex tissues may reveal the presence of telomerase activity in subsets of cells.

TELOMERASE REGULATION AND CELL CYCLING

Cells can cycle with or without telomerase

As illustrated in Figure 2, some cells replicate and divide in the absence of telomerase. Yeast cells lacking the essential RNA component and enzyme activity continue to cycle in the absence of telomerase for at least 40 generations (65, 66). Many normal, mortal mammalian cells and benign hyperplasia also proliferate in the absence of telomerase activity (29, 84, 88, 89, 91, 92). There are also examples of immortal cells that proliferate in the absence of telomerase (27, 84, 93, 94). Thus, telomerase activity is not required for cell cycling.

Eighty percent of tumor-derived cells and tumor samples express telomerase activity (reviewed in 95-98). Some non-tumorigenic mammalian cells also express telomerase (99-102). The presence of telomerase in these cells may provide a long-term replicative advantage.

Telomerase activity in all phases of the cell cycle

Telomerase activity has been detected in G1, S, and G2/M phase of tumor-derived and normal cells. Similar levels of telomerase activity were observed in phase-specific fractions of cervical carcinoma cells blocked with either S-phase or M-phase inhibitors, promyelocytic leukemia and fibrosarcoma cells sorted by fluorescence activated cell sorting (FACS), and primary normal lymphocytes synchronised by drugs and separated by FACS (103, 104); A. Avilion, C. Greider, unpublished data). *Xenopus* S-phase extracts with no cdc2 kinase activity and M-phase extracts with high cdc2 activity both have similar

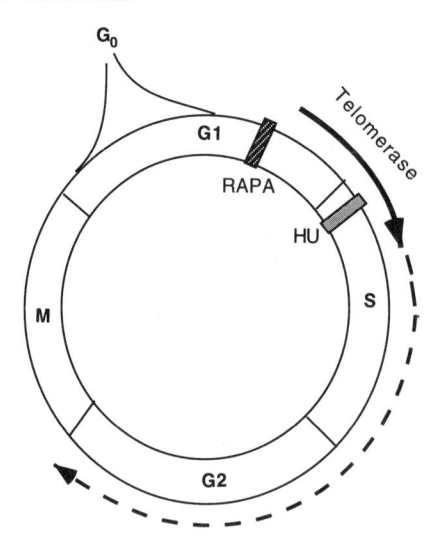

Figure 3. Telomerase activity and RNA increase as T cells progress from G_0 to S phase. When quiescent T cells are immunologically activated, telomerase activity increases at approximately the same time as T cells enter S phase. Hydroxyurea (HU), an inhibitor of DNA synthesis, does not block the induction of telomerase activity; whereas rapamycin (RAPA), an immunosuppressant that interferes with G1 phase signaling pathways, does block telomerase (107). High "activated" levels of enzyme activity have been detected in G_1, S, and G_2/M phases of cycling CD4+ lymphocytes (103).

levels of telomerase activity (105). These data suggest that telomerase activity is constant throughout the cell cycle, although additional analysis is required to confirm this.

Regulation as cells enter and exit the cell cycle

In some types of cells telomerase activity changes as the cells enter and exit the cell cycle (Figure 2). When quiescent T lymphocytes are immunologically activated to enter the cell cycle, telomerase activity increases 10-1000 fold (101, 103, 106-108) and the steady state level of the telomerase RNA component increases 10-20 fold (107, 108). Although telomerase activity increases at approximately the same time as the T cells enter S phase, DNA synthesis is not required for telomerase induction, because inhibition of DNA synthesis by aphidicolin and hydroxyurea does not block telomerase induction (107, 108). The immunosuppressant rapamycin, which binds a protein related to the ataxia telangiectasia gene product and blocks late G1-phase cell cycle events (phosphorylation of RB, down regulation of $p27^{Kip1}$, activation of cdk2; reviewed in 109, 110), does inhibit telomerase induction during T cell immunological activation (107). Thus, telomerase activity is regulated in G1 phase as quiescent T cells enter the cell cycle (Figure 3). How tightly telomerase regulation and cell cycling are linked in these cells remains to be determined. Results from Igarashi and Sakaguchi (1996) (106) suggest that continual T cell receptor stimulation, and not simply T cell proliferation, is required for maximal telomerase expression.

As in T lymphocytes, an increase in telomerase expression correlates with G0 to G1 phase progression in an immortal line of human lung fibroblasts that express SV40 T Antigen (TAg) under the control of a dexamethasone-inducible promoter (104). Cells grown in the absence of dexamethasone, do not express detectable TAg, do not synthesise DNA, and express low levels of telomerase (<10% level of dexamethasone-treated cells). Addition of dexamethasone results in an increase in TAg and telomerase activity that precedes the onset of DNA synthesis; thus telomerase is induced as the quiescent cells progress through G1 phase. These cells may prove useful in dissecting signalling pathways that regulate telomerase.

There are several examples of an increase in telomerase activity or RNA correlating with an increase in cell proliferation: 1) telomerase activity increases as cells from mouse mammary and skin primary explants begin to proliferate in culture (111); 2) telomerase activity is highest in the most highly proliferative cells of the epidermis from human skin samples, (90); 3) the level of telomerase activity in benign tonsils and lymph nodes correlates with the percentage of S phase cells in the samples (112); 4) increases in telomerase activity correspond with increased ^{3}H-thymidine incorporation in T cells treated with a variety of stimuli and inhibitors (103); and 5) the expression levels of telomerase RNA correlates strongly with histone H4 mRNA (a marker of S phase) in normal mammary gland and mammary tumors of transgenic mice (89). In contrast, telomerase activity decreases as proliferation increases in some populations of hematopoietic stem cells and leukemia cells (102)(113). In the latter examples, changes in enzyme levels correlate with changes in differentiation state.

Telomerase activity decreases as cells differentiate in culture. Following treatment with a variety of differentiating agents, telomerase activity declines in human leukemia cells, mouse and human embryonic cells, and mouse myoblasts (104, 113, 114); A. Kass-Eisler, C. Greider, personal communication). Since data on the timing and extent of cell cycle withdrawal during differentiation was not provided with many of these experiments, it is difficult to determine if telomerase down-regulation correlated with cell cycle withdrawal in many cases. In an experiment designed to distinguish between loss of telomerase activity due to differentiation and loss of telomerase due to proliferative arrest, mouse myoblasts were grown under conditions that lead to proliferative arrest in the absence of differentiation (104). Telomerase activity declined to the same extent in the arrested, undifferentiated myoblasts as in the arrested, differentiated myoblasts, suggesting a correlation between proliferative arrest and declining enzyme activity (104).

In conclusion, an increase in telomerase activity correlates with cell cycle entry and a decrease in telomerase activity correlates with cell cycle exit in some cells, but the extent to which telomerase regulation and cell cycling are linked in these cells remains to be determined. Telomerase is regulated independently of cell cycling in some cells by as yet uncharacterised signal transduction pathway(s).

ACKNOWLEDGEMENTS

I thank Carol Greider for the opportunity to work in her laboratory and for comments on the manuscript. Chantal Autexier, Siyuan Le, Alyson Kass-Eisler, Helena Yang, and Philip Sass are thanked for helpful discussions and critical reading of this manuscript. I thank Carolyn Price for reviewing the section on telomere processing and Victoria Lundblad for providing information prior to publication.

REFERENCES

1. McClintock, B. (1941) *Genetics* **26**, 234-282.
2. McClintock, B. (1942) *Proc. Natl. Acad. Sci.* **28**, 458-463.
3. Muller, H. J. (1938) *Collecting Net* **13**, 181-198.
4. Blackburn, E. H. (1991) *Nature* **350**, 569-573.
5. Henderson, E. (1995) in *Telomeres*. (E. H. Blackburn and C. W. Greider, eds), pp. 11-34, Cold Spring Harbor Laboratory Press.
6. Moyzis, R. K. *et al.* (1988) *Proc. Natl. Acad. Sci. USA* **85**, 6622-6626.
7. Brown, W. R. A. (1989) *Nature* **338**, 774-776.
8. Cross, S. H. *et al.* (1989) *Nature* **338**, 771-774.
9. de Lange, T. *et al.* (1990) *Mol. Cell. Biol.* **10**, 518-527.
10. Hastie, N. D. *et al.* (1990) *Nature* **346**, 866-868.
11. Walmsley, R. M. and Petes, T. D. (1985) *Proc Natl. Acad. Sci. USA* **82**, 506-510.
12. Cooke, H. J. and Smith, B. A. (1986) *Cold Spring Harbor Symp. Quant. Biol.* **51**, 213-219.
13. Allshire, R. C., Dempster, M. and Hastie, N. D. (1989) *Nucleic Acid Res.* **17**, 4611-4627.
14. Prowse, K. R. and Greider, C. W. (1995) *Proc. Natl. Acad. Sci.* **92**, 4818-4822.
15. Lansdorp, P. M. *et al.* (1996) **5**, 685-691.
16. Henderson, S. *et al.* (1996) *J. Biol. Chem.* in press.
17. Rhodes, D. and Giraldo, R. (1995) *Curr Opin. Struct. Biol.* **5**, 311-332.
18. Klobutcher, L. A., Swanton, M. T., Donini, P. and Prescott, D. M. (1981) *Proc. Natl. Acad. Sci. USA* **78**, 3015-3019.
19. Henderson, E. and Blackburn, E. H. (1989) *Mol. Cell. Biol.* **9**, 345-348.
20. Vermeesch, J. R. and Price, C. M. (1994) *Mol. Cell Biol* **14**, 554-566.
21. Greider, C. W., Collins, K. and Autexier, C. A. (1996) in *DNA replication in eukaryotic cells*. (M. DePamphlis, eds), pp. 619-638, Cold Spring Harbor Laboratory Press.
22. Greider, C. (1995) in *Telomeres*. (E. H. Blackburn and C. W. Greider, eds), pp. 35-68, Cold Spring Harbor Laboratory Press.
23. de Lange, T. (1996) *Sem. Cell Biol.* **7**, 23-29.
24. Fang, G. and Cech, T. R. (1995) in *Telomeres*. (E. H. Blackburn and C. W. Greider, eds), pp. 69-105, Cold Spring Harbor Laboratory Press.
25. Greider, C. W. (1990) *Bioessays* **12**, 363-369.
26. Lingner, J., Cooper, J. P. and Cech, T. R. (1995) *Science* **269**, 1533-4.
27. Rogan, E. M. *et al.* (1995) *Mol. Cell Biol.* **15**, 4745-4753.
28. Harley, C. B. (1990) in *Methods in Mol. Biol.* (J. W. Pollard and J. M. Walker, eds), pp. 25-32, The Humana Press Inc.
29. Counter, C. M. *et al.* (1992) *EMBO J.* **11**, 1921-1929.
30. Klingelhutz, A. J. *et al.* (1994) *Mol. Cell Biol.* **14**, 961-969.
31. Allsopp, R. C. *et al.* (1992) *Proc. Natl. Acad. Sci. USA* **89**, 10114-10118.
32. Lindsey, J. *et al.* (1991) *Mut. Res.* **256**, 45-8.
33. Greider, C. W. and Harley, C. B. (1996) in *Cellular aging and cell death*. (N. J. Holbrook, G. R. Martin and R. A. Lockshin, eds), pp. 123-138, Wiley-Liss, Inc.
34. Greider, C. W. (1996) *Ann. Rev. Biochem.* **65**, 337-365.
35. McEachern, M. J. and Blackburn, E. H. (1996) *Gen. Devel.* **10**, 1822-1834.
36. Walmsley, R. W., Chan, C. S. M., Tye, B.-K. and Petes, T. D. (1984) *Nature* **310**, 157-160.
37. Wang, S.-S. and Zakian, V. A. (1990) *Nature* **345**, 456-458.
38. Lundblad, V. and Blackburn, E. H. (1993) *Cell* **73**, 347-360.
39. Villeponteau, B. (1996) *Cell Devel. Biol.* **7**, 15-21.
40. Greider, C. W. and Blackburn, E. H. (1985) *Cell* **43**, 405-413.
41. Melek, M., Greene, E. C. and Shippen, D. E. (1996) *Mol. Cell. Biol.* **16**, 3437-3445.
42. McCormick-Graham, M. and Romero, D. P. (1996) *Mol. Cell Biol.* **14**, 1871-1879.
43. Gilley, D. and Blackburn, E. H. (1996) *Mol. Cell Biol.* **16**, 66-75.
44. Prescott, D. M. (1994) *Microbio. Rev* **58**, 233-267.
45. Greider, C. W. and Blackburn, E. H. (1987) *Cell* **51**, 887-898.
46. Blackburn, E. H. *et al.* (1989) *Genome* **31**, 553-560.
47. Harrington, L. A. and Greider, C. W. (1991) *Nature* **353**, 451-454.
48. Collins, K., Koybayashi, R. and Greider, C. W. (1995) *Cell* **81**, 677-686.

49. Zahler, A. M. and Prescott, D. M. (1989) *Nucl. Acid Res.* **17**, 6299-6317.
50. Adams, A. K. and Holm, C. (1996) *Mol. Cell Biol.* **16**, in press.
51. Carson, M. and Hartwell, L. (1985) *Cell* **42**, 249-257.
52. Waga, S. and Stillman, B. (1994) *Nature* **369**, 207-212.
53. Tsurimoto, T. and Stillman, B. (1991) *J. Biol. Chem.* **266**, 1950-1960.
54. Tsurimoto, T. and Stillman, B. (1991) *J. Biol. Chem* **266**, 1961-1968.
55. Roth, M. and Prescott, D. M. (1985) *Cell* **41**, 411-417.
56. Vermeesch, J. R., Williams, D. and Price, C. M. (1993) *Nucleic Acids. Res.* **21**, 5366-5371.
57. Wellinger, R. J., Wolf, A. J. and Zakian, V. A. (1993) *Mol Cell Biol* **13**, 4057-65.
58. Wellinger, R. J., Wolf, A. J. and Zakian, V. A. (1993) *Cell* **72**, 51-60.
59. Wellinger, R. J., Ethier, K., Labrecque, P. and Zakian, V. A. (1996) *Cell* **85**, 423-433.
60. Hicke, B. et al. (1990) *Proc Natl Acad Sci U S A* **87**, 1481-5.
61. Gray, J. T., Celander, D. W., Price, C. M. and Cech, T. R. (1991) *Cell* **67**, 807-14.
62. Gottschling, D. E. and Zakian, V. A. (1986) *Cell* **47**, 195-205.
63. Hicke, B. J. et al. (1995) *Nucl. Acids Res.* **23**, 1887-1893.
64. Harley, C. B., Futcher, A. B. and Greider, C. W. (1990) *Nature* **345**, 458-460.
65. McEachern, M. J. and Blackburn, E. H. (1995) *Nature* **376**, 403-409.
66. Singer, M. S. and Gottschling, D. E. (1994) *Science* **266**, 404-409.
67. Yu, G.-L., Bradley, J. D., Attardi, L. D. and Blackburn, E. H. (1990) *Nature* **344**, 126-132.
68. Feng, J. et al. (1995) *Science* **269**, 1236-1241.
69. Shay, J. W., Wright, W. E., Brasiskyte, D. and Van der Hagen, B. A. (1993) *Oncogene* **8**, 1407-1413.
70. Wright, W. E., Pereira-Smith, O. M. and Shay, J. W. (1989) *Mol. Cell. Biol.* **9**, 3088-3092.
71. Harley, C. B. (1995) in *Telomeres*. (E. H. Blackburn and C. W. Greider, eds), pp. 247-263, Cold Spring Harbor Laboratory Press.
72. Harley, C. B. and Villeponteau, B. (1995) *Curr. Opin. in Genet. and Dev.* **5**, 249-255.
73. Vaziri, H. et al. (1993) *Am. J. Hum. Genet* **52**, 661-667.
74. Harley, C. B. et al. (1994) *Cold Spring Harbor Symp. Quant. Biol.* **59**, 307-315.
75. Levy, M. Z. et al. (1992) *J. Mol. Biol.* **225**, 951-960.
76. Harley, C. B. (1991) *Mut. Res.* **256**, 271-282.
77. Wright, W. E., Brasiskyte, D., Piatyszek, M. A. and Shay, J. W. (1996) *EMBO J.* **15**, 1734-41.
78. Wright, W. E. and Shay, J. W. (1995) *Trends Cell Biol* **5**, 293-297.
79. Hara, E. et al. (1991) *Biochem. Biophys. Res. Commun.* **179**, 528-534.
80. Shay, J. W., Pereira-Smith, O. M. and Wright, W. E. (1991) *Exp. Cell Res.* **196**, 33-39.
81. Weinert, T. A. and Hartwell, L. H. (1988) *Science* **241**, 317-22.
82. Garvik, B., Carson, M. and Hartwell, L. (1995) *Mol. Cell Biol.* **15**, 6128-6138.
83. Lydall, D. and Weinert, T. (1995) *Science* **270**, 1488-1491.
84. Kim, N. W. et al. (1994) *Science* **266**, 2011-2014.
85. Wright, W. E. et al. (1996) *Dev. Genet.* **18**, 173-179.
86. Chadeneau, C. et al. (1995) *Cancer Recearch* **55**, 2533-2536.
87. Blasco, M., Funk, W., Villeponteau, B. and Greider, C. W. (1995) *Science* **269**, 1267-1270.
88. Blasco, M., Rizen, M., Greider, C. W. and Hanahan, D. (1996) *Nature Genetics* **12**, 200-204.
89. Broccoli, D. et al. (1996) *Mol. Cell Biol.* **16**, 3765-3772.
90. Harle-Bachor, C. and Boukamp, P. (1996) *Proc. Natl. Acad. Sci. USA* **93**, 6476-6481.
91. Shay, J. W., Tomlinson, G., Piatyszek, M. A. and Gollahon, L. S. (1995) *Mol Cell Bio* **15**, 425-432.
92. Piatyszek, M. A. et al. (1995) *Methods. Cell Sci* **17**, 1-15.
93. Murnane, J. P., Sabatier, L., Marder, B. A. and Morgan, W. F. (1994) *EMBO J.* **13**, 4953-62.
94. Bryan, T. R. et al. (1995) *EMBO J.* **14**, 4240-4248.
95. Harley, C. B. and Kim, N. W. (1996) *Imp. Adv. Oncol.* in press.
96. Autexier, C. and Greider, C. W. (1996) *Trends in Biochem. Sci.* in press.
97. Bacchetti, S. and Counter, C. M. (1995) *Intl. Journal of Oncology* **7**, 423-432.
98. Shay, J. W. (1995) *Mol. Med. Today* **1**, 376-382.
99. Broccoli, D., Young, J. W. and de Lange, T. (1995) *Proc. Natl. Acad. Sci., USA* **92**, 9082-9086.
100. Counter, C. M. et al. (1995) *Blood* **85**, 2315-2320.
101. Hiyama, K. et al. (1995) *J. Immun.* **155**, 3711-3715.
102. Chiu, C.-P. et al. (1996) *Stem Cells* **14**, 239-248.
103. Weng, N.-P., Levine, B. L., June, C. H. and Hodes, R. J. (1996) *J. Exp. Med.* **183**, 2471-2479.
104. Holt, S. E., Wright, W. E. and Shay, J. W. (1996) *Mol. Cell Biol.* **16**, 2932-2939.
105. Mantell, L. L. and Greider, C. W. (1994) *EMBO J.* **13**, 3211-3217.
106. Igarashi, H. and Sakaguchi, N. (1996) *Biochem. and Biophys. Res. Commun.* **219**, 649-655.
107. Buchkovich, K. and Greider, C. W. (1996) *Mol. Biol. Cell* in press.
108. Bodnar, A. G., Kim, N. W., Effros, R. B. and Chiu, C.-P. (1996) submitted.
109. Chou, M. M. and Blenis, J. (1995) *Cur. Opin. Cell Biol.* **7**, 806-814.
110. Wiederrecht, G. J. et al. (1995) in *Progress in Cell Cycle Research*. (L. Meijer, S. Guidet and H.

Y. Lim Tung, eds), pp. 53-57, Plenum Press, New York, USA.
111. Chadeneau, C. *et al.* (1995) *Oncogene* **11**, 893-898.
112. Norrback, K.-F., Dahlenborg, K., Carlsson, R. and Roos, G. (1996) *Blood* in press.
113. Albanell, J. *et al.* (1996) *Cancer Research* **56**, 1503-1508.
114. Sharma, H. W. *et al.* (1995) *Proc. Natl. Acad. Sci.* **92**, 12343-12346.
115. Counter, C. M. *et al.* (1994) *J. Virol.* **68**, 3410-3414.
116. de Lange, T. (1994) *Proc. Natl. Acad. Sci.* **91**, 2882-2885.
117. Counter, G. M., Hirte, H. W., Bacchetti, S. and Harley, C. B. (1994) *Proc. Natl. Acad. Sci.* **91**, 2900-2904.

chapter 19
The cyclin C/Cdk8 kinase

Vincent Leclerc and Pierre Léopold[1]

URA 671 CNRS, BP28, 06230 Villefranche-sur-mer, France
[1]To whom correspondence should be addressed

Cyclin C was originally identified in a genetic screen for metazoan cDNAs that complement a triple knock-out of the *CLN* genes, involved in G1/S progression in *S. cerevisiae*. Unlike cyclin Ds and cyclin E, also identified in this screen, cyclin C has not been found to have a cell-cycle role in metazoa. Identified as the catalytic partner of cyclin C, Cdk8 is a novel protein-kinase of the Cdk family structurally related to the yeast Srb10 kinase. Cyclin C, Cdk8 and RNA polymerase II are found in a large multi-protein complex that shows structural as well as functional homologies with the yeast polymerase II holoenzyme. These observations and the sequence similarity to the kinase/cyclin pair Srb10/Srb11 in *S. cerevisiae*, suggest that cyclin C and Cdk8 control RNA polymerase II function.

INTRODUCTION

The cyclins are a growing family of structurally related proteins involved in specific activation of serine-threonine protein kinases termed Cdks (Cyclin-dependent-kinases) (for review: 1). Different kinase activities associated to different cyclin/Cdk pairs have been implicated in controlling the main transitions of the eukaryotic cell cycle through the phosphorylation of specific cellular targets (reviewed in: 2-5). More recently, studies on other regulatory mechanisms have led to the characterisation of Cdk/cyclin pairs that were not exclusively or directly involved in the control of the cell cycle.

CYCLINS AND CDKS WITH NON CELL CYCLE-RELATED FUNCTIONS

Cyclin-dependent kinase 5 (Cdk5) was isolated through structural homology to human Cdc2. It is found associated with D-type cyclins in fibroblasts, but no kinase activity or cell cycle function has yet been attributed to these complexes. In addition, p35, a brain-specific regulatory subunit that has no homology with the cyclin family, activates Cdk5 kinase in post-mitotic neurons. Several lines of evidence indicate that Cdk5 is responsible for the phosphorylation of neurofilament proteins and of the neurone-specific microtubule-associated protein tau in the brain (for review: 6).

In budding yeast, the Cyclin-dependent kinase Pho85 interacts with the Pho80 cyclin to form a complex necessary for a transcriptional control mechanism that senses inorganic phosphate in the cell (7, 8). Alternatively, Pho85 can play a cell cycle regulatory role associating in late G1 with two other cyclins, Pcl1 and Pcl2, possibly to integrate cell cycle events with metabolism in the cell (9, 10).

Like the Pho85 and the Cdk5 kinases, Cdk7 is suspected to participate in multiple regulatory events in the cell. It was first identified with two regulatory subunits, cyclin H and Mat1, as a Cdk-activating kinase (CAK) capable of phosphorylating a conserved threonine present in most Cdks (11-19). Quite surprisingly, Cdk7 complexes were subsequently identified as components of a general transcription factor called TFIIH (20, 21). Cdk7, cyclin H and Mat1 co-purify with other subunits of the TFIIH factor; furthermore, they are associated with the kinase activity of TFIIH that *in vitro* phosphorylates RNA Pol II large subunit on its C-terminal domain (CTD) (22, 23). The CTD of RNA pol II is a repetition of a conserved heptapeptide with the consensus sequence YSPTSPS (for review: 24, 25). Studies in mouse, yeast and flies have demonstrated that it is essential for the *in vivo* function of Pol II (26-28). Its phosphorylation state oscillates during the transcription cycle. It is in a hypophosphorylated form prior to the onset of transcription initiation and then follows massive phosphorylation during transcription elongation. Although many aspects of the function of the CTD remain to be elucidated, it has been suggested that heavy phosphorylation, concomitant with the transition to the elongation step, provokes promoter clearance and dissociation of the polymerase from the initiation complex. The identification of Cdk7/cyclin H as the kinase component of TFIIH suggest that this kinase plays a role in CTD-phosphorylation *in vivo* (see Figure 1).

While accumulating evidence implicates the Cdk7 kinase in the control of the transcription machinery, it is not clear whether Cdk7 is responsible for CAK activity *in vivo* as originally suspected. Three major arguments suggest that the Cdk7 trio might be the only source of CAK activity

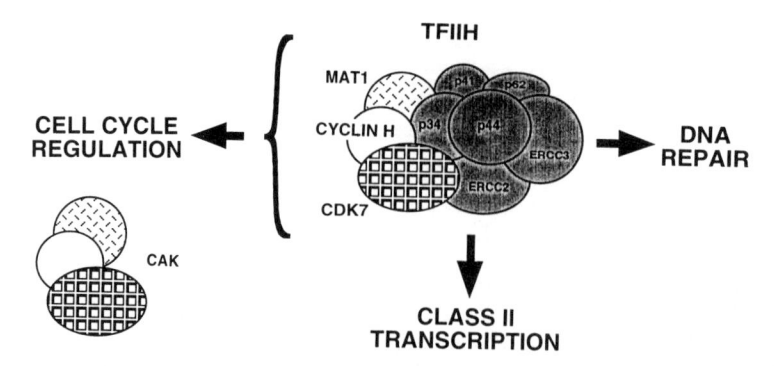

Figure 1. Description of the CAK complex and its possible involvement in multiple regulatory pathways. Cdk7, associated with cyclin H and Mat1, participates to the TFIIH transcription factor. As such, it is found in the Pol II Holoenzyme and is potentially involved in the control of initiation of Pol II transcription. TFIIH participates also in excision-repair mechanisms through the function of ERCC2 and ERCC3. Independently of TFIIH, the Cdk7/cyclin H/Mat1 trio could play an important role in the control of cell proliferation through its ability to activate different Cdks in the cell (CAK function).

in metazoan cells: (i) biochemical purification of CAK enzyme performed in three independent laboratories led to the identification of the same three partners, Cdk7, cyclin H and Mat1; (ii) immunodepletion of *Xenopus* egg extracts of any of these three subunits abolishes all of the CAK activity in the extract (M. Dorée, unpublished results); (iii) in human cells, Cdk7 is found in two different complexes, half being associated with TFIIH in a large nuclear complex while the rest is present in the form of a smaller complex in the cytoplasm, where CAK activity is expected to reside (18). In budding yeast, however, other kinases are responsible for the CAK activity in the cell. Kin 28/Ccl1 kinase, the homolog of Cdk7/cyclin H, participates to the TFIIH transcription factor complex and can phosphorylate the CTD of Pol II *in vitro*. Furthermore, studies of a *kin28-* mutant have established that Kin28 kinase has no CAK activity *in vivo* or *in vitro* (29). Recent work including biochemical purification has identified a novel kinase that possesses CAK activity in budding yeast (P. Kaldis and M.J. Solomon, unpublished results). These observations suggest that, in organisms with more developed regulatory circuits, CAK and CTD-kinase functions might be separated as well, and that other kinases with CAK activity have still to be found in metazoa. Consistent with this, a recent paper by Yee et al. describes the biochemical purification of a second cyclin subunit, most related to the yeast cyclin *mcs2*, which is found associated to Cdk7 in human osteosarcoma cells (30). It is not yet known whether this new cyclin is associated to CAK function *in vivo*, but the possibility that subunit composition plays a role in the multifunctionality of this enzyme is now open.

Recent studies on the control of the transcriptional machinery have led to the characterisation of a new Cdk/cyclin pair in yeast whose activity seems to control RNA polymerase II function. In yeast, partial deletions of the CTD of RNA pol II lead to slow growth and a cold-sensitive phenotype. A genetic screen designed to identify genes involved in the CTD function led to the isolation of nine mutations that revert the cold-sensitive phenotype due to CTD truncations (31). The genes corresponding to these mutations, called *SRBs* (for suppressor of RNA pol b), encode a series of proteins that, together with a subset of general transcription factors (TFIIB, TFIIF, TFIIH) and the RNA Pol II core enzyme, participate in a very large complex called the holoenzyme (32, 33)(for review, 34). It is thought that the holoenzyme represents the form of Pol II that initiates transcription at promoters *in vitro* and directly interacts with and responds to transcriptional activators (see Figure 2).

Among the different *SRBs*, *SRB10* encodes a Cdk-related protein kinase and *SRB11* encodes a distant member of the cyclin family. Alleles of these genes were isolated in two other searches for genes involved in transcriptional repression: *SRB10* (alias *ARE1*) as a repressor of a-specific genes in α-cells (35) and *SRB10* and *SRB11* as repressors of genes in response to glucose (36, 37). Mutants in either of the two genes fail to respond properly to galactose induction and to glucose or α2 repression. Genetic and biochemical studies indicate that the two proteins form a kinase pair *in vivo* and *in vitro*. Holoenzymes purified from *srb10-* strains have a substantially reduced CTD-kinase activity *in vitro* (38). The Srb10/11 kinase is thus, like Kin 28/Ccl1, a Cdk/cyclin complex present in the Pol II holoenzyme possibly involved in the control of CTD phosphorylation *in vivo*.

Another player recently joined the CTD-kinase team. By using CTD-phosphorylation as an assay, a CTD-kinase activity named CTDK-I, consisting of three subunits, was previously purified from yeast

CYCLIN C AND CDK8 FORM A KINASE PAIR *IN VITRO* AND *IN VIVO*

Both human and *Drosophila* cyclin Cs were originally isolated in a screen for genes capable of complementing a G1-cyclin defect in yeast (42-44). Three new cyclins came out from the screen. Two of them, cyclin D and E, are now considered as *bona fide* G1-cyclins in metazoans (for review: 4).

The cyclin C family consists so far of several members from human (44), *Drosophila* (43), *Xenopus* (J. Gautier and P. Léopold, unpublished data) and rat (45). They all share strikingly high levels of sequence identity, varying from 72% to 98%, which is not observed among other cyclin sub-families. The overall sequence similarity between cyclin C and B-type cyclins is only 40%. It reaches 58% (28% identity) with the yeast Srb11 cyclin, recently identified as a partner of the Srb10 kinase (38). The function of cyclin C has remained mysterious until the recent identification of its catalytic partner, Cdk8, itself structurally related to Srb10.

Cdk8 was identified in a PCR-based screen designed to identify new kinases implicated in cell cycle control (46). The initial PCR-fragment was used to isolate a human cDNA called K35 that encoded a putative Cdk of 464 aminoacids (47). Possible association of this new kinase with a cyclin subunit was tested by mixing reticulocyte translation products from K35 and cyclin A, B1, C, D1, E or H clones *in vitro*. After immunoprecipitation with specific anti-K35 antibodies, only cyclin C remained associated with the K35 kinase. These *in vitro* results were then followed by coimmunoprecipitation from cellular extracts. K35 from ^{35}S-labelled cells immunoprecipitated in association with a large number of cellular proteins including a 30 kDa protein subsequently identified as cyclin C by Western blotting and V8 protease digestion pattern. This demonstrated that cyclin C and K35 form a stable complex in the cell. Consequently, the K35 kinase was renamed Cdk8. These results were soon confirmed by two complementary studies. Using anti-human cyclin C antibodies, Rickert *et al.* were able to immunoprecipitate efficiently the Cdk8 kinase from human cell extracts (48). Leclerc *et al.* probed nuclear fractions from *Drosophila* embryos extracts with anti human Cdk8 antibodies and detected a 51 kDa band that co-fractionated along sucrose gradients with cyclin C in a complex of high molecular mass (49). Molecular cloning of a *Drososphila* CDK8 cDNA allowed the identification of this 51 kDa polypeptide as the product of the *Drosophila* CDK8 gene, thus reinforcing the idea that cyclin C and Cdk8 are associated in a common complex in the cell.

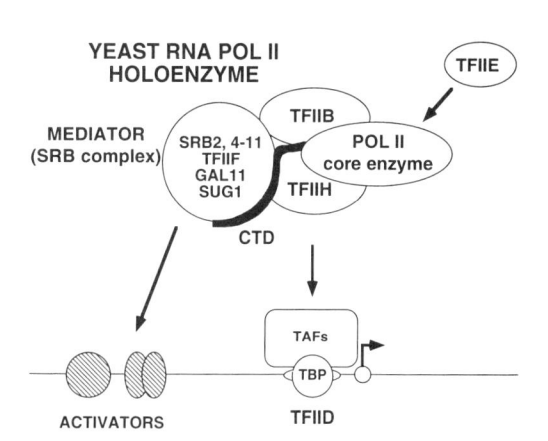

Figure 2. Structure of the yeast RNA polymerase II (Pol II) holoenzyme. A pre-formed multisubunit RNA pol II complex, consisting of several general transcription factors and a group of proteins called the Mediator, forms multiples interactions with transcription factor TFIID bound to the TATA region of the promoter, and the activator proteins bound to upstream activating sequences, thus facilitating formation of a stable initiation complex. SRB: suppressor of RNA polymerase b; CTD: C-terminal domain of Pol II; TBP: TATA-binding protein; TAFs: TBP-associated factors.

extract (39). The genes encoding these subunits have now all been cloned (40, 41). These are a new Cdk/cyclin pair called Ctk1/Ctk2 and a third protein, Ctk3, with no homology to other known proteins. Null mutations in all three genes confer similar growth defects and cold-sensibility. Furthermore, phospho-CTD specific antibodies revealed that null mutant *ctk1* strain were deficient in phosphorylation of the CTD *in vivo*. Ctk1-fusion proteins localise in the nucleus but neither Ctk1 nor Ctk3 are detected in the holoenzyme, suggesting that CTDK-I might not be involved in an initiation-associated CTD phosphorylation event. It will be important to know whether any functional interactions exist between these kinase pairs and what the relative contributions of the three kinase activities are regarding the control of CTD phosphorylation *in vivo*.

All of the cyclins involved in these CTD-kinase complexes (cyclin H, Ccl1, Mcs2, Srb11 and Ctk2) are very distantly related to the mitotic cyclins. They all show most similarities with human and *Drosophila* cyclin Cs, long suspected of playing a G1-cyclin function. We will discuss now the possibility that cyclin C and its newly identified kinase partner also play a role in the control of the transcription machinery in metazoa.

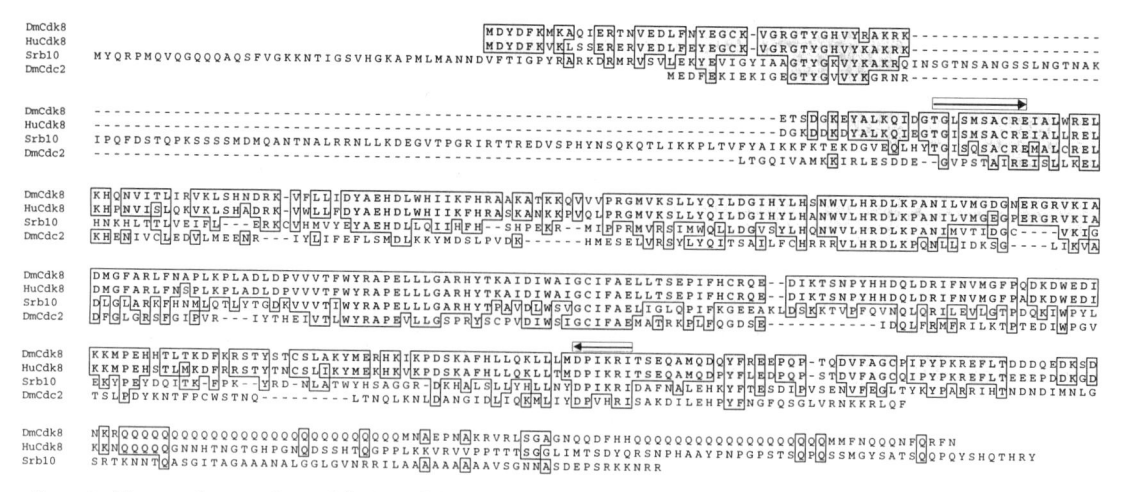

Figure 3. Alignment between *Drosophila* (*Dm*) Cdk8, human Cdk8, Ume5/Srb10 and *Drosophila* Cdc2 sequences. Boxed residues correspond to identities between *Dm* Cdk8 and the other sequences (Reproduced from *Molecular Biology of the Cell*, 1995. Volume 7, pp. 505-513 by copyright permission of the American Society for Cell Biology).

STRUCTURE OF THE CDK8 KINASE

The sequences of human and *Drosophila* Cdk8 show some common features that distinguish them from other members of the Cdk family (47, 49)(see Figure 3). Remarkably, they share 72% identity with each other, a higher score than the Cdc2 homologs among themselves. Their closest homolog is the Srb10 kinase (43% identity). Furthermore, both Cdk8 and Srb10 show a similar variation of the PSTAIRE motif of Cdc2 implicated in cyclin binding (S(M/Q)SACRE), whose conservation could reflect the presence of a contact region for related cyclins like cyclin C and Srb11. The most invariant stretch of residues in the catalytic domain, the DFG triplet, is involved in ATP binding (50). Remarkably, this triplet is changed in both Cdk8 and in Srb10 for DMG and DLG, respectively. All three kinases possess a Y_{15}- and a T_{14}-equivalent as potential regulatory phosphorylation sites in ATP-binding domain I. By contrast, none of the Cdk8 kinases have a potential phosphorylation site in the "T-loop" region, between domains VII and VIII. The "T-loop" of the Cdks has a regulatory function depending on its phosphorylation state: in the cyclin-bound form of the Cdk, phosphorylation of a conserved Thr residue leads to a conformation where the active site of the kinase is unmasked. In the unbound form, the "T-loop" is not phosphorylated and it blocks the active site (1, 51). Since, it seems that both human and *Drosophila* Cdk8 do not require activating phosphorylation in the "T-loop". Although Srb10 has a T within its "T-loop", it is not clear whether it corresponds to the phosphorylated residue of other Cdks. The ambiguity arises because a three aminoacids insertion is present in both Cdk8 and in Srb10 sequences within the conserved "T-loop". As depicted in Figure 4, two different versions of the alignment can be drawn. We favour version I because it preserves a correct alignment of the conserved sequences with no gap up to the W that marks the beginning of subdomain VIII. In this alignment, the T in the Srb10 sequence does not correspond to the phosphorylated T of other Cdks. Instead, at the position normally occupied by T, Srb10 and both Cdk8s have a D residue whose negative charge could substitute for phosphorylation. If indeed neither Srb10 nor Cdk8 require phosphorylation of the "T-loop" for their activation, there are important consequences for the regulatory pathways implicating these kinases (discussed below).

A CTD-KINASE ACTIVITY IS ASSOCIATED WITH CDK8/CYCLIN C IMMUNE COMPLEXES

The efficient co-precipitation of a Cdk8/cyclin C complex with either Cdk8 or cyclin C antibodies allowed associated kinase activity to be tested. None of the usual Cdk/cyclin substrates (histone H1, β-casein, phosvitin, HMG-I, myelin basic protein, pRb) were phosphorylated *in vitro* by immunoprecipitated Cdk8/cyclin C. In contrast, a GST-CTD fusion protein or CTD peptides containing tri or tetrameric repeats of the CTD heptapeptide consensus are phosphorylated efficiently *in vitro*, supporting the idea that the Cdk8/cyclin C , kinase is a CTD-kinase (48, 49). Moreover, autophosphorylation performed on Cdk8 immunoprecipitates from fly embryo extracts produced the labelling of a ~240 kDa polypeptide that co-migrated in SDS-PAGE with the phosphorylated form of Pol II. This phosphorylated form disappeared upon depletion of the extracts with anti-CTD antibodies prior to immunoprecipitation, suggesting that the large subunit of Pol II is part of the immune complex and

Version I

```
                                    #1  #2
DmCdk8      DMGFARLFNAPLKPLADLDPVVVTFWYRAPELLLG
HuCdk8      DMGFARLFNSPLKPLADLDPVVVTFWYRAPELLLG
Srb10       DLGLARKFHNMLQTLYTGDKVVVTIWYRAPELLLG
DmCdc2      DFGLGRSFGIPVR---IYTHEIVTLWYRAPEVLLG
DmCdc2c     DFGLARAFNVPMR---AYTHEVVTLWYRAPEILLG
HuCdk3      DFGLARAFGVPLR---TYTHEVVTLWYRAPEILLG
RaCdk4      DFGLARIYSYQM----ALTPVVVTLWYRAPEVLLQ
HuCdk6      DFGLARIYSFQM----ALTSVVVTLWYRAPEVLLQ
XlCdk7      DFGLAKSFGSPNR---IYTHQVVVTRWYRSPELLFG
ScKin28     DFGLARAIPAPHE---ILTSNVVTRWYRAPELLFG
ScCtk1      DFGLARKMNSRA----DYTNRVITLWYRPPELLLG

ScPho85     DFGLARAFGIPVN---TFSSEVVTLWYRAPDVLMG
DmCdk5      DFGLARAFGIPVK---CYSAEVVTLLYRPPDVLFG
```

Version II

```
                                    #1  #2
DmCdk8      DMGFARLFNAPLKPLADLDPVVVTFWYRAPELLLG
HuCdk8      DMGFARLFNSPLKPLADLDPVVVTFWYRAPELLLG
Srb10       DLGLARKFHNMLQTLYTGDKVVVTIWYRAPELLLG
DmCdc2      DFGLGRSFGIPVR-IYTH--EIVTLWYRAPEVLLG
DmCdc2c     DFGLARAFNVPMR-AYTH--EVVTLWYRAPEILLG
HuCdk3      DFGLARAFGVPLR-TYTH--EVVTLWYRAPEILLG
RaCdk4      DFGLARIYSYQM--ALT--PVVVTLWYRAPEVLLQ
HuCdk6      DFGLARIYSFQM--ALT--SVVVTLWYRAPEVLLQ
XlCdk7      DFGLAKSFGSPNR-IYTHQ--VVTRWYRSPELLFG
ScKin28     DFGLARAIPAPHE-ILT--SNVVTRWYRAPELLFG
ScCtk1      DFGLARKMNSRA--DYTNR--VITLWYRPPELLLG
ScPho85     DFGLARAFGIPVN---TFSSEVVTLWYRAPDVLMG

DmCdk5      DFGLARAFGIPVK-CYS--AEVVTLLYRPPDVLFG
```

Figure 4. Two possible alignments of Cdk8 and Srb10 sequences with nine other Cdks in the region between kinase domains VII and VIII, corresponding to the "T-loop". Version I aligns all the sequences from the conserved W residue in domain VIII (boxed) without gap back to the second D residue in the Cdk8 sequences (marked by #1). The high degree of conservation in the short stretch of sequence between these two residues suggest that no gap should be created in this region. In this case, a D residue (#2) is found in both Cdk8s and in Srb10 at the equivalent position of the phosphorylated T residue in other Cdks. Version II aligns this T residue with D#1 of both Cdk8s. In this case, Srb10 sequence shows a T residue aligned with Ts of other Cdks, but gaps are created on both sides.

that it is phosphorylated efficiently by a CTD-kinase associated with Cdk8/cyclin C. In addition, RNA pol II could be detected directly in Cdk8 immunoprecipitates (49).

Using extracts from elutriated human cells, Rickert et al. showed that neither Cdk8/cyclin C complex levels nor associated kinase activity fluctuate as a function of cell cycle progression, therefore suggesting that, as for Cdk7/cyclin H, the cyclin C/Cdk8 kinase is constitutively active in the cell.

CDK8 AND CYCLIN C PARTICIPATE IN A LARGE COMPLEX WITH RNA POL II

Initial observations suggested strongly that Cdk8, cyclin C and RNA pol II are found in a common complex in the cell. Size fractionation of either embryo extracts or cell lysates indicate that Cdk8 and cyclin C are present in at least two large multiprotein complexes of 180 and 500-600 kDa in size respectively (48, 49). As expected, anti-pol II antibodies detected the presence of RNA pol II in the fractions spanning through the larger complex in sucrose gradients. Moreover, labelled RNA pol II could be efficiently re-immunoprecipitated with an anti-pol II antibody from sizing column fractions, after cyclin C immunoprecipitate-kinase assay. These observations demonstrated that cyclin C, Cdk8 and Pol II are associated in a common complex and suggested that Pol II could be a natural substrate of the Cdk8/cyclin C kinase.

The idea that Srb10/11 and Cdk8/C kinases are functionally related, based initially on sequence similarity, was thus reinforced by the characterisation of a related CTD-kinase activity in vitro. Functional homology between the two kinases still awaited the biochemical definition of a metazoan RNA pol II holoenzyme. This was recently achieved by three independent groups. Ossipow et al. used anti-Cdk7 antibodies to co-precipitate a large group of transcription factors from rat liver extracts that included TFIID, TFIIB, TFIIH, TFIIF and TFIIE, together with RNA Pol II in a complex capable of promoter-specific transcription initiation in vitro. Apart from the identification of the general transcription factors (GTFs) participating in the complex, there was no evidence for Srb homologs in the complex. More recently, Chao et al. identified a human homologue of the yeast SRB7 gene and used anti-Srb7 antibodies in a first step for the purification of a Pol II holoenzyme from calf thymus (52). This pol II complex has an estimated size of M_r ~2000K and contains at least the general transcription factors TFIIH and TFIIE as well as the human Srb7 homolog. The identification of a human holoenzyme was also reported by Maldonado et al. (53). Purified fractions from human cell extracts contain a Pol II complex of approximately the same size than the one described by Chao et al. Western blot demonstrated the presence of stoechiometric quantities of TFIIE, TFIIF and limiting amounts of TFIIH associated with the core subunits of RNA Pol II. This complex is capable of supporting basal as well as activated transcription, responding to the GAL4-VP16 activator. Furthermore, Srb7, Cdk8 and cyclin C were also found in the complex, suggesting the existence of a homologue of the yeast SRB complex called the "mediator of activation" (32). In accordance with this idea, a "co-activator" fraction could be separated from the human holoenzyme complex by fractionation on an anti-CTD antibody column (in this experiment, anti-CTD antibodies were used to provoke the dissociation of CTD-binding proteins from the Pol II complex). The flow-through fraction did not contain RNA pol II but was capable of mediating transcriptional activation when added to core RNA pol II. It is not yet known whether Cdk8, cyclin C and Srb7 are effectively found in the "co-activator" fraction. However, this observation goes in line with the results establishing in yeast that a complex formed of all the SRB members binds to RNA pol II through the CTD and forms direct contacts with activators like VP16 (32).

Attempts at functionally complementing *srb10*⁻ or *srb11*⁻ mutants in yeast by overexpression of the Cdk8/cyclin C kinase have failed so far (D. Balciunas, J.-P. Tassan, E. Nigg and H. Ronne, unpublished results), suggesting that in these experimental conditions, the kinases cannot simply substitute for each other. Whether Cdk8/cyclin C is the kinase of a putative metazoan SRB complex remains to be shown. The recent finding that a human holoenzyme can be separated into a Pol II subcomplex and an activator fraction by chromatography on anti-CTD antibody column opens up the route to such an analysis.

WHAT ROLES FOR THE DIFFERENT CTD-KINASES IN THE HOLOENZYME?

Cdk7/H and Cdk8/C are two potential *in vivo* CTD-kinases but it is difficult to imagine specific functions for each of these kinases in the CTD-phosphorylation process. In order to reconcile the presence of Cdk7/H in the transcription factor TFIIH with its previously described CAK function, it was hypothesised that Cdk7/H could activate through its CAK activity a second kinase present in the holoenzyme that would be the real CTD-kinase. Nevertheless, as previously discussed, sequence comparisons suggest that neither Srb10 nor Cdk8 require phosphorylation of the "T-loop" for their activation. Furthermore, Cdk7/H has a strong CTD-kinase activity by itself, suggesting that it could act directly in the CTD-phosphorylation process *in vivo*. Based on a series of converging observations, it was recently proposed that CTD phosphorylation occurs in two separate steps (53). Following recruitment of the holoenzyme to the promoter region via gene-specific activators, partial phosphorylation of the CTD may occur and provoke SRB complex release, as well as formation of a stabley engaged initiation complex. An SRB kinase like the Srb10/11 and possibly the Cdk8/C pairs could be involved in this process. This first step would be followed by massive CTD phosphorylation, the transition of engaged Pol II into an elongation-competent enzyme and the release of transcriptional cofactors. TFIIH which is proposed to act at a late stage of the transcription initiation cycle contains the Cdk7/H pair, whose kinase activity could be responsible for this second phosphorylation step (see Figure 5). This involvement of several Cdk/kinase activities in controlling the transcriptional machinery represents a new vision of how the initiation events may be ordered and regulated. In addition to the previously described protein-protein interaction and complex recruitment mechanisms, the field is now pointing to a series of enzymatic activities directed towards a CTD-substrate that could act as a molecular integrator.

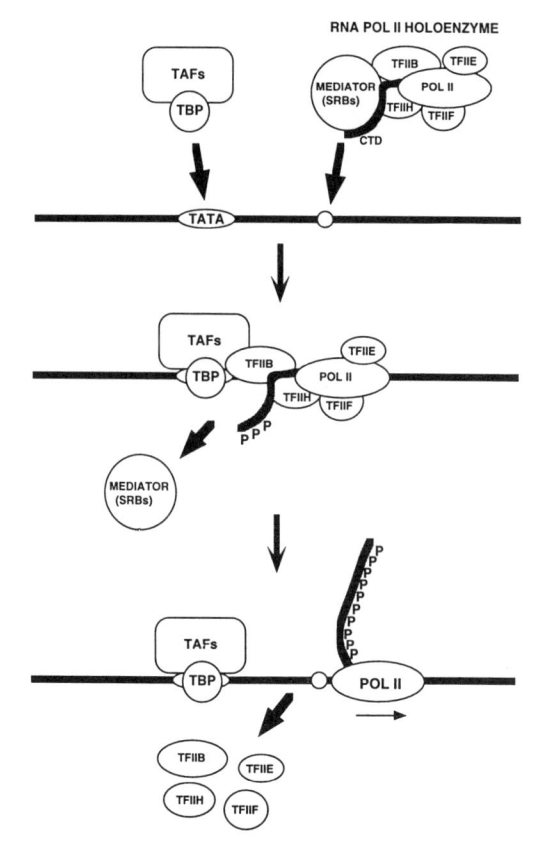

Figure 5. A model for the sequence of events occurring at the promoter site during transcription initiation. The RNA pol II holoenzyme is recruited at the initiation site via interactions with the TFIID complex, consisting of TBP and TAFs. Subunit rearrangements lead to stabilisation of the initiation complex (in particular through specific association between TFIIB and TBP), partial phosphorylation of the CTD and release of the SRB complex. Massive CTD phosphorylation, dissociation of the initiation complex and transition to an elongation mode follows (modified from McKnight, 1996).

COULD THE CDK8/C PAIR BE INVOLVED IN OTHER CELLULAR MECHANISMS?

The kinase/cyclin complex seems to be a rather flexible system that allows a single kinase to participate in different activities depending on the cyclin subunit and other modulators it associates with. The immediate advantage for the cell would be that the kinase subunit then serves as a link between the different processes it is involved in. For example, Pho85 controls phosphate metabolism and cell cycle progression depending on the cyclin subunit it binds to. This could be a way for the cell of coordinating these two processes. The same could apply to Cdk7/H and the control of Cdk activation, transcription and DNA repair.

Cyclin C was first proposed to play a cell cycle role on the basis of the initial complementation screen in yeast. Nevertheless, a reevaluation of this hypothetical G1 function is possible in the light of Cdk8/C's role in transcriptional control. While it is possible that cyclin C can play a cell cycle role in a foreign context, it is possible that cyclin C function in yeast bypassed the G1 block indirectly by disturbing the transcriptional controls of bona fide cell cycle control genes. One such could be the cyclin-coding *CLB5* gene, since a very modest increase in copy number of this gene rescues a triple knock-out of the *CLN* genes (55). A small increase of expression of this gene, provoked by massive overexpression of cyclin C, could thus push the cells through the G1 block. Therefore, there is still no direct evidence for a cell cycle regulatory function for the Cdk8/cyclin C kinase.

Two recent papers bring new data on the biology of cyclin C. Li *et al.* reported that the cyclin C gene is localised in 6q21, a region which was found deleted in a series of human acute lymphoblastic leukaemia cell lines (56). Over 13 patient samples examined, 12 showed a deletion of one allele of cyclin C. However, single strand conformational polymorphism established that the second cyclin C allele was not altered in these cell lines. These results suggested that either haploinsufficiency of the cyclin C protein can promote tumorigenesis, or that an important tumour-suppressor gene shares physical linkage with the cyclin C locus. While studying expression of the cyclin C gene in an avian B-cell line, the same research group isolated a widely expressed alternatively spliced cyclin C cDNA (57). This spliced version of the mRNA is a consequence of the insertion of a unique exon within the cyclin box region. It encodes a truncated protein that contains only the amino terminus and part of the cyclin box. Intriguingly, expression of the spliced mRNA is cell cycle-regulated, peaking in G2/M and early G1. The function of these spliced cyclin C mRNAs is still mysterious, but their cyclical appearance suggest that they could participate in a cell cycle-regulated mechanism like, for instance, preventing transcription during late G2/M by modulating the function of the Cdk8 kinase. The functional significance of these two lines of results is still unclear. While a role for cyclin C in the control of transcription could account for these results, one cannot definitely exclude the possibility that Cdk8 kinase is involved independently in other cellular mechanisms.

CONCLUSION

Characterisation of the function of the Cdk8/cyclin C kinase has recently advanced significantly. The findings allow us to consider the function of the cyclin/Cdk complexes in a new light. Initially proposed as keys in the control of cell cycle progression, it was admitted that some of them can also, in parallel, participate to different cellular functions. It appears now that some of these kinases are implicated in other functions like transcription control without even being suspected of having a cell cycle role. It is likely that this very flexible two-subunit system will appear more ubiquitously involved in the multiple post-translational controls used by the cell.

ACKNOWLEDGEMENTS

We thank Evelyn Houliston for careful reading of the manuscripts and comments. This work was supported by grants from the Centre National de la Recherche Scientifique, the Association pour la Recherche contre le Cancer and the Groupement de Recherches et d'etudes sur le génome.

REFERENCES

1. Morgan, D.O. (1995) *Nature* **374**, 131-134.
2. King, R.W., Jackson, P.K. and Kirschner, M.W. (1994) *Cell* **79**, 563-571.
3. Nurse, P. (1994) *Cell* **79**, 547-550.
4. Sherr, C.J. (1994) *Cell* **79**, 551-555.
5. Nigg, E.A. (1995) *Bioessays* **17**, 471-480.
6. Lew, J. and Wang, J.H. (1995) *Trends Biochem Sci* **20**, 33-37.
7. Hirst, K., Fisher, F., McAndrew, P.C. and Goding, C.R. (1994) *Embo J* **13**, 5410-5420.
8. Kaffman, A., Herskowitz, I., Tjian, R. and O'Shea, E.K. (1994) *Science* **263**, 1153-6.
9. Espinoza, F.H., Ogas, J., Herskowitz, I. and Morgan, D.O. (1994) *Science* **266**, 1388-1391.
10. Measday, V., Moore, L., Ogas, J., Tyers, M. and Andrews, B. (1994) *Science* **266**, 1391-1395.
11. Fesquet, D., Labbe, J., Derancourt, J., Capony, J., Galas, S., Girard, F., Lorca, T., Shuttleworth, J., Doree, M. and Cavadore, J. (1993) *EMBO J.* **12**, 3111-3121.
12. Poon, R., Yamashita, K., Adamczewski, J., Hunt, T. and Shuttleworth, J. (1993) *EMBO J.* **12**, 3123-3132.
13. Solomon, M.J., Harper, J.W. and Shuttleworth, J. (1993) *Embo J* **12**, 3133-42.
14. Fisher, R.P. and Morgan, D.O. (1994) *Cell* **78**, 713-724.
15. Makela, T.P., Tassan, J.P., Nigg, E.A., Frutiger, S., Hughes, G.J. and Weinberg, R.A. (1994) *Nature* **371**, 254-257.
16. Tassan, J.P., Schultz, S.J., Bartek, J. and Nigg, E.A. (1994) *J Cell Biol* **127**, 467-478.
17. Devault, A., Martinez, A.M., Fesquet, D., Labbe, J.C., Morin, N., Tassan, J.P., Nigg, E.A., Cavadore, J.C. and Doree, M. (1995) *Embo J* **14**, 5027-5036.
18. Fisher, R.P., Jin, P., Chamberlin, H.M. and Morgan, D.O. (1995) *Cell* **83**, 47-57.

19. Tassan, J.P., Jaquenoud, M., Fry, A.M., Frutiger, S., Hughes, G.J. and Nigg, E.A. (1995) *Embo J* **14**, 5608-5617.
20. Feaver, W.J., Svejstrup, J.Q., Henry, N.L. and Kornberg, R.D. (1994) *Cell* **79**, 1103-9.
21. Roy, R., Adamczewski, J.P., Seroz, T., Vermeulen, W., Tassan, J.P., Schaeffer, L., Nigg, E.A., Hoeijmakers, J.H. and Egly, J.M. (1994) *Cell* **79**, 1093-1101.
22. Serizawa, H., Makela, T.P., Conaway, J.W., Conaway, R.C., Weinberg, R.A. and Young, R.A. (1995) *Nature* **374**, 280-2.
23. Shiekhattar, R., Mermelstein, F., Fisher, R.P., Drapkin, R., Dynlacht, B., Wessling, H.C., Morgan, D.O. and Reinberg, D. (1995) *Nature* **374**, 283-287.
24. Corden, J.L. (1990) *Trends Biochem Sci* **15**, 383-7.
25. Young, R.A. (1991) *Annu Rev Biochem* 689-715.
26. Nonet, M., Sweetser, D. and Young, R.A. (1987) *Cell* **50**, 909-15.
27. Bartolomei, M.S., Halden, N.F., Cullen, C.R. and Corden, J.L. (1988) *Mol Cell Biol* **8**, 330-9.
28. Zehring, W.A., Lee, J.M., Weeks, J.R., Jokerst, R.S. and Greenleaf, A.L. (1988) *Proc Natl Acad Sci U S A* **85**, 3698-702.
29. Cismowski, M.J., Laff, G.M., Solomon, M.J. and Reed, S.I. (1995) *Mol Cell Biol* **15**, 2983-2992.
30. Yee, A., Wu, L., Liu, L., Kobayashi, R., Xiong, Y. and Hall, F.L. (1996) *J Biol Chem* **271**, 471-477.
31. Thompson, C.M., Koleske, A.J., Chao, D.M. and Young, R.A. (1993) *Cell* **73**, 1361-75.
32. Hengartner, C.J., Thompson, C.M., Zhang, J., Chao, D.M., Liao, S.M., Koleske, A.J., Okamura, S. and Young, R.A. (1995) *Genes Dev* **9**, 897-910.
33. Koleske, A.J. and Young, R.A. (1994) *Nature* **368**, 466-9.
34. Koleske, A.J. and Young, R.A. (1995) *Trends Biochem Sci* **20**, 113-116.
35. Wahi, M. and Johnson, A.D. (1995) *Genetics* **140**, 79-90.
36. Kuchin, S., Yeghiayan, P. and Carlson, M. (1995) *Proc Natl Acad Sci U S A* **92**, 4006-4010.
37. Balciunas, D. and Ronne, H. (1995) *Nucleic Acid Res.* **23**, 4421-4425.
38. Liao, S.M., Zhang, J., Jeffery, D.A., Koleske, A.J., Thompson, C.M., Chao, D.M., Viljoen, M., Hj, v.V. and Young, R.A. (1995) *Nature* **374**, 193-6.
39. Lee, J.M. and Greenleaf, A.L. (1989) *Proc Natl Acad Sci U S A* **86**, 3624-8.
40. Lee, J.M. and Greenleaf, A.L. (1991) *Gene Expr* **1**, 149-67.
41. Sterner, D.E., Lee, J.M., Hardin, S.E. and Greenleaf, A.L. (1995) *Mol Cell Biol* **15**, 5716-5724.
42. Lahue, E.E., Smith, A.V. and Orr-Weaver, T.L. (1991) *Genes and Dev.* **5**, 2166-2175.
43. Léopold, P. and O'Farrell, P.H. (1991) *Cell* **66**, 1207-1216.
44. Lew, D.J., Dulic, V. and Reed, S. (1991) *Cell* **66**, 1197-1206.
45. Tamura, K., Kanaoka, Y., Jinno, S., Nagata, A., Ogiso, Y., Shimizu, K., Hayakawa, T., Nojima, H. and Okayama, H. (1993) *Oncogene* **8**, 2113-8.
46. Schultz, S.J. and Nigg, E.A. (1993) *Cell Growth Differ* **4**, 821-30.
47. Tassan, J.-P., Jaquenoud, M., Léopold, P., Schultz, S.J. and Nigg, E.A. (1995) *Proc. Natl. Acad. Sci* **92**, 8871-8875.
48. Rickert, P., Seghezzi, W., Shanahan, F., Cho, H. and Lees, E. (1996) *Oncogene* in press.
49. Leclerc, V., Tassan, J.-P., O'Farrell, P.H., Nigg, E.A. and Léopold, P. (1995) *Mol. Biol. Cell* **7**, 505-513.
50. Hanks, S.K., Quinn, A.M. and Hunter, T. (1988) *Science* **241**, 42-52.
51. Clarke, P.R. (1995) *Curr Biol* **5**, 40-42.
52. Chao, D.M., Gadbois, E.L., Murray, P.J., Anderson, S.F., Sonu, M.S., Parvin, J.D. and Young, R.A. (1996) *Nature* **380**, 82-85.
53. Maldonado, E., Shiekhattar, R., Sheldon, M., Cho, H., Drapkin, R., Rickert, P., Lees, E., Anderson, C., Linn, S. and Reinberg, D. (1996) *Nature* **381**, 86-89.
54. McKnight, S.L. (1996) *Genes Dev.* **10**, 367-381.
55. Epstein, C.B. and Cross, F.R. (1992) *Genes Dev* **6**, 1695-706.
56. Li, H., Lahti, J.M., Valentine, M., Saito, M., Reed, S.I. and Kidd, V.J. (1996) *Genomics* **32**, 253-259.
57. Li, H. and Kidd, V.J. (1996) *Oncogene* in press.

chapter 20
Cyclin-dependent kinase 5 (Cdk5) and neuron-specific Cdk5 activators

Damu Tang and Jerry H. Wang[1]

Department of Biochemistry, The Hong Kong University of Science and Technology,
Clear Water Bay, Kowloon, Hong Kong
[1]To whom correspondence should be addressed

While cyclin-dependent kinase 5 (Cdk5) is widely distributed in mammalian tissues and in cultured cell lines, Cdk5-associated kinase activity has been demonstrated only in mammalian brains. An active form of Cdk5, called neuronal cdc2-like kinase (Nclk) has been purified from mammalian brain and shown to be a heterodimer of Cdk5 and a 25 kDa protein, which is derived proteolytically from a 35 kDa brain and neuron-specific protein. The protein is essential for the kinase activity of Cdk5 and is therefore designated neuronal Cdk5 activator, p25/35^{Nck5a}. Nclk appears to have important neuronal functions. The changes in Cdk5 and Nck5a expression appear to correlate with the terminal differentiation of neurons of the mouse embryonic brain. Transfection of cultured cortical neurons with dominant negative cdk5 mutants or Nck5a antisense DNA may reduce neurite growth, suggesting that Nclk plays an active role in neuron differentiation. A number of cytoskeletal proteins including neurofilament proteins, the neuron-specific microtubule associated protein tau, and the actin binding protein caldesmon are *in vitro* substrates of Nclk. Although Nck5a has cyclin-like activity, it shows minimal amino acid sequence identity to members of cyclin family proteins. The mechanism of activation of Cdk5 by Nck5a differs from that of cyclin activation of Cdks in that full Cdk5 kinase activity can be achieved in the absence of phosphorylation of Cdk5. An isoform of Nck5a, a 39 kDa protein has been cloned and shown to share extensive amino acid identity and the mechanism of Cdk5 activation with Nck5a. These proteins may represent a subfamily of Cdk activators distinct from cyclins.

While cdc2 protein family members are noted mainly for their pivotal roles in cell division cycle (1-9), Cdk5 does not have any known cell cycle function. On the other hand, evidence has been accumulating to suggest that Cdk5 has important neuronal functions (10-13). The Cdk5 protein is enriched in neurons of adult brain (14, 12) whereas Cdc2 and Cdk2 are essentially absent (15). To date, Cdk5-associated kinase activity has been demonstrated only in mammalian brains (16, 17, 18), although Cdk5 protein appears to be ubiquitously distributed in mammalian tissues and cultured cells (12). This may be attributed, at least in part to the existence of the central nervous system neuron-specific Cdk5 activator proteins (19, 20). In addition to the specific tissue distribution, the neuronal Cdk5 activators display structural and regulatory properties distinct from cyclins. Thus, a new principle governing the regulation and function of Cdk family proteins appears to have emerged, whereby a Cdk is adapted to perform specific and unique functions unrelated to cell cycle control by using an activator protein distinct from cyclins.

NEURONAL CDC2-LIKE KINASE-AN ACTIVE FORM OF CDK5

A highly active protein kinase containing Cdk5 has been identified and purified in two different laboratories. Lew et al. (18) used a synthetic peptide derived from one of the cdc2 phosphorylation sites on pp60^{src}, thr 46, to demonstrate in bovine brain extract the existence of a protein kinase with cdc2 kinase-like properties and to purify the protein kinase to homogeneity. Since the protein kinase has the phosphorylation site specificity strictly dependent of the proline residue carboxyl side of the phosphorylation serine or threonine, the kinase was initially called brain proline-directed kinase. Ishiguro and co-workers (16, 17), during their long term investigation of neuro-cytoskeletal proteins, identified and purified two bovine brain protein kinases which catalysed the phosphorylation of the neuron-specific microtubule associated proteins, tau proteins. The two protein kinases were designated as tau kinase I and II (17). Subsequent studies have established that brain proline-directed kinase and tau kinase II are the same protein. Both are heterodimer of a 33 (or 32) kDa and a 25 (or 23) kDa subunits (17, 18, 21). Molecular cloning has identified the 33 kDa subunit of the brain protein kinase as Cdk5 (22, 23). Cloning of the 25 kDa subunit has revealed the protein as a proteolytic derivative of a 35 kDa protein (19, 24). As will be detailed later, the protein p25/35 has been shown to be essential for Cdk5 kinase activity and to be specifically expressed in neurons of central nervous system (10, 11), it is designated neuronal

Cdk5 activator, p25/35^{Nck5a}. The holoenzyme is called neuronal cdc2-like kinase, Nclk, on account of its neuron-specific regulatory subunits and the cdc2 related catalytic subunit.

Nclk has been characterised in some detail in terms of its catalytic properties. Its enzymological properties such as substrate specificity and susceptibility to inhibition by certain purine analogues (25, 26, 27) are similar to that of cdc2 kinase. The best *in vitro* substrate of the enzyme identified to date is histone H1. A peptide derived from the phosphorylation site sequence of histone H1 is as good a substrate as histone H1 itself (26) suggesting that the substrate activity determinants are restricted to the primary sequence around the phosphorylation sites. In comparison with the pro-src peptide, histone H1 peptide is over 50 fold more efficient as a Nclk substrate. Using a set of synthetic peptides systematically modified from the histone H1 peptide, the substrate determinants of the peptide, both positive and negative, have been elucidated as schematically represented in Fig. 1.

Shetty et al. (28) have identified and isolated an active form of Cdk5 from rat spinal cord that has similar catalytic properties as Nclk. While the existence of Cdk5 has been suggested by immunological criteria (18), the regulatory subunit of the enzyme was not identified.

IDENTIFICATION AND CLONING OF CDK5

Cdk5 was identified and cloned independently in four different laboratories. Using degenerated oligo-nucleotide corresponding to conserved regions of cdc2 as primers, Meyerson et al (29) isolated by PCR cloning a number of clones encoding distinct human cdc2 homologous kinase subunits, one of these, called PSSALRE was the first Cdk5 cloned. The name indicated the unique amino acid sequence of the protein at the highly conserved PSTAIRE regions of many cdc2-family members. Hellmich et al (14) used a cdc2 cDNA probe and low-stringency procedures to obtain a similar clone from an adult rat brain library. The encoded protein was called neuronal cdc2-like kinase since analyses by Northern blot and in situ hybridisation indicated that the protein was highly enriched in terminally differentiated neurons. Lew et al. (22) identified and cloned Cdk5 as the catalytic subunit of the "brain proline-directed kinase" which is now called neuronal cdc2-like kinase(see above). The name Cdk5 was first used in a publication by Xiong et al (30) who identified and cloned the protein as a D cyclin-associated protein.

Both human and mouse cdk5 genes have been localised on chromosome 5 at 7q36 and the

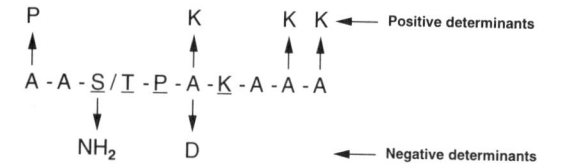

Figure 1. Structural determinants of CDK5 substrate peptide. Positive or negative determinants are the amino acid residues whose substitution for alanine results in an increase or a decrease in substrate activity of the peptide, respectively. Substitution of any of the underlined amino acid residues by alanine abolishes the substrate activity of the peptide.

centromeric region, respectively (31, 32). While the location, 12q13, for human cdk2 and cdk4 genes is known to be associated with chromosome alteration in solid tumours, the location for human cdk5 gene, 7q36, is not a major site of chromosome alteration in tumours (31). Genomic clone of mouse Cdk5 has recently been obtained and a 5.8 kb fragment has been characterised. (33, 32). The mouse Cdk5 gene contains at least 12 exons. The promoter region contains at least two negative and two positive regulatory elements. In addition to Cdk5 clones from the mammalian species, a genomic clone from *Drosophila* melanogaster (34) and *C. elegans* cDNA clone (Tang and Wang, unpublished observation) that encode protein kinases with amino acid sequences highly homologous to the mammalian Cdk5 sequences have been obtained. The *Drosophila* gene shows that the protein kinase is contained on 4 exons thus distinct from the intron/exon organisation of the mouse gene. Sequence alignment (Fig. 2) of the various Cdk5 homologous proteins revealed highly conserved regions of these proteins and some of these appear to be unique and can be used to distinguish themselves from other Cdk family proteins. While the amino acid sequence identity between mammalian Cdk5 proteins are over 99%, the protein kinases of the lower animal species are much less conserved. For example, the sequence of the *Drosophila* protein displays 77% amino acid identity to the mammalian Cdk5 sequences. Certain conserved regions, however, appear to contain amino acid sequences characteristic of Cdk5. Examples are the totally conserved regions of residues 40 to 60 containing the PSSALRE sequence and of residues 159 to 179 starting with Ser159, the residue corresponding to the positive regulatory phosphorylation site of Cdc2, Thr161. Since the proteins encoded by the *Drosophila* and *C. elegans* clones have not yet been expressed and tested for catalytic and functional properties, the assignment of these proteins as Cdk5 homologues is tentative.

IDENTIFICATION AND CLONING OF NCK5A AND RELATED PROTEINS

Nck5a was cloned as the regulatory subunit of bovine brain Nclk (21, 19) or as a 35 kDa Cdk5

```
HCDK5   : mqkyeklekigegtygtvfkaknretheivalkrvrlddddegvpssalr      50
XENOCD  : mqkyeklekigegtygtvfkaknrdtheivalkrvrlddddegvpssalr      50
DROPCD  : mqkydkmekigegtygtvfkgrnratmeivalkrvrldeddegvpssalr      50
CECDK5  : mlnydkmekigegtygtvfkarnknsgeivalkrvrlddddegvpssalr      50
          *  * * *************  *       ********** **********

HCDK5   : eicllkelkhknivrlhdvlhsdkkltlvfefcdqdlkkyfdscngdldp     100
XENOCD  : eicllkelkhknivrlhdvlhsdkkltlvfefcdqdlkkyfdscngdldp     100
DROPCD  : eicllkelkhknivrlidvlhsdkkltlvfehcdqdlkkyfdslngeidm     100
CECDK5  : eicilrelkhrnvvrlydvvhsenkltlvfeycdqdlkkffdslngymda     100
          ****  ****  *  ***  **  ** ******* ******* *** **   *

HCDK5   : eivksflfqllkglgfchsrnvlhrdlkpqnllinrngelkladfglara     150
XENOCD  : eivksfmyqllkglafchsrnvlhrdlkpqnllinrngelkladfglara     150
DROPCD  : avcrsfmlqllrglafchshnvlhrdlkpqnllinkngelkladfglara     150
CECDK5  : qtarslmlqllrglsfchahhvlhrdlkpqnllintngtlkladfglara     150
              *   ***  ** *** *    ************** **  **********

HCDK5   : fgipvrcysaevvtlwyrppdvlfgaklystsidmwsagcifaelanagr     200
XENOCD  : fgipvrcysaevvtlwyrppdvlfgaklystsidmwsagcifaelanagr     200
DROPCD  : fgipvkcysaevvtlwyrppdvlfgaklyttsidmwsagcilaeladagr     200
CECDK5  : fgvpvrcfsaevvtlwyrppdvlfgaklyntsidmwsagcifaeisnagr     200
          ** **  * ********************** **********  **   ***

HCDK5   : plfpgndvddqlkrifrllgtpteeqwpsmtklpdykpypmypattslvn     250
XENOCD  : plfpgndvddqlkrifrllgtpteeqwpamtklpdykpypmypatmslvn     250
DROPCD  : plfpgsdvldqlmkifrvlgtpnedswpgvshlsdyvalpsfpaitswsq     250
CECDK5  : plfpgadvddqlkrifkqlgspsednwpsitqlpdykpypiyhptltwsq     250
          *****  ** ***  **    ***       **      * **        *

HCDK5   : vvpklnatgrdllqnllkcnpvqrisaeealqhpyfsdfcpp             292
XENOCD  : vvpklnatgrdllqnllkcnpvqricadealqhpyfadfcpp             292
DROPCD  : lvprlnskgrdllqkllicrpnqrisaeaamqhpyftdssssgh           294
CECDK5  : ivpnlnsrgrdllqkllvcnpagridadaalrhayfadtsdv             292
          ** **  ******* **  * *  **  *   * ** *   *
```

Figure 2. Alignment of the amino acid sequences of human CDK5, Xenopus CDK5, Drosophila melanogaster CDK5, and Caenorhabditis elegans CDK5. Asterisks indicate the conserved amino acid residues. HCDK5: human CDK5; XENOCD: Xenopus CDK5; DROPCD: D. melanogaster CDK5; CECDK5: C. elegans s CDK5.

associated protein detected by co-immunoprecipitating with Cdk5 (20). While the regulatory subunit of the purified bovine brain Nclk is a 25 kDa protein, the cDNA clone contains an open reading frame of a 34 kDa protein (19, 24), indicating that the 25 kDa protein is a truncated form of the protein. The deduced amino acid sequences show that the conversion of the full length protein to the 25 kDa form involves the removal of 98 amino acid residues from the amino terminal region (19). In crude bovine extract, the intact 35 kDa form of the protein is the predominant form (35). As the regulatory subunit of Nclk, the 25 kDa protein is essential for the kinase activity. Bacterial expressed Cdk5 or Cdk5 monomeric form isolated from bovine brain can be converted from a totally inactive protein to an active kinase upon mixing with a bacterially expressed p25 or a truncated form, p21. Nclk reconstituted from bacterial expressed recombinant subunits with specific kinase activity as high as, or higher than those of homogeneous bovine brain Nclk preparations has been obtained (36).

Although Nck5a has cyclin-like activity in that it is an essential activator of a Cdk, the protein shows no significant amino acid homology to members of cyclins (19, 20, 21). Only the sequence of 17 amino acid residues, residues 222 to 238 displays any similarity to equivalent region of cyclin box consensus sequence (19). Mammalian Nck5a is a highly conserved protein; amino acid sequences of the protein from bovine and human show more than 99% amino acid identity (19, 20). Northern blot analysis of human brain mRNA using a Nck5a probe has revealed two RNA transcripts, a 4 kb and 2.4 kb species (19). While the 4 kb species is derived from the Nck5a gene, the smaller mRNA species is the

A CEP39: mganltsplphhhrtqttsvcnhlfspgdggptfapqrdsntssrsssna   50
   HP39I: mgtvlslspassakgrrpgglpeekkkappagdealggygappvgkggkg   50
   HP35 : mgtvlslsps--------------yrkatlfedgaatvghytavqnskna   36
           **   *   *

   CEP39: kesvlmqgwnwskrniqpvmsrrslpksgssseatsskssdslvsftrnv   100
   HP39I: esrlkrpsvlisaltwkrlvaasakkkkgsk-------------------   81
   HP35 : kdknlkrhsiisvlpwkrivavsakkknskk-------------------   67
                       *   *

   CEP39: ststssqygkisisldrnqnyksvprpsdtttiipnyyslreefrrglqi   150
   HP39I: --------------kvtpkpastgpdplvqqrnrenllrkgrdppdgggt   117
   HP35 : ---------------------------------vqpnssyqnnithln   82

   CEP39: ntrqdlvnnnlasndlsvigspkhvprprsvlrddnancdispiaeqenv   200
   HP39I: akplavpvptvpaaaatceppsggsaaaqppgsgggkpppppppapqvap   167
   HP35 : nenlkkslscanlstfaqpppaqppappasqlsgsqtggsssvkkaphpa   132

   CEP39: psekrgtkktiiqastsellrglgifisnnc-dvsdfdpahlvtwlrsvd   249
   HP39I: pvpggsprrvivqastgellrclgdfvcrrcyrlkelspgelvgwfrgvd   217
   HP35 : vtsagtpkrvivqastsellrclgeflcrrcyrlkhlsptdpvlwlrsvd   182
                * **** **** ** *   *   *   * * * **

   CEP39: rslllqgwqdiafinpanlvfifllvrdvlpderhlihtleelhawilsc   299
   HP39I: rslllqgwqdqafitpanlvfvyllcreslrgd--elasaaelqaafltc   265
   HP35 : rslllqgwqdqgfitpanvvflymlcrdvisse---vgsdhelqavlltc   229
           ********** ** *** **   * *   ** * * *

   CEP39: lyvsysymgneisyplkpfligndrntfwnrcvamvtshsrqmlllnsss   349
   HP39I: lylaysymgneisyplkpflvepdkerfwqrclrliqrlspqmlrlnadp   315
   HP35 : lylsysymgneisyplkpflvesckeafwdrclsvinlmsskmlqinadp   279
           ** **************   ** **   *   * *** *

   CEP39: tffsevftdlkhcssse   366
   HP39I: hfftqvfqdlknegeaaasgggppsggapaassaardscaagtkhwtmnl   365
   HP35 : hyftqvfsdlknesgqedkkrllllgldr   307
            * ** ***

   CEP39:
   HP39I: dr   367
   HP35 :

                      *   *
B cyclin consensus   L Q L V G - - A M F L A S K Y E E
   CEP39   292   L H A W I L S C L Y V S Y S Y M G 308
   HP39I   258   L Q A A F L T C L Y L A Y S Y M G 274
   HP35   222   L Q A V L L T C L Y L S Y S Y M G 238
   CLN1   A K L V V G T C L W L A A K T W G
   ORFD   R H R I F L G C L I L A A K T L N
   HCS26   I H R I F L A C L I L S A L F H N
   PH080   A H R F L L T A T T V A T K G L C

Figure 3. A. alignment of amino acid sequences of a putative *C. elegans* CDK5 activator p39, an isoform of human CDK5 activator p39i, and human CDK5 activator p35. CEP39: a putative *C. elegans* CDK5 activator; HP39I: human p39[Nck5ai]; HP35: human p35[Nck5a]. B. comparison of amino acid sequences of *C. elegans* p39, HP39I, and HP35 with cyclin and cyclin-like sequences. The cyclin consensus sequence is derived from cyclins A, B, D, and E with the highly conserved Leu and Lys residues highlighted with asterisks. Bolded residues indicate matches to *C. elegans*, HP35, and HP35I. Number on both side indicates the location of the pepetide sequences of these CDK5 activators used in the comparison.

transcript of a distinct gene, which has been cloned from a human hippocampus library (37). The cDNA clone of the second gene contains an open reading of a 39 kDa homologous protein. This protein is designated neuronal Cdk5 activator isoform, p39[Nck5ai]. Like p35[Nck5a], p39[Nck5ai] is capable of activating monomeric Cdk5. The two Cdk5 activators appear to have similar kinase

```
HP39I:  mgtvlslspassakgrrpgglpeekkkappagdealggygappvgkggkg    50
HP35 :  mgtvlslsps--------------yrkatlfedgaatvghytavqnskna    36
YORF :                        lealmdillcyqklfsqfindhil       554


HP39I:  esrlkrpsvlisaltwkrlvaasakkkkgskkvtpkpastgpdplvqqrn    100
HP35 :  kdknlkrhsiisvlpwkrivavsakkknskk-------------------    67
YORF :  ftktfifiykkvlkekdvpaynvtsfmpfwkffmknfpfvlkvdndlrie    604
                         *

HP39I:  renllrkgrdppdgggtakplavpvptvpaaaatceppsggsaaaqppgs    150
HP35 :  --vqpnssyqnnithlnnenlkkslscanlstfaqpppaqppappasqls    115
YORF :  lqsvyndeklkteklkndksevlkvysminnsnqavgqtwnfpevfqvni    654


HP39I:  gggkpppppppapqvappvpggsprrvivqastgellrclgdfvcrrcyr    200
HP35 :  gsqtggsssvkkaphpavtsagtpkrvivqastsellrclgeflcrrcyr    165
YORF :  rfllhnseiidtntskqfqkarnnvmlliatnlkeynkfmsiflkrkdft    704
                                                *       *

HP39I:  lkelspgelvgwfrgvdrslllqgwqdqafitpanlvfvyllcreslrgd    250
HP35 :  lkhlsptdpvlwlrsvdrslllqgwqdqgfitpanvvflymlcrdvisse    215
YORF :  nknliqlislklltfevtqnvlgleyiirllpinlenndgsyglflkyhk    754
         *  *               *

HP39I:  elasaaelqaafltclylaysymgne--------------isyplkpflve   287
HP35 :  -vgsdhelqavlltclylsysymgne--------------isyplkpflve   252
YORF :  eqfiksnfekilltcyelekkyhgneceinyyeillkilitygsspklla    804
             ***   *    * ***                    *  *   * *

HP39I:  pdkerfwqrclrliqrlspqmlrlnadphfftqvfqdlknegeaaasggg    337
HP35 :  sckeafwdrclsvinlmsskmlqinadphyftqvfsdlknesgqedkkrl    302
YORF :  t--------------stkiimlllndsvenssniledilyystcpsetdl    840
                       **   *                   *

HP39I:  ppsggapaassaardscaagtkhwtmnldr    367
HP35 :  llgldr                          307
YORF :  ndiplgsgqpdn                    852
```

Figure 4. Comparison amino acid sequences of the yeast protein which can bind to human CDK5 with high affinity to human CDK5 activator and an isoform of human CDK5 activator. HP39I: human p39[Nck5ai]; HP35: human p35[Nck5a]; YORF: the yeast open reading frame which is know as yeast SRB8 (39).

activating efficiency (37). An open reading frame on chromosome III of *C. elegans* encoding a 39 kDa protein homologous to the Cdk5 activators is among the *C. elegans* genome sequences deposited in the gene bank (NewEMBL). The *C. elegans* protein shows similar degrees of sequence identity to the two mammalian activators. A fragment of the *C. elegans* protein equivalent to the p25 region of Nck5a has been expressed in *E. coli* and shown to activate the bacterial expressed bovine Cdk5. The maximal kinase activity achieved by the *C. elegans* protein, however, is only a few percent that achieved with the mammalian Nck5a or Nck5ai (Tang and Wang, unpublished observation). The sequence alignment of the homologous Cdk5 activators is shown as Fig. 3. These proteins may represent a subfamily of Cdk activators that are specific for Cdk5, or a subfamily of Cdks more closely related to Cdk5 than to other Cdks.

A yeast gene on chromosome III of *S. cerevisiae* contains an open reading frame encoding a 144 kDa protein has been identified to have a region homologous to Nck5a, albeit with a very low percentage of sequence similarity (Fig. 4). When a fragment of this protein corresponding to the p25 region of Nck5a was cloned and expressed in *E. coli*, the expressed protein showed high affinity and specific binding to Cdk5. The protein did not activate Cdk5, and it prevented the kinase activation by Nck5a (38). These observations suggest that this protein or the Nck5a homologous domain of the protein mimics Nck5a in binding to Cdk5. That this protein domain functions as an activator for a Cdk5-like yeast kinase is an intriguing notion worthy of exploring. This large yeast protein has been identified recently as one of the suppressors of RNA polymerase B(SRB), called SRB8 (39). Yeast RNA polymerase II holoenzyme is a protein complex consisting of RNA polymerase II, several general transcription factors and up to 9 SRBs. It is interesting to note that among the SRBs, SRB10 and SRB11 display high sequence homology to a 35 kDa human Cdk called Cdk8 and cyclin C respectively, and that cyclin C and Cdk8 are also present in the human RNA polymerase II complex (40, 41, 42). The possibilities of a human homologue of SRB8 and of its existence in RNA polymerase complex should therefore also be considered.

MECHANISM OF CDK5 ACTIVATION BY NCK5A

As indicated above, Cdk5 shows extensive homology to Cdc2 and Cdk2, but it has not been shown to be activated by any cyclin and nor is Cdk5 capable of rescuing yeast cdc2 mutants. On the other hand, Nck5a and Nck5ai have little similarity to cyclins in amino acid sequence but can activate Cdk5. In spite of lacking sequence similarity to cyclins, the Cdk5 activators may share with cyclins similar protein folding characteristics. Nck5a is capable of activating Cdk2, although the level of activation is only a fraction of that of cyclin A activation. Recent crystallographic structures of cyclin A and cyclin A-Cdk2 complex have demonstrated the presence in cyclin A two repeating domains with similar protein folding (43, 44). Comparison of amino acid sequences of the two domains revealed very low homology. Furthermore, using Threader program which calculates the energy of a protein when folded in a particular conformation, $p25^{Nck5a}$ has been suggested to assume a cyclin-like structure (44). The minimal size of kinase activating Nck5a matches closely to that of cyclin A required to activate Cdk2(Tang and Wang, unpublished observation).

The activation of several Cdks, Cdc2, Cdk2 and Cdk4 by their respective cyclins have been shown to depend on the phosphorylation of a specific threonine residue, Thr161 of Cdc2 or equivalents in other cdks, by an exogenous kinase called Cdk activating kinase, Cak (45-49). In contrast, the activation of Cdk5 by Nck5a or Nck5ai does not seem to depend on the phosphorylation of Cdk5. Highly active Nclk has been reconstituted from the bacterial expressed Cdk5 and $p25^{Nck5a}$ in the absence of any additional kinases. Addition of Cak to the reconstitution reactions was found to have no effect on the Nclk kinase activity (36).

The kinase Cak is itself a cdc2-like kinase in that it is composed of a cdc2-related catalytic subunit, Cdk7 and a specific cyclin, cyclin H (48). The activation of Cdk7 by cyclin H has been suggested to be enhanced by autophosphorylation on threonine 170 of Cdk7. The possibility that activation of Cdk5 by Nck5a or Nck5ai involves autophosphorylation on Cdk5 has been examined and ruled out (36). The activation of Cdk7 by cyclin H may be facilitated by an alternative mechanism which entails the formation of a trimeric protein complex of Cdk7, cyclinH and a third protein subunit (50). The trimeric protein complex has full Cak activity without the phosphorylation of the catalytic subunit. Thus, Nclk resembles Cak in its phosphorylation-independent kinase activation except this is achieved in the absence of an additional protein subunit.

TISSUE DISTRIBUTION AND EMBRYONIC DEVELOPMENT OF CDK5 AND NCK5A

On the basis of Northern analysis, Cdk5 has a ubiquitous tissue distribution in mammals with brain containing the highest amount of the transcript (12, 20, 19). On the other hand, the expression of Nck5a, as well as Nck5ai, is strictly confined to brains. Analysis by in situ hybridisation has revealed that Cdk5 mRNA is expressed in neurons throughout adult rat brain with a high level of expression in pyrimidal cell layer of hippocampus (10). There is no apparent Cdk5 mRNA expression in glia. Nck5a distribution has also been analysed by in situ hybridisation as well as by immuno-histochemistry (11). Like Cdk5, Nck5a appears to be expressed at very high levels in pyrimidal cell layer of hippocampus. The protein is restricted to neurons of central nervous system. Although both Cdk5 and Nck5a may exist in same neurons, they do not share common subcellular localisation in adult rat brain. Immunohistochemical analysis of adult rat brains localised Cdk5 and Nck5a to axon and cell body respectively (10, 11). Although the significance of the observation is not clear, it is important to examine the subcellular localisation of the isoform of the Cdk5 activator, $p39^{Nck5ai}$.

The expression of Cdk5 in mouse embryonic brain during development was found by both Western immunoblot and Northern analyses to undergo progressive increase from E 11 up to birth (12). In parallel with the change in protein and mRNA levels, Cdk5 associated histone H1 kinase also increased progressively. While Cdk5 expression is demonstrated in many tissues and proliferating cells in culture, Cdk5 kinase activity has only been detected in brain (20, 19, 21), suggesting the Cdk5 activity change in developing brain is an indication of the change in Nck5a expression. Indeed, the expression of Nck5a during embryonic development of rat brain was found to increase from E12 up to birth (11). The high level of Nck5a expression was maintained for about two weeks after birth and then underwent slow decline. Significantly, the change in Nck5a mRNA expression paralleled the change in Cdk5 kinase activity (24). The possibility that the parallel development of Nck5a and Cdk5 kinase activity is a fortuitous observation cannot be ruled out. Immuno-histochemical study has revealed a developmental change in subcellular location of Cdk5 in neurons that was independent of the localisation of Nck5a (10, 11). In the early neonatal stage, Cdk5 is strongly expressed in the cell bodies of neurons. With neuronal maturation, the protein changes from the cell body to axon (10).

During development, the neuronal precursor cells of the embryonic brain differentiate into neurons; this is under rigourous spatial and temporal control. This process of neuron differentiation appears to be well correlated with the changes in the expression and activity of Cdk5 and its activator Nck5a. Of particular interest is the observation of an inverse relationship between the changes of Cdk5 expression and activity with those of the expressions and activities of cell cycle-related cdc2-like kinases such as Cdc2 and Cdk2 as well as cyclins A and B during embryonic development (12, 51). Thus, as cells enter into terminally differentiated neurons, cell cycle kinases and regulatory proteins disappear whereas Cdk5 and Nck5a appear.

The neuron specificity of Nclk and the correlation of the appearance of the protein subunits with neuron differentiation and maturation has raised the possibility that Nclk may be an essential factor for neuronal differentiation. Nikolic et al. (13) have obtained evidence to support such a suggestion. They demonstrated that transfection of cultured primary cortical neurons from E17-18 rats with dominant negative Cdk5 mutants or Nck5a antisense cDNA resulted in much reduced neurite growth of the cells. The effects could be prevented by co-transfection with wild type Cdk5 or sense Nck5a cDNA. In these neurons, Cdk5 and Nck5a were found to co-localize with microtubules in the cell body and axon, whereas at growth cones of the neurons, the two proteins were also present at the periphery where actin filaments are located. The growth cone localisation of Nclk supports a role of the enzyme in neurite outgrowth.

P67, A 67 KDA NEURONAL CDK5 ACTIVATOR

Shetty et al. (52) identified a protein activator of Cdk5 with an apparent molecular mass of 62 kDa during the purification of Cdk5 kinase from rat spinal cord. Partial sequencing of a purified sample of the protein suggested that the protein was identical to a previously cloned 67 kDa protein called Munc-18, the mammalian homolog of the C. elegans unc-18 gene. The suggestion was confirmed by the observation that a bacterially expressed His-tagged fusion protein of p67 could activate a sample of Cdk5 purified from rat spinal cord by several fold. P67 has no sequence homology to any of the cyclins, nor any of the Nck5a related proteins.

The significance of the two types of Cdk5 activator proteins is not clear. In situ hybridisation studies localised p67 in neurons of both central nervous system and peripheral nervous system E18 rat embryo (52). Immuno-histochemical studies of a primary hippocampus neuron culture showed that p67 and NH-F were co-localized in axons and cell bodies but not in dendrites. Western blot analysis has detected p67 or p67-related proteins in non-neuronal tissues such as heart and liver in addition to nervous tissues. Thus, p67 appears to have a much wider distribution than Nck5a type of Cdk5 activators in mammals, and therefore may serve as the primary Cdk5 activator in tissues lacking Nck5a and Nck5ai.

Since direct comparison of the Cdk5 activation activities of p67 and Nck5a has not been made, the relative efficiencies of these two activators are not known. It is also unclear whether or not the totally inactive monomeric form of Cdk5, such as the bacterially expressed Cdk5 can be activated by p67 and whether or not Cdk5's activation by p67 and by Nck5a/Nck5ai are additive. Further studies are required to clarify the relationship between the two types of Cdk5 activator in the kinase activation.

THE ROLE OF NCLK IN NEURO-CYTOSKELETON DYNAMICS

The suggestion that Nclk plays a role in neurite outgrowth is in agreement with the large amount of evidences from biochemical studies implicating the enzyme in the regulation of neuro-cytoskeleton dynamics (22, 53-60, 28). Proteins involved in all three classes of cytoskeleton systems: microtubules, intermediate filaments and actin filaments, have been found to be phosphorylated *in vitro* by Nclk (22). Thus, neuron-specific microtubule associated proteins, tau proteins and MAP 2 (61, 17, 23, 53, 62, 63, 64), and the subunits of neurofilaments (neuronal intermediate filaments), neurofilaments M(NF-M) and neurofilaments H(NF-H) (22, 28, 57) have been shown to be *in vitro* substrates of Nclk and the sites of phosphorylation of these proteins corresponding to some of the *in vivo* phosphorylation sites. Caldesmon, an actin filaments binding protein has been suggested to be a physiological substrate of cdc2 kinase. Nclk has also been found to phosphorylate caldesmon *in vitro* and the phosphorylated caldesmon showed drastically reduced binding to actin filaments(A. Mak, J. Lew and J.H.Wang, unpublished observation).

Tau protein exists in adult neurons as a set of six isoforms derived from a single gene by alternative mRNA splicing (65, 66, 67). It promotes *in vitro* microtubule assembly from the tubulin dimers (68, 69). Expression of tau in insect cells has been found to cause the cells to form neurite-like processions suggesting that the protein takes part in the maintenance of neuron polarity (70). Tau is especially rich in "proline-directed" serine/threonine residues (55). Recent studies have implicated the phosphorylation of tau at some of these proline-directed sites in Alzheimer pathology (71, 53). Tau is the major protein components of paired helical filaments (PHFs), a structure found in abundance in brains of Alzheimer's Disease patients (72, 60, 73, 74, 65). Tau proteins isolated from PHFs are hyperphosphorylated on many of the proline-directed sites (75, 55). Most of these sites have been found to be phosphorylated by Nclk in $vitro$ (53, 62). Although other kinases, such as MAP kinase and glycogen synthase kinase 3 are also capable of phosphorylating tau proteins at proline-directed sites (76, 77, 55), Nclk appears to be the most selective in targeting the phosphorylation sites of Alzheimer tau (62). Mass spectrometry has identified eight serine/threonine residues as the sites phosphorylated or potentially phosphorylated in tau proteins isolated from Alzheimer brains (56). In one study, phosphorylation of purified bovine tau by purified Nclk was shown to result in the phosphorylation of seven of the eight Alzheimer sites while none of the other proline directed sites in tau protein was phosphorylated (53).

Neurofilaments, the neuron-specific intermediate filaments represent the major cytoskeletal organelle in axons in terms of mass and volume (78, 79, 80). In mature mammalian brains, neurofilaments are composed of three subunits of different molecular weights which are referred to as low (68 kDa), medium (95 kDa) and high molecular weight (115 kDa) neurofilament proteins: NF-L, NF-M and NF-H respectively (81, 82). Each of the subunits contains an alpha-helical rich conserved central core which is involved in the coil-coil assembly of the filamentous structure (83). The amino side of the core domain is a conserved globular domain which contains multiple second messenger regulated phosphorylation sites and is therefore suggested to play a role of regulating the filament assembly (84-91). The carboxyl terminal regions of the neurofilaments subunits are the regions discriminating the three neurofilament subunits (92, 80, 54). NF-M and NF-H but not NF-L contain long caboxyl-terminal extensions which are rich in proline-directed serine/threonine residues (93-96). This is especially true for NF-H which has more than 50 such sites (96). These sites are heavily phosphorylated in neurons. Several studies have shown that NF-H and NF-M can be phosphorylated by Nclk or related kinases (80, 28, 57, 14, 63, 97).

The proline-directed phosphorylation appears to play important roles in neuro-skeleton structure and function. Dephosphorylated NF-H display specific binding to microtubules in $vitro$ whereas the phosphorylated form of the protein does not bind microtubules (98). The phosphorylation is under strict spatial and temporal control. The assembly of neurofilaments occurs in the cell body (99, 100); once assembled, the filaments are transported along the axon. During the axonal transport, the subunits NF-M and NF-H become heavily phosphorylated on the proline-directed sites (101, 102). It has been suggested that the phosphorylation depends on the presence of myelin sheath (102). A mouse strain with defective myelin, trembling mouse, has been found to have greatly reduced neurofilaments phosphorylation. Abnormality in NF-H and NF-M phosphorylation has also been implicated in amylotrophic lateral sclerosis.

Although it is yet to be established that the proline-directed tau and neurofilament phosphorylations are catalysed in neurons by Nclk, several observations support such a suggestion. Nclk or its subunits have been found to co-localize with neurofilaments and tau proteins in the cells (10, 11) or in association with these proteins in the cell extract (63, 53). The major tau kinase activity or proline-directed kinase activity in mammalian brain extracts has been shown to be with Nclk or Cdk5 (63). However, it should be noted that the phosphorylation of tau and neurofilaments by Nclk and the phosphorylation of tau and neurofilaments by other proline-directed kinases in neurons do not have to be mutually exclusive.

MULTIPLE PROTEIN ASSOCIATION STATES OF CDK5

Western immunoblot analysis in combination with protein kinase activity and protein fractionation analyses of bovine brain extracts has suggested that Cdk5 in brain existed in at least three protein association states (35): the free monomeric Cdk5, Cdk5 in complexing with p25^{Nck5a} and Cdk5 in complexing with the intact p35^{Nck5a}. Most of Cdk5 is present in the monomeric state which can be activated by the bacterially expressed p25^{Nck5a}. A small amount of Cdk5 is present as the heterodimer of Cdk5-p25^{Nck5a} and this form is the only Cdk5 form in the brain extract with intrinsic histone H1 peptide kinase activity. The complex containing Cdk5 and intact p35^{Nck5a}, on the other hand, exhibits no kinase activity (35). This protein complex has a molecular mass over 600 kDa and appears to be composed of many additional protein components. Since the complex is devoid of kinase activity, the existence of an inhibitory protein in the complex has been considered. When the protein complex was chromatographed on a gel filtration column in the presence of 10% ethylene glycol, it became active towards histone H1 peptide, though with relatively low activity. The observation is compatible with the suggestion that gel filtration under the condition had removed a low molecular weight inhibitor (35). This low molecular weight inhibitor may not be the well known Cdk inhibitory proteins. The p21^{cip} has been shown to be a poor inhibitor for Cdk5/p35 complex (103). A 27 kDa general Cdk protein inhibitor called p27^{inkp} is present in high levels in brains, but this protein is a very weak inhibitor for Nclk (104).

Although the significance of the existence of multiple association states of Cdk5 is not known, it seems reasonable to suggest that most Cdk5 in brain exists in an inactive monomeric states, i.e., the kinase activity of Cdk5 is limited by the amount of Nck5a. Since the Cdk5-p35^{Nck5a} is inactive, it may be further suggested that Cdk5 activity is controlled by additional factors, such as the specific inhibitor alluded above. The assembly of the >600 kDa protein complex may be dependent on the 98 residue aminoterminal region of Nck5a since no 25 kDa Nck5a species was detected in the complex. In the presence of an excess amount of the inhibitor protein, therefore, the Nclk may escape the inhibition by conversion of the intact Nck5a to the 25 kDa truncated form of Nck5a. Amino acid sequence comparison between Cdk5 and Cdc2 has revealed that the negatively regulatory phosphorylation sites thr 14 and tyr 15 of Cdc2 are conserved in Cdk5 and the positive regulatory phosphorylation site, thr 161 of Cdc2 is conservatively substituted by a serine (105). As indicated above, the activation of Cdk5 by p25^{Nck5a} is independent of the phosphorylation of Cdk5 (36). On the other hand, the negative regulatory phosphorylation mechanism may be conserved in the Cdk5 kinase.

The protein composition of the >600 kDa protein complex is not known at present. However, a number of proteins which have been reported to be "Cdk5-associated", may exist as such components. For example, p67 and Nck5ai may be present in the complex. Proteins which have been shown to co-immunoprecipitate with Cdk5 should also be considered as candidates, these include cyclin D (30, 106), cyclin E (107), a 60 kDa protein and a 180 kDa protein (20). Neurofilament proteins (28), a p13 suc1 related protein p15 (108) and Tau proteins (53) which have been demonstrated to undergo high affinity association with Cdk5 should also be tested for their possible presence in the >600 kDa protein complex. The protein complex may represent a functional unit of Cdk5 kinase. The elucidation of the protein composition of the protein complex will shed light on the regulatory as well as functional properties of Nclk.

ACKNOWLEDGEMENT

The authors wish to acknowledge the financial support of grants from the Research Grants Council of Hong Kong and Research Infrastructure Grants of The Hong Kong University of Science and Technology.

REFERENCES

1. Draetta, G. (1990) *Trends Biochem. Sci.* **15**, 378-383.
2. Murray, A.W. and Kirschner, M.W. (1989) *Science* **246**, 614-621.
3. Hunt, T. (1989). *Curr. Opin. Cell Biol.* **1**, 268-274.
4. Pines, J. and Hunter, T. (1990) *New Biol.* **2**, 389-401.
5. Massague, J. and Roberts, J.M. (1995) *Curr. Opin. Cell Biol.* **7**, 769-772.
6. Nurse, P., (1990) *Nature* **344**, 503-507.
7. Pines, J. (1992) *Curr. Opin. Cell Biol.* **4**, 144-148.
8. Pines, J. (1994) *Can. Biol.* **15**, 305-313.
9. Hunt, T. and Kirschner, M. (1993) *Curr. Opin. Cell Biol.* **5**, 163-165.
10. Matsushita, M., Matsui, H., Itano, T., Tomizawa, K., Tokuda, M., Suwaki, H., Wang, J.H., and Hatase, O. (1995) *NeuroReport* **6**, 1267-1270.
11. Tomizawa, K., Matsui, H., Matsushita, M., Lew, J., Tokuda, M., Itano, T., Konishi, R., Lew, J., Tokuda, M., Itano, T., Konishi, R., Wang, J.H., and Hatas, O. (1996) **In Press**.
12. Tsai, L.-H., Takahashi, T., Caviness Jr, V.S., and Harlow, Ed (1993) *Development* **119**, 1029-1040.
13. Nikolic, M., Dudek, H., Kwon, Y.T., Ramos, Y.F.M., and Tsai, L.-H. (1996) *Genes and Development* **10**, 816-825.
14. Hellmich, M.R. Pant, H.C., Wada, E., and Battey, J.F. (1992) *Proc. Natl. Acad. Sci. USA* **89**, 10867-10871.
15. Hayes, T.E., Valtz, N.L.M., and McKay, R.D.G. (1991) *New Biol.* **3**, 259-269.
16. Ishiguro et al. (1988) *J. Biochem.* **104**, 319-321.
17. Ishiguro, K., Takamatsu, M., Tomizawa, K., Omori, A., Takahashi, M., Arioka, M., Uchida, T., and Imahori, K. (1992) *J. Biol. Chem.* **267**, 10897-10901.
18. Lew, J., Beaudette, K., Litwin, C.M.E., and Wang, J.H. (1992a) *J. Biol. Chem.* **267**, 13383-13390.
19. Lew, J., Huang, Q.-Q., Qi, Z., Winkfein, R.J., Aebersold, R., Hunt, T., and Wang, J.H. (1994) *Nature* **371**, 423-426.
20. Tsai, L.-H., Delalle, I., Caviness Jr, V.S., Chae, T., and Harlow, Ed (1994) *Nature* **371**, 419-423.

21. Ishiguro, K., Kobayashi, S., Omori, A., Takamatsu, M., Yonekura, S., Anzai, K., Imahori, K., and Uchida, T. (1994) *FEBS Lett.* **342**, 203-208.
22. Lew, J., Winkfein, R.J., Paudel, H.K., and Wang, J.H. (1992b) *J. Biol. Chem.* **267**, 25922-25926.
23. Kobayashi, S., Ishiguro, K., Omori, A., Takamatsu, M., Arioka, M., Imahori, K., and Uchida, T. (1993) *FEBS Lett.* **335**, 171-175.
24. Uchida, T., Ishiguro, K., Ohnuma, J., Takamatsu, M., Yonekura, S., and Imahori, K. (1994) *FEBS Lett.* **355**, 35-40.
25. Lew, J., Qi, Z., Huang, Q.-Q., Paudel, H., Matsuura, I., Matsushita, M., Zhu, X., and Wang, J.H. (1995) *Neurobiol. Aging* **16**, 263-270.
26. Beaudette, K.N., Lew, J., and Wang, J.H. (1993) *J. Biol. Chem.* **268**, 20825-20830.
27. Vesely, J., Havlicek, L., Strnad, M., Blow, J.J., Donella-Deana, A., Pinna, L., Letham, D.S., Kato, J.-Y., Detivaud, L., Leclerc, S., and Meijer, L. (1994) *Eur. J. Biochem.* **224**, 771-786.
28. Shetty, K.T., Link, W.T., and Pant, H.C. (1993) *Proc. Natl. Acad. Sci. USA* **90**, 6844-6848.
29. Meyerson, M., Enders, G.H., Wu, C.-L., Su, L.-K., Gorka, C., Nelson, C., Harlow, Ed, and Tsai, L.-H. (1992) *EMBO J.* **11**, 2909-2917.
30. Xiong, Y., Zhang, H., and Beach, D. (1992) *Cell* **71**, 505-514.
31. Demetrick, D.J., Zhang, H., and Beach D.H. (1994) *Cytogenet. Cell Genet.* **66**, 72-74.
32. Ohshima, T., Nagle, J.W., Pant, H.C., Joshi, J.B., Kozak, C.A., Brady, R.O., and Kulkarni, A.B. (1995) *Genomics* **28**, 585-588.
33. Ishizuka, T., Ino, H., Sawa, K., Suzuki, N., and Tatibana, M. (1995) *Gene* **66**, 267-271.
34. Hellmich, M.R., Kennison, J.A, Hampton, L.L., and Battey, J.F. (1994) *FEBS Lett.* **356**, 317-321.
35. Lee, K.-Y., Rosales, J.L., Tang, D., and Wang, J.H. (1996) *J. Biol. Chem.* **271**, 1538-1542.
36. Qi, Z., Huang, Q.-Q., Lee, K.-Y., Lew, J., and Wang, J.H. (1995a) *J. Biol. Chem.* **270**, 10847-10854.
37. Tang, D., Yeung, J., Lee, K.-Y., Matsushita, M., Matsui, H., Tomizawa, K., Hatase, O., and Wang, J.H. (1995) *J. Biol. Chem.* **270**, 26897-26903.
38. Huang, Q.-Q., Lee, K.-Y, and Wang, J.H. (1996) *FEBS Lett.* **378**, 48-50.
39. Hengartner, C.J., Thompson, M.C., Zhang, J., Chao, D.M., Liao, S.-M., Koleske, A.J., Okamura, S., and Young, R.A. (1995) *Genes and Development* **9**, 897-910.
40. Tassan, J.P., Jaquenoud, M., Leopold, P., Schultz, S.J. and Nigg, E.A. (1995) *Proc. Natl. Acad. Sci. USA* **92**, 8871-8875.
41. Maldonado, E., Shiekhattar, R., Sheldon, M., Cho, H., Drapkin, R., Richert, P., Lees, E., Anderson, C.W., Linn, S., and Reinberg, D. (1996) *Nature* **381**, 86-89.
42. Chao, D.M., Gadbois, E.L., Murray, P.J., Anderson, S.F., Sonu, M.S., Parvin, J.D., and Young, R.A. (1996) *Nature* **380**, 82-85.
43. Jeffrey, P.D., Russo, A.A., Polyak, K., Gibbs, E., Hurwitz, J., Massague, J., and Pavletich, N.P. (1995) *Nature* **376**, 313-320.
44. Brown, NR, Noble, M., Endicott, JA, Garman, EF, Wakatsuki, S., Mitchell, E., Rasmussen, B., Hunt, T., and Johnson, LN (1995) *Structure* **3**, 1235-1247.
45. Solomon, M.J., Harper, J.W., and Shuttleworth, J. (1993) *EMBO J.* **12**, 3133-3142.
46. Poon, R.Y.C., Yamashita, K., Adamczewski, J.P., Hunt, T., and Shuttleworth, J. (1993). *EMBO J.* **12**, 3123-3132.
47. Fesquet, D., Labbe, J.C., Derancourt, J., Capony, J.P., Galas, S., Girard, F., Lorca, T., Shuttleworh, J., Doree, M., and Cavadore, J.C. (1993) *EMBO J.* **12**, 3111-3121.
48. Fisher, R.P. and Morgan, D.O. (1994) *Cell* **78**, 713-724.
49. Kato, J.-Y., Matsuoka, M., Strom, D.K., and Sherr, C.J. (1994) *Mol. Cell. Biol.* **14**, 2713-2721.
50. Devault, A., Martinez, A.-M., Fesquet, D., Labbe, J.-C., Morin, N, Tassan, J.P., Nigg, E.A., Cavadore, J.-C., and Doree, M. (1995) *EMBO J.* **14**, 5027-5036.
51. Ino, H., Ishizuka, T., Chiba, T., and Tatibana, M. (1994) *Brain Research* **661**, 196-206.
52. Shetty, K.T., Kaech, S., Link, W.T., Jaffe, H., Flores, C.M., Wray, S., Pant, H.C., and Beushausen, S. (1995) *J. Neurochem.* **64**, 1988-1995.
53. Paudel, H. K., Lew, J., Ali, Z., and Wang, J.H. (1993) *J. Biol. Chem.* **268**, 23512-23518.
54. Nixon, R.A. and Sihag, R.K. (1991) *TINS* **14**, 501-506.
55. Mandelkow, E.-M. and Mandelkow, E. (1993) *Trends Biochem. Sci.* **18**, 480-483.
56. Hasegawa, M., Morishima-kawashima, M., Takio, K., Suzuki, M., Titani, K., and Ihara, Y. (1992) *J. Biol. Chem.* **267**, 17047-17054.
57. Hisanaga, S., Kusubata, M., Okumura, E., and Kishimoto, T. (1991) *J. Biol. Chem.* **266**, 21798-21803.
58. Guan, R.G., Hall, F.L., and Cohlberg, J.A. (1992) *J. Neurochem.* **58**, 1365-1371.
59. Mawal-Dewan, M., Sen, P.C., Abdel-Ghany, M., Shalloway, D., and Racker, E. (1992) *J. Biol. Chem.* **267**, 19705-19709.
60. Vulliet, R., Halloran, S., Braun, R., Smith, A., and Lee, G. (1992) *J. Biol. Chem.* **267**, 22570-22574.

61. Ishiguro, K., Omori, A., Sato, K., Tomizawa, K., Imahori, K., and Uchida, T. (1991) *Neurosci. Lett.* **128**, 195-198.
62. Baumann, K., Mandelkow, E.-M., Biernat, J., Piwnica-Worms, H., Mandelkow, E. (1993) *FEBS Lett.* **336**, 417-424.
63. Hosoi, T., Uchiyama, M., Okumura, E., Saito, T., Ishiguro, K., Uchida, T., Okuyama, A., Kishimoto, T., and Hisanaga, S.-I. (1995) *J. Biochem.* **117**, 741-749.
64. Mandelkow, E.-M., Biernat, J., Drewes, G., Gustke, N., Trinczek, B., and Mandelkow, E. (1995) *Neurobiology of Aging* **16**, 355-363.
65. Lee, V.M.Y., Balin, B.J.; Otvos, L., and Trojanowski, J.Q. (1991) *Science* **251**, 675-678.
66. Himmler, A., Drechsel, D., Kirschner, M., and Martin, D. (1989) *Mol. Cel. Biol.* **9**, 1381-1388.
67. Goedert, M., Spillantini, M., Jakes, R., Rutherford, D., and Crowther, R.A. (1989) *Neuron* **3**, 519-526.
68. Goedert, M., Crowther, R.A., and Garner, C.C. (1991) *TINS* **14**, 193-199.
69. Ksiezak-Reding, H., Liu, W.K., and Yen, S.H. (1992) *Brain Res.* **597**, 209-219.
70. Knops, J., Kosik, K.S., Lee, G., Pardee, J.D., Cohen-Gould, L., and McConlogue, L. (1991) *J. Cell Biol.* **114**, 725-735.
71. Kosik, K.S., and Greenberg, S.M. (1994) in *Alzheimer Disease.* (R.D. Terry, R. Katzman, and K.L. Bick, eds.) Raven Press. New York, 335-344.
72. Anderton, B.H. (1993) *Hippocampus* **3**, 227-237.
73. Will, H., Drewes, G., Biernat, J., Mandelkow, E.M., and Mandelkow, E. (1992) *J. Cell Biol.* **118**, 573-584.
74. Crowther, R.A., Olesen, O.F., Jakes, R., and Goedert, M. (1992) *FEBS Lett.* **309**, 199-202.
75. Kosik, K.S. (1992) *Science* **256**, 780-783.
76. Drewes, G., Lichtenberg-Kraag, B., Doring, F., Mandelkow, E.-M., Biernat, J., Goris, J., Doree, M., and Mandeklow, E. (1992) *EMBO J.* **11**, 2131-2138.
77. Mandelkow, E.-M., Drewes, G., Biernat, J., Gustke, N., Van Lint, J., Vandenheede, J.R., and Mandelkow, E., (1992) *FEBS Lett.* **3**, 314-321.
78. Liem, R.K.H. (1993) *Curr. Opin. Cell Biol.* **5**, 12-16.
79. Steinert, P.M. and Roop, D.R. (1988) *Annu. Rev. Biochem.* **57**, 593-625.
80. Pant, H.C. and Veeranna (1995) *Biochem. Cell Biol.* **73**, 575-592.
81. Shaw, G. (1991) in *The Neuronal Cytoskeleton*, (Burgoyne, R.D. ed.) New York, 186-214.
82. Fliegner, K.H. and Liem, R.K.H. (1991) *Int, Rev. Cytol.* **131**, 109-167.
83. Parry, D.A.D. and Steinert, P.M. (1992) *Curr. Opin. Cell Biol.* **4**, 94-98.
84. Gill, S.R., Wong, P.C., Monteiro, M.J., and Cleveland, D.W. (1990) *J. Cell. Biol.* **111**, 2005-2019.
85. Sihag, R.K. and Nixon, R.A. (1989) *J. Biol. Chem.* **264**, 457-464.
86. Sihag, R.K. and Nixon, R.A. (1990) *J. Biol. Chem.* **265**, 4166-4171.
87. Sihag, R.H. and Nixon, R.A. (1991) *J. Biol. Chem.* **266**, 18861-18867.
88. Hisanaga, S., Gonda, Y., Inagaki, M., Ikai, A., and Hirokawa, N. (1990) *Cell Regul.* **1**, 237-248.
89. Nakamura, Y., Takeda, M., Angelides, H.J., Tanaka, T., Tada, K., and Nishimura, T. (1990) *Biochem. Biophys. Res. Commun.* **169**, 744-750.
90. Dosemeci, A., Floyd, C., and Pant, H.C. (1990) *Cell. Mol. Neurobiol.* **10**, 369-382.
91. Dosemeci, A. and Pant, H.C. (1992) *Biochem. J.* **282**, 477-481.
92. Pant, H.C. (1990) in *Advances in Physiological Sciences.* (Manchanda, S.K., Murthy, W.S., and Mohankumar, V. eds.) MacMillan India Ltd., Delhi. pp. 96-100.
93. Lee, V.M.Y., Carden, M., Schlaepfer, W., and Trojanski, J. (1987) *J. Neurosci.* **7**, 3474-3488.
94. Lee, V.M.Y., Otvos., L., Carden, M.J., Hollosi, M., Dietzschold, B., and Lazzarini, R.A. (1988) *Proc. Natl. Acad. Sci. USA* **85**, 1998-2002.
95. Xu, Z.-S., Liu, W.-S., and Willard, M.B. (1992) *J. Biol. Chem.* **267**, 4467-4471.
96. Elhanany, E., Jaffe, H., Link, W.T., Sheeley, D.M., Gainer, H., and Pant, H.C. (1994) *J. Neurochem.* **63**, 2324-2335.
97. Qi, Z., Tang, D., Matsuura, I., Lee, K.-Y., Zhu, X., Huang, Q.-Q., and Wang, J.H. (1995b) *Mol. Cell. Biochem.* **149**(150), 35-39.
98. Miyasaka H., Okabe, S., Ishiguro, K., Uchida, T., and Hirokawa, N. (1993) *J. Biol. Chem.* **268**, 22695-22702.
99. Lasek, R.J., Garner, J.A., and Brady, S.T. (1984) *J. Cell Biol.* **99**, 212s-221s.
100. Hollenbeck, P.J. (1989) *J. Cell Biol.* **108**, 223-227.
101. Ksiezak-Reding, H., and Yen, S.H. (1987) *J. Neurosci.* **7**, 3554-3560.
102. Waegh, S.M.D., Lee, V.M-Y., and Brady, S.T. (1992) *Cell* **68**, 451-463.
103. Harper, J.W., Elledge, S.J., Keyomarsi, K., Dynlacht, B., Tsai, L.-H., Zhang, P., Dobrowolski, S., Bai, C., Connell-Crowley, L., Swindell, E., Fox, M.P., and Wei, N. (1995) *Mol. Biol. Cell* **6**, 387-400.
104. Lee, M.-H., Nikolic, M., Baptista, C.A., Lai, E., Tsai, L.-H., and Massague, J. (1996) *Proc. Natl. Acad. Sci. USA* **93**, 3259-3263.
105. Lew, J. and Wang, J.H. (1995) *TIBS* **20**, 33-37.

106. Bates, S., Bonetta, L., MacAllan, D., Parry, D., Holder, A., Dickson, C., and Peters, G. (1994) *Oncogene* **9**, 71-79.
107. Miyajima, M., Nornes, H.O., and Neuman, T. (1995) *NeuroReport* **6**, 1130-1132.
108. Azzi, L., Meijer, L., Ostvold, A.-C., Lew, J., and Wang, J.H. (1994) *J. Biol. Chem.* **269**, 13297-13288.

chapter 21
Role of Ca++/Calmodulin binding proteins in *Aspergillus nidulans* cell cycle regulation

Nanda N. Nanthakumar, Jennifer S. Dayton and Anthony R. Means[1]

Department of Pharmacology, Duke University Medical Center, Durham NC 27710, USA
[1] To whom correspondence should be addressed

The goal of this review is to summarise the current knowledge concerning the targets of Ca++/calmodulin that are essential for cell cycle progression in lower eukaryotes. Emphasis is placed on *Aspergillus nidulans* since this is the only organism to date shown to posses essential Ca++ dependent calmodulin activated enzymes. Two such enzymes are the calmodulin activated protein phosphatase, calcineurin and the calmodulin dependent protein kinase. These proteins, each the product of a unique gene, are required for progression of quiescent spores into the proliferative cycle and also for execution of the nuclear division cycle in exponentially growing germlings.

INTRODUCTION

Calcium is absolutely required for cell cycle progression of eukaryotic cells. However, the order of the cascade of events responsible has yet to be elucidated. Mitogenic stimulation frequently results in cytoplasmic free Ca++ transients that are generated by the influx of extracellular Ca++ through Ca++ channels on the plasma membrane or release from IP-3 sensitive intracellular pools (1-5). Calmodulin (CaM) is one of the primary intracellular receptors for Ca++ and is the target required for mediation of cell cycle effects (6,7). Ca++ and CaM not only participate in cell cycle control but are also essential for a variety of other physiological functions including maintenance of the structural integrity of all eukaryotic cells (8). Because of the wide array of requirements for Ca++/CaM, it has been difficult to identify specific roles for CaM in higher eukaryotes. This is made doubly difficult due to the presence of multiple genes encoding the identical protein (9-11). Frequently, all three genes are expressed and differentially regulated which makes attempts to examine the effects of gene knockout quite frustrating. Therefore, a number of investigators have turned to three genetically tractable ascomycetes fungi: filamentous fungi *Aspergillus nidulans*, budding yeast *Saccharomyces cerevisiae* and fission yeast *Schizosaccharomyces pombe*, to elucidate the role of CaM in cell cycle control (12-15). Unlike vertebrate cells, these fungal organisms each have a single CaM gene, the disruption of which is lethal and causes arrest at various stages in the cell cycle.

The purpose of this review is to summarise the role of Ca++/CaM binding proteins in cell cycle progression of lower eukaryotes. Our discussion will be focused primarily on *Aspergillus nidulans*, but information obtained from the study of budding yeast and fission yeast will be included. Comprehensive reviews on the role of Ca++ in yeast (16) and vertebrates (1), and the regulation of the cell cycle by Ca++ (17), Ca++/CaM in general (18) or in yeast (19) are available.

THE ROLE OF CAM IN CELL CYCLE PROGRESSION

Our aim is to understand the role of Ca++ action mediated by CaM in cell cycle control. CaM is essential in *S. cerevisiae*, although the essential functions of CaM do not require Ca++ binding (20). This is unique in eukaryotes and is probably due to the considerable difference in intracellular Ca++ homeostasis between vertebrate cells and budding yeast (21,22). Furthermore, *S. cerevisiae* CaM is only 59% identical to vertebrate CaM (13), the lowest homology for any sequenced CaM. In addition *S. cerevisiae* CaM is the only known CaM which binds three instead of four Ca++ ions (20). Therefore, to elucidate the role of Ca++-dependent CaM in cell cycle regulation, our choice of a model system was *A. nidulans*. The CaM of this filamentous fungus is 84% identical to vertebrate CaM (12), and it can activate vertebrate CaM-dependent enzymes in a Ca++ dependent manner similar to that achieved by CaM from higher eukaryotes. Furthermore, mutant *A. nidulans* CaM which is unable to bind to Ca++ fails to activate these enzymes (23). Therefore, it appeared that *A. nidulans* could be a relevant model system in which to delineate the role(s) of Ca++/CaM in cell cycle progression.

The growth and nuclear division cycles are morphologically well characterised in *A. nidulans*

(Reviewed in 24). Vegetative growth is initiated when the haploid uninuclear conidia (spores) are deposited in a suitable environment. Spores germinate by absorbing water and swelling. As they grow, the symmetrical spores become bipolar by forming a growing tip or cell apex (25-27). Once polarised growth is established, the cell apex leads to development of hyphae. Unlike yeast, this fungus grows as a multinucleated syncitium and cytokinesis occurs when a septum is formed after the third nuclear division cycle (28,29). The 100 minute nuclear division cycle has been well characterised, with G_1, S, G_2, and M phases requiring 15, 40, 40, and 5 minutes respectively, when grown in enriched medium at 37°C (30). Spore nuclei are arrested perhaps analogously to the quiescent state (G_0) in vertebrate cells. When spores are germinated, they progress into the nuclear division cycle via G_1 and enter the first S phase at 3 to 4 hours, rather than the 15 minutes it takes to proceed from M to S in exponentially growing germlings.

In addition to sophisticated genetics, the availability of a number of temperature sensitive (ts) mutants that cause arrest of the nuclear division cycle (31) or defects in septum formation (31,29) make this organism an attractive system in which to identify the molecules that control these events. Several essential proteins required to mediate the G_2/M transition in *A. nidulans* have been defined through the characterisation of ts mutants defective in cell cycle progression. The corresponding genes include *nim*A, *nim*T, *nim*E and *nim*X (32-35). Similar to other organisms, the activation of the cyclin-dependent kinase, NIMX ($p34^{cdc2}$), is essential for the G_2/M transition. This $p34^{cdc2}$ homologue is activated by the binding of cyclin B (NIME) when dephosphorylated by the tyrosine phosphatase, NIMT^{cdc25} (34). In addition, the activation of another protein Ser/Thr kinase, NIMA, is also essential for the G_2/M transition (32,35). NIMX activity is also required for the G_1/S transition (34), however, the regulatory components involved have yet to be defined in *A. nidulans*. Recent identification of a ts mutant, *nim*R, which arrests in G_1 at the restrictive temperature should help to identify additional regulators of the G_1/S transition (36). Detailed descriptions of these proteins and their functions are presented in other chapters of this volume.

The CaM gene from *A. nidulans* was cloned using the chicken CaM cDNA as a probe. It was shown that the CaM gene was unique and its essential nature was demonstrated by disrupting the gene by site specific recombination as shown in Figure 1A (12). Conidia of the null mutants lack a germ tube and arrest at multiple points in the nuclear division cycle resulting in germlings with one or two nuclei (12). Similarly, when wild type conidia were grown in the presence of the anti-CaM drug, W-7, germlings did not enter the first S-phase (37) confirming that CaM was essential for progression through the nuclear division cycle.

To determine which phases of the nuclear division cycle required CaM in *A. nidulans*, a strain was created in which the endogenous promoter of the CaM gene was replaced by the regulatable *alcA* promoter by homologous recombination as shown in Figure 1B (38). Overexpression of CaM in this strain accelerated cell cycle progression and decreased the requirement for extracellular Ca^{++}. Repression of the CaM gene in exponentially growing cells resulted in both G_2 and G_1 arrest (38). Data were generated using W-7 that suggested CaM was also necessary for completion of mitosis in *A. nidulans* (37). To date, however, appropriate experiments have not been performed in the conditional strain to conclusively demonstrate a role for CaM during mitosis in this organism. Calmodulin is also essential for the re-entry of quiescent conidia into the cell cycle in a manner that may be similar to the requirement for progression of vertebrate cells from G_0 through G_1 and into S in response to mitogenic stimuli (39).

Since elements required for the G_2/M transition have been characterised in this filamentous fungi using ts mutants, the next aim was to determine whether Ca^{++}/CaM is involved in the cascade leading to activation of either NIMX or NIMA. To answer this question, strains conditional for CaM expression were constructed in the background of *nim*A^{ts} or *nim*T^{ts} mutants (40,41). In these strains the presence of both Ca^{++} and CaM was shown to be essential for the activation of both NIMA and NIMT^{cdc25}, and thus were required for the G_2/M transition (42). A detailed review of the experiments that led to the suggestion of a role for Ca^{++}/CaM in regulating the G_2/M transition is available elsewhere (18,42). However, the targets of CaM which are involved in the regulation of NIMA and NIMT activities have yet to be defined in molecularly precise terms.

Another approach to examine the roles of CaM in cell cycle progression is to create CaM ts mutants. Using site directed mutagenesis a number of temperature sensitive CaM mutants were derived in budding yeast (43-45). Despite the fact that *S. cerevisiae* CaM does not require the binding of Ca^{++}, such mutants were useful in defining cellular functions that require CaM. These CaM^{ts} mutants were categorised into four complementation groups which cause growth arrest due to defects in mitosis, bud emergence, actin organisation, or CaM localisation respectively. Two essential CaM target proteins have been identified in budding yeast, Nuf1/Spc110 and Myo2. Their role in growth and mitosis will be discussed later in this chapter.

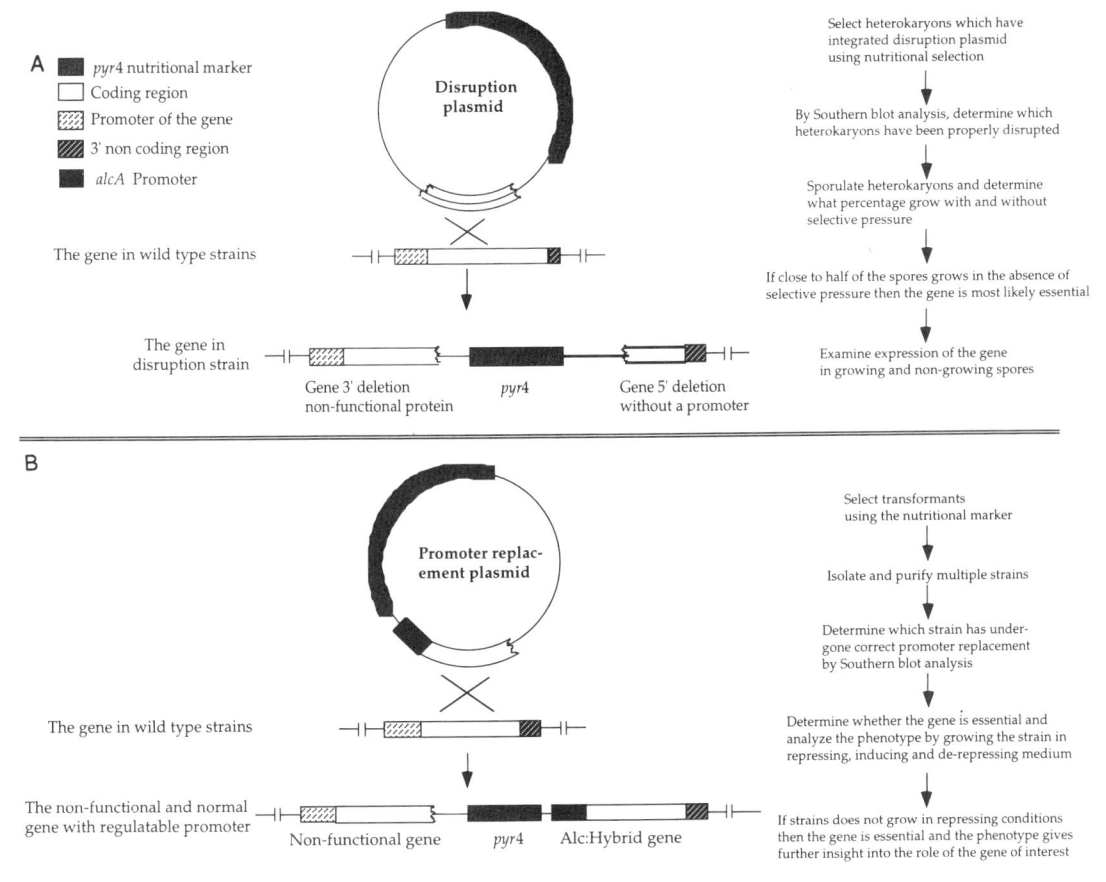

Figure 1. Schematic diagrams of gene disruption or promoter replacement and analysis of the phenotype.
A. The disruption is accomplished by site specific homologous recombination of the disruption plasmid into the locus of interest. A specific internal fragment of the gene was cloned into the plasmid which after integration creates a non-functional gene product with a 3' deletion and a second non-functional, promoterless gene with 5' deletion. These two non-functional genes are now separated by a nutritional marker (*pyr4*).
B. Promoter replacement is the result of site specific homologous recombination with a plasmid containing the alcA promoter ligated to the 5' end of the gene of interest without the wild type promoter. The result is from 5' to 3': a wild type promoter and gene with a 3' truncation, the nutritional marker, and the alcA promoter adjacent to a wild type gene.

TARGETS OF Ca^{++}/CaM INVOLVED IN THE REGULATION OF CELL CYCLE PROGRESSION

Once the CaM requirements for cell cycle progression were clearly established, the next logical question was to identify the CaM-dependent proteins that mediate Ca^{++} signalling. The CaM overlay technique was used to examine the number of cellular proteins in *A. nidulans* that bind to CaM in a Ca^{++}-dependent manner (37). This technique demonstrated the existence of two major CaM-binding proteins and a number of less abundant species. The two major proteins have been identified as the Ca^{++}/CaM dependent protein phosphatase A subunit (CnA or calcineurin) and protein kinase (CaMK). Each of these proteins had been previously implicated in cell cycle regulation in other organisms. For example, mating factor-arrested *S. cerevisiae* cells require the activity of CnA to re-enter the cell cycle (46,47). Similarly, mitogenic activation of quiescent T-cells requires calcineurin activity (48-50). In the later case, an understanding of the function of calcineurin activity in cell cycle progression has been greatly advanced by the characterisation of the mechanism of action of two immunosuppressive drugs, FK506 and Cyclosporin A. FK506 binds to the FK506-binding protein (FKBP), whereas cyclosporin A binds to cyclophilin, but both drug binding protein complexes inhibit calcineurin activity (48-50). Inhibition of CnA prevents dephosphorylation of the cytoplasmic component of the NF-AT transcription factor. This dephosphorylation is necessary for nuclear entry of the NF-AT protein which is an essential component of the complex required for activation of the cytokine gene, IL-2 (48-50). An analogue of FK506, L-683-818, binds to the same intracellular receptor, FKBP, with similar affinity but has no effect on the phosphatase activity (51-52), and hence does not prevent the activation of T cells which is required for entering the proliferative cell cycle. A role for CaMK in cell cycle progression has been proven for the metaphase/anaphase transition in Xenopus oocytes (53). In addition, a reasonably specific CaMK inhibitor, KN-93, prevents quiescent fibroblasts from reaching S phase (54). Finally the expression of a constitutively active CaMK arrests both vertebrate (55) and *S. pombe* cells (56) in G_2 suggesting the possibility that a CaMK initiated phosphorylation/dephosphorylation cycle may be required for the G_2/M transition. Collectively these studies demonstrated the importance of CaMK and CnA in cell cycle progression of a number of cell types and prompted us to investigate whether either or both enzymes were Ca^{++}/CaM-dependent targets involved in cell cycle control in *A. nidulans*.

CALMODULIN ACTIVATED PHOSPHATASE IN CELL CYCLE CONTROL

The *A. nidulans* CnA gene was cloned by a degenerate PCR strategy based on the known sequence of *Neurospora crassa* CnA (57). Molecular analysis demonstrated that CnA is a unique gene with a single 2.5 kb mRNA. The putative coding sequence predicts a 530 amino acid protein with a calculated molecular weight of 61 kDa. Single copy CnA genes have also been reported for fission yeast (58,59) and *Neurospora crassa* (60), however two genes for CnA have been cloned from budding yeast (61,47). The predicted amino acid sequence of the *A. nidulans* CnA is most similar to the *Neurospora crassa* sequence. However, the catalytic core of *A. nidulans* CnA is more identical to vertebrate CnA (70-80%) than to yeast CnA (48%). The CnB binding site is the most highly conserved region across all species including *A. nidulans* and the crystal structure demonstrates that CnB holds the substrate in the catalytic site of the phosphatase (62). In addition, the CaM binding site is well conserved.

The CnA protein and mRNA in *A. nidulans* undergo changes during the cell cycle. The expression of the steady state level of the 2.5 kb mRNA oscillates in a cell cycle dependent manner with highest expression at the G_1/S transition, suggesting that the CnA gene could be transcriptionally regulated in a cell cycle dependent manner (37). Similar cell cycle dependent regulation has been shown for CnA in fission yeast (58). Transcriptional activation of S phase specific genes by the transcription factor E2F in vertebrate cells (63,67) and their functional homologs in yeast (64-66) is a limiting event in the G_1/S transition. Likewise, transcriptional regulation of the *A. nidulans* CnA gene with highest levels of mRNA just prior to DNA replication is consistent with a regulatory function for CnA at the G_1/S transition. Since we have cloned the 5' regulatory region of the *A. nidulans* CnA gene it will be possible to determine if it is transcriptionally regulated in G_1 and, if so, which transcription factor(s) is responsible for the cell cycle dependent expression. In addition, the activity of this phosphatase depends on a regulatory B subunit and CaM. Therefore, it will be interesting to examine whether the expression of CnB and enzyme activity also show cell cycle dependent changes.

To demonstrate that CnA is essential for growth, gene disruption experiments were performed in a heterokaryon (57) as described in Figure 1A. A heterokaryon with one disrupted and one wild type CnA gene was recovered and confirmed by Southern blot analysis. The haploid conidia (spores) were allowed to grow with and without selective pressure. Only the spores containing the wild type gene (46%) were able to grow, whereas 95% of the spores from a wild type strain grew under identical conditions, suggesting that CnA is essential in *A. nidulans*. The nuclear morphology of the germlings in which the CnA gene was disrupted showed that nuclear division was arrested after completion of one or in a few cases two cycles in 11 hours. On the other hand, germlings of a wild type strain grown under identical conditions completed 4-5 nuclear division cycles in 11 hours. These results confirmed that a functional CnA gene was essential for cell cycle progression in *A. nidulans*. In contrast, calcineurin null mutants are sterile but viable in both fission and budding yeast (59,46) suggesting the presence of a functionally redundant phosphatase in these organisms. However, in the presence of a high salt environment or in mutants defective in ion homeostasis, CnA becomes essential in budding yeast (68-71).

To further characterise the essential nature of this gene in *A. nidulans*, a strain conditional for the expression of CnA (AlcCnA) was created as shown in Figure 1B (72). As expected, CnA mRNA and protein were undetectable in repressing conditions, overexpression resulted in a 5-10 fold elevation, whereas de-repressing conditions resulted in expression but at a level lower than that found in wild type cells. When AlcCnA spores were grown in repressing medium, growth was arrested. The induction of CnA in AlcCnA had no effect on growth, whereas this low expression in de-repressing medium resulted in slowed growth. Growth of AlcCnA was assessed by measuring radial colony diameter on solid plates, dry weight in liquid medium, or by counting the number of nuclei per cell. It is interesting to note that the overexpression of CnA did not alter growth relative to that of wild type strains. This is in contrast to the AlcCaM strain grown in inducing conditions which shows an accelerated rate of nuclear division. Perhaps the lack of effect of excess CnA should not be surprising, because an elevation of calcineurin activity also requires both the regulatory subunits and Ca^{++}. Thus, the enzyme activity may not have increased as, at present, there is no evidence for concomitant elevation of either CnB or CaM. Creation of a strain in which both CaM and CnA genes are inducible may provide further insight into the mechanism by which elevated CaM enhances the rate of nuclear division as well as allows determination of the consequences of elevated calcineurin activity.

There were three striking features of AlcCnA grown in repressing conditions. First, the nuclear division cycle was inhibited, second, growth was no longer polarised, and third, the cellularisation of germlings was disrupted. Regarding nuclear division, after 9 hours of CnA repression, AlcCnA germlings contained only one or two nuclei. DNA synthesis was measured in spores grown under repressing or non-repressing conditions to determine whether spores of AlcCnA strains entered S phase. Under repressing conditions, the AlcCnA strain undergoes only one round of DNA synthesis. This result suggests that either CnA is not the target of CaM that is essential for re-entry into the cell cycle, or the low levels of CnA present in the spores provided enough activity to allow entry into S phase but as this CnA is turned over the cells arrest. The nuclear morphology of AlcCnA germlings grown under repressing conditions suggests that growth is arrested in G_1 with a small percentage at G_2 and mitosis. This suggests that CnA may play an important role in G_1 but also may be involved in other phases of the cell cycle. Due to the lack of an antibody to *Aspergillus* CnA, determination of the half life of CnA *in vivo* has not been possible. Once we know the turnover rate of CnA, reciprocal growth arrest and shift experiments can be done to demonstrate whether CnA is essential only in G_1 or in other phases of the nuclear division cycle as well.

A similar growth arrest phenotype was observed when wild type spores were grown in the presence of the specific calcineurin inhibitors, FK506 and Cyclosporin A (72). FK506 was found to be a very potent inhibitor of *A. nidulans* radial growth, with an IC-50 of 0.1μg/ml. However, L-683-818 had no effect on growth even at 100μg/ml (72). When used together in high concentrations, the analogue begins to relieve growth inhibition caused by FK506, suggesting that the inhibitory effect of the later drug is mediated by FKBP and, most likely, the target is calcineurin. When wild type spores were grown in the presence of FK506 nuclear division was inhibited, arresting the cells with either one, two, or four nuclei. Examination of nuclear morphology revealed that FK506 can cause arrest at G_1, G_2 and mitosis, similar to what was observed for the AlcCnA strain grown in repressing conditions. When similar experiments were performed in the presence of the CaM antagonist, W-7, growth arrested prior to the first S phase (37), a finding compatible with the possibility that the target of CaM essential for re-entry into the cell cycle may not be CnA.

The second phenotype of AlcCnA grown under repressing conditions was lack of polarised growth. This phenotype is consistent with that exhibited by the CnA null mutant of *S. pombe* (59). In the absence of CnA cell polarity was disturbed and rather than bipolar growth, a multi-hyphal branched phenotype was observed. It has been shown in another filamentous fungi that the extracellular level of Ca^{++} and the local concentration of Ca^{++} at the hyphal tip were critical for normal extension (25). Changes in the Ca^{++} concentration at the tip resulted in cessation of polarised growth and the development of hyphae in a new direction (73). In addition two yeast strains containing CaM^{ts} mutations cause inhibition of bud emergence and altered CaM localisation (44) which are phenotypes compatible with a role for a CaM-dependent protein in regulation of polarised growth. In vertebrate neurons, calcineurin is also enriched in dendritic growth cones where it seems to mediate Ca^{++}/CaM effects on cytoskeletal integrity (74). Inhibition of calcineurin activity disrupts the long bipolar growth characteristic of the developing neuron and results in short, branched dendrites displaying depolarised growth, a phenotype similar to that observed when CnA is repressed in the AlcCnA strain of *A. nidulans*. The phosphorylation states of τ-1, tubulin and MAP-2 have been shown to be altered by calcineurin (75). It has also been shown that cytoskeletal rearrangement and deposition of F-actin and tubulin are necessary for process extension (25). Taken together, there is significant evidence from several organisms that CnA plays a role in regulating polarised growth. Whether or not this occurs by the same pathway by which the phosphatase influences nuclear division remains to be determined.

Lastly, the most unusual feature of repression of CnA in AlcCnA is the formation of multiple septa creating cells without nuclei. The germlings also have wide hyphae and thin cell walls. This phenotype is similar to that observed in a CnA null mutant of fission yeast (59). Nine temperature sensitive septation mutants (Sep) have been isolated in *A. nidulans* and were shown to fall into three classes each epistatic to one another (29,31,76). Septum formation in this filamentous fungus is closely associated with the third nuclear division and actin organisation (29). The morphology of one mutant, (SepA), at the nonpermissive temperature, shows wide hyphae with extensive branching, and appears to be similar to the AlcCnA strain grown in repressing conditions. The downshift of SepA to the permissive temperature results in normal septation, suggesting a defect in cell wall biosynthesis. Glucan is a major component of the fungal cell wall and 1,3–ß-D-glucan synthase is responsible for the biosynthesis of this polymer (77). In budding yeast this enzyme is transcriptionally regulated and the induction is dependent on Ca^{++} and calcineurin activity (77). Thus it will be interesting to determine whether glucan biosynthesis is essential in *A. nidulans* and whether the SepA gene product and CnA interact with each other to form the septa that are required for normal cell wall deposition.

CALMODULIN ACTIVATED KINASE IN CELL CYCLE CONTROL

Ca^{++}/calmodulin-dependent kinase purified from *A. nidulans* (78) is a monomer with a calculated molecular weight of 41 kDa and migrates at 51 kDa on an SDS polyacrylamide gel. It undergoes autophosphorylation which makes it more acidic, however this does not generate Ca^{++}/calmodulin-independent activity as occurs with other members of this family of enzymes (79). The *A. nidulans* enzyme phosphorylates glycogen synthase, microtubule associated protein 2, synapsin, tubulin, gizzard myosin light chain, and myelin basic protein. These substrates can also be phosphorylated by mammalian CaMK II isoforms which led to the classification of *A. nidulans* CaMK as a "multifunctional" enzyme. The cDNA for *A. nidulans* CaMK was isolated and Southern blot analysis suggested that the gene was unique (80). The cDNA contains an ORF which predicts a 414 amino acid protein. The predicted amino acid sequence shares 29% identity with rat brain CaMKIIα and 40 and 44 % identity, respectively, with the *S. cerevisiae* CaMKs: YCMK1 and YCMK2 (81,82). The *A. nidulans* CaMK and YCMK1 sequences lack the regulatory RQET phosphorylation site present in mammalian CaMKII that generates CaM-independent activity upon autophosphorylation. This might explain why these fungal CaMKs do not acquire autonomous activity when autophosphorylated. The YCMK2 sequence contains the sequence RVET at the same site as the RQET in mammalian CaMKIIα and it acquires some autonomous activity (83). The *S. cerevisiae* and *A. nidulans* CaMKs lack the carboxy-terminal association domain found in mammalian CaMKII, however the yeast enzymes may have an amino-terminal association domain (81).

As shown in Figure 1B, an *A. nidulans* strain conditional for the expression of CaMK (AlcCaMK) was created in order to study the effects of repression of CaMK expression on the nuclear division cycle (84). As expected, the conditional strain showed markedly decreased or undetectable CaMK mRNA and very little protein by western blot analysis when grown in repressing conditions. In inducing conditions, the AlcCaMK strain showed markedly increased mRNA and protein which corresponded to a 10-to 18-fold increase in Ca^{++}/calmodulin-dependent protein kinase activity in extracts. The AlcCaMK strain showed no difference in any growth parameter: radial colony growth on solid plates, dry weight in liquid culture, or nuclear division in inducing medium compared to a control strain overexpressing an inactive *A. nidulans* CaMK (created by mutating the lysine essential for ATP binding to alanine). However, repression of CaMK expression caused a marked slowing of growth evident in all three of the following assays. The radial growth was slowed significantly (75%) during the first 24 hours, and the increase in total dry weight during the first 15 hours of growth from spores was slowed by 60 to 80 % compared to the growth of the control strain which contained a normal CaMK gene and a repressible inactive CaMK cDNA. The most impressive effect of repression of CaMK was on nuclear division. After 9 hours of germination in repressing medium, the control strain had an average ten nuclei per germling, whereas, the AlcCaMK strain had an average of 2 nuclei per germling. Thus CaMK was clearly required for optimal growth, it did not appear to be essential from analysis of the conditional strain. Although the *alcA* promoter is tightly regulated, low levels of expression under repressing conditions have been documented (85). For this reason we assayed CaMK activity in extracts prepared from AlcCaMK grown in repressing medium for 16 hours. Although there was no detectable CaMK activity in extracts, we were able to detect CaMK activity in extracts purified by CaM-sepharose affinity chromatography. The eluted activity corresponded to approximately 10% of the activity recovered on CaM-sepharose from extracts of a control strain grown in identical condition. Perhaps the low level of expression of CaMK explained the partial phenotype of the AlcCaMK strain and thus, indeed, the gene was essential. This could only be verified by disrupting the CaMK gene in *A. nidulans*.

The CaMK gene was disrupted in a heterokaryon as described in Fig 1A. The disruption was performed as we have previously described for the CaM (12) and CnA (57) genes by transformation of GR5 strain The GR5 strain carries the *pyrG89* nutritional marker and thus requires exogenous pyrimidines (uridine and uracil) for growth. A plasmid containing an internal fragment of the CaMK gene was used to disrupt the endogenous

CaMK gene and insert the *Neurospora pyr4* gene which complements the *pyrG89* nutritional requirement and thus allows growth in the absence of pyrimidines. After verification of successful recovery of heterokaryons with a wild type and a disrupted copy of CaMK by Southern blot analysis, spores from the heterokaryons were analysed for their ability to grow in the presence or absence of selective pressure (uridine and uracil). The heterokaryon would be expected to contain nuclei and ultimately spores that were untransformed (*pyr4⁻*CaMK⁺) as well as nuclei and spores that were transformed (*pyr4⁺*CaMK⁻). In the absence of uridine and uracil, none of the spores progressed through nuclear division. Thus, as expected the untransformed *pyr4⁻*CaMK⁺ spores could not grow in the absence of exogenous pyrimidines, but in addition the transformed *pyr4⁺*CaMK⁻ spores also did not grow. In the presence of uridine and uracil approximately 47-65% of the germlings have greater than 4 nuclei and extended germtubes, corresponding to the *pyr4⁻*CaMK⁺ spores. The remaining 35-50% of the spores representing the *pyr4⁺*CaMK⁻ population fails to undergo nuclear division, send out germtubes, or enlarge in preparation for either of these events. The phenotype of the CaM disruption is not quite as complete. In the CaM disruption, approximately 50% of the disrupted spores undergo one round of nuclear division (12), but they do not send out germtubes. CaM and CaMK proteins are detectable in spores and the difference in half-life or cellular requirement for these proteins is the likely explanation for the differences in phenotype. Presumably the spores have CaM or CaMK from the heterokaryon and the effect of the disruption is manifest when this level of protein is insufficient for some process required for germination. The disruption of CaMK suggests that CaMK is essential and required for germtube extension and for nuclear division. Alternatively, CaMK may be required for some early event in germination and the absence of CaMK arrests cells before they are capable of germtube extension or nuclear division. The former possibility seems more likely based on the data from the conditional strain. In the conditional strain both of these processes are inhibited, however, the effect on nuclear division is more pronounced perhaps suggesting differential requirements for CaMK in these two processes.

Additional insight into the role of CaMK in cell cycle progression has been obtained from the overexpression of a constitutively active form of CaMK. It was demonstrated in C127 mouse cells that the expression of a constitutively active CaMK under the control of the MMTV promoter caused a G_2 arrest (55). It was a nearly complete block as 85% of the cells arrested in G_2 after 12 hours. These cells arrested with elevated H1 kinase activity which indicates active complex of $p34^{cdc2}$. Secondly, a constitutive form of mouse CaMKIIα expressed in *S. pombe* also causes a G_2 arrest, but this arrest occurs without elevated H1 kinase activity (56). In the first study the constitutive enzyme was created by truncation, removing the regulatory CaM-binding domain. In the latter, the constitutive enzyme resulted from the single point mutation of Thr286 to Asp to mimic a phosphorylated Thr residue. This difference in the protein used to generate constitutive activity could affect the precise mechanism causing the arrest and thus may explain the differences observed in H1 kinase activity.

A constitutive CaMK (CaMKct) was created by truncation of the *A. nidulans* CaMK at Ile292 removing the CaM-binding regulatory domain. It was subcloned into an *A. nidulans* expression vector behind the alcohol dehydrogenase promoter and transformed into a wild type strain (86). The resulting strain was designated AlcCaMKct. Attempts to grow AlcCaMKct spores in inducing medium resulted in the inhibition of germtube extension and inhibition of entry into the first S phase, similar to the effects of the CaMK disruption. Further examination showed that the overexpression of CaMKct in spores leads to a premature activation of H1 kinase by 1 hour of germination. Normally, the increase in H1 kinase activity is seen about the time of the first entry into S phase at 3 to 4 hours. The elevated H1 kinase activity is likely due, at least in part, to premature activation of NIMX, as an anti-NIMX antibody also immunoprecipitated protein that exhibited increased H1 kinase activity. The NIMX protein is detectable in spores using the anti-PSTAIR antibody, but using an anti-NIME antibody it is difficult to determine whether NIME is present in spores. However, the mRNA for both *nimX* and *nim*E are detectable by northern blot analysis of total RNA from *A. nidulans* spores (86). Thus, it is possible that this premature H1 kinase activity is the result of the NIME-dependent NIMX kinase. It is not clear why premature activation of H1 kinase arrests spores early in germination, however, it has been shown that the presence of cyclin-dependent kinase (CDK) activity inhibits the formation of origins of replication (87). Perhaps *A. nidulans* spores do not have pre-formed origins of replication and the premature H1 kinase activity prevents their formation. This early arrest is completely reversible up to 9 hours after the induction of CaMKct demonstrating that the checkpoint causing the arrest protects the cells from further damage for some period of time. After longer incubations increasing numbers of cells fail to recover.

If AlcCaMKct spores are allowed to enter the cell cycle before induction of CaMKct, the cells arrest in G_2 prior to mitosis as seen in C127 mouse cells (55) and in *S. pombe* (56). It should be mentioned that these experiments must be interpreted carefully because the overexpression of a constitutive kinase may not be reflecting a physiologically relevant event. We do not know the target(s) for CaMK and thus cannot verify that CaMKct is phosphorylating the normal *in vivo* substrate. However, since the phenotype of the disruption of CaMK is the same as the phenotype of the overexpression of CaMKct in spores, i.e., the inhibition of nuclear division and germtube extension, these data argue for a role for CaMK at both the G_1/S and G_2/M transitions in the nuclear division cycle of *A. nidulans*. Furthermore, the data support a requirement for orderly phosphorylation and dephosphorylation of a substrate(s) for progression through the cell cycle. It is interesting to note that nuclear division and cellular growth in *A. nidulans* are separable processes. For example, treatment of *A. nidulans* with hydroxyurea arrests nuclear division in S phase but the hyphae continue to grow (29). Also, the growth of the $nimT^{ts}$ strain at restrictive temperature arrests nuclear division in G_2 but does not inhibit hyphal growth. Thus the fact that disruption of CaMK or the expression of CaMKct prevents germtube extension suggests that CaMK may also have a specific role in the growth process in *A. nidulans*.

We have demonstrated that the CaM-dependent kinase inhibitors KN-62 and KN-93 inhibit purified *A. nidulans* CaMK *in vitro* with approximate IC_{50}s of 0.6 and 1μM, respectively. However there is no effect of KN-62 on radial colony growth of *A. nidulans* and only partial inhibition of growth is achieved with KN-93. This may be due to the poor solubility of these compounds (88) or to inefficient transfer across the *A. nidulans* cell wall. The eventual availability of the more soluble KN-93 phosphate salt could provide a powerful tool for delineating the functions of CaMK in *A. nidulans*.

OTHER CaM-BINDING PROTEINS AFFECTING CELL CYCLE PROGRESSION

Nuf1/Spc110 is part of spindle pole body complex and binds to CaM in *S. cerevisiae*. It was first identified as a component of the spindle pole body (89,90) and was demonstrated to be an essential target of CaM required for mitosis in budding yeast (91). Ca^{++} binding is not essential for the CaM mediated function of Nuf1/Spc110 protein and deletion of the CaM binding domain of Nuf1/Spc110 relieves the CaM requirement during mitosis in *S. cerevisiae*. In vertebrate cells, CaM had been suggested to play a role in chromosome segregation (92) and had been immunolocalised to the mitotic spindle (93). CaM was demonstrated to be required for mitosis in mouse C127 cells by expressing CaM anti-sense RNA (94). Taken together, these observations suggest that CaM is necessary for proper assembly of the mitotic spindle and may mediate its function through binding to a functional homologue of Nuf1/Spc110 protein in all organisms. It will be of interest to determine whether a homologue of Nuf1/Spc110 exists in *A. nidulans* and, if so, whether the CaM interaction requires Ca^{++}.

In *A. nidulans* MyoA, the CaM-binding unconventional myosin, was shown to be an essential gene necessary for hyphal elongation (95). However, the nuclear division cycle was unaffected by the absence of this unconventional myosin. Although, MyoA is essential for normal growth, it is not required for nuclear division. A similar CaM binding protein Myo2 exists in budding yeast (96). Myo2 is also a member of the class of unconventional myosins, and had been shown to be a target of CaM. Myo2 mediates CaM function at the site of bud emergence and helps to transport vesicles to the site of growth. Myo2 is an essential gene in yeast and binds CaM primarily in a Ca^{++} independent manner. Similarly in vertebrate brain, CaM binds to an unconventional myosin, p190, independent of Ca^{++} binding (97). Taken together, MyoA, Myo2 and p190 might be functional homologs that mediate CaM function universally in specific cytoskeletal rearrangements at the site of growth.

HYPOTHETICAL MODEL FOR THE ROLE OF CNA AND CAMK IN CELL CYCLE PROGRESSION

The demonstration that both CnA and CaMK are essential in *A. nidulans* and may be involved in growth and in nuclear division is a step forward in our effort to understand the regulatory roles of Ca^{++} and CaM in these processes. The precise point at which they affect the cell cycle could be demonstrated by creating conditional strains in the background of critical ts mutants such as *nim*A or *nim*T at G_2 and *nim*R at G_1. Evidence indicates that *nim*R is upstream of CaM during G_1 (37). First it might be of interest to determine whether the level of CaM doubles at the restrictive temperature in the *nim*R strain and if not whether the *nim*R strain could be rescued by elevating the level of CaM by creating an AlcCaM strain in the genetic background of *nim*R. If nimR could be rescued by elevating CaM then using the same strategy one could question whether CnA or CaMK is the downstream target of *nim*R. Similarly creating conditional strains of CnA and/or CaMK in the background of *nim*T and *nim*A may help to determine which CaM binding protein regulates *nim*A or *nim*T *in vivo*. In order to understand why these processes require CaMK and CnA it is necessary to identify their requisite specific substrates.

One means to identify potential substrates of the CaM-dependent enzymes is a biochemical approach. We have shown (84,86) that the expression of CaMK protein does not vary with cell cycle progression in exponentially growing cells of *A. nidulans*, thus post-translational modification seems a likely mechanism to explain the G_2/M and perhaps the G_1/S requirement(s) for CaMK. Biochemical or two-hybrid strategies may enable the isolation of specific substrates for these two proteins. It has been suggested that NIMA may be a substrate for CaMK because CaMK can phosphorylate and partially re-activate previously dephosphorylated and inactive NIMA (98). In addition, the mammalian homologue of NIMT^{cdc25} has been shown to be an *in vitro* substrate for mammalian CaMKII, with phosphorylation causing increased cdc25 activity (99,100). However, demonstrating that these *in vitro* substrates mediate the function of CaMK and CnA at various stages in cell cycle progression *in vivo* necessitates the usage of genetic means such as creating conditional strains and in disruption of these candidate proteins in the background of critical ts mutants as well as using drugs that inhibit CaM, CnA, and CaMK.

Alternatively, genetic approaches may be useful to isolate substrates. Unfortunately, there are no conditional ts mutants reported for any CaM-dependent enzymes in any organism. Since CnA and CaMK are unique and essential for growth in *A. nidulans* we can attempt to create ts mutants of CaMK or CnA. However, this requires gene replacement with copies of the relevant gene containing random mutations in the coding sequence followed by a search for any strain that displays the desired phenotype. This time consuming and tedious approach could be very risky. Alternatively, it might be possible to create a conditional ts mutant of CnA by creating a point mutation at the homologous position that led to generation of a ts strain of the protein phosphatase 1 gene, *bimG* (101). It might also be possible to identify suppressers of AlcCnA. Such genetic interactions may not identify CnA substrates but might provide a downstream target that mediates the function of CnA in the nuclear division cycle.

There is evidence in other systems for roles of CaM-dependent enzymes in the activation of transcription factors (102). Precedent for this exists in yeast where a transcription factor functions as a dosage dependent suppresser of one ts mutant of CaM (103). Recently it has been shown that both mammalian CaMKII and CaMKIV can phosphorylate the transcription factors CREB (102), CREMτ (104), SRF (105), ets (106), and c-jun (107). In several instances such phosphorylation has been shown to affect transcriptional activation.

Perhaps *A. nidulans* CaMK phosphorylates a transcription factor that induces the transcription of genes required for germination. It is known that the activation of T cells requires the dephosphorylation of the cytoplasmic component of the transcription factor, NF-AT, by CnA (47-50). Dephosphorylation of NF-AT$_c$ by CnA results in nuclear translocation and transactivation of early response genes such as IL-2. Therefore, it is possible that transcription factors involved in regulation of gene expression at G_1/S may be targets for both CnA and CaMK in *A. nidulans*.

Since a summary of critical proteins required for the G_2/M transition and a theoretical discussion of how they could be regulated by CaM binding proteins has been presented elsewhere (18,42), we will only consider a potential mechanism for the requirement of CnA and CaMK in the G_1/S transition as shown schematically in Figure 2. Whereas this is a purely a hypothetical model, it can serve as a conceptual framework for delineating the role of CaM binding protein(s) during the G_1/S transition. An essential step for execution of the G_1/S transition in vertebrates is the activation of a transcription factor, E2F and its functional equivalents in yeast (63-67). E2F is anchored in the cytoplasm by a member of the retinoblastoma family of proteins (Rb). A two step process is necessary for E2F function. First E2F is released from sequestration by phosphorylation of Rb. Once E2F is released from Rb, then the transcription factor is activated by phosphorylation, as in vertebrate cells (63,67), or association with other factors of the transcription initiation complex, as in yeast (64-66) or activtion and association. CnA and/or CaMK could directly or indirectly mediate sequestration, release from sequestration and/or activation of a transcription factor and thereby control an essential event necessary for G_1/S progression. Both of these CaM binding proteins would be regulated by the availability of CaM at G_1. It has been shown that a doubling of the CaM concentration occurs in late G_1 and may be essential for the G_1/S transition. This increase in CaM could result in the activation of CnA and/or CaMK the activity of which would be critical for this process.

Precedent for activation of transcription factors by CnA is well described in T-cell signalling (47-50). It has been shown that CnA is directly responsible for activating NF-AT$_c$ by dephosphorylation after mitogenic stimulation during T-cell activation. Dephosphorylation of NF-AT$_c$ results in nuclear translocation where it associates with its nuclear components (c-fos and c-jun family members) to transactivate the IL-2 promoter. Although, nuclear transport of NF-AT$_c$ depends on dephosphorylation rather than by sequestration by another protein, as suggested in the model, it is a variation of the basic

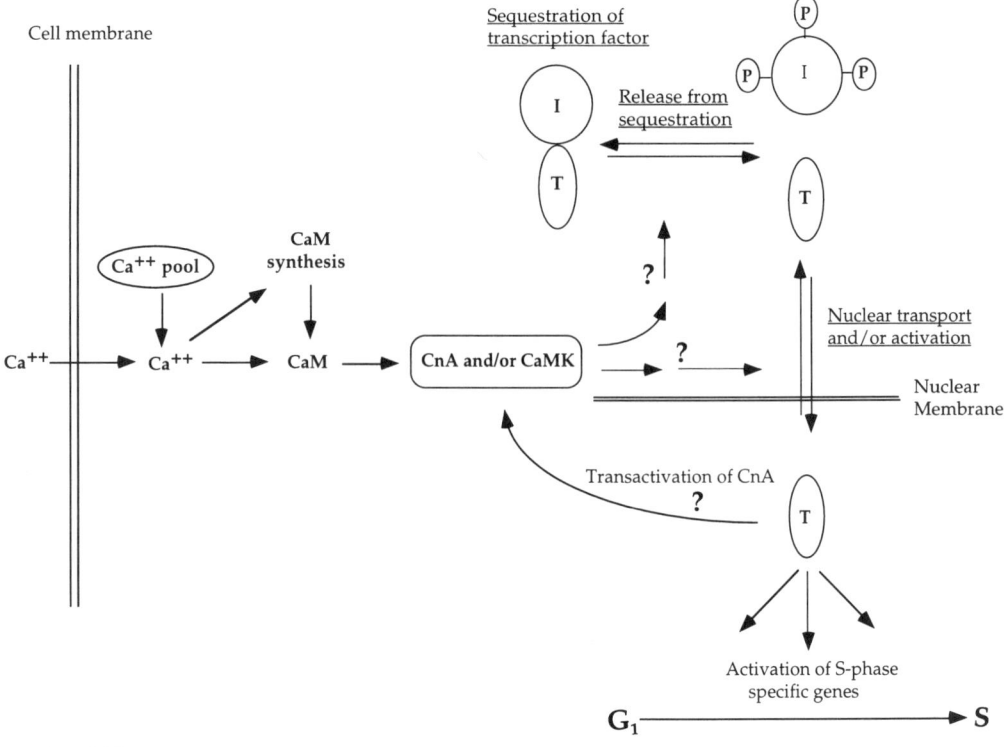

Figure 2. A model for potential regulation of a rate limiting transcription factor by CnA and CaMK at the G_1/S transition. I = inhibitor, T = transcription factor, P = phosphate group. The function of the inhibitor is to anchor the transcription factor in the cytoplasm. Sequestration of the transcription factor is released by phosphorylation and/or dephosphorylation. The next step is to allow the transport of the transcription factor, T in to the nucleus either by phosphorylation and/or dephosphorylation. Finally this factor transactivates genes required for the G_1/S transition in association with other factors. In the case of transcription factor E2F, the inhibitor is either Rb or RB family of proteins but in the case of NF-AT$_c$, the inhibitor is unknown. The activity of the putative transcription factor(s) could in turn, upregulate CnA expression at this stage of the cell cycle in *Aspergillus nidulans*.

principle described in Figure 2. Clearly the model is based on transcription factors known to be involved in the G_1/S transition in other systems that are also substrates for Ca^{++}/CaM dependent enzymes. At this point, analogy to the requirements for CaMK and CnA in *Aspergillus nidulans* is entirely speculative but does point to what may well be a fruitful avenue of investigation.

ACKNOWLEDGEMENTS

We would like to thank the members of Dr. Means' laboratory for constructive criticism during the course of these studies, and particularly Jim Joseph for assistance in the revision of this manuscript. The work reported in this review was supported by PHS NRSA GM-16922 to N. N. N from NIGMS and an American cancer society grant BE-185Y to A. R. M.

REFERENCES

1. Clapham, D.E. (1995) *Cell* **80**, 259-268.
2. Short, A.D., Bian, J., Ghosh, T.K., Waldron, R.T., Rybak, S.L., and Gill, D.L. (1993) *Proc. Natl. Acad. Sci. USA* **90**, 4986-4990.
3. Poenie, M., Alderton, J., Tsien, R.Y., and Steinhardt, R.A. (1985) *Nature* **315**, 147-149.
4. Allbritton, N.L., and Meyer, T. (1993) *Cell Calcium* **14**, 691-697.
5. Berridge, M.J. (1995) *BioEssays* **17**, 491-500.
6. Means, A.R. (1994) *FEBS Ltr.* **347**, 1-4.
7. Whitaker, M., and Patel, R. (1990) *Development* **108**, 525-542.
8. Means, A.R., VanBerkum, M.F.A., Bagchi, I.C., Lu. K.P., and Rasmussen, C.D. (1991) *Pharmacol. Ther.* **50**, 255-270.
9. Fischer, R., Koller, M., Flura, M., Mathews, M., Strehler-Page, M.A., Krebs, J., Penniston, J.T., Carafoli, E., and Strehler, E.E. (1988) *J. Biol. Chem.* **263**, 17055-17062.
10. Nojima, H. (1989) *J. Mol. Biol.* **208**, 269-282.
11. Christensen, M.A., and Means, A.R. (1993) *J. Cell Physiol.* **154**, 343-349.

12. Rasmussen, C.D., Means, R.L., Lu, K.P., May, G.S., and Means, A.R. (1990) *J. Biol. Chem.* **265**, 13767-13775.
13. Davis, T.N., Urdea, M.S., Masiarz, F.R., and Thorner, J. (1986) *Cell* **47**, 423-431.
14. Takeda, T., and Yamamoto, M. (1987) *Proc. Natl. Acad. Sci. USA* **84**, 3580-3584.
15. Moser, M.J.,Lee, , S.Y., Klevit, R.E., and Davis, T.N. (1995) *J. Biol. Chem.* **270**, 20643-20652.
16. Davis, T.N. (1995) in *Adv. Sec. Mess. Phosphoprot. Res.* (Means, A.R. ed.) Vol **30**, pp. 339-358, Raven Press, N.Y.
17. Whitaker, M. (1995) in *Adv. Sec. Mess. Phosphoprot. Res.* (Means, A.R. ed.) Vol **30**, pp. 299-310, Raven Press, N.Y.
18. Lu, K.P., and Means, A.R. (1993) *Endocrine. Rev.* **14**, 40-58.
19. Anraku, Y., Ohya, Y., and Iida, H. (1991) *Biochem. Biophys. Acta* **1093**, 169-177.
20. Geiser, J.R., van Tuinen, D., Brockerhoff, S.E., Neff, M.M., and Davis, T.N. (1991) *Cell* **65**, 949-959.
21. Iida, H., Yagawa, Y, and Anraku, Y. (1990) *J. Biol. Chem.* **265**, 13391-13399.
22. Halachmi, D., and Eliam, Y. (1993) *FEBS Ltr.* **316**, 73-78.
23. Lu, K.P., and Means, A.R. unpublished data
24. Doonan, J.H. (1992) *J. Cell Sci.* **103**, 599-611.
25. Heath. I.B. (1995) *Can. J. Bot.* **73**(Suppl. 1), S131-S139.
26. Kropf. D.L., Money, N.P., and Gibbon, B.C. (1995) *Can. J. Bot.* **73**(Suppl. 1), S126-S130.
27. Sietsma. J.H., Wosten, H.A.B., and Wessels, J.G.H. (1995) *Can. J. Bot.* **73**(Suppl. 1), S388-S395.
28. Clutterbuck, A.J. (1977) *J. Gen. Microb.* **60**, 133-135.
29. Harris. S.D., Morrell, J.L., and Hamer, J.E. (1994) *Genet.* **136**, 517-532.
30. Bergen, L.G., and Morris, N.R. (1983) *J. Bacteriol.* **156**, 155-160.
31. Morris, N.R. (1976) *Genet. Res.* **26**, 237-254.
32. Osmani, S.A., Pu, R.T., and Morris, N.R. (1988) *Cell* **53**, 237-244.
33. O'Connell, M.J., Osmani, A.H., Morris, N.R., and Osmani, S.A. (1992) *EMBO J.* **11**, 2139-2149.
34. Osmani, A.H., van Peij, N., Mischke, M., O'Connell, M.J., and Osmani, S.A. (1994) *J. Cell Sci.* **107**, 1519-1528.
35. Ye, X.S., Xu, G., Pu, R.T., Fincher, R.R., Osmani, A.H., and Osmani, S.A. (1995) *Can. J. Bot.* **73**(Suppl. 1), S359-S363.
36. James, S.W., Mirabito, P.M., Scacheri, P.C., and Morris, N.R. (1995) *J. Cell Sci.* **108**, 3485-3499.
37. Nanthakumar, N. N., and Means, A. R. Unpublished data
38. Lu, K.P., Rasmussen, C.D., May, G.S., and Means, A.R. (1992) *Mol. Endocrinol.* **6**, 365-374.
39. Chafouleas, J.G., Bolton, W.E., Boyd III, A.E., and Means, A.R. (1984) *Cell* **36**, 73-81.
40. Lu, K.P., Osmani, S.A., Osmani, A.H., and Means, A.R. (1993) *J. Cell Biol.* **121**, 621-630.
41. Lu, K.P., and Means, A.R. (1994) *EMBO J.* **13**, 2103-2113.
42. Lu, K.P., Nanthakumar, N.N., Dayton, J.S., and Means, A.R. (1995) in *Adv. Mol. Cell Biol.* (Bitter, E.E., and Whitaker, M. eds.) Vol. **13**, pp.89-136, JAI Press, Greenwich, CN.
43. Davis, T.N. (1992) *J. Cell Biol.* **118**, 607-617.
44. Ohya, Y., and Botstein, D. (1994) *Science* **263**, 963-966.
45. Sun, G.-H., Hirata, A., Ohya, Y., and Anraku, Y. (1992) *J. Cell Biol.* **119**, 1625-1639.
46. Cyert, M.S., and Thorner, J. (1992) *Mol. Cell. Biol.* **12**, 3460-3469.
47. Liu, Y., Ishii, S., Tokai, M., Tsutsumi, H., Ohke, O., Akada, R., Tanaka, K., Tsuchiya, E., Fukui, S., and Miyakawa, T. (1991) *Mol. Gen. Genet.* **227**, 52-59.
48. Crabtree, G.R., and Clipstone, N.A. (1994) *Annu. Rev. Biochem.* **63**, 1045-1083.
49. Schreiber, S.L. (1992) *Cell* **70**, 365-368.
50. Cardenas, M.E., andHeitman, J. (1995) in *Adv. Sec. Mess. Phosphoprot. Res.* (Means, A.R. ed.) Vol **30**, pp. 281-298, Raven Press, N.Y.
51. Becker, J.W., Rotonda, J., McKeever, B.M., Chan, H.K., Marcy, A.I., Wiederrecht, G., Hermes, J.D., and Springer, J.P. (1993) *J. Biol. Chem.* **268**, 11335-11339.
52. Dumont, F.J., Staruch, M.J., Koprak, S.L., Siekierka, J.J., Lin, C.S., Harrison, R., Sewell, T., Kindt, V.M., Beattie, T.R., Wyvratt, M., and Sigal, N.H. (1992) *J. Exp. Med.* **176**, 751-760.
53. Lorca, T., Cruzalegui, F.H., Fesquet, D., Cavadore, J.-C., Méry, J., Means, A., and Dorée, M. (1993) *Nature* **366**, 270-273.
54. Tombes, R.M., Grant, S., Westin, E.H., and Krystal, G. (1995) *Cell Growth Diff.* **6**, 1063-1070.
55. Planas-Silva, M.D., and Means, A.R. (1992) *EMBO J.* **11**, 507-517.
56. Rasmussen, C., and Rasmussen, G. (1994.) *Mol. Biol. Cell.* **5**, 785-795
57. Rasmussen, C., Garen, C., Brining, S., Kincaid, R.L., Means, R.L., and Means, A.R. (1994) *EMBO J.* **13**, 2545-2552.
58. Plochocka-Zulinska, D., Rasmussen, G., and Rasmussen, C.D.. (1995) *J. Biol. Chem.* **270**, 24794-24799.
59. Yoshida, T., Toda, T., and Yanagida, M. (1994) *J. Cell Science* **107**, 1725-1735.

60. Higuchi, S., Tamura, J., Giri, P.R., Polli, J.W., and Kincaid, R.L. (1991) *J. Biol. Chem.* **266**, 18104-18112.
61. Cyert, M.S., Kunisawa, R., Kaim, C., and Thorner, J. (1991) *Proc. Natl. Acad. Sci. USA* **88**, 7376-7380.
62. Griffith, J.P., Kim, J.L., Kim, E.E., Sintchak, M.D., Thomson, J.A., Fitzgibbon, M.J., Fleming, M.A., Caron, P.R., Hsiao, K., and Navia, M.A. (1995) *Cell* **82**, 507-522.
63. DeGregori, J., Kowalik, T., and Nevins, J.R. (1995) *Mol. Cell. Biol.* **15**, 4215-4224.
64. Vemu, S., and Reichel, R.R. (1995) *J. Biol. Chem.* **270**, 20724-20729.
65. Dirick, L., Moll, T., Auer, H., and Nasmyth, K. (1993) *Nature* **357**, 508-510.
66. Koch, C., Moll, T., Neuberg, M., Ahorn, H., and Nasmyth, K. (1993) *Science* **261**, 1551-1554.
67. Muller, R. (1995) *Trends Genet.* **11**, 173-178
68. Cunningham, K.W., and Fink, G.R. (1994) *J. Cell Biol.* **124**, 351-363.
69. Garrett-Engele, P., Moilanen, B., and Cyert, M.S. (1995) *Mol. Cell. Biol.* **15**, 4103-4114.
70. Mendoza, I., Rubio, F., Rodriguez-Navarro, A., and Pardo, J.M. (1994) *J. Biol. Chem.* **269**, 8792-8796.
71. Breuder, T., Hemenway, C.S., Movva, N.R., Cardenas, M.E., and Heitman, J. (1994) *Proc. Natl. Acad. Sci. USA* **91**, 5372-5376.
72. Nanthakumar, N.N. and Means, A.R. Manuscript in preparation
73. Gooday, G.W., and Schofield, D.A. (1995) *Can. J. Bot.* **73**(Suppl. 1), S114-S121.
74. Ferreira. A., Kincaid, R., and Kosik, K.S. (1993) *Mol. Biol. Cell.* **4**, 1225-1238.
75. Goto. S, Yamamoto, H., Fukunage, K., Iwasa, T., Matsukodo, Y., and Miyamoto, E. (1985) *J. Neurochem.* **45**, 276-283.
76. Momany M., Morrell, J.L., HArris, S.D., and Hammer, J.E. (1995) *Can. J. Bot.* **73**(Suppl. 1), S396-S399.
77. Mazur. P., Morin, N., Baginsky, W., El-Sherbeini, M., Clemas, J.A., Nielsen, J.B., and Foor, F. (1995) *Mol. Cell. Biol.* **15**, 5671-5681.
78. Bartlet, D.C., Fidel, S., Farber, L.H., Wolff, D.J., and Hammell, R.L. (1988) *Proc. Natl. Acad. Sci. USA* **85**, 3279-3283.
79. Hanson, P.I., and Schulman, H. (1992) *Annu. Rev. Biochem.* **61**, 559-601.
80. Kornstein, L.B., Gaiso, M.L., Hammell, R.L., and Bartelt, D.C. (1992) *Gene* **113**, 75-82.
81. Pausch, M.H., Kaim, D., Kunisawa, R., Admon, A., and Thorner, J. (1991) *EMBO J.* **10**, 1511-1522.
82. Ohya, Y., Kawasaki, H., Suzuki, K., Londesborough, J., and Anraku, Y. (1991) *J. Biol. Chem.* **266**, 12784-12794.
83. Londesborough, J., and Nuutinen, M. (1987) *FEBS Ltr.* **219**, 249-253.
84. Dayton, J.S. and Means, A.R. Manuscript in Preparation.
85. Waring,R.B., May, G.S., and Morris, N.R. (1989) *Gene* **79**, 119-130.
86. Dayton, J.S., M. Sumi and Means, A.R. Manuscript in preparation.
87. Su, T.T., Follette, P.J. and O'Farrell, P. H. (1995) *Cell* **81**, 825-828.
88. Sumi, M., Kazutoshi, K., Ishikawa, T., Ishii, A., Hagiwara, M., Nagatsu, T., and Hidaka, H. (1991) *Biochem. Biophys. Res. Commun.* **181**, 968-975.
89. Mirzayan, C., Copeland, C.S., and Snyder, M. (1992) *J. Cell Biol.* **116**, 1319-1332.
90. Stirling, D.A.,Welch, K.A., and Stark, M.J.R. (1994) *EMBO J.* **13**, 4329-4342.
91. Geiser, J.R., Sundberg, H.A., Chang, B. H., Muller, E.G.D., and Davis, T.N. (1993) *Mol. Cell. Biol.* **13**, 7913-7924.
92. Welsh, M.J., Dedman, J.R., Brinkley, B.R., and Means, A.R. (1978) *Proc. Natl. Acad. Sci. USA* **75**, 1867-1871.
93. Sweet, S.C., and Welsh, M.J. (1987) *Eur. J. Cell Biol.* **47**, 88-93z.
94. Rasmussen, C.D., and Means, A.R. (1989) *EMBO J.* **8**, 73-82.
95. McGoldrick, C.A., Gruver, C., and May, G.S. (1995) *J. Cell Biol.* **128**, 577-587.
96. Brockerhoff, S.E., Stevens, R.C., and Davis, T.N. (1994) *J. Cell Biol.* **124**, 315-323.
97. Espinola, F.S., Espreafico, E.M., Coelho, M.V., Martins, A.R., Costa, F.R.C., Mooseker, M.S., Larson, R.E. (1992) *J. Cell Biol.* **118**, 359-368.
98. Lu, K.P., Osmani, S.A., and Means, A.R. (1993) *J. Biol. Chem.* **268**, 8769-8776.
99. Izumi, T., and Maller, J.L. (1995) *Mol. Biol. Cell.* **6**, 215-226.
100. Patel, R., Philipova, R., Moss, S., Schulman, H., Hidaka, H., and Whitaker, M. submitted.
101. Doonan, J.H., and Morris, N.R. (1989) *Cell* **57**, 987-996.
102. Rosen,L.B., Ginty, D.D., and Greenberg, M.E. (1995) in *Adv. Sec. Mess. Phosphoprot. Res.* (Means, A.R. ed.) Vol **30**, pp. 225-254, Raven Press, N.Y.
103. Zhu. G., Muller, E.G.D., Amacher, S.L., Northrop, J.L., and Davis, T.N. (1993) *Mol. Cell. Biol.* **13**, 1779-1787.
104. Sun, Z, and Means, A.R. (1995) *J. Biol. Chem.* **270**, 20962-20967.
105. Miranti, C. K., Ginty, D. D., Huang, G., Chatila, T, and Greenberg, M. E. (1995) *Mol. Cell. Biol.* **15**, 3672-3684.
106. Rabult, B., and Ghysdael, J. (1994) *J. Biol. Chem.* **269**, 28143-28151.
107. Anderson, K., and Means, A.R. Unpublished data.

chapter 22
The roles of DNA topoisomerase II during the cell cycle

Annette K. Larsen[1,*], Andrzej Skladanowski[1,2] and Krzysztof Bojanowski[3]

Department of Structural Biology and Pharmacology,
Institut Gustave Roussy PR2, Villejuif 94805 cedex, France[1]
Department of Pharmaceutical Technology and Biochemistry,
Technical University of Gdansk, Narutowicza St 11, 80-952 Gdansk, Poland[2]
Department of Surgery and Pathology, Children's Hospital
and Harvard Medical School, 300 Longwood Ave., Boston MA 02115, USA[3]
*To whom correspondence should be sent

DNA topoisomerase II (topo II) is essential for survival of all eukaryotic cells. Topo II is both an enzyme and a structural component of the nuclear matrix. It regulates the topological states of DNA by transient cleavage, strand passing and re-ligation of double-stranded DNA resulting in decatenation of intertwined DNA molecules and relaxation of supercoiled DNA. Topo II plays an important role in DNA replication and is required for condensation and segregation of chromosomes. The expression of topo II is cell cycle dependent with both protein levels and catalytic activity peaking at G2/M. Phosphorylation/dephosphorylation of topo II may be a part of regulatory checkpoints at the entry and progression of mitosis.

INTRODUCTION

DNA topoisomerase II (topo II) is an enzyme that regulates the topology of DNA and which is the principal target for a number of important antitumor agents (1). The enzyme catalyses many types of interconversions between DNA topological isomers in an ATP-dependent fashion through transient cleavage, strand passing, and re-ligation of double-stranded DNA, resulting in relaxation of supercoiled DNA, catenation, decatenation, knotting or unknotting (2). Eukaryotic topo II exists as a homodimer consisting of two polypeptides of ~170kD (1) and is essential for the viability of proliferating cells (3-6). The enzyme is required for chromosome condensation and separation of intertwined chromosomal DNA molecules during mitosis (3,4,7-10) and for chromosomal segregation in meiosis (11). Topo II has been implicated in the periodic spacing of nucleosomes (12,13). Furthermore, the enzyme can relieve DNA supercoiling introduced by chain elongation during transcription and replication (14-17). In addition to its catalytic activity, topo II has been reported to play a structural role in maintenance of the chromatin structure of interphase cells (18) and mitotic chromosomes, where the enzyme is associated with the DNA at the matrix attachment regions, which maintain the DNA loops or domains (19-23).

Higher eukaryotes contain two isoforms of topoisomerase II, termed α and β, which are coded by two different genes (24-29). Yeast and *Drosophila* contain only one topo II (30,31) which, like topo IIα (32,33), is expressed and regulated in a cell cycle dependent manner (34,35). Topo IIβ is predominantly localised in the nucleolus of interphase cells, which suggests a role for this isoform in transcription of ribosomal RNA (33,36).

In this paper, we review the growth state and cell cycle-dependent fluctuations of topo II as well as its various biological functions. The different methods available to study this essential nuclear protein will also be discussed.

REGULATION OF THE AMOUNT OF TOPO II THROUGHOUT THE CELL CYCLE

Topo IIα, but not topo IIβ is a proliferation - dependent enzyme (32,37-39). The average level of topo IIα monomers is around 10^2 to 10^3 per cell in serum-deprived and in differentiated cells and 10^5 to 10^6 per cell in proliferating, non synchronised populations (21,40-43). In proliferating cells, the level of topo IIβ does not change significantly throughout the cell cycle and accounts for approximately 20 to 30% of the total topo II (44-46). It has been suggested that the topo IIβ isoform is involved in the very early stages of lymphocyte activation where cells pass from G0 into and through G1 (47). Otherwise, topo IIβ does not seem to be involved in activities specific for a given phase of the cell cycle as is the case for topo IIα.

During G1 when intensive transcription of housekeeping genes takes place, the amount of topo IIα is at its lowest. Therefore, topo IIα does not seem to be required for transcription. The level of topo IIα protein starts to raise before the S phase, remains stable during most of S and starts to increase again in late S (44,48-52, K. Bojanowski and D. Ingber-unpublished results). This is consistent with a role for topo IIα in release of the torsional stress generated by progression of the replication forks and with its localisation in the nuclear matrix, where it is associated with newly synthesised DNA. The level of topo IIα peaks in late G2. During mitosis, the enzyme plays an essential role in chromatin condensation, chromosome assembly and segregation of sister chromatids. In *Drosophila* embryos (where only one form of topo II has been identified), the amount of chromatin-bound enzyme starts to decrease at prophase and continues to decrease until its near disappearance in early G1 (35). In other systems, the decrease of topo IIα is observed in early G1 (50, K. Bojanowski and D. Ingber-unpublished results).

The regulation of topo IIα occurs both at the mRNA and at the post-translational level. The dissection of the 5'-flanking region of the human topo II gene shows that the regulatory elements are common to other S phase-related proteins such as thymidine kinase and DNA polymerase α (53). However, the inhibition of DNA synthesis in synchronised 3T3 cells has no effect on the increase in topo IIα protein levels (51). In higher eukaryotes, the increase of topo IIα protein is due to transcriptional activation of the topo IIα gene and an increased half-life of its mRNA and protein. Similarly, the abrupt decrease of topo II following exit from M can be explained both by down-regulation of transcription and by a 7-8 fold decrease in mRNA and protein stability (50,52,54). The mechanism responsible for the decreased protein stability is not known. Topo IIα might be proteolysed by an ubiquitin-dependent pathway similar to the mitosis-related cyclins (55) and has been shown to possess a cyclin destruction box motif in the N-terminal end (56). Alternatively, topo II might be degraded by proteases activated by the transient increase in the level of intracellular free Ca^{2+}, which occurs during anaphase (57). Finally, a different way to regulate the level of nuclear topo II has been described for *Drosophila* embryos, where the cell cycle-dependent fluctuations in the amount of nuclear enzyme are, at least in part, due to cyclic translocations between the nuclear and the cytoplasmic compartments (35).

POST-TRANSLATIONAL MODIFICATIONS OF TOPO II

Topo II can modified by poly-ADP ribosylation and phosphorylation. Poly-ADP ribosylation has been shown to inhibit the catalytic activity of topo II *in vitro* (58) and was also detected *in vivo* (59). Although poly-ADP ribosylation of proteins usually takes place following DNA damage, this does not appear to be the case for topo IIα, which is poly-ADP ribosylated even in the absence of DNA lesions. Furthermore, the occurrence of DNA damage does not increase the level of poly-ADP ribosylation of topo IIα (59). We are not aware of studies concerning any cell cycle dependence of poly-ADP ribosylation. In contrast, the phosphorylation of topo II during the cell cycle has been well characterised.

In vitro studies show that phosphorylation of purified topo II by serine/threonine protein kinases such as casein kinase II and protein kinase C stimulates its ATPase activity, and thereby its catalytic functions (60,61). Moreover, completely dephosphorylated topo II from budding yeast (62) and from Swiss 3T3 cells (63) is almost totally inactive. However, this is not the case for fission yeast, where phosphorylation does not seem to stimulate topo II activity, but rather enhances its transport into the nucleus (64). *In vivo*, both topo II α and β are present as phosphoproteins in proliferating but not in serum-deprived cells (51,65-68). The total level, as well as the sites of phosphate incorporation, vary in a cell cycle-dependent manner for topo IIα. The level of topo II phosphorylation is low during the G1 phase, intermediary during DNA replication and high in the G2 phase and during mitosis (34,51,69). Although the increase in the phosphorylation of topo II during the cell cycle is similar to the increase in topo II protein levels, their regulation may differ since studies with synchronised 3T3 cells show that the phosphorylation, but not the increase in protein levels, is affected by aphidicolin treatment (51).

The protein kinases interacting with topo II *in vivo* are the same as those which has been shown to phosphorylate the enzyme *in vitro* (65,66). Tryptic phosphopeptide mapping reveals that topo II is predominantly phosphorylated by casein kinase II in yeast, *Drosophila* and in human cells (34,66,70). The major phosphorylation sites for casein kinase II in human topo IIα are ser 1376 and ser 1524 (70). It has also been shown, that topo II and casein kinase II form molecular complexes which remain associated even after the phosphorylation of topo II by the kinase. An unusual property of these complexes is that both enzymes keep their respective catalytic activities (71).

In addition to an increased frequency of phosphorylation of old sites, new sites are being phosphorylated during the G2/M phase (72-74). In human cells, this involves ser 1212 and ser 1246 which are phosphorylated by a proline-directed kinase such as $p34^{cdc2}$ (73) and ser 29 which is phosphorylated by protein kinase C (74). Furthermore, in mitotic cells, topo IIα is the major chromosomal protein recognised by MPM-2, an antibody that recognises a phosphorylated epitope found predominantly in mitotic cells (46). This mitosis-specific phosphorylation is due to phosphorylation of threonine residues by unknown kinases (46,75) which is consistent with the studies of Burden et al. (76) who detected phosphoserine in all phases of the cell cycle, with a contribution of phosphothreonine only in M-phase cells. Hyperphosphorylation of topo IIα during G2 and M might be necessary to maximise its catalytic activity, which is required for chromosome assembly and disjunction of sister chromatids. The hyperphosphorylation may also stimulate multimerisation of chromatin-bound topo II, which could be necessary for chromatin condensation (77). In addition, the phosphorylation of topo II may allow or modify its interactions with other proteins which play a role in mitosis.

Finally, cell cycle-dependent modifications have also been reported for topo IIβ, where M-phase specific phosphorylation leads to an altered electrophoretic migration of this isoform. The phosphorylation takes place on serine, and to a lesser extent, on threonine residues (46,78). So far, the physiological role of these modifications as well as the protein kinases involved remain unknown.

THE ROLE OF TOPO II IN DNA REPLICATION

During the S phase, the separation of the two DNA strands by replication helicases induces torsional stress in terms of positive superhelical turns, building up in front of the moving replication fork. This torsional stress stops fork movement unless it is relaxed by DNA topoisomerases (14).

Studies with topo I and topo II mutants of budding and fission yeast show that initiation of DNA synthesis can occur in the absence of both topo I and topo II. However, DNA synthesis stops when the DNA fragments have reached a length of a couple of thousands base-pairs, and at least one of the two topoisomerases has to be present for further elongation (7,14,17). It is suggested that in wild-type yeast it is topo I rather than topo II that is the major replication swivel (17). In contrast, several studies suggest that the topo IIα isoform is required for DNA synthesis in higher eukaryotes. The topo II inhibitor teniposide induces a block of DNA synthesis of SV40 minichromosomes in cellular extracts *in vitro*, which prevents the final replication steps leading from late Cairns structures of replicative intermediates to monomeric mini-chromosomes (79). A different topo II inhibitor, ICRF-193, has similar effects on the replication of SV40 *in vivo*. In addition, ICRF-193 seems to interfere with DNA synthesis in the final stage of SV40 replication where the length of the DNA increases from ~1400 nucleotides to 2700 nucleotides (80). The topo II inhibitors etoposide and teniposide have a similar effect on the synthesis of genomic DNA, since drug treatment prevents the formation of large intermediates, such as 10-kilobase DNA without affecting the formation of small intermediates i.e. Okazaki fragments (81).

Some models of chromatin assembly suggest a biphasic molecular mechanism. The first phase, nucleosome formation, comprises the formation of histone-DNA complexes which mature into a canonical nucleosome structure. The second phase represents the process by which these nucleosomes become properly spaced with a regular periodicity on the DNA. *In vitro* models for chromatin assembly using cell-free *Xenopus* extracts suggest that topoisomerase activity (usually topo I) is needed for the first phase of chromatin assembly. This is probably due to the requirement for relaxation of the topological constraints generated by chromatin assembly rather than for the process of assembly by itself. Topo II seems to be specifically required for the second phase since topo II inhibitors significantly inhibit the periodic spacing of nucleosomes without affecting nucleosome formation (12,13).

Finally, topo II is an integral part of the nuclear scaffold, which associates preferentially with newly synthesised DNA (82,83). More recently it has been shown that nascent DNA is associated with topo IIα, whereas the β isoform is only associated with bulk DNA not undergoing replication (84). Taken together, these results suggest that topoisomerase II is needed for DNA replication which, at least in part, could be due to a specialised role for topo IIα in the nuclear matrix.

G2 CATENATION CHECKPOINT

One of the well characterised biological effects of topo II inhibitors, both those which stabilise cleavable complexes and those which inhibit the catalytic activity of the enzyme, is arrest of cell cycle progression in the G2 phase. This effect has recently been attributed to a catenation-sensitive checkpoint (85). This checkpoint exists at least in some mammalian cells and depends on the decatenation state of DNA during prophase. This is in line with earlier reports concerning inhibition of cell cycle progression by topo II-interacting agents about 30-60 min. before metaphase (86,87) associated with an absence of nuclear lamina depolymerisation, lack of metaphase-specific phosphorylation of histone H3 and the presence of elongated, partially condensed chromosomes (88). This checkpoint is apparently lacking in yeast since even in the absence of topo II, top2 mutant cells attempt chromosome segregation resulting in chromosome loss and breakage (6,89,90). The catenation checkpoint is leaky and may be overcome by kinase and phosphatase inhibitors such as caffeine (85,91), fostriecin, okadaic acid, and interestingly, by inactivation of the RCC1 protein (92). The RCC1 protein is believed to be a part of the mitotic entry checkpoint and prevents entry into mitosis of cells with not fully replicated or damaged DNA (93). This suggests that the phosphorylation/dephosphorylation status is an important factor regulating G2/M progression. Consistent with this notion, cells treated with topo II-interacting agents show reduced $p34^{cdc2}$ kinase activity and decreased phosphorylation of histone H1 (88,94,95). The $p34^{cdc2}$ kinase by itself is indirectly involved in chromatin condensation by its histone H1 kinase activity and inhibition of $p34^{cdc2}$ results in G2 arrest (96). The inhibitory effect of topo II inhibitors toward $p34^{cdc2}$ is indirect since these agents have no effect on H1 kinase activity *in vitro*. In addition, it is not related to altered synthesis of cdc25 nor inhibition of protein phosphatase activity (95). Both caffeine and okadaic acid were shown to activate $p34^{cdc2}$ in G2-arrested cells and this activation seems to require a minimal amount of cyclin B (97). It is interesting to note, that some topo II inhibitors induce cyclin B1 accumulation (95). The effect of okadaic acid is post-translational and does not require *de novo* protein synthesis. This is in contrast to what is observed for caffeine and the tsBN2 mutant (where RCC1 is inactivated), suggesting that the tsBN2 phenotype and caffeine induce (a) protein(s) with okadaic acid equivalent activity (97 and references therein).

Although cleavable complexes are also formed under physiological conditions, they may be recognised, when stabilised by topo II inhibitors or if transformed into frank DNA breaks, as DNA damage and therefore activate the DNA damage-dependent checkpoint, which would also lead to G2 arrest. It is not clear at present whether the catenation-sensitive checkpoint is different from the DNA-damage-dependent one in terms of molecular mechanisms (85,92).

THE ROLE OF TOPO II IN CHROMATIN CONDENSATION AND CHROMOSOMAL ASSEMBLY

As eukaryotic cells progress from G2 into mitosis, they undergo a programmed series of morphological changes that culminate in the faithful segregation of each chromosome pair and the formation of two individual cells. One of the most dramatic events that occurs during this process is the compaction of interphase chromatin to generate metaphase chromosomes. The involvement of topo II in this process is well documented, although its precise mechanism of action is still unknown. Genetic studies in *S. pombe* show that at the restrictive temperature, thermosensitive topo II mutants which lack ATPase activity produce extended, entangled chromosomes. These elongated prophase-like structures become fully condensed when the topo II activity is restored (89). Immunodepletion of topo II from *Xenopus* egg extracts or incubation of these extracts with topo II inhibitors suggests that topo II is required from the very beginning of chromatin condensation (8,9,98). However, results from other systems show that topo II catalytic activity is not required for the early steps of chromatin condensation. Mitotic cell extracts can induce initial chromosome condensation in purified interphase nuclei regardless of the topo II content, but the further resolution into distinct chromosomes does not take place in nuclei lacking topo II (43). Elongated, entangled chromosomes were observed following treatment of cells with topo II inhibitors of the ICRF series (ICRF-159, ICRF-187 and ICRF-193) or with etoposide, merbarone and aclarubicin (85,92,99,100). This suggests that topo II activity is required at a late stage of chromosome condensation. A similar conclusion was obtained from studies in which the catalytic activity of topo II in living *Drosophila* blastoderm embryos was disrupted by microinjection of anti-topo II antibodies or with the topo II inhibitor teniposide (101). A possible explanation for all these findings could be that topo II has a structural, but not a catalytic role in the early stages of chromatin condensation.

Chromatin condensation may require the formation of multiprotein complexes, analogous to those which enable DNA replication. Recent data suggest that topo II forms molecular complexes with the Sc II protein during mitosis (102,103). Sc II belongs to the family of SMC proteins which may play a role as motor proteins involved in chromosome compaction (for review see 104). It is possible that topo II can serve as a receptor for Sc II which, in contrast to topo II, is only associated with the nuclear matrix during mitosis, but not in interphase (103).

Another possibility is that the hyperphosphorylation of topo II, which occurs during the G2/M transition, could induce conformational changes in the structure of topo II leading to its multimerisation (77). This would be similar to another structural protein, histone H1, for which multimerisation serve to regulate the degree of chromatin folding (105,106).

Recent work shows that catalytically inactive purified topo II can recondense proteolysed chains of metaphase chromosomes removed microsurgically from living mitotic cells as shown in Fig. 1. Further studies are necessary to demonstrate whether the

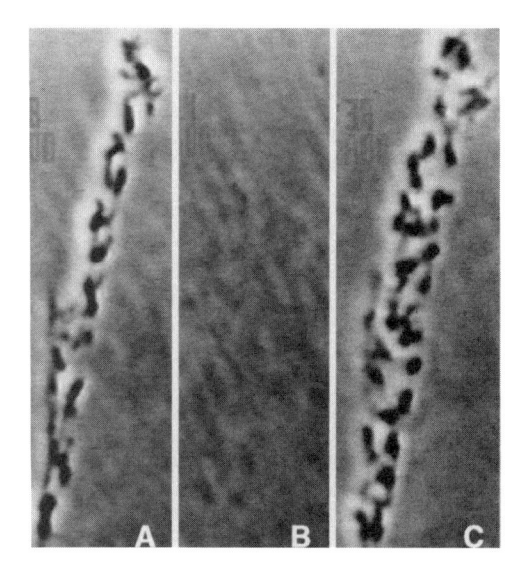

Figure 1. Chromosomes isolated from a living mitotic bovine endothelial cell are placed onto a Petri dish (A) and decondensed with a trypsin/proteinase K mixture (B). Addition of purified human topo IIα results in recondensation of chromosomes, even in the absence of ATP (C). This suggests that the chromatin condensation induced by topo II is at least partially independent from its catalytic activity (K. Bojanowski, A.K. Larsen and D. Ingber, unpublished results).

results obtained with metaphase chromosomes are also applicable to cells during the early phase of chromosome condensation.

Results from experiments where topo II inhibitors, that induce topo II-associated DNA strand breaks, were added to either interphase or mitotic cells suggest that the distribution of topo II on the chromatin may differ between mitotic and interphase cells. Treatment of non-synchronised cells with etoposide resulted in the formation of >600 kilobasepair (kbp) DNA fragments (107) whereas teniposide treatment resulted in the formation of DNA fragments between 200 and 300 kbp (108). In contrast, treatment of metaphase chromosomes with teniposide generated a smear of DNA fragments ranging from 50 to 900 kbp (88). Similar results were found by alkaline elution where amsacrine or etoposide induced 4 to 15-fold more cleavage in mitotic HeLa cells that in S phase cells (109).

SEPARATION OF SISTER CHROMATIDS

DNA replication results in the newly replicated chromosomes being intertwined. This is why the decatenation activity of topo II is required for faithful chromosome segregation. Studies with mutant topo II protein in yeast as well as the use of topo II inhibitors in mammalian cells all show that the catalytic activity of topo II is required for the final stages of chromosome condensation prior to the metaphase-anaphase transition (10,85,92,100). The final decatenation is probably carried out by topo IIα (110) and may require association with Sgs1, an eukaryotic homolog of the E. coli RecQ helicase (111). It is not yet known, whether topo II activity is required for the final condensation process as such or only for sister chromatid decatenation, the chromatids being unable to condense completely while still intertwined.

The decatenation of DNA during the late stages of DNA condensation is an extremely delicate process. On one hand, the sister chromatids need to be kept in close physical proximity to facilitate identification of those chromatids that should be segregated from one another during mitosis. This might in part explain why the cell waits to the very last moment to separate the sister chromatids. On the other hand, sister chromatids need to be totally disentangled when chromosome segregation takes place since, in contrast to many other essential processes, there appears to be no regulatory checkpoint for DNA entangling. The lack of final chromosome condensation and segregation is not coupled to other mitotic events such as activation of $p34^{cdc2}$ kinase, spindle formation and disassembly of the nuclear envelope (99). If entangling is prevented by mutation of topo II or by drug treatment, chromosome segregation nonetheless proceeds leading to chromosome breakage, non-disjunction and cell death (6,7,10,89,90,99,112,113).

METAPHASE/ANAPHASE TRANSITION

It has been shown that addition of the topo II inhibitor teniposide to BHK cells during the G2 phase of the cell cycle prevents histone H1 phosphorylation. In cells that were arrested in mitosis, teniposide induced dephosphorylation of histones H1 and H3, DNA breaks and partial chromosome decondensation (88). It is likely that the signal for dephosphorylation is a change in the mechanical properties of the chromosomes due to the following series of events. Teniposide-induced DNA-topo II complexes might be converted into frank DNA strand breaks through the action of mitosis-specific helicases such as Sgs1 (111). This would allow DNA unwinding around the cleavage site resulting in long-range changes in chromosome tension.

Changes in mechanical properties also seem to be involved in a different checkpoint during mitosis. Some cells have a control system which can detect a single misattached chromosome and delay the onset of anaphase thus allowing time for error correction. The checkpoint signal appears to be tension since tension from a microneedle on a misattached chromosome leads to anaphase (114). The mechanical error in attachment must somehow be linked to the chemical regulation of the cell cycle.

The 3F3 antibody detects phosphorylated kinetochore proteins that might serve as the required link since tension, whether from a micromanipulation needle or from normal mitotic forces, causes dephosphorylation of the kinetochore proteins recognised by 3F3 (115). Recently, it has been shown that topo IIα from mitotic cells carries a phosphoepitope which is recognised by the 3F3 antibody (116). However, it is not known at this time if dephosphorylation of topo IIα is directly linked to the spindle checkpoint.

TOPO II AFTER METAPHASE/ANAPHASE TRANSITION

After resolving residual intertwinings between sister chromatids just before anaphase, the active role of topo II in the cell cycle is apparently completed and the amount of cellular topo II decreases in late mitosis reaching its minimum in G1 (50). The data are scarce concerning changes in the distribution and amount of topo II where discrimination between nuclear and cytoplasmic pools of the enzyme have been reported. The elegant studies of Swedlov et al. (35) have shown that about 70% of the chromosomal topo II leaves the chromosomes during mitosis in *Drosophila*, part of it after the completion of chromosome condensation, and the other after chromosome segregation. It is not clear why topo II is no longer necessary, especially if we consider its structural role. It may be that after condensation and segregation, the availability and/or affinity of topo II to DNA is modified causing the enzyme to leave the chromosomes and diffuse out of the nucleus. In theses studies (35), it was not possible to determine whether the enzyme that leaves the nucleus during mitosis is degraded. However, the proteolytic system which degrades cyclins also recognises other proteins, especially those for which destruction is required for sister chromatid segregation (117). Topo IIα possesses a cyclin B-type destruction box in its N-terminal part (118) which makes it prone to ubiquitination and degradation by the proteasome system during the general proteolysis that leads to metaphase-anaphase transition (117). Topo II may be also degraded by calcium-dependent proteases which are activated by the increase of intracellular Ca^{+2} which occurs during anaphase (57) and which has also been postulated to participate in the destruction of the mitotic cyclins (119). The rapid decrease in the amount of topo II following exit from mitosis may also be related to down-regulation of its transcription and decreased mRNA and protein stability (50,52,54). Finally, given the role of topo II in chromosome condensation, degradation of the enzyme in late mitosis may also, at least in part, lead to decondensation of DNA before entering a new round of the cell cycle. The different modifications and functions of topo IIα during the cell cycle are summarised in Fig. 2.

REGULATION OF TOPO II IN CANCER/TRANSFORMED CELLS COMPARED TO NORMAL CELLS

Many types of tumors are characterised by an altered pattern of topo II regulation, which makes this enzyme an important target for anticancer therapy. Cancer cells often have a higher level of topo II protein and topo II activity as compared to normal cells, during all stages of the cell cycle (120). This makes them more sensitive to chemotherapeutic agents which kill cells by stabilising the topo II in covalent complexes with DNA (121). Transformed cell lines also exhibit higher topo II level and activity, as compared to non-transformed cells (50,122,123). In addition, the topo II content in cancer and transformed cells is less regulated by growth conditions (42,124), which might explain why the sensitivity of different tumors to topo II inhibitors is usually correlated with topo II protein expression and activity, but not necessarily with the cell doubling time (125,126). An analysis of the effect of transformation on the topo II isoforms shows that although the relative amounts of both isoenzymes are higher in transformed cells, the ratio of topo IIα (replication-oriented) to topo IIβ (transcription-oriented) is higher in transformed than in non-transformed cells (124).

EXPERIMENTAL SYSTEMS FOR STUDIES OF TOPO II

In higher eukaryotes, no suitable topo II mutants are available and our knowledge concerning the genetics and functions of topo II is based mostly on experiments with yeast. However, yeast systems have undoubtedly some limitations since at least one major difference exists concerning functions of DNA topoisomerases, which is the case of topo I. Although topo I is essential for the viability of multicellular organisms (127), topo I mutants of both budding and fission yeast, which exhibit less than 1% of normal topo I activity grow normally (4,128). The same is true for *E. coli* (129). It has been proposed that the topo I deletion can be complemented by topo II since top1top2 double mutants have defects in DNA replication and transcription not seen for either of the single mutants (128). Topo I can also complement some of the functions of topo II since a mitotic block can prevent killing of temperature-sensitive top2 mutants at the non-permissive temperature, whereas the same treatment is ineffective in preventing death of top1top2 double mutants (5). Many of these findings are probably true for higher eukaryotes as well but the exact relationship

Modifications of topo IIα ## Functions of topo IIα

G₁: proteolysis of topo II (possible involvement of cyclin destruction box) decrease in stability of the protein and mRNA — no particular function reported

S: induction of topo II expression, increased topo II mRNA and protein stability, ser/thr phosphorylation on the C-terminal domain, predominantly by casein kinase II — possible role in DNA replication through the release of torsional stress, possible structural role in the organization of the chromatin at the nuclear matrix

G₂: hyperphosphorylation, modification of phosphorylation pattern, maximal level of the protein — essential catalytic and possible structural role in chromatin condensation and chromosome assembly

M: — possible structural role in the organization of chromosome loops, essential role in segregation of sister chromatids

Figure 2. Modifications and functions of topo IIα during the cell cycle.

between topo I and topo II functions in multicellular organisms is still an open question.

Many of the important findings concerning the mitotic checkpoints come from genetic studies in fungi, *Drosophila* and *Xenopus* systems. However, it should be mentioned that although, for example, *S. cerevisiae* possess a clear metaphase checkpoint, there seems not to be any catenation-sensitive one (6,90). Another important difference when comparing the metaphase checkpoint in yeast and mammalian cells is that in mammals multiple microtubules interact with a single kinetochore, in contrast to the presence of a single kinetochore microtubule in budding yeast (98). Some of the functional aspects of spindle assembly are also different even between higher eukaryotes, such as animals, *Drosophila* and *Xenopus* (130).

Apart from its enzymatic activity, topo II also plays a structural role in chromosome organisation. In this respect, it is interesting to note that immunolocalisation studies with *Drosophila* embryos (35) and *Xenopus* extracts (98) suggest that topoisomerase II is not specifically associated with axial filaments but rather is distributed throughout chromosomes, in contrast to what was found earlier (20). Furthermore, topo II can be extracted from condensed chromosomes of *Xenopus* mitotic extracts without causing major changes in their morphology (98). This result may be however somewhat artifactual since subsequent studies show that a component of the mitotic extracts, β-glycerophosphate, has a marked effects on the enzymatic activity of topo II and its ability to multimerize *in vitro* (77). Some of the structural aspects of chromatin organisation are clearly different in various cell models. In yeast, genomic DNA is only about 1% of the amount of genomic DNA present in mammalian cells. Therefore, the topological problems associated with DNA replication and transcription as well as those posed during chromosome segregation during mitosis are certainly less important. For example, small chromosomes in yeast apparently can be resolved "passively" during mitosis without any topo II activity, by falling off from one end of the chromosome when the sister chromatids separate (112). This phenomenon is absent in mammalian cells.

Many different topo II inhibitors, such as epipodophyllotoxins (etoposide and teniposide), amsacrine, anthracyclines (doxorubicin, aclarubicin), anthracenediones (mitoxantrone) and bis-dioxopiperazine derivatives (ICRF-159, 187 and 193), have been applied to study topo II functions in various biological systems. However, the specificity of many of these compounds in their interactions with topo II may be quite limited. Some of these agents (epipodophyllotoxins, anthracyclines, anthracenediones, amsacrine), in addition to inhibition of topo II, produce DNA single- and double-strand breaks *via* topo II-mediated cleavable complex formation. DNA intercalators (anthracyclines, anthracenediones) may prevent not only topo II but also other proteins from binding to DNA (131), whereas etoposide inhibits thymidine transport (132). ICRF analogs are believed to inhibit the catalytic activity of topo II without measurable DNA damage. However, these agents induce formation of a stable DNA-topo II complex in which the enzyme is immobilised around the DNA as a closed clamp (133). Mechanistically, these 'closed clamp' complexes are quite different from a simple inhibition of catalytic activity. Finally, no drug has been identified so far which is able to inhibit the structural functions of topo II.

ACKNOWLEDGEMENTS

This work was supported in part by the Association pour la Recherche sur le Cancer (ARC) Villejuif, France.

REFERENCES

1. Liu, L.F. (1989) *Annu. Rev. Biochem.* **58**, 351-375.
2. Wang, J.C. (1985) *Annu. Rev. Biochem.* **54**, 665-697.
3. DiNardo, S., Voelkel, K. and Sternglanz, R. (1984) *Proc. Natl. Acad. Sci. USA* **81**, 2616-2620.
4. Uemura, T. and Yanagida, M. (1984) *EMBO J.* **3**, 1737-1744.
5. Goto, T. and Wang, J.C. (1984) *Cell* **36**, 1073-1080.
6. Holm, C., Goto, T., Wang, J.C. and Botstein, D. (1985) *Cell* **41**, 553-563.
7. Uemura, T. and Yanagida, M. (1986) *EMBO J.* **5**, 1003-1010.
8. Newport, J. and Spann, T. (1987) *Cell* **48**, 219-230.
9. Adachi, Y., Luke, M. and Laemmli, U.K. (1991) *Cell* **64**, 137-148.
10. Downes, C.S., Mulliger, A.M. and Johnson, R.T. (1991) *Proc. Natl. Acad. Sci. USA* **88**, 2616-2620.
11. Rose, D., Thomas, W. and Holm, C. (1990) *Cell* **60**, 1009-1017.
12. Almouzni, G. and Mechali, M. (1988) *EMBO J.* **7**, 4355-4365.
13. To, R.Q. and Kmiec, E.B. (1990) *Cell Growth Diff.* **1**, 39-45.
14. Brill, S. J., DiNardo, S., Voelkel-Meiman, K. and Sternglanz, R. (1987) *Nature* **326**, 414-416.
15. Wu, H.Y., Shy, S.H., Wang, J.C. and Liu L.F. (1988) *Cell* **53**, 433-440.
16. Brill, S. J. and Sternglanz, R. (1988) *Cell* **54**, 403-411.
17. Kim, R.A. and Wang, J.C. (1989) *J. Mol. Biol.* **208**, 257-267.
18. Berrios, M., Osheroff, N. and Fisher, P.A. (1985) *Proc. Natl. Acad. Sci. USA* **82**, 4142-4146.
19. Earnshaw, W.C., Halligan, B., Cooke, C.A., Heck, M.M.S. and Liu, L.F. (1985) *J. Cell Biol.* **100**, 1706-1715.
20. Earnshaw, W.C. and Heck, M.M. (1985) *J. Cell Biol.* **100**, 1716-1725.
21. Gasser, S.M., Laroche, T., Falquet, J., Boy de la Tour, E. and Laemmli, U.K. (1986) *J. Mol. Biol.* **188**, 613-629.
22. Gasser, S.M. and Laemmli, U.K. (1986) *EMBO J.* **5**, 511-518.
23. Razin, S.V., Petrov, P. and Hancock, R. (1991) *Proc.Natl.Acad.Sci.*USA **88**, 8515-8519.
24. Drake, F.H., Zimmerman, J.P., McCabe, F.L., Bartus, H.F., Per, S.R., Sullivan, D.M., Ross, W.E., Mattern, M.R., Johnson, A.K., Crooke, S.T. and Mirabelli, C.K. (1987) *J. Biol. Chem.* **262**, 16739-16747.
25. Tsai-Pflugfelder, M., Liu, L.F., Liu, A.F., Tewey, K.M., Whang-Peng, J., Knutsen, T., Huebner, K., Croce, C.M. and Wang, J.C. (1988) *Proc. Natl. Acad. Sci. USA* **85**,7177-7181.
26. Chung, T.D.Y., Drake, F.H., Tan, K.B., Per, S.R., Crooke, S.T. and Mirabelli, C.K. (1989) *Proc. Natl. Acad. Sci. USA* **86**, 9431-9435.
27. Austin, C.A. and Fisher, L.M. (1991) *FEBS Lett.* **266**, 115-117.
28. Jenkins, J.R., Ayton, P., Jones, T., Davies, S.L., Simmons, D.L., Harris, A.L., Sheer, D. and Hickson, I.D. (1992) *Nucl. Acids Res.* **20**, 5587-5592.
29. Austin,C.A., Sng, J.-H., Patel, S. and Fisher, L.M. (1993) *Biochim. Biophys. Acta* **1172**, 283-291.
30. Giaever, G., Lynn, R., Goto, T. and Wang, J.C. (1986) *J. Biol. Chem.* **261**, 12448-12454.
31. Nolan, J.M., Lee, M.P., Wyckoff, E. and Hsieh, T.S. (1986) *Proc. Natl. Acad. Sci. USA* **83**, 3664-3668.
32. Drake, F.H., Hofmann, G.A., Bartus, H.F., Mattern, M.R., Crooke, S.T. and Mirabelli, C.K. (1989) *Biochemistry* **28**, 8154-8160.
33. Negri, C., Chiesa, R., Cerino, A., Bestagno, M., Sala, C., Zini, N., Maraldi, N.M. and Astaldi-Ricotti, G.C.B. (1992) *Exp. Cell Res.* **200**, 452-459.
34. Cardenas, M.E., Dang, Q., Glover, C.V.C. and Gasser, S.M. (1992) *EMBO J.* **11**, 1785-1796.
35. Swedlow, J.R., Sedat, J.W. and Agard, D.A. (1993) *Cell* **73**, 97-108.
36. Zini, N., Martelli, A.M., Sabatelli, P., Santi, S., Negri, C., Astaldi-Ricotti, G.C. and Maraldi, N.M. (1992) *Exp. Cell Res.* **200**, 460-466.

37. Capranico, G., Tinelli, S., Austin, C.A., Fisher, M.L. and Zunino, F. (1992) *Biochim. Biophys. Acta* **1132**, 43-48.
38. Tsutsui, K., Tsutsui, K., Okada, S., Watanabe, M., Shohmori, T., Seki, S. and Inoue, Y. (1993) *J. Biol. Chem.* **268**,19076-19083.
39. Kimura, K., Nozaki, N., Saijo, M., Kikuchi, A., Ui, M. and Enomoto, T. (1994) *J. Biol. Chem.* **269**, 24523-24526.
40. Heck, M.M. and Earnshaw, W.C. (1986) *J.Cell Biol.* **103**, 2569-2581.
41. Fairman, R. and Brutlag,D.L. (1988) *Biochemistry* **27**, 560-565.
42. Hsiang, Y.H., Wu, H.Y. and Liu, L.F. (1988) *Cancer Res.* **48**, 3230-3235.
43. Wood, E.R. and Earnshaw, W.C. (1990) *J. Cell Biol.* **111**, 2839-2850.
44. Woessner, R.D., Mattern, M.R., Mirabelli, C.K., Johnson, R.K. and Drake, F.H. (1991) *Cell Growth Diff.* **2**, 209-214.
45. Prosperi, E., Sala, E., Negri, C., Oliani, C., Supino, R., Astraldi-Ricotti, G.B.C., Bottiroli, G. (1992) *Anticancer Res.* **12**, 2093-2100.
46. Taagepera, S., Rao, P.N., Drake, F.H. and Gorbsky, G.J. (1993) *Proc. Natl. Acad. Sci. USA* **90**, 8407-8411.
47. Daev, E., Chaly, N., Brown, D.L., Valentine, B., Little, J.E., Chen, X. and Walker, P.R. (1994) *Exp. Cell Res.* **214**, 331-342.
48. Miskimins, R., Miskimins, W.K., Bernstein, H. and Shimizu, N. (1983) *Exp. Cell Res.* **146**, 53-62.
49. Chow, K.-C. and Ross, W.E. (1987) *Mol. Cell. Biol.* **7**, 3119-3123.
50. Heck, M.M., Hittelman, W.N. and Earnshaw, W.C. (1988) *Proc. Natl. Acad. Sci. USA* **85**, 1086-1090.
51. Saijo, M., Ui, M. Enomoto, T. (1992) *Biochemistry* **31**, 359-363.
52. Goswami, P.C., Roti, J.L. and Hunt, C.R. (1996) *Mol. Cell. Biol.* **16**, 1500-1508.
53. Hochhauser, D.C., Stanway, C.A., Harris, A.L. and Hickson, I.D. (1992) *J. Biol. Chem.* **267**, 18961-18965.
54. Ramachandran, C., Mead, D., Wellham, L.L., Sauerteig, A. and Krishan, A. (1995) *Biochem. Pharmacol.* **49**, 545-552.
55. Sudakin, V., Ganoth, D., Dahan, A., Heller, A., Hershko, J., Luca, F.C., Ruderman, J.V. and Hershko, A. (1995) *Mol. Biol. Cell* **6**, 185-197.
56. Nakajima, T., Ohi, N., Arai, T., Nozaki, N., Kikuchi, A. and Oda, K. (1995) *Oncogene* **10**, 651-652.
57. Poenie, M. J., Alderton, R., Steinhardt, R. and Tsien, R. (1986) *Science* **233**, 886-889.
58. Darby, M.K., Schmitt, B., Jongstra-Bilen, J. and Vosberg, H.P. (1985) *EMBO J.* **4**, 2129-2134.
59. Scovassi, A.I., Mariani, C., Negroni, M., Negri, C. and Bertazzoni, U. (1993) *Exp.Cell Res.* **206**, 177-181.
60. Corbett, A.H., DeVore, R.F. and Osheroff, N. (1992) *J. Biol. Chem.* **267**, 20513-20518.
61. Corbett, A.H., Fernald, A.W. and Osheroff, N. (1993) *Biochemistry* **32**, 2090-2097.
62. Alghisi, G.C., Roberts, E., Cardenas, M.E. and Gasser, S.M. (1994) *Cell Molec. Biol. Res.* **40**, 563-571.
63. Saijo, M., Enomoto, T., Hanaoka, F. and Ui, M. (1990) *Biochemistry* **29**, 583-590.
64. Shiozaki, K. and Yanagida, M. (1992) *J. Cell Biol.* **119**, 1023-1036.
65. Rottmann, M., Schröder, H.C., Gramzow, M., Renneisen, K., Kurelec, B., Dorn, A., Freise, U. and Müller, W.E.G. (1987) *EMBO J.* **6**, 3939-3944.
66. Ackerman, P., Glover, C.V. and Osheroff, N. (1988) *J. Biol. Chem.* **263**, 12653-12660.
67. Kroll, D.J. and Rowe, T.C. (1991) *J. Biol. Chem.* **266**, 7957-7961.
68. Bernardi, R., Negri, C., Donzelli, M., Guano, F. and Scovassi, A.I. (1995) *Int. J. Oncol.* **6**, 203-208.
69. Heck, M.M., Hittelman, W.N. and Earnshaw, W.C. (1989) *J. Biol. Chem.* **264**, 15161-15164.
70. Wells, N.J., Addison, C.M., Fry, A.M., Ganapathi, R. and Hickson, I.D. (1994) *J. Biol. Chem.* **269**, 29746-29751.
71. Bojanowski, K., Filhol, O., Cochet, C., Chambaz, E.M. and Larsen, A.K. (1993) *J. Biol. Chem.* **268**, 22920-22926.
72. Burden, D.A. and Sullivan, D.M. (1994) *Biochemistry* **33**, 14651-14655.
73. Wells, N.J. and Hickson, I.D. (1995) *European J. Biochem.* **231**, 491-497.
74. Wells, N.J., Fry, A.M., Guano, F., Norbury, C. and Hickson, I.D. (1995) *J. Biol. Chem.* **270**, 28357-28363.
75. Taagepera, S., Campbell, M.S. and Gorbsky, G.J. (1995) *Exp. Cell Res.* **221**, 249-260.
76. Burden, D.A., Goldsmith, L.J. and Sullivan, D.M. (1993) *Biochem. J.* **293**, 297-304.
77. Vassetzky, Y.S., Dang, Q., Benedetti, P. and Gasser, S.M. (1994) *Mol. Cell. Biol.* **14**, 6962-6974.
78. Kimura, K., Saijo, M., Ui, M. and Enomoto, T. (1994) *J. Biol. Chem.* **269**, 1173-1176.
79. Richter, A. and Strausfeld, U. (1988) *Nucl. Acids Res.* **16**, 10119-10129.
80. Ishimi, Y., Sugasawa, K., Hanaoka, F., Toshihiko, E. and Hurwitz, J. (1992) *J. Biol. Chem.* **267**, 462-466.
81. Lonn, U., Lonn, S., Nylen, U. and Winblad, G. (1989) *Cancer Res.* **49**, 6202-6207.
82. Nelson, W.G., Liu, L.F. and Coffey, D.S. (1986) *Nature* **322**, 187-189.

83. Fernandes, D.J., Smith-Nanni, C., Paff, M.T. and Neff, T.A. (1988) *Cancer Res.* **48**, 1850-1855.
84. Qiu, J., Catapano, C.V. and Fernandes, D.J. (1996) *Proc. Am. Assoc. Cancer Res.* **37**, 2964.
85. Downes, C.S., Clarke, D.J., Mullinger, A.M., Gimenez-Abian, J.F., Creighton, A.M. and Johnson R.T. (1994) *Nature* **372**, 467-470.
86. Tobey, R.A. (1972) *Cancer Res.* **32**, 2720-2725.
87. Rowley, R. and Kort, L. (1989) *Cancer Res.* **49**, 4752-4257.
88. Roberge, M., Th'ng, J., Hamaguchi, J. and Bradbury, E.M. (1990) *J. Cell Biol.* **111**, 1753-1762.
89. Uemura, T., Ohkura, H., Adachi, Y., Morino, K., Shiozaki, K. and Yanagida, M. (1987) *Cell* **50**, 917-924.
90. Holm, C., Stearns, T. and Botstein, D. (1989) *Mol. Cell. Biol.* **9**, 159-168.
91. Lock, R.B., Galperina, O.V., Feldhoff, R.C. and Rhodes, L.J. (1994) *Cancer Res.* **54**, 4933-4939
92. Anderson, H. and Roberge, M. (1996) *Cell Growth Diff.* **7**, 83-90.
93. Roberge, M. (1992) *Trends Cell Biol.* **2**, 277-281.
94. Lock, R.B. and Ross, W.E. (1990) *Cancer Res.* **50**, 3761-3766.
95. Ling, Y-H., El-Naggar, A.K., Priebe, W. and Perez-Soler, R. (1996) *Mol. Pharmacol.* **49**, 832-841.
96. Pagano, M., Pepperkok, R., Lukas, J., Baldwin, V., Ansorge, W., Bartek, J. and Draetta, G. (1993) *J. Cell Biol.* **121**, 101-111.
97. Yamashita, K., Yasuda, H., Pines, J., Yasumoto, K., Nihitani, H., Ohtsubo, M., Hunter, T., Sugimura, T. and Nishimoto, T. (1990) *EMBO J.* **9**, 4331-4339.
98. Hirano, T. and Mitchison, T.J. (1993) *J. Cell. Biol.* **120**, 601-612.
99. Ishida, R, Sato, M., Narita, T., Utsumi, K.R., Nishimoto, T., Morita, T., Nagata, H. and Andoh, T. (1994) *J. Cell Biol.* **126**, 1341-1351.
100. Gorbsky, G.J. (1994) *Cancer Res.* **54**, 1042-1048.
101. Buchenau, P., Saumweber, H. and Arndt-Jovin, D.J. (1993) *J. Cell Sci.* **104**, 1175-1185.
102. Ma, X., Saitoh, N. and Curtis, P.J. (1993) *J. Biol. Chem.* **268**, 6182-6188.
103. Saitoh, N., Goldberg, I.G., Wood, E.R. and Earnshaw, W.C. (1994) *J. Cell. Biol.* **127**, 303-318.
104. Peterson, C.L. (1994) *Cell* **79**, 389-392.
105. Lennard, A.C. and Thomas, O.J. (1985) *EMBO J.* **4**, 3455-3462.
106. Kamakaka, R.T. and Thomas, J.O. (1990) *EMBO J.* **9**, 3997-4006.
107. Beere, H.H., Chresta, C.M., Alejo-Herberg, A., Skladanowski, A., Dive, C., Larsen, A.K. and Hickman, J.A. (1995) *Mol. Pharmacol.* **47**, 986-996.
108. Charron, M. and Hancock, R. (1990) *Biochemistry* **29**, 9531-9537.
109. Estey, E., Adlakha, R.C., Hittelman, W.N. and Zwelling, L.A. (1987) *Biochemistry* **26**, 4338-4344.
110. Giménez-Abián, J.F., Clarke, D.J., Mullinger, A.M., Downes, C.S. and Johnson, R.T. (1995) *J. Cell Biol.* **131**, 7-17.
111. Watt, P.M., Louis, E.J., Borts, R.H. and Hickson, I.D. (1995) *Cell* **81**, 253-260.
112. Spell, R.M. and Holm, C.(1994) *Mol. Cell. Biol.* **14**, 1465-1476.
113. Sumner, A.T. (1995) *Exp. Cell Res.* **217**, 440-447.
114. Li, X. and Nicklas, R.B. (1995) *Nature (London)* **373**, 630-632.
115. Nicklas, R.B., Ward, S.C. and Gorbsky, G.J. (1995) *J. Cell Biol.* **130**, 929-939.
116. Gorbsky, G.J., Campbell, M.S. and Daum, J.D. (1996) *Proc. Am. Assoc. Cancer Res.* **37**, 32.
117. Holloway, S.L., Glotzer, M., King, R.W. and Murray, A.W. (1993) *Cell* **73**, 1393-1402.
118. Nakajima, T., Ohti, N., Arai, T., Nozaki, N. Kikuchi, A. and Oda, K. (1995) *Oncogene* **10**, 651-662.
119. Lorca, T., Galas, S., Fesquet, D., Devault, A., Cavadore, J-C. and Dorée, M. (1991) *EMBO J.* **10**, 2087-2093.
120. Nelson, W.G., Cho, K.R., Hsiang, Y.H., Liu, L.F. and Coffey, D.S. (1987) *Cancer Res.* **47**, 3246-3250.
121. Zwelling, L.A., Estey, E., Silberman, L., Doyle, S. and Hittelman, W. (1987) *Cancer Res.* **47**, 251-257.
122. Crespi, M.D., Mladovan, A.G. and Baldi, A. (1988) *Exp. Cell Res.* **175**, 206-215.
123. Rainwater, R. and Mann, K. (1991) *Virology* **181**, 408-411.
124. Woessner, R.D., Chung, T.D.Y., Hofmann, G.A., Mattern, M.R., Mirabelli, C.K., Drake, F.H. and Johnson, R.K. (1990) *Cancer Res.* **50**, 2901-2908.
125. Fry, A.M., Chresta, C.M., Davies, S.M., Walker, M.C., Harris, A.L., Hartley, J.A., Masters, J.R. and Hickson, I.D. (1991) *Cancer Res.* **51**, 6592-6595.
126. Giaccone, G., Gazdar, A.F., Beck, H., Zunino, F. and Capranico, G. (1992) *Cancer Res.* **52**, 1666-1674.
127. Lee, M.P., Brown, S.D., Chen, A. and Hsieh, T.-S. (1993) *Proc. Natl. Acad. Sci. USA* **90**, 6656-6660.
128. Trash, C., Voelkel, K., Di Nardo, S., Sternglanz. (1984) *J. Biol. Chem.* **259**, 1375-1377.
129. Sternglanz, R., DiNardo, S., Voelkel, K.A., Nishimura, Y., Hirata, Y., Becherer, K., Zumstein, L. and Wang, J.C. (1981) *Proc. Natl. Acad. Sci. USA* **78**, 2747-2751, 1981.

130. Therkauf, W.E. and Hawley, R.S. (1992) *J Cell Biol.* **11**, 1167-1180.
131. Bartkowiak, J., Kapuscinski, J., Melamed, M.R., Darzynkiewicz, Z. (1989) *Proc. Natl. Acad. Sci. U.S.A.* **86**, 5151-5154.
132. Yalowich, J.C. and Goldman, I.D. (1984) *Cancer Res.* **44**, 984-989.
133. Roca, J., Ishida, R., Berger, J.M., Andoh, T. and Wang, J.C. (1994) *Proc. Natl. Acad. Sci. USA* **91**, 1781-1785.

chapter 23
Circadian rhythm of cell division

Rune Smaaland

Department of Oncology, Haukeland Hospital,
University of Bergen, N-5021 Bergen, Norway

The existence of circadian oscillations in the level of hormones, in numerous physiological parameters, in toxicity and in behavior is now fully recognized in all living organisms. In contrast, the synchronisation and regulation of cell proliferation by circadian rhythms *in vivo* is only starting to be appreciated. This article reviews the experimental evidence for circadian synchronisation of cell division in different mammalian tissues (mainly the gastro-intestinal tract and hemapoietic system), including tumoral tissues. The possible causes of this coupling of the cell cycle phases to the circadian rhythm are discussed. Testing of novel anti-tumour agents using murine models should take into consideration the temporal difference between murine and human circadian control of proliferation (the peak of DNA synthesis occurs during the activity period, i.e. during daytime in man, and at night-time in mice and rats). Experimental and clinical data clearly support the important implications of the circadian control of the cell cycle in the optimisation of cancer chemotherapy, both for reducing toxicity and increasing the antitumour effects.

INTRODUCTION

It has been commonly believed that the evolution of any physiochemical system invariably leads to a state of maximum disorder as a result of the effect of the second law of thermodynamics. As a consequence, any coherent behaviour, such as self-sustained oscillations, has been thought to be ruled out. However, stable oscillations are a common phenomenon in animal and human biology, with positive or negative feedback being a necessary physical prerequisite, sometimes combined with some form of cross-coupling. Such a theoretical notion is embodied in the biological periodicities prevalent at every level of structural and physiological organisation throughout the animal and plant kingdoms: Oscillatory behaviour is the rule rather than the exception (1-3). The period (t) of known biological oscillators (or "clocks") ranges from milliseconds, minutes, and hours (ultradian, t < 20 h) to approximately one day (circadian, 20 h < t < 28 h), to longer, infradian (t > 28 h) time spans. For example, circadian rhythms of transcription of clock genes lead to increased abundance of mRNA, which is followed by peak availability of clock proteins some hours later (Figure 1). The proteins are able, directly or indirectly and mostly through unknown mechanisms, to inhibit transcription of their encoding gene. This transcriptional inhibition cannot be sustained, because in the absence of transcription, levels of clock protein will inevitably decline; transcription will therefore restart and a self-sustaining oscillation is generated. The stable circadian period is presumed to arise from kinetics of the constituent events, for example the rates of synthesis, modification and degradation of mRNA and protein, and rates of transfer between cytoplasm and nucleus (4).

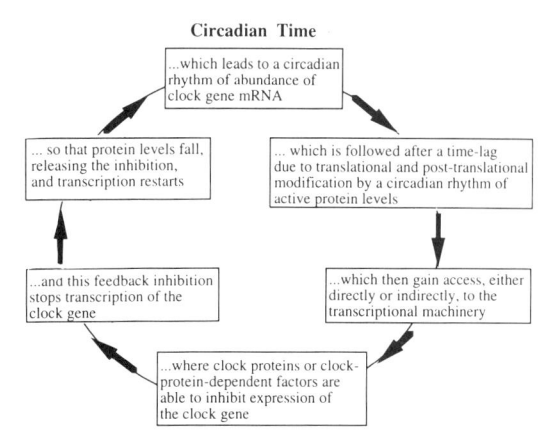

Figure 1. **Circadian clocks: a never ending sentence**. Circadian rhythms of transcription of clock genes lead to increased abundance of mRNA, which is followed by peak availability of clock proteins some hours later. The proteins are able, directly or indirectly and through unknown mechanisms, to inhibit transcription of their encoding gene. This transcriptional inhibition cannot be sustained because in the absence of transcription, levels of clock protein will inevitably decline; transcription will therefore restart and a self-sustaining oscillation is generated. The stable circadian period is presumed to arise from kinetics of the constituent events, for example the rates of synthesis, modification and degradation of mRNA and protein, and rates of transfer between cytoplasm and nucleus (Modified after Hastings, M. *Nature* **376**,296-297,1995).

Since their identification 40 years ago, we have known that our circadian clocks have an intrinsic period of about 25 solar hours. They therefore need frequent resetting by the environment, i.e., entrainment, if they are to be effective. That organisms can and do measure astronomical time in some manner is reflected in the ubiquitous persistent rhythms having solar daily, lunar monthly, and yearly (circannual) periods. Thus, the different types of time-keeping have external correlates

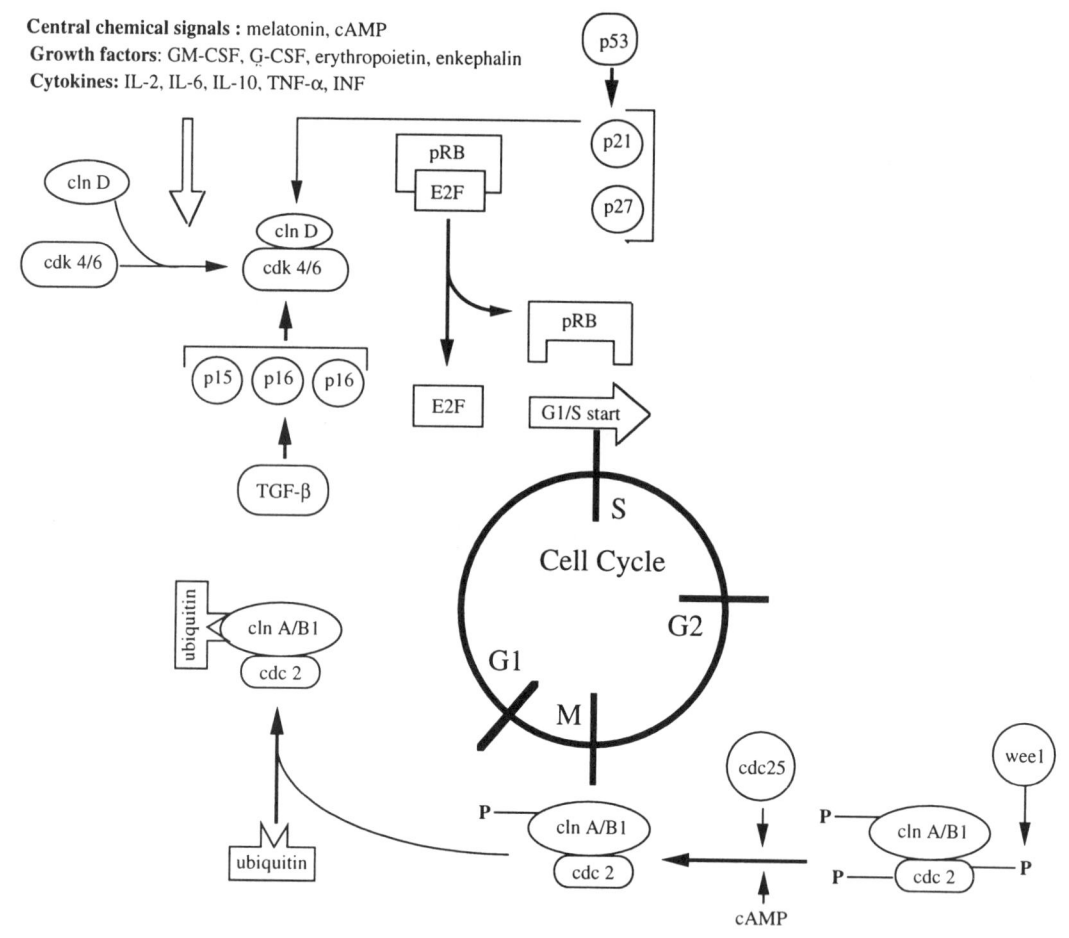

Figure 2. Schematic representation of functional interrelationships between cell cycle-regulatory proteins. The figure shows the cascade of regulatory mechanisms controlling activation of cdks by the cell cycle-dependent regulation of specific cyclins and by a series of cdk inhibitors. The contributions of ubiquitin and phosphatases (e.g., cdc25) to activation and downregulation of cell cycle-regulatory complexes are indicated. Also designated are interactions of cyclin-related proteins with growth factors, cytokines, tumour suppressors (pRB), TGF-beta and cAMP. Competency for cell-cycle progression is reflected by convergence of growth factor (tumour promoters and suppressors) signalling pathways with cyclin-related proteins. Cell-cycle progression is supported by transcriptional control (activation and suppression) of genes at a series of checkpoints (e.g., G1/S and G2/M) by the cyclin-dependent phosphorylation of transcription factors. Several of the effectors which affect the regulation of the cell cycle progression have been shown or will have the potential to do this in a circadian stage-dependent manner (Modified after Stein, G.S., Stein, J.L., van Wijnen, A.J., Lian, J.B., Quesenberry, P.J., *Exp. Hematol.*, **23**, 1053, 1995.)

(generated by the movements of the earth, moon and sun) to which the organism has adapted. An understanding of the physiological and biochemical bases of simpler autogenous oscillators and clocks underlying biological periodicities, such as those characteristic of most cell types, may be crucial to the understanding of the more complex time-keeping displayed at more advanced levels, i.e., organ, organismal and even population levels. Thus, since the metabolic system changes rhythmically in time, it follows that an organism such as man is biochemically and physiologically a different entity at different circadian stages; therefore it reacts differently to an identical stimulus given at different times.

In this chapter, documentation, general aspects and control of circadian rhythm of the cell division will be discussed, on the basis of *in vitro* studies, animal studies and human studies. Especially will circadian variations in the highly proliferating cells of the GI-tract and hematopoietic system be discussed. Data from studies on humans will be specifically addressed in relation to their significance for cancer chemotherapy.

In our laboratory we have performed extensive studies on hematopoiesis in general, and on temporal growth control and proliferation in particular; in recent years we have performed extensive temporal studies of the human bone marrow and non-Hodgkin lymphomas. To a large extent we have used flow cytometry for measuring DNA distribution in the evaluation of circadian cell proliferation, but have also performed bone marrow culturing. Results from time sequence studies on bone marrow cell kinetics in both animals and humans will be presented and compared.

Finally, it will be discussed rather extensively how these rhythmic variations in cell cycle division can be exploited in anticancer therapy.

MOLECULAR AND CELLULAR ASPECTS OF BIOLOGICAL RHYTHMS

The cell division cycle (CDC) of eukaryotes is a complex cascade of developmental and morphogenetic events that culminate in cell duplication (5). Rapid progress has been made within the past several years towards dissecting the regulatory network that provides precise coordination of these processes. A unifying view has emerged: the cell division cycle consists of transitions from one regulatory state to another (6, 7), the cell cycle progression being regulated at several irreversible transition points (8-10). These transitions initiate the modification of substrates that determine the physical state of the cell and are themselves feedback controlled. The basic mechanisms, as elucidated by a combination of genetic and biochemical analyses, appear to have been conserved in organisms ranging from unicellular organisms, such as yeast, to clams, and to cultured human cells. This work has led to the general conclusion that the cyclin-dependent protein kinases (CDKs) are central to controlling the cell cycle, and that the basic elements of these controls are common to all eukaryotic cells (Figure 2).

Three types of cell cycle control have been described as being important for regulating the orderly progression through S-phase and mitosis. The first is a commitment control (Start) acting at the beginning of the cell cycle which begins the programme of events eventually leading to cell division. The second determines the overall cell cycle timing of the onset of S-phase and mitosis. The third type of cell cycle control ensures that S-phase and mitosis occur in the correct sequence and are dependent one upon the other, which is an example of the checkpoint controls proposed by Hartwell and Weinert, whereby at specific points in the cell cycle the cell checks whether an essential early event has been completed before proceeding to the next event.

Two crucial transition control points exist, namely in the G1 phase and at the G2/M boundary. In the fission yeast the commitment control Start acts in G1 and determines the cell cycle timing of S-phase. At this phase of the cell cycle p34cdc2, encoded by cdc2 gene, is complexed with the B-cyclin cig2. Activation of the p34cdc2/cig2 protein kinase is necessary to bring about the onset of the S-phase. In mammalian cells the situation with respect to the G1/S transition is more complex, with other kinases and cyclins being involved as well (Figure 2) (5, 6). Cells undergoing DNA replication generate a signal that S-phase is in progress. This signal restraints mitotic p34cdc2/p56cdc13 protein kinase by maintaining the inhibitory Y15 phosphorylation on p34cdc2. Only when S-phase is complete can Y15 become dephosphorylated allowing mitosis to take place. Progression from G1 to S-phase in mammalian cells is regulated by the accumulation of cyclins D, E, and A, which bind to and activate different CDK catalytic subunits (9). However, cyclin accumulation and CDK binding do not constitute the only levels of regulation of CDK activity. CDK activity is also regulated by both positive and negative phosphorylation events (10), as well as by association with inhibitory proteins (11, 12). Two major classes of CDK inhibitors have recently been identified in mammalian cells. Whereas p15, p16 and p18 specifically inhibit CDK4 and CDK6 by binding to the CDK subunit alone, p21, p27 and p57 can bind to and inhibit a broad range of CDK-cyclin complexes. The p27Kip1 protein is implicated in the negative regulation of G1 progression in response to a number of antiproliferative signals (13). For example, studies in macrophages have linked cyclic AMP-induced growth arrest to an increase amount of p27Kip1 protein, whereas the drug rapamycin abrogates a small reduction in p27 abundance observed after colony-stimulating factor-1 stimulation (14). Hengst and Reed have recently demonstrated that translational control is in part or predominantly responsible for the regulation of p27 abundance under various conditions during the cell cycle. A rapid increase in translation of p27 may be essential for negative regulation of G1 progression in response to antiproliferative signals. Alternatively, regulation of the half-life of the protein could be important for maintaining the arrested state (15). The rapid change in the level of p27 may well be brought about by growth factors, i.e., interleukins, hematopoietic growth factors. In this context it is therefore of relevance and interest that several authors have demonstrated significant circadian changes in *in vitro* pharmacodynamic effects of recombinant IL-3, GM-CSF and G-CSF on murine myeloid progenitor cells (16-19). Perpoint *et al.* have also demonstrated that in the absence of any added CSF, the number of clusters of myeloid progenitor cells also varied significantly according to sampling time, showing that circadian stage-dependent variation in the level of growth factors may induce circadian variation in proliferative activity, and that progenitor cells may themselves exhibit temporal variation due to some endogenous growth promoting signal which varies during the 24-hour time span (19). Their conclusion was that both proliferation, circulation and functions of hematopoietic cells are under circadian control. These findings also indicate that the circadian variations of hemopoiesis are not a passive adaptive process, but are actively regulated at the cellular level.

The M-phase is characterised by the activation of a kinase (MPF, or maturation (M-phase)-promoting factor), which consists of two component subunits. One is the p34cdc2 protein (the product of the *Schizosaccharomyces pombe* gene cdc2), which is able to phosphorylate histone H1 *in vitro* and has maximal kinase activity at mitosis. The other is the mitotic B-cyclin p56cdc13, whose cellular concentration fluctuates during the cell cycle: it accumulates gradually during interphase, forms a complex with p34cdc2, and activates the protein kinase. It is then degraded during mitosis, turning off kinase and MPF activity. The system thus behaves like an oscillator, or "clock", which is reset to its interphase state during mitosis.

The improved knowledge of cell cycle controls and the importance of CDKs in these controls has implications for clinical applications in two ways. The first is that it identifies a whole new series of proteins which have to be activated to allow a cell to reproduce, and so are potential targets for therapy. However, this will be a relatively crude approach as normal cells also have to activate the same functions to proliferate, but further work may identify more specific changes to these proteins in cancerous cells which perhaps can be exploited to develop more specific therapies. However, these signals may exhibit a different phasing between normal and malignant cells. This is in accordance with the fact that the final effect of many different cancers will be to disturb overall cell cycle control. If these changes can be identified, whether they be phenotypical differences or differences in temporal phasing, then one might be able to therapeutically differentiate normal cells from cancerous cells. One possibility is the control which prevents cells from undergoing mitosis until S-phase is complete. Mammalian cells may be like fission yeast in having an alternative mechanism of p34cdc2 activation in early G1 compared with late G1 and S-phase. Because many cancerous cells are unable to become properly quiescent in early G1, the mechanism usually restraining mitosis may be the one which operates in late G1 or S-phase cells. Interfering with this control at specific circadian stages may have differing effects in normal and cancerous cells. Thus, these differences could be exploited for therapy since premature activation of p34cdc2 would bring about the death of the cancerous cell.

The hormone melatonin secreted by the pineal organ has been found to have varying proliferative effects; for example *in vitro* on MCF-7 cells, depending on whether the cells are exposed to hormone concentrations which remain constant in culture media or varying at 12-h intervals, thus simulating a diurnal rhythm. The highest antiproliferative effect has been demonstrated to be obtained by sequential exposure of melatonin, which mimics the physiological rhythm of serum melatonin concentration (Figure 2) (20)

There is now also strong evidence that an autonomous oscillator(s) (independent of either p34cdc2, cyclins, or both) is part of, or is coupled to, the cell division cycle. Thus, when the *S. pombe* cell cycle is arrested by a temperature-sensitive cdc2 mutation, the activity of the enzyme nucleotide diphosphonate kinase and the rate of CO_2 production still undergo periodic oscillations, with a period that approximates the length of the CDC (21, 22). The view, therefore, that the cell cycle duration would depend only on the accumulation of cellular components such as cyclins may be too simplistic.

The circadian clock has been shown to underlie persisting cell division rhythmicity in numerous and diverse systems (1). The group of Edmunds has discovered that adenylate cyclase and phosphodiesterase, the two enzymes whose dynamic balance determines the intracellular level of cAMP, also exhibit inverse, bimodal circadian changes in their own levels in both dividing and nondividing cultures. Effectors such as calcium, calmodulin, and cyclic GMP may modulate these rhythmicities.

The mechanism whereby the circadian oscillator and other higher-frequency oscillators interact with the p34cdc2/cyclin pathway for the control of cell division is an important question to be addressed. Cyclic AMP, known to play a pivotal role in cellular regulation (23), if not an element in the clock itself, may effect such coupling between oscillator and CDC, participating in the gating of CDC events to specific phases of the circadian cycle. In fact, cAMP seems to have the capacity to control some of the rate-limiting steps of cell cycle progression in many cell types (23, 24). It has been found to stimulate the proliferation of some cells, but to have the opposite effect, or no effect at all, on others. Such contradictory findings might be explained by a variation in different systems of the cAMP concentration required for optimal stimulation. Pharmacological studies in mammalian cells (23, 25) and genetic experiments in *Saccharomyces* (26) have shown that a transient rise and the ensuing fall in the cAMP level are necessary for the initiation of DNA synthesis, and that a second cAMP surge is correlated with the onset of mitosis. Similar changes of cAMP concentration at different phases of the CDC have been shown to occur in Euglena (27), this unicell system exhibiting an extensively characterized circadian rhythm of cell division (1). A clock-controlled variation of cAMP level - that is, the periodic repetition of a cAMP signal - thus may participate in the "gating" of DNA synthesis and cell division to a certain phase of the circadian cycle. The cellular circadian oscillator of Euglena has been shown to modulate the progression of cells through the different phases of their CDC (28). The results suggest a possible role for cAMP, either as an element of the coupling pathway for the control of the CDC by the circadian oscillator, or as a "gear" of the clock itself, although the latter possibility is less likely. If the first hypothesis holds true, conditions that would override the control of the cAMP level by the oscillator should also cause perturbations of the cell division rhythm (29). Indeed, it has been found that cAMP seems to cause either a shortening or a lengthening of the CDC, depending on the phase of the CDC when the molecule is given (30, 31). It should be noted, however, that cAMP in *Euglena* oscillates with the same periodicity in nonproliferating cells as in actively growing, synchronously dividing cells, but at a slightly higher level. Thus, changes in cAMP levels do not always serve as signals for cell cycle progression. Rather, cAMP may modulate other biochemical pathways that are essential for cell cycle progression, such as the activation of the cdc2 protein kinase (Figure 2) (32).

By adding cAMP to a cell system, Edmunds *et al.* found that this artificially increased cAMP resulted in the temporary uncoupling of the CDC from the circadian timer. Measurement of cellular DNA content by flow cytometry indicated that cAMP pulses at circadian time 06-08 hours delayed progression through S-phase (and perhaps also through mitosis), but at circadian time 18-20 hours accelerated the G2/M transition. The circadian oscillator was not perturbed by the cAMP pulse: The division rhythm soon returned to its original phase. Further, if the periodic cAMP signals generated by the oscillator were suppressed by the addition of the plant drug forskolin, an activator of adenylate cyclase that maintained cAMP at an abnormally high level, division rhythmicity was lost in DD (Dark/Dark), and even in LD:12,12 (Light/Dark:12 hours, 12 hours). These findings indicate that although cAMP signals do not reset the circadian oscillator itself in *Euglena* (and, thus, that cAMP is unlikely to represent a "gear" of the clock), they do regulate CDC progression and may be sufficient to generate cell division rhythmicity (33).

In the near future we will know in even more detail how the cell cycle is regulated. The effects of both positive and negative growth factors on cell cycle control genes probably will become clearer, as will the correlation between the cell cycle and cell function. Further discoveries on the inter-relationship between proto-oncogenes, tumour suppressor genes and cell cycle control genes should not only lead to an elucidation of their roles in the cell, including the control of temporal variations, but also to a greater understanding of the molecular biology of neoplastic disease. The pace of discovery does not seem to be decreasing and the next few years are expected to add new important information.

EXPERIMENTAL EVIDENCE FOR CIRCADIAN VARIABILITY OF THE CELL CYCLE IN DIFFERENT TISSUES

Circadian rhythms in the mitotic index have been known for a long time (34). Circadian fluctuations in DNA synthesis were however first demonstrated in 1963 (35). Also, already by the late 1950´s Halberg *et al.* (36) demonstrated the existence of circadian rhythms in hepatic cellular DNA and RNA content and also in mitotic activity, in addition to showing that the timing could be shifted by altering the lighting regimen. Since then, a vast literature on circadian rhythms in mitosis and other cycle phases has emerged. Some general examples and features of cell cycle rhythms will be described.

The epidermis has in particular been used as a model for the study of the circadian variations of cell proliferation (37). One reason is that the different subclasses of cells can clearly be identified in layers and that maturation only occurs in one direction, i.e., by keratinisation. It was increasingly recognised that the numbers of cells in certain phases of the cell cycle varied with 24 hour periods, i.e., more and more studies demonstrated circadian variations in the mitotic index of different tissues, e.g., in the epidermis (38-42). The same circadian dependence was found for the DNA synthetic phase (37).

The ascertainment of consistent diurnal variations in human epidermal cell proliferation may have important implications for the treatment of many skin diseases.

For the evaluation of diurnal rhythms in the growth of human epidermis, skin biopsies were taken every 4 hour for 48 hours from each of two persons under synchronised living conditions. The epidermal cell proliferation was assessed by the fraction of cells in S- and in G2/M-phase as determined by measurements of the DNA content in the individual cells in single-cell suspensions. The fraction of cells in G2/M-phase indicated circadian rhythmicity for the first 32 hours of the test period. However, no regular variation according to time of day could be established in the fraction of cells in S-phase in this study (43).

Tvermyr found, by use of a double labelling technique, that there were circadian variations both of the influx to and efflux from the S-phase (44). In addition, subpopulations of S-phase cells had differing rates of DNA synthesis (37, 45). The duration of the S-, G- and M-phases underwent strong circadian variations, with amplitudes up to 3 times the minimum value (37). In general, the time of high proliferative activity was accompanied by low cell cycle phase duration. In the oral mucosa, i.e., the hamster cheek pouch epithelium, similar variations have been found (46, 47).

Brown has re-analysed data from published studies of circadian rhythms in epidermal cell proliferation in mice, rats, and humans (48). Composite circadian rhythm curves were plotted from the combined data for each species for S-phase and M-phase. Each group of studies showed a general consensus on timing, and the composite curve showed a regular sinusoidal pattern. The rhythms in mice and rats were the same, whereas those in humans were in the opposite phase. In rodents, the S-phase peaked at about 03.30 hours and M-phase peaked at about 08.30 hours. In humans S-phase peaked at about 15.30 hours and M-phase peaked at about 23.30 hours.

Circadian variations of liver cell proliferation have long been known (36, 49); in the mouse both hepatocyte DNA synthesis and mitotic activity have been studied. In contrast, the sinusoid littoral cells showed a bimodal pattern of the mitotic index, indicating a different type of regulation. In line with this, it was found that the timing of partial hepatectomy was of importance for the onset of regeneration in hepatocytes (50).

In the cornea of the eye, strong circadian variations of the cell proliferation have also been found. These are so constant, that they can be used as a reliable marker for circadian rhythmicity in mice and rats (51, 52). Indeed, rat corneal epithelium has been chosen as a model for studying growth regulation. In this epithelium a large single cohort of cells enters the S-phase during a fairly short time period once a day. The factor responsible for this wave of cell proliferation is unknown, but it may be a chemical signal from the central nervous system (the suprachiasmatic nucleus or the pineal gland). The mature cell compartment of the corneal epithelium is assumed to produce a negative feedback factor on the local cell proliferation (chalone), counteracting the effect of the circadian proliferative factor on the local cell proliferation.

This interaction between the circadian proliferative factor and the negative feedback factor for regulation of proliferation with its accompanying stimulatory effect on maturation, may represent a general mechanism in the regulation of cell proliferation in any tissue. Since in at least some organs virtually all cells entering the S-phase do this as a single wave once a day, this mechanism may be enough to explain the regulation of cell proliferation during both normal and regenerative conditions (53).

The hematopoietic and lymphopoietic systems have also been studied with regard to circadian variations. However, the complex system with different maturation lines and stages as well as distribution all over the body, have made such studies in the human difficult (54). However, such data are now available from our laboratory (see later). It is also very important and noteworthy that the number of different classes of lymphocytes in peripheral blood has also been shown to undergo strong circadian variations, both in experimental animals and in man (55, 56).

In mice it has long been known that both the stem, progenitor and precursor cells of the different classes undergo circadian variations (57-59). The pattern in mice seems to vary strongly from time to time, probably reflecting a labile equilibrium between cell production, migration and loss (60). Furthermore, both the phasing and the amplitude of the different maturation stages undergo alterations with increasing age of mice (61, 62). There seems to be an increase of mature cells along myelopoiesis, and lower numbers of stem and progenitor cells at high age. In consequence, comparisons between young and old mice at only one time point may lead to erroneous conclusions.

Over several decades there was a controversy as to whether epithelial cells in the alimentary tract of the rodent divide with a circadian frequency. As early as 1947, Klein and Geisel reported a rhythmicity for the mitotic rate in the duodenal epithelium of rodents (63). Several other authors also reported a circadian rhythm of different parts of the alimentary tract in the rat and mouse (39, 64-70).

In spite of the above evidence, reports persisted that there was no cell proliferation rhythm in the gut epithelium (35, 71-74). Unfortunately, these latter reports had a great deal of influence on many scientists, because they have been the most frequently cited and often without any reference to any of the positive findings mentioned above. It is now quite clear that much of the controversy arose from technical variation (75). However, there also were those who would rather carry out their research without having to deal with an oscillating variable. This was easier to do if they had a reference to cite claiming no rhythm. As late as in 1981, a monograph on cell proliferation in the gastrointestinal tract did not deal with circadian rhythms (76). However, strong circadian variations have been found in cell proliferation both of the small intestinal and colon epithelium (37, 77-79).

In 1972, Scheving et al. reported that there was a circadian rhythm in both the mitotic index and DNA synthesis phase of the cell cycle in the mouse duodenum (75). This study involved sampling at 2-hour intervals and, therefore, represented a far more extensive study than any of the previous ones. From these studies it became clear that the duodenum, which had been the most extensively studied region of the gut from the viewpoint of cell kinetics, showed a daily rhythm in cell proliferation. Indeed, in six subsequent investigations these findings were confirmed. These investigations demonstrated that a reproducible low-amplitude rhythm in cell proliferation existed in the duodenum. The percentage of change from the lowest to the highest mean value in the duodenum was 30-60%, depending upon the study. The essential finding was the documentation of similar phasing of the rhythms (80).

In this same study it was demonstrated that the rhythm was functionally important because a reported mitotic stimulant, isoproterenol (IPR), when given as a single intraperitoneal injection, would bring about an increase in DNA synthesis if given at one circadian stage, a decrease at another stage, while at other stages there was no difference between the responses to IPR and saline (75). Thus, if such diverse responses in the duodenum could be produced subsequent to the perturbation at different stages of a rhythm by a single dose of IPR, then a naturally occurring oscillation should be an important variable to consider, irrespective of whether its amplitude was small or large.

Scheving et al. also examined cell proliferation in several regions in the digestive tract in the same animals that had been used for the duodenum study (81). Thus, in addition to the duodenum, five other organs were evaluated (tongue, oesophagus, stomach, jejunum, and rectum) over a 48-hour span (Figure 3). These data systematically documented for the first time, in the same animals, the dramatic variation encountered in cell proliferation from one region to another. The major variations among the different regions were seen in the amplitude of the rhythms and in the overall 24-hour means. However, the phasing in the rhythms in the different regions of the gut were quite similar. In these animals, standardised to 12 hours of light (06.00 to 18.00) alternating with 12-hour

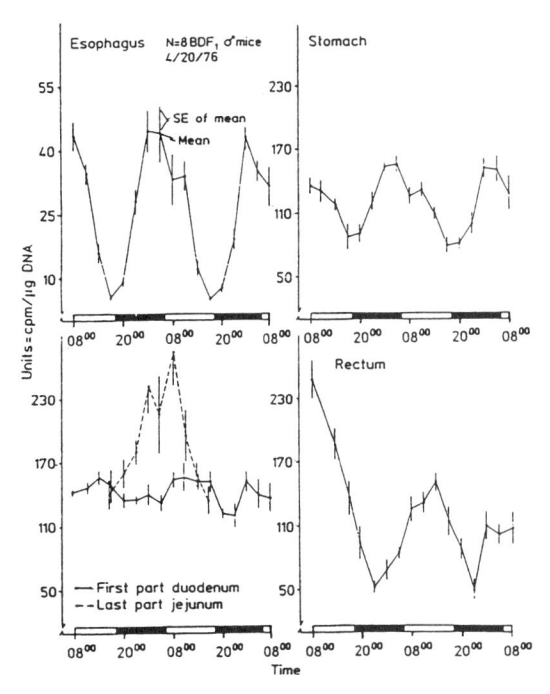

Figure 3. Chronograms illustrating the rhythmic patterns of 3H-thymidine ([3H]TdR) incorporation into DNA in five different regions of the alimentary canal of male mice. (From Scheving, L.E., Burns, E.R., Pauly, J.E., Tsai, T.H., Anat. Rec., 191, 479, 1978. With permission).

darkness, it appeared that the peak in DNA synthesis occurred around the transition from dark to light (daily sleep onset) and the trough occurred around the transition from light to dark (daily activity onset). These rhythms were also reproducible from one study to another for the different organs (tongue, oesophagus, glandular stomach, and rectum). Thus it was demonstrated that cell proliferation is rhythmic throughout the gut and the once-prevailing view of randomness had to be abandoned (82).

It should be noted that such fluctuations are evident whether mitotic figures are counted, the incorporation of tritiated thymidine [3H]TdR into DNA is measured, or cytofluorometric technique (FCM) of analysing cell proliferation is utilised (83-86). Rubin et al have carried out extensive studies pointing out that the biochemical technique used to measure [3H]TdR uptake into DNA has been confirmed by the FCM technique to be a highly reliable method for analysing circadian variation *in vivo* in those tissues that can be reliably compared (87-89). The above data clearly establish that cell proliferation in the entire intestinal tract of the rodent is remarkably coordinated along the 24-hour time scale. It should be underlined that these rhythms are not simply a response to meal timing and evidence supporting this has been documented (90). Although nutrition influences the

24-hour mean DNA synthesis level, nutrition is normally not responsible for phasing the circadian pattern of gut DNA synthesis.

In the 1960s and 1970s there were numerous publications determining cell-cycle time in various tissues. The technique used was the "frequency of labelled mitosis method" (FLM). This technique was designed in 1959 with the assumption that the cells such as those in the intestinal tract divided at random (91). Such studies gave rise to the once popular view that the duration of the synthesis phase was constant within the cell cycle. Many of these studies involved measuring the duration of each stage or total generation time of the cell cycle. Since many of these studies *in vivo* did not consider the circadian timing of tissue sampling, these results are questionable. Therefore, Burns *et al.* studied generation time in the corneal epithelium of the rodent by simply initiating the FLM technique at two appropriately different stages of the mouse circadian system (09.00 or 21.00 hours), instead of only at one time (92, 93). They were able to show dramatic statistically significant differences in the duration of the S-phase.

Thus, significant circadian changes occur in intestinal epithelial cell proliferation/DNA synthesis at all levels of the GI tract - from the tongue to the rectum - in the murine/rodent model (94). Such variations have even been found in adenomas and carcinomas of the mice. While the cell proliferation was faster in the normal epithelium at night, the highest activity in adenomas and carcinomas was found during the day (78). This may indicate that the regulatory mechanisms involved in circadian rhythmicity are disturbed during malignant development.

Both the amplitude and the phasing of cell cycle kinetics have been found to vary from tissue to tissue. Thus, after it had been demonstrated that all phases of the cell cycle underwent such circadian variations, it was recognised that the circadian variations were not synchronous in different tissues. The epidermis tended to have the highest proliferative activity during the night in mice (44, 85).

Although circadian variations seem to be the dominating pattern of proliferative rhythms in various tissues, other rhythms have also been described. However, the study of infradian aspects of the cell cycle has been hampered by both statistical and practical shortcomings, especially in labile tissues. This is not surprising, since the purpose of cell division in tissues is to maintain a constant number of cells according to the actual needs of cell renewal.

CIRCADIAN CYTOKINETICS OF THE HUMAN GI-TRACT

In man, the data on circadian GI cytokinetic characteristics are much less complete, mostly due the technical difficulty in obtaining intestinal tissue samples at frequent intervals and the attendant ethical considerations of such invasive procedures. Nevertheless, Warnakulasuriya and MacDonald have demonstrated a 24-hour periodicity in the labelling index (LI) of human buccal epithelium by using *in vitro* pulse labelling with tritiated deoxythymidine. The mean peak labelling index obtained at 22.00 hours was significantly different from means obtained at rest of the times, the lowest labelling index however being found later during night (95). The same group also used a double labelling technique *in vitro* to examine fluctuations in the time spent in S-phase (Ts) and the rate of cell entry to S-phase (S-influx). They interestingly observed that while the range in Ts was limited (5.1-6.9h), influx to S-phase ranged considerably, from 0-1.26% per hour at six different time periods. An increase in the S influx was apparent around 18.00 h leading to a peak LI 4h later. Thus, these data indicate that the most likely kinetic mechanism related to circadian variations in LI is a variable G1-S influx at this site (96).

Furthermore, a circadian rhythm of cellular proliferation in the human rectal mucosa has been documented by ex vivo measurements of tritiated thymidine ([3H]TdR incorporation into mucosal tissue samples; also, *in vivo* bromodeoxyuridine (BrdUrd) labelling of human colonic mucosa illustrates that such studies in the human may be feasible (97-99).

In 16 healthy men, and altogether 24 studies, the proliferative activity along the circadian scale was measured in rectal mucosa measured as thymidine incorporation (98). Both fasted and fed subjects showed significant rhythms in thymidine incorporation, i.e., DNA synthesis, which peaked about 07.00 hours.

A corresponding study of the human rectal mucosa was performed on samples from 23 subjects by Marra *et al.* (100). [3H]thymidine histoautoradiography was used to determine total labelling index in the rectal crypt. Normal subjects and patients with polyps displayed similar circadian behaviours. They found that circadian fluctuation in proliferation was confined to the area of the crypt normally associated with replication. In spite of inter-individual variations, in this study proliferation was found to be generally higher at night and lower during the afternoon.

CIRCADIAN CYTOKINETICS OF MURINE AND HUMAN BONE MARROW

Until now, little has been known about the biological rhythms of the different parameters of human bone marrow. It is therefore not so surprising to see the systematic neglect of consideration of biological rhythms in the haematological literature. Thus, the patterns of proliferation of bone marrow cells have been thought to be characterised either by: a) a steady-state cytokinetic rate equalling the rate at which cells in the peripheral blood are removed, or b) a rise and fall occurring only under certain abnormal physiological or pathological conditions (101). The homeostatic notion that production rates precisely equal destruction rates is so pervasive that several large surveys on hematopoiesis did not even consider time-dependent variations (102, 103). However, since the production and migration of mature granulocytes into peripheral blood is both dependent on the actual needs of the body, and on several hormonal and regulatory factors (reactive homeostasis) (104), the proliferative pool in the marrow may vary considerably from time to time (105). For example, physical exercise and cortisone/cortisol are strong mobilisers of granulocytes, as are acute bacterial infections. In addition, there are strong endogenous rhythmic variations in these proliferative and mobilising processes, which further complicate the picture, but are also a part of the organism´s homeostasis (predictive homeostasis) (104). Indeed, in classic textbooks it is taught, but usually not put to practical use, that the cell division cycle is about 24 hours in length, with phase M (the period of mitosis) lasting 0.5-1 hr, phase G1 (the post-mitotic/pre-synthetic gap) lasting 10 hours, phase S (the period of DNA synthesis) lasting about 9 hours, and phase G2 (the post-synthetic/pre-mitotic gap) lasting about 4 hours (a resting cell is in phase G0) (101). New information related to proliferation and the redox cycle has recently emerged, and it has become increasingly clear that hematopoiesis is not a temporally fixed phenomenon (106).

Hematopoiesis (hemato = blood; poiesis = to make) is the multi-phase process of proliferation, differentiation and development resulting in mature functional blood cells which develop from the pluripotent stem cell in the bone marrow. The continuous, extremely high proliferative capacity of the bone marrow is rivalled only by the skin and the intestinal mucosa, both of which have been shown to exhibit circadian rhythms in humans (40, 98, 107). It has been estimated that every second, up to 2 million red cells, 2 million platelets and 700,000 granulocytes are produced in the bone marrow (108). The bone marrow, an extremely complex tissue comprising approximately 4.5% of an adult's body weight (a mass comparable to the liver) (109), is found in the ends of flat bones (sternum, ribs, skull, vertebrae and innominates) and contains hematopoietic stem cells which give rise to the many developing functional blood cell lineages within the marrow spaces. The production occurs as a combination of cell proliferation and gradual maturation, until the end stage is reached with a population of mature cells that can exert their specialised functions, but are no longer capable of cell proliferation (110).

Under normal circumstances it takes approximately 14 days from immature stem cell proliferation onset and 7 days from the myeloblast or pronormoblast stage until mature cells are released into the circulating blood (111). Proliferation of stem cells, i.e., CFU-S, appears to be under the control of competing glycoprotein inhibitory and stimulatory factors, the first blocking entry to DNA synthesis and accumulating cells in the G1-phase, while the other triggers cells from G1 rapidly into DNA synthesis.

Circadian aspects in murine bone marrow

Several studies in mice have demonstrated circadian variations in the proliferation of total bone marrow cells or specific subpopulations by measurement of DNA synthesis and/or mitotic index/activity or duration of mitosis, either by using [3H]TdR labelling, percent of labelled mitoses or flow cytometry (81, 112-116). For example, studying erythropoiesis, Dörmer et al. reported that DNA synthesis in mice underwent circadian variations, with the acrophase during the dark (activity) period (117). In addition, with the introduction of assay methods for the various classes of stem and progenitor cells, data have become available from the stem cell proliferation of animals. Several authors have demonstrated circadian and seasonal variations of multi-potent and committed stem cells in mice (57, 86, 118-120). For instance, the circadian rhythms of multi-potent stem cells, myelopoietic progenitor cells (CFU-GM), and recognisable myelopoietic cells have been measured in parallel in female C3H mice. An interesting observation was partial synchronisation of these rhythms (121). Although large differences in proliferation according to circadian time have been found, the results have not been consistent. An explanation to this variation may be investigation on different species, inter-strain differences, animals of different age and sex, different lighting schedules, and possibly infradian variations in rhythm characteristics (mesor [rhythm-adjusted average], amplitude [the extent of predictable change above or below the mesor] and/or acrophase [peak of a fitted curve indicating location of high values]) at different times of the year.

Figure 4. Peak times of circadian variations in murine bone marrow DNA synthesis (S-phase) from review of the literature. The 95% confidence interval (CI) for the peak of the best fitting 24 hour cosine (acrophase =closed dot) can be compared with the actual peak in each time series (macrophase = open square). Following overall summary of acrophase by population mean cosinor, the highest values for DNA occur during the animals activity (dark span) (95% CI = 12.40 to 22.36 HALO). (Modified from Smaaland, R. and Sothern, R.B., *Circadian Cancer Therapy*, Hrushesky, W.J.M. (ed.), CRC Press, **131**, 1994.)

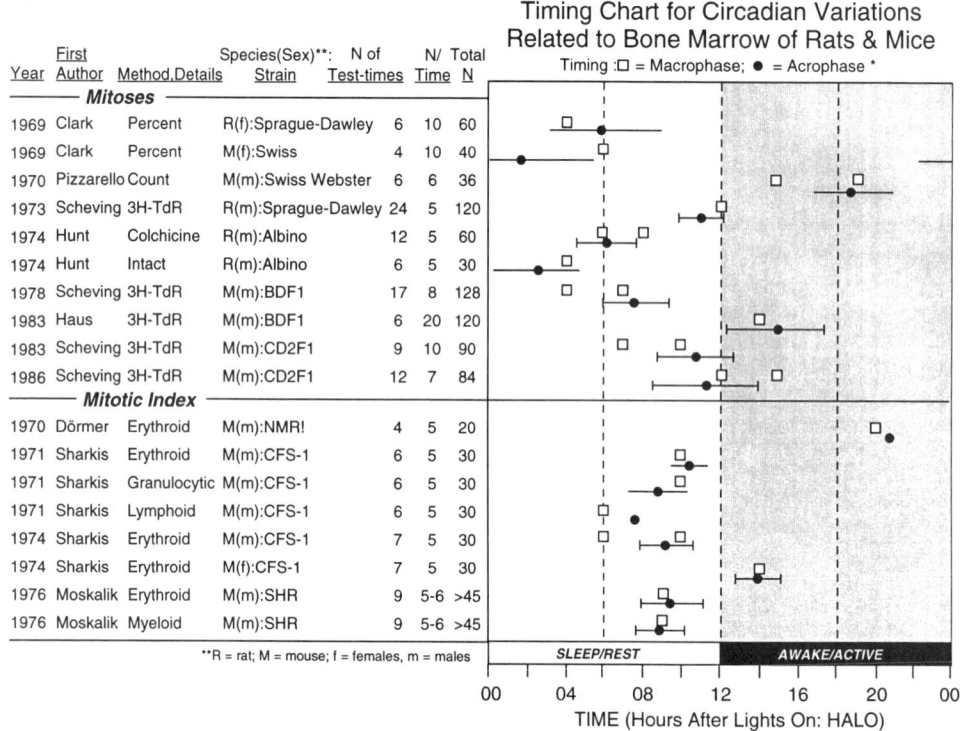

Figure 5. Peak times of circadian variations in indices of murine bone marrow from review of the literature. The 95% confidence interval (CI) for the peak of the best fitting 24 hour cosine (acrophase=closed dot) is presented for comparison with the actual peak in each time series (macrophase=open square). Following overall summary of acrophases by population mean cosinor, the maximal number of mitoses and highest mitotic index were computed to occur during the second half of the resting (light) span (95% CI = 05.20 to 13.08 HALO (hours after lights on). (From Smaaland, R. and Sothern, R.B., *Circadian Cytokinetis of Murine and Human Bone Marrow and Human Cancer*, **131** in *Circadian Cancer Therapy*, Hrushesky, W.J.M. (ed.), CRC Press, 1994. With permission.)

We have done a literature search for all reports on circadian variations related to bone marrow of healthy rats and mice (122). This includes reports on DNA synthesis (S-phase), mitoses and mitotic index (57, 59, 81, 102, 112, 114-118, 123-132).

Two timing charts allow visualisation of the documentation of macrophases and acrophases (using HALO only) between studies and variables for DNA S-phase (Figure 4) and mitoses and mitotic index (Figure 5). As shown in Figure 4, it is clear that the DNA synthesis in 6 out of 6 studies is highest during the activity (dark) span in mice. On the other hand, the maximal number of mitoses and highest mitotic index are generally found during the second half of the resting phase (Figure 5), i.e., almost 12 hours apart from the location of peak DNA synthesis. This large difference in circadian phasing may seem somewhat puzzling until one considers that studies of synchronised cells in culture have shown that the total generation time of a typical proliferating bone marrow cell is about 24 hours, as mentioned above, with mitoses beginning about 12 hours after the onset of the DNA S-phase stage.

On the basis of these data it may be inferred that S-phase specific drugs or drugs with their major effect on cells in S-phase would be least toxic to the bone marrow of rats and mice when administered during the resting (light) part of the 24-hour period. By extrapolation, least toxicity may also be achieved for M-phase specific drugs if they are administered during the daily activity (dark) span.

It should be kept in mind that age-related changes in physiological variations, including hematopoiesis, may be a modifying factor influencing the circadian stage dependent effect of cytotoxic therapy (61, 62).

Circadian aspects related to human bone marrow

Circadian stage dependent variations in proliferative activity of bone marrow cells are highly relevant in the context that cytotoxic therapy of cancer could be properly timed to take advantage of the proliferation-related increase and decrease in sensitivity of bone marrow cells to drug-induced toxicity.

Until recently few data have been available concerning temporal variations of parameters in the human bone marrow, even though Goldeck in 1948 showed a marked circadian variation in reticulocytes sampled in the bone marrow and blood at 3 times of the day (133, 134). Our review of the literature on circadian variations related to human bone marrow is summarised by study details and acrophases in a timing chart of macrophases and acrophases in Figure 6 (135-140). Already in 1962 and 1965, two limited studies, both now considered classics, were published in which DNA synthesis (by use of [3H]TdR technique) and mitotic index of the human bone marrow were measured at different circadian stages (135, 136). By sampling bone marrow at four different times (6 AM and 6 PM, noon and midnight) during a 42-hour period in 1965, Mauer found that the [3H]TdR labelled cells of the myeloid lineage were clearly higher during the day (at 6 AM and noon) as compared to midnight in 3 of 4 individuals, and with a trend towards lower DNA synthesis at midnight in the fourth individual

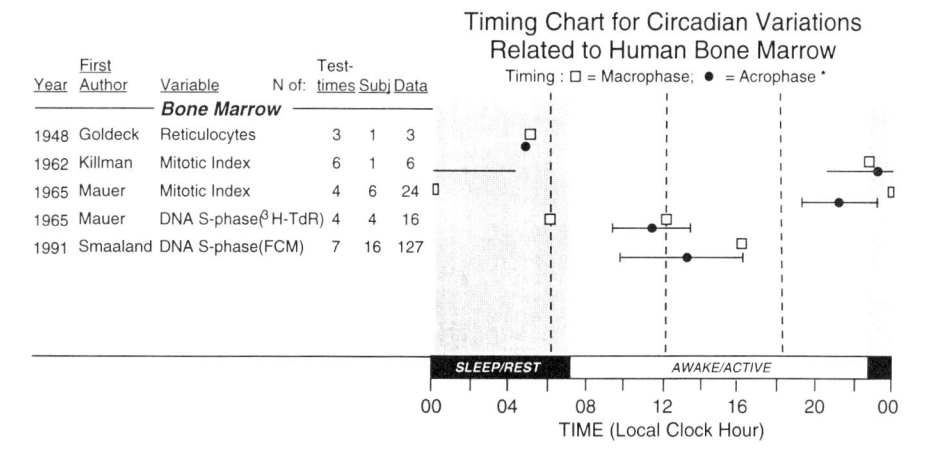

Timing Chart for Circadian Variations Related to Human Bone Marrow

Timing : □ = Macrophase; ● = Acrophase *

Year	First Author	Variable	N of: Test-times	Subj	Data
Bone Marrow					
1948	Goldeck	Reticulocytes	3	1	3
1962	Killman	Mitotic Index	6	1	6
1965	Mauer	Mitotic Index	4	6	24
1965	Mauer	DNA S-phase(³H-TdR)	4	4	16
1991	Smaaland	DNA S-phase(FCM)	7	16	127

* Literature summary. Macrophase = clock hour of actual maxima; Acrophase = peak of best-fitting cosine determined from least-squares fit of a 24-hour cosine to mean values or all data. Confidence limits for acrophase: ±95% limits (close-ended bars) if p-value ≤0.05; or ± 2 SE's (open-ended bars) if p-value ≤0.20.

Figure 6. Circadian phase chart showing peak times derived from review of the literature for temporal variations in human bone marrow. Single cosinor-derived acrophase (closed dot) and a 95% confidence interval (CI) for the peak of the best fitting 24-hour cosine shown if three or more timepoints available for analysis. The actual peak in each time series (macrophase) is represented by an open square. DNA synthesis is maximal during the daily activity span, while mitosis are greatest later in the evening.

(136). However, he was unable to find any circadian variation of the [3H]TdR labelling index of erythropoiesis. The mitotic index was found to be highest at 6 PM or at midnight and lowest at 6 AM in 5 of 6 individuals. Our single cosinor analysis of these data computed the acrophase for DNA synthesis at 11:21 h, and at 21:09 h for the mitotic index, i.e., the highest proliferative activity (S and G2/M) was found to be during day and early evening. Thus, the time with greatest percent of cells incorporating the labelled thymidine was computed to precede the time of greatest number of mitotic figures by 10 hours. Killman et al. made corresponding observations in one human volunteer in an earlier study reported in 1962 with regard to mitotic indices, demonstrating an increase in this proliferative parameter during the day until late evening (135).

Recently, we have done serial sampling of bone marrow from 16 human volunteers to investigate circadian variations of cell cycle distribution of bone marrow cells by using flow cytometric technique (FCM). This rather extensive study was initiated because there was a general lack of data on biological rhythms in DNA synthetic activity and mitosis, as well as for stem and progenitor cell activity of human bone marrow, with the exception of the above-mentioned limited data of Killman, et al. (135) and Mauer (136). Our studies of the human bone marrow will be presented in some detail.

The cell cycle distribution of bone marrow cells from 16 healthy male volunteers (mean age = 33.7 years; range 19 - 47 years) was investigated, five of these persons undergoing the sampling procedure twice, making altogether twenty-one 24-hour periods. Bone marrow was sampled seven times by puncturing the sternum and anterior iliac crests in a randomised sequence every four hours during a single 24-hour period. Two parallel samples were stained at each time point. The two single cell suspensions were analysed on a Cytofluorograph 50 H (Ortho), interfaced to a Model 2150 Computer. For details concerning the method see Smaaland, et al., 1991 (100).

Venous blood was obtained from the same subjects at the same time as bone marrow sampling to determine peripheral blood parameters, including total and differential blood cell counts, in addition to cortisol measurements. The blood was obtained as the initial procedure or immediately after the anaesthesia of periost before the bone marrow puncture. In this way an artificially increased level of cortisol as a result of the puncture procedure itself was avoided (106). All individuals but two had followed a regular diurnal activity schedule for at least 3 weeks before the experiment. They continued their usual activities during the study period, apart from the sampling periods. They went to sleep after the 00:00 h sample was taken, but were awake for 15-30 minutes during the 04:00 h sampling.

The mean value of percent of cells in DNA synthesis (S-phase) of the two differently stained samples of bone marrow cells harvested at each time point showed a large variation along the circadian scale for all twenty-one 24-hour periods. The range of change from lowest to highest value during the 24-hour period for each subject varied between 29% and 339%, with a mean difference of 118%. The mean values of the periods with lowest and highest S-phase were 8.9%±0.5% (SE) and 17.6%±0.6%, respectively, i.e., a difference of nearly 100%.

Differences in phasing along the 24-hour period between the subjects were observed. Six examples of individual circadian stage-dependent variations of fraction of cells in DNA synthesis are shown in Figure 7. Lowest values are always seen between 20:00 and 04:00 h (late evening to midsleep). When pooling the data for all subjects for the mean S-phase values, a consistent pattern is seen, with a lower DNA synthesis around midnight as compared to the day (Figure 8). The observation period goes over 32 hours because the time of sampling started either at 08:00, 12:00 or at 16:00 h and continued for 24 hours. This makes it possible to measure the DNA synthesis for the pooled data for two consecutive day-periods, which corroborates the pattern of DNA synthesis values measured during daytime. Due to different phasing between the subjects, the difference between the lowest and highest average value is smaller than found for the individual subjects. However, the circadian stage-dependent variation is statistically significant, analysed both by analysis of variance (ANOVA) and Cosinor (85) methods; $p = 0.018$ and $p = 0.016$, respectively. The mean S-phase value of the 24-hour sampling period varied from 10.9% to 16.6% for the different individuals, i.e., a difference of 52.3%. The time of highest DNA synthesis (acrophase) estimated by cosinor analysis was at 13:16 h.

DNA analyses of samples from both sternum and iliac crests demonstrated the same circadian pattern of the S-phase. This finding rules out the possibility of different sampling sites being the reason for the observed circadian stage dependence of DNA synthesis and contradicts the possibility that the overall circadian rhythm detected could be attributed to a difference in level of S-phase dependent on sampling site. Also, the finding of nearly the same mean value of DNA synthesis from samples of the left and right iliac crests strongly indicates that the total red bone marrow must be

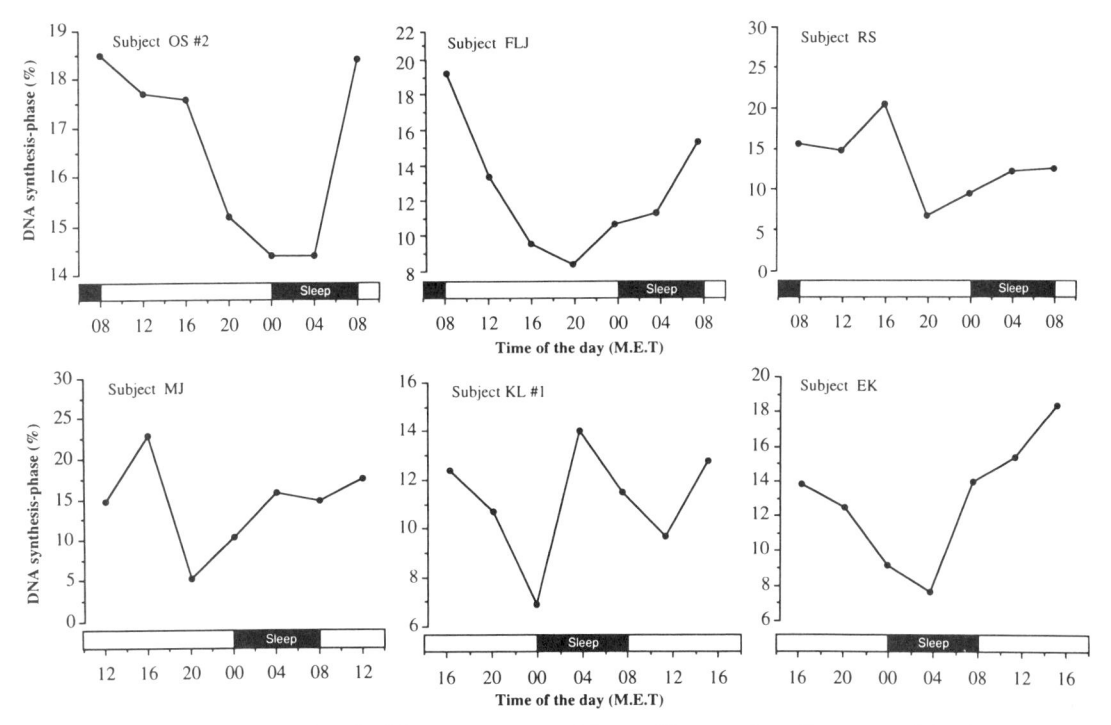

Figure 7. DNA synthesis variation along the 24 hour time span in six different subjects, sampling of bone marrow being done every four hours (N = 7 samples/subject). Results are expressed as the mean of two parallel analyses. The time of starting the experiments was randomised to 08.00, 12.00 and 16.00 hours. (From Smaaland, R., Laerum, O.D., Lote, K., Sletvold, O., Sothern, R.B. and Bjerknes, R. (1991) *Blood* **77**, 2603-2611. With permission.)

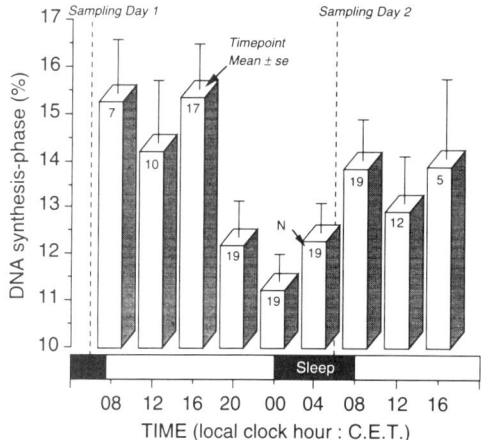

Figure 8. Circadian variation in human bone marrow DNA synthesis for *nineteen* 24 hour periods (Total N = 127).Timepoint means and standard errors are given. The time scale is extended for 32 hours due to the different times of starting each individual study (see text). (From Smaaland, R., Laerum, O.D., Lote, K., Sletvold, O., Sothern, R.B. and Bjerknes, R. (1991) *Blood* **77**, 2603-2611. With permission.)

looked upon as a functional entity. The demonstration of the same circadian variation in the bone marrow of the sternum as in the iliac crests further corroborates this functional homogeneity.

Our studies demonstrated a circadian stage-dependent variation in the proliferative activity of total bone marrow cells, i.e., DNA synthesis. The data corroborate the earlier findings of Mauer, who measured DNA synthetic activity by [3H]TdR uptake in myeloid cell and demonstrated highest and lowest [3H]TdR uptake in 3 of 4 individuals during day and evening, respectively (136). Thus, these two studies of direct measurement of circadian proliferative activity in the human bone marrow demonstrate the same circadian variation with regard to DNA synthetic activity.

At the present time we are studying subpopulations in the human bone marrow by use of multi-parameter flow cytometric analysis (141) in order to see if this all-over pattern is reflected in all lineages as well as in different stages of maturation. Preliminary data have verified that the various sub-populations of myelo-and erythropoiesis undergo similar variations, although their phasing may be slightly different (Figure 9) (142).

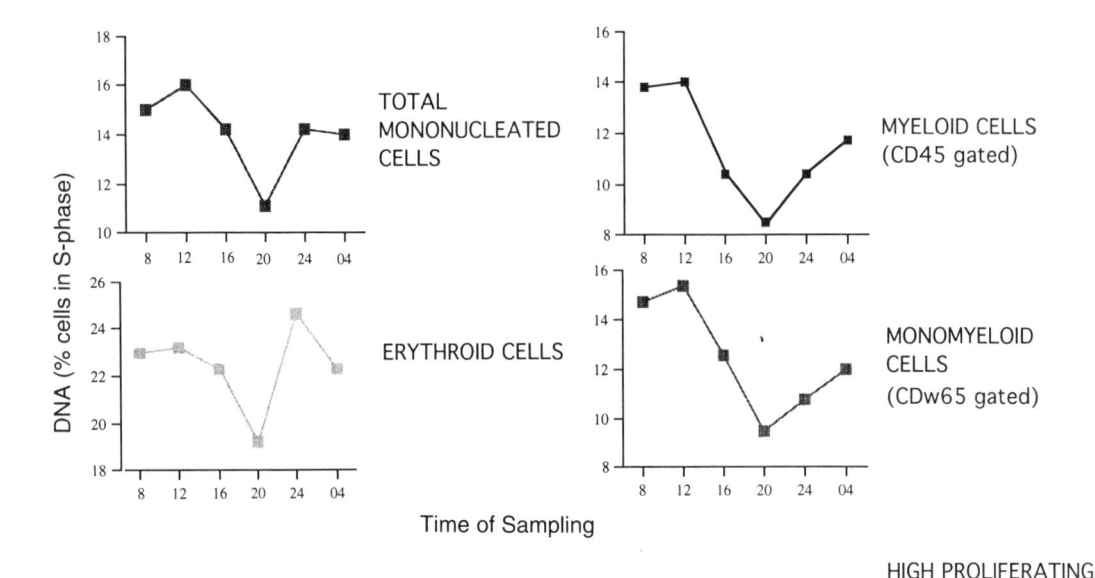

Figure 9. Different subpopulations of the bone marrow in a healthy person demonstrating some phase difference between cell subpopulations. A drop in the S-phase of the myeloid cells occurs before the drop in the erythroid cells.

A comparison of murine and human bone marrow rhythms

Based on natural evolutionary phylogenetic relationships, a certain degree of similarity should be expected in the timing of any rhythmic pattern between mice and men with regard to the rest/activity cycle. While an identical phasing of the same parameters in murine and human bone marrow relative to the rest/activity schedule is not a prerequisite, comparable circadian phasing would increase the importance of the argument in favour of extrapolating results on murine biological rhythms to the treatment of human cancer.

An important study in this regard was published in 1988 by Lévi et al. (59). He and his co-workers demonstrated a circadian variation in CFU-GM and CFU-F (fibroblastoid stroma cells) in murine bone marrow, with highest values of these two parameters of proliferation late in the activity (dark) span. In the same study a circadian toxicity rhythm of the anticancer agent 4'-O-tetrahydropyranyl (THP) adriamycin was demonstrated, with the lowest toxicity when CFU-GM and CFU-F were at the lowest, i.e., during the rest (light) span.

A population mean cosinor summary (143) for murine DNA and mitoses was run in order to allow better comparison of the overall timing of these bone marrow parameters in rodents and humans. Summary of the acrophases from the 6 studies on DNA resulted in a statistically-significant group rhythm ($p = 0.025$), with acrophase (±95% limits) at 17:08 (12:40, 22:36) HALO. This is approximately 5 hours after the beginning of the daily activity span, which is between 12-24 HALO, and is comparable to the timing reported for the two human studies available. Our cosinor analysis of data from Mauer (136) showed a significant circadian rhythm ($p = 0.009$) in bone marrow DNA of 4 men, with an acrophase at 11:21 h (09:16, 13:28 h). This acrophase was slightly more than 5 hours after the onset of activity in these men, who reportedly arose at 6 AM. Similarly, we have reported an acrophase at 13:04 h (09:32, 16:04 h) for a group of 16 men ($p = 0.004$), about 6 hours after the subjects' arising timing of 7 AM (138).

The animal data on circadian timing in mitoses and mitotic index seem to be more at variance with the limited corresponding human data. From the population mean cosinor summary of acrophases from 18 series, the group acrophase of mitoses/mitotic index in mice and rats was found during the rest span at 09:12 (05:20, 13:08) HALO ($p = 0.023$). This is more than 16 hours after or 8 hours before the group acrophase for DNA. In a study of 6 men by Mauer (138), we computed the acrophase for mitotic index to be at 21:09 h (19:08, 23:12 h) ($p = 0.005$), some 10 hours after the DNA measured in the same subjects, but still during late activity.

In summary, data for comparing the timing of mitoses and mitotic index in animals with that in humans are limited, with acrophases from human

CHAPTER 23 / CIRCADIAN RHYTHM OF CELL DIVISION

Figure 10. Distribution of cells in S-, G2/M- and S+G2/M-phase according to circadian stage (10.00 hours vs. 24.00 hours) for cancer patients with normal cortisol pattern and no bone marrow infiltration. The circadian variation in S-phase of bone marrow cells in these cancer patients suggests a circadian pattern similar to that of healthy male subjects sampled every 4 hours and shown for reference (Modified from Smaaland, R., Abrahamsen, J.F., Svardal, A.M., Lote, K., Ueland, P.M., *Br. J. Cancer*, **66**, 39, 1992.)

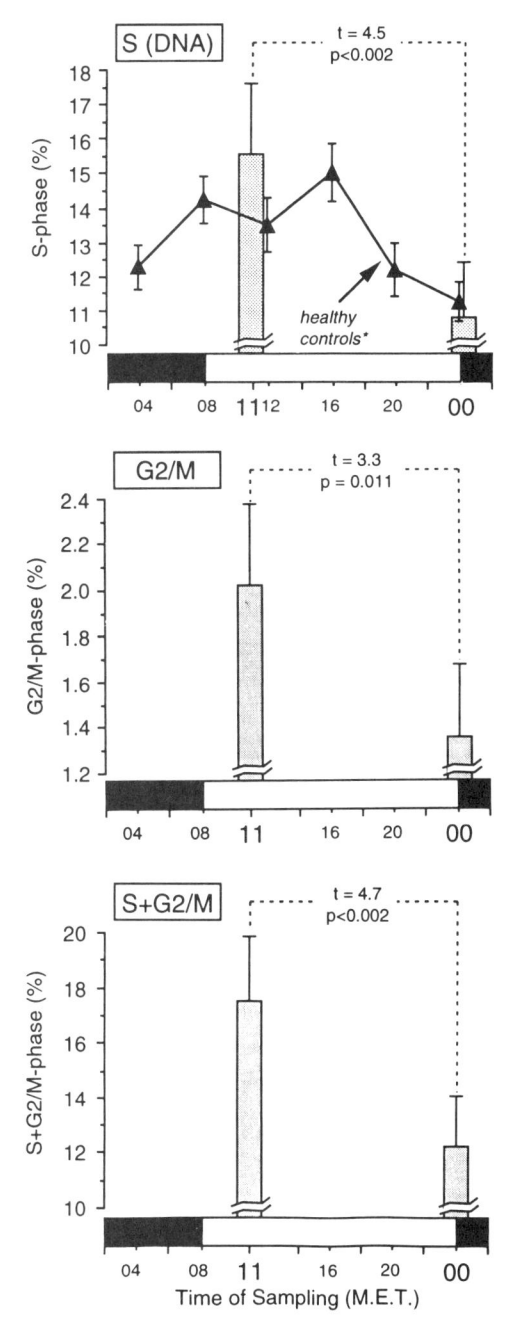

*6 men and 3 women with various malignancies. The control circadian variation in S-phase of bone marrow cells derived from 19 series in 16 healthy men sampled every 4-hrs for 24-hrs.

data of Killman and Mauer seemingly at variance with acrophases in mice and rats in relation to the rest/activity schedule.

Nevertheless, the accumulated data on the timing of the circadian variation in DNA synthesis of total bone marrow cells in animals and in humans strongly suggest a similarly-timed rhythm, with lowest activity during the rest span and, thereby, possibly a lower sensitivity to cytotoxic drug effects if given during this cell cycle phase.

Circadian variation in bone marrow DNA synthesis in cancer patients

In order to find out whether these results in healthy subjects were also valid in cancer patients in whom the circadian rhythmicity might be disturbed due to the malignant disease, we performed a study of 15 patients (6 women, 9 men; mean age 49.4 years, range 27-70 years) with various malignancies. All patients had a regular diurnal rest-activity schedule for at least 3 weeks prior to bone marrow sampling at two different times of the day, i.e., at 11 hours and midnight. A statistically significant higher fraction of cells in S-phase and G2/M-phase was found during the day as compared to midnight when excluding patients with an abnormal circadian variation in cortisol (Figure 10) (144). Thus, these data indicate that patients that are not too afflicted with disease do have a circadian variation in DNA synthesis which corresponds to that in healthy subjects.

REGULATION OF HAEMOPOIETIC CIRCADIAN RHYTHMS

Generally, studies on rhythmic variations in haemopoiesis have mainly been descriptive. Very little is known about regulatory aspects underlying these phenomena, although the regulation of erythropoiesis seems to follow a rhythmic pattern through the circadian variation of erythropoietin (16, 145). A temporal relation of DNA synthesis in marrow to cortisol rhythm in peripheral blood with a certain phase difference must be considered as a descriptive pattern and no proof of a causal relation (139). Accumulating data indicate, however, that growth factors may be involved in the temporal regulation of proliferative activity for different types of tissues, including haemopoietic tissue (Figure 2).

The pineal organ and its hormone melatonin play a central role in the control of the circadian organisation of vertebrates including human beings (146). A role for melatonin within vertebrate circadian systems has long been postulated from the largely circumstantial association with circadian patterns of various other physiological parameters (147). Evidence has been presented in the literature indicating that the pineal gland has a physiological role in the proliferation of granulocyte/macrophage colony-forming unit (CFU-GM). By pinealectomy and administration of melatonin to pinealectomised rats the rhythm of CFU-GM was obliterated or changed. Thus, the pineal gland or its main hormone melatonin seems to have a regulatory role in the temporal pattern of proliferation of CFU-GM in rat bone marrow cultures (148). Recent evidence has demonstrated that the circadian rhythms of many vertebrate species, including humans, are synchronised by the administration of exogenous melatonin as the suprachiasmatic nuclei are directly affected by melatonin (149). Indeed, in the superior control of circadian rhythms and control of regulatory cellular functions, the suprachiasmatic nucleus (SCN) is a central and highly specialised structure with distinct efferent and afferent connections and intrinsic anatomy. It also appears to have singular function - to provide temporal control over physiology and behaviour by generating circadian rhythms and integrating cyclic environmental information.

CIRCADIAN RHYTHMS OF CELL DIVISION AND IMPLICATIONS FOR OPTIMISING ANTICANCER TREATMENT

Cancer is overwhelmingly a dynamic disease. Tumour cells replicate and divide within a tissue that is largely of *de novo* origin and novel in its spatial and temporal organisation. The major shortcoming of classic non-circadian cytokinetics has been the inability to translate *in vitro* advances directly from the Petri dish to the cancer patient. This is due to the fact that it has been thought that the *in vitro* synchronised cell population behaved differently from *in vivo* populations, which were classically assumed to be non-rhythmic with regard to cytokinetic parameters. The idea of using cell kinetic information to optimise cancer chemotherapy is not new, but in the absence of compelling evidence to the contrary, it was previously applied with the underlying assumption that the tumour cell population and the dose-limiting normal tissues were growing asynchronously (150, 151). This stochastic cell kinetic view led to increasing dissatisfaction with cell kinetic based treatment strategies and DNA replication specific drugs (152, 153). So pervasive was the notion of random growth in situ, that numerous attempts to study drug perturbation of cytokinetics (154) or to enhance kinetic effects by synchronisation of the tumour (150), were made without the obvious control for natural synchrony.

However, in spite of its recent ontogeny, tumour growth and cell division are not necessarily random or stochastic processes. Cell replication can be stably entrained to the host circadian rhythm. Frequently, when these bursts in replication and division are not circadian in frequency, they may be of an ultradian frequency that is an integral submultiple of the circadian (155). The primary evidence of dynamic heterogeneity lies in the fact that DNA content, seemingly the most clonal and adamantine of cellular traits, changes on a time scale too rapid to be mutational. Largely unexplored is the role of frequency coding in the stimulation or inhibition of cell proliferation. The dynamics of cell growth and division is logically the core process in the development of tumours.

In the late 1960s and early 1970s the question of whether the rhythms in cell proliferation in sensitive high proliferating tissues like the intestinal tract and the bone marrow, could be used to improve experimental cancer chemotherapy, was thoroughly investigated. Badran and Echave-Llanos were among the first to start such investigations (156). They demonstrated the persistence of a circadian rhythm in the mitotic activity of mammary carcinoma in C3H/Mza mice and then administered cyclophosphamide between 12.00 and 16.00 hours, a time when the tumour mitotic index was high. A significant lower tumour volume was found with treatment in this time interval as compared to between 20.00-24.00 hours when the tumour activity was low ($p<0.001$).

Until very recently only four studies existed that specifically addressed the question of the proliferation dynamics of human tumours in situ (157-160). Human solid tumours, particularly in advanced disease, have been viewed as temporally disorganised masses, comprised of largely autonomous cells bearing little resemblance to the tissue of origin. In the early studies by Tähti and Voutilainen, studies that were not until very recently repeated, higher frequency, 8- and 12-hour rhythms in DNA replication and cell division were noted and contrasted with the expected 24-hour rhythms in normal tissues. Subsequent analysis of the data by the Cosinor method yielded an estimated time of peak mitotic index at 15.00 hours and a nadir at 03.00 hours.

Klevecz and co-workers have further demonstrated that tumour cell DNA replication is synchronous and stably entrained in human ovarian and gastrointestinal cancers in a significant fraction of patients studied; tumour cell DNA synthesis was found to take place out of phase or partially out of phase with non-tumour cell DNA synthesis (123, 161). The tumour and the normal cells were harvested by using an indwelling catheter implanted following surgical debulking in patients with ovarian cancer or other advanced cancers with gastrointestinal involvement. Though individual differences in circadian DNA synthesis were demonstrated, when data from all patients in the study were pooled and analysed by spectral analysis, patients with aneuploid or diploid tumours showed maximum S-phase fractions near midday (11.00 to 14.00 hours). The phase of peak DNA synthesis was not affected by surgery or by the use of saline lavage for removal of cells from the abdomen. Following surgery the baseline for the S-phase fraction in tumour cells increased for 5-7 days and then declined to pre-surgery levels. This increase was superimposed on the daily maximum (162). Of particular interest was the observation that in a patient with stage IV, grade 3 ovarian adenocarcinoma, positive at second look, the peak S-phase fraction appeared in the days immediately following surgery to be at 13.00 hours. When this same patient returned to begin therapy 3 weeks later, ascites cells were removed for 2 days prior to beginning treatment. Peak S-phase fraction in this series was at 11.00 hours, i.e., within 2 hours of the time of maximum seen in the first assess-ment (161). When individual patient data sets were analysed, 8- and 12-hours periods were common and these seemed to cluster at certain times of the day.

Klevecz et al. have also reported on recent protocols involving the diagnostic use of BUDR in advance of surgical debulking as well as in a phase-I trial of IUDR, with and without high-dose folinic acid, which have extended the understanding of circadian cell kinetics of human cancer. These studies lead to the conclusions that stable entrainment of tumour cell replication occurs frequently in human tumours in situ and that both circadian and ultradian phase relationships exist. Long-term (10 day), and repeat long-term analyses of halogenated pyrimidine incorporation into tumour cells suggest that a highly ordered spatial and temporal structure exists in the solid tumour (155).

Enlarged pathological lymph nodes in patients with non-Hodgkin's lymphomas are easily palpable and accessible. We have performed an around-the-clock study (109) in which lymphoma cells were sampled by fine-needle aspirations from palpable lymph nodes in either the cervical, axillae or groin regions in 26 patients (20 men and 6 women) with histologically established non-Hodgkin's lymphomas. Median age of the patients was 57 years (range: 23 - 91 years). The same subcutaneous pathological lymph node in each patient was punctured every 4 hours during a single 24-hour time span, resulting in 7 sampling times per subject. From the total of 182 samples collected around-the-clock, 161 (88.5%) gave a reliable histogram. Determination of cell cycle distribution was possible in 25 of the 26 patients and in one of these only two time points could be analysed. Thus, 24 patients were included in the circadian analyses of the different cell cycle phases.

The percentage variation in S-phase within each patient between the lowest and highest S-phase as compared to the lowest value (range of change = ROC) during the 24-hour time span varied from 21 to 353%, with a mean ROC of 128 ± 19% (median: 107%). No statistically significant difference in S-phase was found between the first (7.6 ± 1.7%) and last samples (7.4 ± 1.6%) in the 17 patients where samples were obtained 24 hours apart (p = 0.75). In 18 of 24 patients (75%), location of the acrophase representing higher S-phase was found during evening and night as compared to daytime (20:00 - 04:00 h vs. 08:00 - 16:00 h). This observation was validated when each individual S-phase series was converted to a percentage of the mean and combined for analysis by one-way ANOVA to test for time-effect across two 12-hour time spans (20-08 h $vs.$ 08-20 h). S-phase variation according to circadian stage was found to be statistically significant (p<0.004), with higher values found in the 20-08 h timespan. A oneway ANOVA across the six individual test-times resulted in borderline statistical significance (p = 0.059).

By single cosinor analysis of all data combined, S-phase yielded a marginally significant p-value of 0.069 for the least-squares fit of a 24-hour cosine to all data as percent of mean, with an amplitude of 8.5% and the acrophase found to be near midnight (00:05 h). The circadian pattern of S-phase in NHL thus appears to be out of phase with that reported by us in the bone marrow of healthy men. By analysing the data according to tumour stage, a circadian rhythm in S-phase was detected in lymphomas ≤stage IIB (n = 10, p = 0.046, amplitude = 17.0±6.7%, acrophase at 01:04 h, with confidence limits from 22:20 to 03:48 h). When examining the distribution of individual acrophases around-the-clock for DNA synthesis in malignant lymphomas with acrophases for DNA synthesis in healthy human bone marrow cells in our previous study an apparent difference in the timing of S-phase in the two populations was found. When comparing the number of acrophases for each group within the two 12-hour time spans, a significant difference in temporal distribution was found by a chi-square test - the S-phase

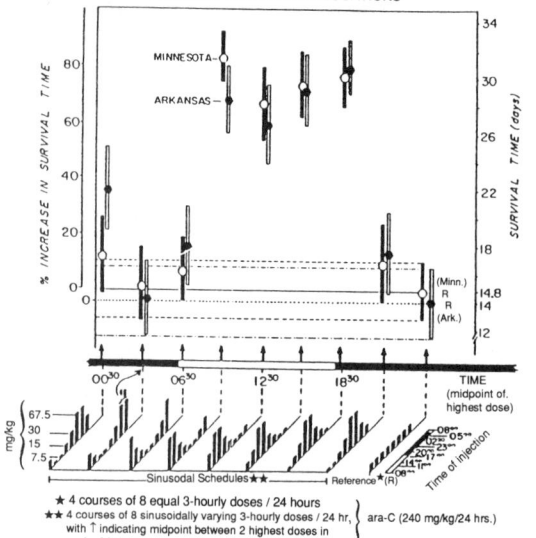

Figure 11. Survival time of CD2F1 mice on different administration schedules (top), and timing of doses of ara-C (bottom) in sinusoidal and reference (R) schedules. For all treatment schedules, four courses were given with a total dosage of 240 mg/kg per 24 hours. When the same total doses of ara-C were given, certain sinusoidal drug administration schedules were definitely better tolerated by mice than were other sinusoids or the conventional reference treatment schedule of eight equal doses over a 24-hour span. Also note the unequivocal reproducibility of chronotoxicity to ara-C in experiments done on the same days in different laboratories in different geographic locations (From Scheving, L.E., Haus, E., Kühl, J.F.W., Pauly, J.E., Halberg, F., Cardosos, S.S., *Cancer Res.*, 36, 1133, 1976. With permission).

acrophase was found between 20:00 - 08:00 h in 15 lymphomas and 3 bone marrow series, while there were 9 lymphomas and 16 bone marrow series with S-phase acrophase between 08:00 - 20:00 h (p = 0.002). Thus, obtaining tissue samples for determination of S-phase several times during a 24-hour period may thus help in the selection of the optimal circadian time of administering cytotoxic drugs for an individual.

In experimental chronotherapy Haus et al. have demonstrated better survival rates with a chronobiological approach of administering the S-phase specific drug arabinofuranosylcytosine (Ara-C) as compared to Skipper's extensive studies on L1210 leukemic mice in the late 1960s (163, 164). Furthermore, Scheving and co-workers have unequivocally demonstrated a circadian stage dependent toxicity of the drug Ara-C with a larger percentage of mice surviving at certain sinusoidal schedules and better than the non-circadian schedule, with all groups receiving the same total dose (Figure 11) (165). An even more impressive tumour effect has been accomplished by combining ara-C with cyclophosphamide in a chrono-modulated regimen, i.e., several fold more animals were cured and unacceptable acute toxicity of the host to the drug was eliminated (166). Also for the long-acting drug adriamycin has a significant circadian stage dependent variation in toxicity been documented (167), i.e., it was unequivocally demonstrated that maximum mortality to adriamycin occurred during the middark and that the host was more resistant to the drug during midlight.

Thus, the response to both short-acting and long-acting cytotoxic anticancer drugs is circadian-stage dependent. These differences in response to different drugs may be interpreted as reflecting the rhythmic changes in variables in the principal target organs of the drugs, for instance DNA synthesis in the cells of the bone marrow and intestinal tract, as well as rhythmic changes in the biochemistry of a large number of metabolic variables.

In addition to reduced mortality due to acute toxicity, preclinical studies have shown that an increase in effect on tumour or cure rate can also be obtained (164, 168-172), or that it is possible to eliminate or reduce drug-induced death due to toxicity, while still using an effective dose (166). Clinical studies have also demonstrated a circadian dependence of drug cytotoxicity to the bone marrow, resulting in fewer dose reductions, fewer treatment-related complications and fewer postponements of treatment courses when drugs have been administered at certain times of the day (173, 174), as well as increased long term survival (175). A reduced chance of relapse was found for acute lymphoblastic leukemia when giving maintenance cytotoxic therapy with 6-mercaptopurine (6-MP) in the evening (after 17:00 h) as compared to the morning (before 10:00 h) in 118 children (176). The original 8-year observation that the risk of relapsing was 4.6 times greater among children on the morning schedule as compared to those on the evening schedule has recently been confirmed in a 15-year disease-free survival analysis (177). The relative risk of relapse was still 2.6 times greater for children taking maintenance 6-MP in the morning (46% disease-free) as compared to the evening (64% disease-free). These studies strongly suggest that an increased dose intensity, as well as an increased therapeutic index, may be feasible by taking circadian rhythms of bone marrow cell proliferation into consideration when delivering myelosuppressant chemotherapy.

Unfortunately, it took almost 50 years from the time of the first observation of circadian variation in cell proliferation to its application on an experimental basis to cancer chemotherapy because conventional treatment strategies generally assumed some stochastic, steady-state process as responsible for bringing cells into S-phase. Clearly, the dynamics of replication in tissues and tumours are far more complex than was once thought (178) and it seems that our understanding would benefit from a molecular definition of the temporal structure of S-phase. More than ever, the critical mass of experimental data leads to the obvious conclusion that attempts must be intensified to bring the concept of chronochemotherapy to the clinic.

We have only now reached the stage when most scientists accept the existence and the potential importance of circadian susceptibility rhythms, even though they may still disregard them in their own experimental design. The realisation that there is relative synchrony along the circadian time scale within all *in vivo* cell populations is the only solution if cytokinetics is ultimately to become a practical, clinically relevant tool for treating cancer patients.

Thus, a better understanding of the mechanism of both normal and abnormal cell division *in vivo* can be expected only if its underlying rhythmicity is considered. As positive data on human beings accumulates, as it clearly already has in animal studies, it ultimately may become an ethical (and possibly a legal) issue as to whether the cancer patient will be treated with or without chronobiological consideration. As advances in technology such as programmable pumps develop, chronotherapy has become a logistically feasible approach to treatment at optimal circadian stages (179).

What may be of benefit in refining our understanding of the requirements for successful application of timing strategies is the use of anticancer drugs, e.g., antimetabolites or analogues that have more quantifiable molecular and cell biological mechanisms. The use of DNA replication specific drugs such as iododexouridine (IUDR), ara-C, or bromo-deoxyuridine (BUDR) under circum-stances where the times of replication of the major dose-limiting tissues or the tumour, or both, are known, may provide the most direct way of testing the efficacy of this strategy. Drugs such as floxuridine (FUDR), with somewhat more complex pharmakokinetics have yielded results consistent with our understanding of circadian gating of cell proliferation (180).

Some fraction of the population of cells will be replicating DNA even at circadian minimum, but maximally only in a restricted period. In the many studies of the prognostic value of the S-phase fraction, if S-phase fractions from single haphazard time samplings (samplings cannot be considered random around the clock, because they are obtained at clinically convenient times) of solid tumours are used to form the basis for the prognostic value of the S-phase fraction, it would be surprising if such data correlated at all with the outcome. Indeed, contradictions in this area are frequent and may be a result of time sampling differences.

Moreover, one should be aware of the fact that nocturnal rodents have a partially to completely inverted chronotoxicity curve compared to humans. All of the drugs tested and found to be not-overly-toxic and efficacious to rodents would be those that were nontoxic in the light phase of the light/dark cycle (the rodent's sleep phase). It is just possible that new drugs being launched in phase-I or phase-II clinical trials have been administered at the wrong time of day. This leaves the possibility that efficacious and useful drugs have not been entered into the clinic due to the failure of the scientific community to take circadian aspects of biology, e.g., cell division, into consideration.

The notion that we can extrapolate cell kinetic information directly from rodent systems to human cancer treatment encounters several problems. Rodents are nocturnal, requiring at the least a 12-hour inversion of all results. However, the phase relationship for DNA replication in the two systems may not be taken care of by a simple 12-hour phase shift in scheduling. Moreover, the data for normal tissue proliferation are not adequate if the phase relationship between toxicities is subtle, close, or complicated significantly by pharmacokinetics/ pharmacodynamics. Even more problematical is the possibility that trans-plantable tumours in rodents often lack circadian rhythmicity and display difficult to reproduce high-frequency rhythms or no significant rhythm, whereas human tumours seem in many cases to be entrained.

Thus, biological rhythms along ultradian, circadian, and infradian time scales characterise the cytokinetics of both healthy and tumour tissue (181), as well as the susceptibility of tumour and target organs to cytotoxic agents and their pharmacokinetics (182). Circadian time keeping mechanisms may also be present in tumour cells, although they are not always synchronised and/or exhibit a phasing which differs from that of normal cells (157, 183, 184). All of these aspects will eventually need to be considered when attempting to optimise the administration schedule of an anticancer agent. Better use of anticancer agents may result from integrating a chronopharmacologic strategy at an early stage of drug development.

Most cytotoxic drugs interact with the DNA of target cells. As a result of these interactions, the replicative and/or transcriptional functions of DNA are impaired (185), thereby causing cell death or reduced proliferative rate. It is only relatively recently that analyses of drug effects on cells at various phases of the cell cycle became a common practice in programs of drug evaluation. Results of these studies indicate that the lethal effects of short term exposure of cells to various drugs are correlated with the cell position in the cell cycle during the exposure (186-188).

In our extensive studies, the proliferative activity of all nucleated bone marrow cells were measured (138, 189, 190), including erythroid cells, which have been shown to have a high proliferative activity (142, 191). Therefore, by scheduling the administration of cytotoxic drugs to the circadian variation in DNA synthesis of these cells as well, fewer patients may require blood transfusions, as anemia is an increasing problem with increasing treatment courses.

Bone marrow still remains the primary normal target biosystem for most cytostatic drugs. In animal studies it has been shown again and again that protection of host normal tissues, including the bone marrow, can be achieved for different cytotoxic agents by determining host toxicity rhythms along the 24-hour time scale. Optimal circadian therapeutic schedules can then be devised for the administration of cytotoxic drugs (192). Thus, the ultimate goal of circadian studies of haematopoiesis in humans is to employ cancer treatment in a similar chronomodulated way (106).

By taking circadian stage dependent variations in DNA synthesis into account it may be possible to reduce bone marrow toxicity of S-phase cytotoxic drugs by administering the drugs or the major dose of a continuous drug infusion during the time of lowest proliferative activity, i.e., late in the evening or during the night (during the late activity or resting/sleeping spans for diurnally-active humans). Cells in the DNA synthesis phase will then be less susceptible, and cells in the G0/G1 phase will have more time for repairing damage before entering into the DNA synthesis phase. This point is underscored by the studies of Haus (164) and Scheving (165) demonstrating in mice that cytosine arabinosyl (Ara-C), an S-phase specific drug, has its lowest toxicity during the light span, i.e., during the time of resting/sleeping for nocturnally-active rodents. This contention is additionally corroborated by the study of Lévi et al., mentioned previously (59), which demonstrated that the lowest DNA synthesis and CFU-GM during the resting phase in mice coincided with the time of lowest toxicity to the drug THP-adriamycin, a drug mainly affecting DNA synthesis.

Certain drugs, however, may have their least toxic effects at other times during the 24 hour period, i.e. alkylating agents (185), but this may easily be adjusted for when knowing the circadian variation of the DNA synthesis, as the distribution of cells within the cell cycle will be interdependent. It has been suggested that cells alkylated during G1/S-phase could be more sensitive because they have a lower probability of repairing potentially lethal damage before the next phase of synthesis than those in G2/M-phase (193). Another aspect of DNA repair processes which could contribute to such cell cycle effects is the possibility that repair rates themselves vary in the different phases of the cell cycle.

It is of interest that the time of highest DNA synthesis in non-Hodgkin's lymphomas coincides with the time of lowest DNA synthesis, as well as lowest number of CFU-GM in healthy human bone marrow (184). Recently, a circadian rhythm in DNA synthesis in human rectal mucosa has been reported in healthy men (98). In that study, highest proliferative activity, as reflected by in vitro [3H]TdR uptake, was found near the time of awakening at 07:00 h. Together, these findings on differences in the circadian timing of proliferation in healthy and tumourous cells suggest the possibility of achieving the specification of a therapeutic window in time within which an increase in dosage and cytotoxic effect, together with a decrease in undesired toxicity, might occur. The results imply that a chronotherapeutic administration of all or the major portion of an S-phase s pecific drug (or drug with its main effect on S-phase) during the time of highest DNA synthesis in lymphoma cells and lowest DNA synthesis in healthy tissues such as the human bone marrow and the gastrointestinal mucosa, i.e., during late evening and night, could result in a maximised therapeutic index.

Nevertheless, as tumour cell proliferation is likely to be more autonomous, depending on differentiation grade, heterogeneity of the tumour and tumour burden, the timing of cancer therapy should still primarily be adjusted to the temporal pattern of proliferation in normal tissues, e.g., the bone marrow and the gastrointestinal tract, which are the major dose limiting tissues today relative to cytotoxic therapy. This approach is supported by the fact that circadian variation in normal cell proliferation is coupled to a broad spectrum of circadian rhythmicity in the body, which may remain stable even as treatment continues.

Not only proliferative rhythms of normal and tumour tissue, but also metabolic and/or pharmacokinetic circadian changes must be considered when optimising treatment according to the circadian system to improve the therapeutic index (194, 195, 196, 197, 198), e.g., temporal changes in drug absorption and distribution, temporal aspects in drug metabolism, membrane permeability, plasma protein binding, blood flow changes in the target organs, glomelular filtration/renal elimination and pH changes.

The time factor may be even more relevant to the effective use of biological response modifiers. The circadian stage ("time of day") when interleukins, interferons, and growth factors are given will ultimately determine how effective these factors will be (16, 17, 18, 19). It is therefore of importance to define the time structure of the pharmacodynamic effects of biological agents. A potentially important consequence of the circadian pattern and phasing of bone marrow DNA synthesis, as well as CFU-GM, is the possibility of increasing the effect of biological response modifiers like GM-CSF, G-CSF, and regulatory peptides (19, 60, 110, 199) by administering the optimal dose at the time of greatest responsiveness of the bone marrow. This may increase the usefulness and efficacy of these biological substances, and also possibly reduce side effects.

Finally, the temporal variations in bone marrow cytokinetics suggest that the rate of success of bone marrow auto- or allografting may be enhanced with careful selection of not only the sampling time of the bone marrow from the donor, but also of the administration time to the recipient. The use of recombinant hematopoietic growth factors to accelerate the regeneration of granulocyte numbers after conventional or high dose cytotoxic therapy with or without autologous/allogeneic bone marrow transplantation, may further take advantage of biological rhythms of the bone marrow by increasing therapeutic effects or reducing side effects. An optimal use of these growth factors relative to biological rhythms may be potentially beneficial for patients with acquired immunodeficiency syndrome, myelodysplastic syndrome, or aplastic anemia (200).

In conclusion, knowledge about predictable temporal changes of bone marrow proliferation *in vitro* and *in vivo* may not only be useful for understanding the regulation of bone marrow proliferation, but also for screening optimal dosing time of cytotoxic agents, hematopoietic growth factors and for choosing the right time for harvesting and delivering bone marrow cells for auto-/allografting. The possibility of treating patients at optimal circadian time, interval and schedule is today feasible and cost-effective through programmable delivery systems (172, 201, 202, 203).

Circadian cytokinetics, as well as circadian aspects of pharmacokinetics, endocrinology, immunology, and toxicology can no longer be ignored, but must be used to treat human cancer more efficiently.

REFERENCES

1. Edmunds, Jr. L.N. and Laval-Martin, D.L. (1982) in *Cell cycle clocks* (Edmunds Jr., L. N. ed.), pp. 295-324, New York, Marcel Dekker.
2. Touitou, Y. and Haus, E. (1992) *Biologic rhythms in clinical and laboratory medicine* Berlin, Springer-Verlag.
3. Smaaland, R. and Sothern, R.B. (1994) in *Circadian cancer therapy* (Hrushesky, W.J.M. ed.) Boca Raton, CRC Press.
4. Hastings, M. (1995) *Nature* **376**, 296-297
5. Cross. F., Roberts, J. and Weintraub, H. (1989) *Annu Rev Cell Biol.* **5**, 341-395.
6. Murray, A.W. and Kirschner. M.W. (1989) *Science* **246**, 614-621.
7. Lewin, B. (1990) *Cell* **61**, 743-752.
8. King, R.W. Jackson, P.K. and Kirschner, M.W. (1993) *Cell* **79**, 563-571.
9. Sherr, C.J. (1994) *Cell* **79**, 551-555.
10. Coleman, T.R. and Dumphy, W.G. (1994) *Curr. Opin. Cell Biol.* **6**, 877-882.
11. Sherr, C.J. and Roberts, J.M. (1995) *Genes Dev.* **9**, 1149-1163.
12. Peters, G. (1994) *Nature* **371**, 204-205.
13. Polyak, K., Lee, M.H., Erdjument-Bromage, H., Tempst, P. and Massague, J. (1994) *Cell* **78**, 59-66.
14. Kato, J., Matsuoka, M., Polyak, K., Massague, J. and Sherr, C.J. (1994) *Cell* **79**, 487-496.
15. Hengst, L. and Reed, S.I. (1996) *Science* **271**, 1861-1864.
16. Wood, P.A. and Hrushesky, W.J.M. (1994) in *Circadian cancer therapy* (Hrushesky, W. J. M. ed.) Boca Raton, CRC Press.
17. Young, M.R.I., Matthewa, J.P., Kanabrocki, E.L., Sothern, R.B., Roitman-Johnson, B. and Scheving, L.E. (1995) *Int. J. Chronobiol.* **12**, 19-27.
18. Sothern, R.B., Roitman-Johnson, B., Kanabrocki, E.L., Yager, J.G., Roodell, M.M., Weatherbee, J.A., Young, M.R.I., Nemchausky, B.M. and Scheving, L.E. (1995) *J. Allergy Clin. Immunol.* **95**, 1029-1035.
19. Perpoint, B., Le Bousse-Kerdiles, C., Clay, D., Smadja-Joffe, F., Depres-Brummer, P.,

Laporte-Simitsidis, S., Jasmin, C. and Lévi, F. (1995) *Exp. Hematol.* **23**, 362-368.
20. Cos, S. and Sanches-Barcelo, E.J. (1994) *Cancer lett.* **85**, 105-109.
21. Creanor, J. and Mitchison, J.M. (1986) *J. Cell Sci.* **86**, 207-215.
22. Novak, B. and Mitchison, J.M. (1986) *J. Cell Sci.* **86**, 191-206.
23. Whitfield. J.F., Durkin, J.P., Franks, D.J., Kleine, L.P., Raptis, L., Rixon, R.H., Sikorska, M. and Walker, P.R. (1987) *Cancer Metast. Rev.* **5**, 205-250.
24. Dumont, J.E., Jauniaux, J.C. and Roger, P.P. (1989) *Trends Biochem. Sci.* **14**, 67-71.
25. Boynton, A.L. and Whitfield, J.F. (1983) *Adv. Cyclic Nucleotide Res.* **15**, 193-294.
26. Matsumoto, K., Uno, I. and Ishiakawa, T. (1985) *Yeast* **1**, 15-24.
27. Carell, E.F. and Deardfield, K.L. (1982) *Life Sci.* **31**, 249-254.
28. Carré, I.A., Laval-Martin, D.L. and Edmunds, Jr. L.N. (1989) *J. Cell Sci.* **94**, 267-272.
29. Edmunds, Jr. L.N., Carré, I.A., Tamponnet, C. and Tong, J. (1992) *Chronobiol. Int.* **9**, 180-200.
30. Carré, I.A. and Edmunds, Jr. L.N. (1990) Cyclic AMP, cell division cycles, and circadian oscillators in Euglena. Bethesda, MD. ASCB/EMBO Conference on "The Cell Cycle".
31. Edmunds, Jr. L.N. and Carré, I.A. (1990) Cyclic AMP, cell division cycles, and circadian oscillators in Euglena, 37. 2nd. meeting of the Society for Research on Biological Rhythms. Amelia Island, Jacksonville, FL.
32. Draetta, G., Luca, F., Westendorf, J., Brizuela, L., Ruderman, J. and Beach, D. (1989) *Cell* **56**, 829-838.
33. Edmunds, Jr. L. (1993) Signal transduction between circadian clock and cell division cycle. I-4, 21st. Conference of the International Society for Chronobiology. Quebec City.
34. Fortuyn-Van Leyden, C.E.D. (1917) *Proc. kon. nederl. Akad. Wet. (c)* **19**, 38-44.
35. Pilgrim, C., Erb, W. and Maurer, W. (1963) *Nature* **199**, 863-865.
36. Halberg, F., Halberg, E., Barnum, C.P. and Bittner, J.J. (1959) in *Photoperiodism and Related Phenomena in Plants and Animals* (Withrow, R. B. ed.) pp. 803-878, Ed. Publ. No 55 of the Amer. Assoc. Adv. Sci. Washington D.C.
37. Thorud, E., Clausen, O.P.F. and Laerum, O.D. (1984) in *Cell cycle clocks* (Edmunds, L. N. j. ed.) pp. 113-133, New York, Basel, Marcel Dekker, Inc.
38. Cooper, Z.K. and Schiff, A. (1938) *Proc. Soc. exp. Biol. (N.Y.)* **39**, 323-324.
39. Bullough, W.S. (1948) *Proc. roy. Soc. Lond. (Biol.)* **135**, 212-232.
40. Scheving, L.E. (1959) *Anat. Rec.* **135**, 7-19.
41. Evensen, A. (1963) *Bull. WHO* **28**, 513-515.
42. Rubin, N. and Scheving, L.E. (1983) *J. invest. Dermatol.* **80**, 79-80.
43. Frentz, G., Moller, U., Holmich, P. and Christensen, I.J. (1991) *Acta Derm. Venereol.* **71**, 85-87.
44. Tvermyr, E.M.F. (1972) *Virch Arch (Cell Pathol)* **11**, 43-54.
45. Clausen, O.P.F., Thorud, E. and Kaufman, S.L. (1982) *Int. J. Chronobiol.* **8**, 83-93.
46. Møller, U. and Keiding, N. (1982) *Cell Tiss. Kinet.* **15**, 341-350.
47. Møller, U., Keiding, N. and Engel, F. (1982) *Cell Tiss. Kinet.* **15**, 157-168.
48. Brown, W.R. (1991) *J. Invest. dermatol.* **97**, 273-280.
49. Halberg, F., Barnum, C.P., Silber, R.H. and Bittner, J.J. (1958) *Proc. Soc. Exp. Biol. (N.Y.)* **97**, 897-900.
50. Souto, M. and Echavellanos, J.M. (1985) *Chronobiol. Int.* **2**, 169-175.
51. Scheving, L.E. and Pauly, J.E. (1973) *Int. J. Chronobiol.* **1**, 269-286.
52. Scheving, L.E., Pauly, J.E., Von Mayersbach, H. and Dunn, J.D. (1974) *Acta Pathol.* **88**, 411-423.
53. Refsum, S.B., Håskjold, E., Bjerknes, R. and Iversen, O.H. (1991) *Virchows Arch. B. Cell Pathol. Incl. Mol. Pathol.* **60**, 225-230.
54. Benestad, H.B. and Laerum, O.D. (1989) in *Current Topics In Pathology* (Iversen, O.H. ed.) **79**,7-36, Berlin, Springer-Verlag.
55. Canon, C., Lévi, F., Reinberg, A. and Mathé, G. (1985) *Leukemia Res.* **9**, 1539-1546.
56. Lévi, F., Canon, C., Blum, J.P., Mechkouri, M., Reinberg, A. and Mathé, G. (1985) *J. Immunol.* **134**, 217-222.
57. Bartlett, P., Haus, E., Tuason, T., Sackett-Lundeen, L. and Lakatua, D. (1982)in *Proc. 15th international Conference on Chronobiology* (Haus, E. and Kabat, H. F. eds) Basel, Karger, S.
58. Laerum, O.D. and Aardal, N.P. (1985) in *Clinical Aspects of Chronobiology* (Rietveld, W. J. eds.) pp. 85-97, Leiden, Meducation Service Hoechst.
59. Lévi, F., Blazcek, I. and Ferlé-Vidovic, A. (1988) *Exp. Hematol.* **16**, 696-701.
60. Laerum, O.D; and Paukovits, W.R. (1984) *Exp. Hematol.* **12**, 7-17.
61. Sletvold, O., Laerum, O.D. and Riise, T. (1988) *Eur. J. Haematol.* **40**, 42-49.
62. Sletvold, O., Laerum, O.D. and Riise, T. (1988) *Mech. Aging Dev.* **42**, 91-104.

63. Klein, C.S. and Geisel, H. (1947) *Klin. Wochenschr.* **25**, 662-.
64. Raitsina, S.S. (1961) *Bull. Exp. Biol. Med.* **51**, 494-497.
65. Gololobova, M.T. (1958) *Bull. Exp. Biol. Med.* **40**, 1143-.
66. Loisner, L.D., Artemiva, N.S. and Babaeva, A.G. (1962) *Bull. Exp. Biol. Med.* **54**, 77-.
67. Sinha, H.P. (1960) *Patna. J. Med.* **34**, 301-.
68. Clark, R.H. and Baker, B.L. (1963) *Am. J. Physiol.* **204**, 1018-1022.
69. Sigdestad, C.T., Bauman, J. and Lesher, S. (1969) *Exp. Cell Res.* **538**, 159-162.
70. Sigdestad, C.T. and Lesher, S. (1970) *Experientia (Experimenta)* **26**, 1321-1322.
71. Leblond, C.P. and Stevens, C.E. (1948) *Anat. Rec.* **100**, 357-377.
72. Bertalanfy, F.D. (1960) *Acta Anat.* **40**, 130-.
73. Hunt, T.E. (1952) *Anat. Rec. (Abstr.)* **112**, 346.
74. von Muhlemann, H.R., Marthaler, T.M. and Rateitschak, K.H. (1956) *Acta Anat.* **28**, 331.
75. Scheving, L.E., Burns, E.R. and Pauly, J.E. (1972) *Am. J. Anat.* **135**, 311-317.
76. Appleton, D.R., Sunter, J.P. and Watson, A.J. (1981) *Cell proliferartion in the gastrointestinal tract.* London, Pitman Medical Ltd.
77. Scheving, L.E., Tsai, T.H. and Scheving, L.A. (1983) *Amer. J. Anat.* **168**, 433-465.
78. Kennedy, M.F.G., Tutton, P.J.M. and Barkla, D.H. (1985) *Cancer Letters* **28**, 169-175.
79. Kaur, P. and Potten, C.S. (1986) *Cell Tiss. Kinet.* **19**, 591-599.
80. Scheving, L.E., Feuers, R.J., Tsai, T.H. and Scheving, L.A. (1994) in *Circadian cancer therapy* (Hrushesky, W.J.M. ed.) pp. 19-40, Boca Raton.
81. Scheving, L.E., Burns, E.R., Pauli, J.E. and Tsai, T.H. (1978) *Anat. Rec.* **191**, 479-486.
82. Scheving, L.E., Tsai, T.H. and Scheving, L.A. (1983) *Am. J. Anat.* **168**, 433-465.
83. Thorud, E., Clausen, O.P.F. and Bjerknes, R. (1978) in *Pulse Cytometry* (Lutz, D. ed.) **3**, 359, Ghent, European Press.
84. Møller, U. and Larsen, J.K. (1978) *Cell Tissue Kinet.* **12**, 405-413.
85. Clausen, O.P.F., Thorud, E., Bjerknes, R. and Elgjo, K. (1979) *Cell Tissue Kinet.* **12**, 319-337.
86. Laerum, O.D. and Aardal, N.P. (1981) in *11th International Congress of Anatomy, part C, Biological rhythms in structure and function* (von Mayersbach, H., Scheving, L.E.and Pauli, J.E. eds.) pp. 87-97, New York, Alan R. Liss.
87. Rubin, N.H. (1981) in *Biological Rhythms in Structure and Function* (von Mayersbach, H., Scheving, L.E. and Pauly, J.E. eds.) pp. 82-, New York, Alan R. Liss.
88. Rubin, N.H. (1982) *Radiat. Res.* **89**, 65-76.
89. Rubin, N.H., Hokanson, J.A. and Mayshak, J.W. (1983) *Am. J. Anat.* **168**, 15-26.
90. Scheving, L.E., Scheving, L.A., Tsai, T.H. and Pauly, J.E. (1984) *J. Nutr.* **114**, 2160-2166.
91. Quastler, H. and Sherman, F.G. (1959) *Exp. Cell Res.* **17**, 420-.
92. Burns, E.R. and Scheving, L.E. (1975) *Cell Tissue Kinet.* **8**, 61-66.
93. Burns, E.R., Scheving, L.E., Fawcett, D.F., Gibbs, W.M. and Galatzan, R.E. (1976) *Anat. Rec.* **184**, 265-273.
94. Scheving, L.E., Tsai, T.H., Scheving, L.A., Feuers, R.J. and Kanabrocki, E.L. (1992) in *Biologic Rhythms in Clinical and Laboratory Medicine* (Touitou, Y. and Haus, E. eds.) **566**, Berlin, Springer-Verlag.
95. Warnakulasuriya, K.A. and MacDonald, D.G. (1993) *Arch. Oral Biol.* **12**, 1107-1111.
96. Warnakulasuriya, K.A. and MacDonald, D.G. (1995) *Arch. Oral Biol.* **40**, 107-110.
97. Khan, S., Raza, A., Petrelli, N. and Mittleman, A. (1988) *J. Surg. Oncol.* **39**, 114-118.
98. Buchi, K.N., Moore, J.G., Hrushesky, W.J.M., Sothern, R.B. and Rubin, N.H. (1991) *Gastroenterology* **101**, 410-415.
99. Potten, C.S., Chwalinski, S., Swindell, R. and Palmer, M. (1982) *Cell Tiss. Kinet.* **15**, 351-370.
100. Marra, G., Anti, M., Percesepe, A., Armelao, F., Ficarelli, R., Coco, C., Rinelli, A., Vecchio, F.M .and D'Arcangelo, E. (1994) *Gastroenterology* **106**, 982-987.
101. Beck, L., Reick, D.V. and Damann, B. (1959) *Proc. Soc. Exp. Biol. Med.* **97**, 229-231.
102. Lohrman, H-P. and Schreml, W. (1982) in *Cytotoxic drugs and the granulopoietic system* **81**, 155-182, Berlin, Heidelberg, New York., Springer-Verlag.
103. Gordon, M.Y., Barrett, A.J. and Gordon-Smith, E.C. (1985) *Bone marrow disorders. The biological basis of clinical problems* pp. 1-422, Oxford, Blackwell Scientific.
104. Arendt, J., Minors, D.S. and Waterhouse, J.M. (1989) in *Biological rhythms in clinical practice.* (Arendt, J., Minors, D. S. and Waterhouse, J. M. eds.) pp. 3-7, London, Boston, Singapore, Sydney, Toronto, Wellington, Wright.
105. Butcher, E.C. (1990) *Am. J. Pathol.* **136**, 3-11.
106. Laerum, O.D. (1995) *Exp. Hematol.* **23**, 1145-1147.
107. Fisher, L.B. (1968) *Br. J. Dermatol.* **60**, 75.
108. Spivak, J.L. (1984) in *The principles and Practice of Medicine* (Harvey, A. M., Johns, R. J., McKusick, V. A., Owens, A. H. and

Ross, R. S. eds.) 21st ed., 465, Norwalk, Appleton-Century-Crofts.
109. Nathan, D.G. (1987) in *Oxford Textbook of Medicine* (Weatherall, D.J., Ledingham, J.G.G. and Warrell, D.A. eds) 2nd. ed. **2**, Oxford University Press.
110. Laerum, O.D. and Paukovits, W.R. (1989) *Pharmac. Ther.* **44**, 335-349.
111. Gordon, M.Y., Barrett, A.J. and Gordon-Smith, E.C. (1985) *Bone marrow disorders. The biological basis of clinical problems* 1-422, Oxford, Blackwell Scientific.
112. Pizzarello, D.J. and Witcofski, R.L. (1970) *Radiology* **97**, 165-167.
113. Burns, E.R. (1981) *Cancer Res.* **41**, 2795-2802.
114. Sharkis, S.J., LoBue, J., Alexander, P.J., Rakowitz, F., Weitz-Hamburger, A. and Gordon, A.S. (1971) *Proc. Soc. Exp. Biol. Med.* **138**, 494-496.
115. Sharkis, S.J., Palmer, J.D., Goodenough, J., LoBue, J. and Gordon, A.S. (1974) *Cell Tiss. Kinet.* **7**, 381-387.
116. Moskalik, K.G. (1976) *Bull. Exp. Biol. Med. (USSR)* **81**, 594.
117. Dörmer, P., Schmolke, W., Muschalik, P. and Brinkman, W. (1970) *Beitr. Pathol.* **141**, 174-186.
118. Stoney, P.J., Halberg, F. and Simpson, H. W. (1975) *Chronobiologia* **2**, 319-323.
119. Aardal, N.P., Laerum, O.D. and Paukovits, W.R. (1982) *Virch. Arch. B Cell Pathol.* **38**, 253-261.
120. Aardal, N.P. (1984) *Exp. Hematol.* **12**, 61-67.
121. Aardal, N.P. and Laerum, O.D. (1983) *Exp. Hematol.* **11**, 792-801.
122. Smaaland, R. and Sothern, R.B. (1994) in *Circadian Cancer Therapy* (Hrushesky, W.J.M. ed.) pp. 119-163, Boca Raton, CRC Press.
123. Klevecz, R.R., Shymko, R.M., Blumenfeld, D. and Braly, P.S. (1987) *Cancer Res.* **47**, 6267-6271.
124. Hromas, R.A., Hutchins, J.T., Marke, D.E. and Scholes, V.E. (1981) *Chronobiologica* **8**, 369-373.
125. Clark, R.H. and Korst, D.R. (1969) *Science* **166**, 236-237.
126. Scheving, L.E. and Pauly, J.E. (1973) *Int. J. Chronobiol.* **1**, 269-286.
127. Hunt, N.H. and Perris, A.D. (1974) *J. Endocrinol.* **62**, 451-462.
128. Haus, E., Lakatua, D.J., Swoyer, J. and Sackett-Lundeen, L. (1983) *Am. J. Anat.* **168**, 467-517.
129. Scheving, L.A., Tsai, T.H. and Scheving, L.E. (1986) *Chronobiol. Intl.* **3**, 1-15.
130. Vacek, A. and Rotkovska, D. (1970) *Strahlentherapie* **140**, 302-306.
131. Simmons, D.J., Loeffelman, K., Frier, C., McCoy, R., Friedman, B., Melville, S. and Kahn, A.J. (1984) in *Chronobiology* (Haus, E. and Kabat, H. eds.) pp. 37-42, Basel, S.Karger.
132. Halder, C., Häussler, D. and Gupta, D. (1992) *J. Pineal Res.* **12**, 79-83.
133. Goldeck, H. (1948) *Ärztl Forsch* **2**, 22-27.
134. Goldeck, H. and Siegel, P. (1948) *Ärtzl Forsch* **2**, 245-248.
135. Killmann, S-Å., Cronkite, E.P., Fliedner, T.M. and Bond, V.P. (1962) *Blood* **19**, 743-750.
136. Mauer, A.M. (1965) *Blood* **26**, 1-7.
137. Lasky, L.C., Ascencao, J., McCullough, J. and Zanjian, E.D. (1983) *Br. J. Haemotol.* **55**, 615-622.
138. Smaaland, R., Laerum, O.D., Lote, K., Sletvold, O., Sothern, R.B. and Bjerknes, R. (1991) *Blood* **77**, 2603-2611.
139. Smaaland, R., Lote, K., Sletvold, O., Bjerknes, R., Aakvaag, A., Vollset, S.E. and Laerum, O.D. (1991) *Ann. NY Acad. Sci.* **618**, 605-609.
140. Morra, L., Ponassi, A., Caristo, G., Bruzzi, P., Zunino, R., Parodi, G.B. and Sacchetti, C. (1984) *Biomed. Pharmacother.* **38**, 167-170.
141. Lund-Johansen, F., Bjerknes, R. and Laerum, O.D. (1990) *Cytometry* **11**, 610-616.
142. Abrahamsen, J.F., Smaaland, R. and Laerum, O.D. (1994) Amelia Island, Abstract VIIIa-2.
143. Nelson, W., Tong, Y., Lee, J.K. and Halberg, F. (1979) *Chronobiologia* **6**, 305-323.
144. Smaaland, R., Abrahamsen, J.F., Svardal, A.M., Lote, K. and Ueland, P.M. (1992) *Br. J. Cancer* **66**, 39-45.
145. Wide, L., Bengtsson, C. and Birgegard, G. (1989) *Br. J. Haemotol.* **72**, 85-90.
146. Reiter, R.J. (1986) in *Endokrinologie der kindheit und Adolesenz* (Gupta, D. ed.) pp. 53-67, Stuttgart, Thieme Verlag.
147. Aschoff, J. (1979) in *Endocrine rhythms* (Krieger, D. ed.) pp. 1-67, New York, Raven Press.
148. Haldar, C., Häußler, D. and Gupta, D. (1992) *J. Pineal Res.* **12**, 79-93.
149. Cassone,V.M. (1990) *Trends Neurosci.* **13**, 457-464.
150. Lampkin, B.C., McWilliams, N.B. and Mauer, A.M. (1976) *Br. J. Haematol.* **32**, 29-40.
151. Pallavicini, M.G. and Gray, J.W. (1982) *Med. Ped. Oncol.* **1** (Suppl.), 109-123.
152. Tannock I. (1978) *Cancer Treat. Rep.* **62**, 1117.
153. Hill, B.T. (1986) *Rec. Results Cancer Res.* **103**, 41-53.
154. Simpson-Herren, L. (1982) *Ann. NY Acad. Sci.* **397**, 88-100.

155. Klevecz, R.R. and Braly, P.S. (1994) in *Circadian cancer therapy* (Hrushesky, W.J.M. ed.) pp. 165-183, Boca Raton, CRC Press.
156. Badran, A.F. and Echave-Llanos, J.M. (1965) *J. Natl. Cancer Inst.* **35**, 285.
157. Voutilainen, A. (1953) *Acta Path. Microb. Scan.* **99**, 1-104.
158. Tähti, E. (1956) *Acta Path. Microbiol. Scand.* **117**, 1-61.
159. Stoll, B.A. and Burch, W.M. (1968) *Cancer* **21**, 193-196.
160. Wolley-Hart, A., Twentyman, P., Corfeild, I., Joslin, C., Morrison, R. and Fowler, J.F. (1986) *Br. J. Radiol.* **41**, 440.
161. Klevecz, R.R. and Braly, P.B. (1991) *Ann. NY Acad. Sci.* **618**, 257-276.
162. Braly, P.S. and Klevecz, R.R. (1993) *Cancer* **71**, 1621-1628.
163. Skipper, H.E., Schabel, F.M.J. and Walcox, W.S. (1967) *Cancer Chemother. Res.* **51**, 125.
164. Haus, E., Halberg, F., Scheving, L., Cardoso, S., Kuhl, J.F.W., Sothern, R., Shiotsuka, R., Hwang, D. and Pauli, S.E. (1972) *Science* **177**, 80-82.
165. Scheving, L.E., Haus, E., Kühl, J.F.W., Pauly, J.E., Halberg, F. and Cardoso, S.S. (1976) *Cancer Res.* **36**, 1133-1137.
166. Scheving, L.E., Burns, R., Pauly, J.E., Halberg, F. and Haus, E. (1977) *Cancer Res.* **37**, 3648-3655.
167. Sothern, R.B., Nelson, W.L. and Halberg, F.A. (1977) in *12th International Conference of the International Society for Chronobiology* pp. 433-438, Milano, Il Ponte.
168. Kühl, J.F.K., Haus, E., Halberg, F., Scheving, L.E., Pauly, J.E., Cardoso, S.S. and Rosene, G. (1974) *Chronobiologia* **1**, 316-317.
169. Scheving, L.E., Burns, E.R., Halberg, F. and Pauly, J.E. (1980) *Chronobiologia* **17**, 33-40.
170. Scheving, L.E., Burns, E.R., Pauly, J.E. and Halberg, F. (1980) *Cancer Res.* **40**, 1511-1515.
171. Sothern, R.B., Lévi, F., Haus, E., Halberg, F. and Hrushesky, W.J.M. (1989) *J. Natl. Cancer Inst.* **81**, 135-145.
172. von Roemeling, R. and Hrushesky, W. (1989) *J. Clin. Oncol.* **7**, 1710-1719.
173. Hrushesky, W.J.M. (1985) *Science* **228**, 73-75.
174. Kerr, D.J., Lewis, C., O'Neill, B., Lawson, N., Blackie, R.G., Newell, D.R., Boxall, F., Cox, J., Rankin, E.M. and Kaye, S.B. (1990) *Hematol. Oncol.* **8**, 59-63.
175. Hrushesky, W.J.M., Roemeling, R.V. and Sothern, R.B. (1989) in *Biological rhythms in clinical practice* (Arendt, J., Minors, D.S. and Waterhouse, J.M. eds.) pp. 225-252, London, Boston, Singapore, Sydney, Toronto, Wellington., Wright.
176. Rivard, G.E., Infante-Rivard, C., Hoyoux, C. and Champagne, J. (1985) *Lancet* **2**, 1264-1266.
177. Rivard, G.E., Infante-Rivard, C., Dresse, M-F., Leclerc, J-M. and Champagne, J. (1993) *Chronobiol. Int.* **10**, 201-204.
178. Aarnaes, E., Clausen, O.P., Kirkhus, B. and DeAngelis, P. (1993) *Cell Proliferation* **26**, 205-219.
179. Hrushesky, W.J.M., Langer, R. and Theeuwes, F. (1991) *Ann. NY Acad. Sci.* **618**.
180. von Roemeling, R. and Hrushesky, W.J.M. (1990) *J. Natl. Cancer Inst.* **82**, 386-393.
181. Scheving, L.E., Pauly, J.E., Tsai, T.H. and Scheving, L.A. (1983) in *Biological Rhythms and Medicine. Cellular, metabolic, physiopathologic, and pharmacologic aspects* (Reinberg, A. and Smolensky, M.H. eds.) pp. 79-130, Berlin, New York, Springer-Verlag.
182. Lévi, F., Halberg, F., Nesbit, M., Haus, E. and Levine, H. (1981) in *Neoplasms - Comparative pathology of growth in animals, plants and man* (Kaiser, H. ed.) pp. 267-316, Baltimore, Williams & Wilkins.
183. Focan, C. Le rythme nycthéméral de la prolifération tumorale: Aspects expérimentaux et cliniques: Implication pour la chimiothérapie oncolytique. (1985) Thesis. University of Liège, Belgium.
184. Smaaland, R., Lote, K., Sothern, R.B. and Laerum, O.D. (1993) *Cancer Res.* **53**, 3129-3138.
185. Meyn, R.E. and Murray, D. (1986) in *Cell cycle effects of drugs* (Dethlefsen, L. A. ed.) **121**, pp. 170-188, Oxford, Pergamon Press.
186. Bhuyan, B.K., Schedit, L.G; and Fraser, T.J. (1972) *Cancer Res.* **32**, 398-407.
187. Traganos, F., Darzynkiewicz, Z. and Melamed, M.R. (1981) *Cancer Res.* **41**, 4566-4576.
188. Traganos, F., Staiano-Coico, L., Darzynkiewicz, Z. and Melamed, M.R. (1981) *Cancer Res.* **41**, 2728-273.
189. Smaaland, R., Svardal, A.M., Lote, K., Ueland, P.M. and Laerum, O.D. (1991) *J. Natl Cancer Inst.* **83**, 1092-1098.
190. Smaaland, R., Laerum, O.D., Sothern, R.B., Sletvold, O., Bjerknes, R. and Lote, K. (1992) *Blood* **79**, 2281-2287.
191. Brons, P.P.T., Pennings, A.H.M., Haanen, C., Wessels, H.M.C. and Boezeman, J.B.M. (1990) *Cytometry* **11**, 837.
192. Hrushesky, W.J.M. and Bjarnason, G.A. (1993) in *Cancer. Principles and Practice of Oncology* (DeVita, J.V.T., Hellman, S. and Rosenberg, S.A. eds.) Philadelphia., J.B.Lippincott Company.

193. Ludlum, D. (1977) in *Cancer, a comprehensive treatise* (Becker, F.F. ed.) **5**, pp. 285-307, New York, Plenum Press.
194. Petit, E., Milano, G., Lévi, F., Thyss, A., Bailleul, F. and Schneider, M. (1988) *Cancer Res.* **48**, 1676-1679.
195. Mormont, M.C., Boughattas, A.N. and Lévi, F. (1989) in *Chronopharmacology. Cellular and biochemical interactions* (Lemmer, B. ed.) pp. 395-437, New York, Marcel Dekker.
196. Boughattas, A.N., Lévi, F., Fournier, C., Lemaigre, G., Roulon, A., Hecquet, B., Mathé, G. and Reinberg, A. (1989) *Cancer Res.* **49**, 3362-3368.
197. Harris, B., Song, R., Soong, S. and Diasio, R.B. (1989) *Cancer Res.* **49**, 6610-6614.
198. Boughattas, N.A., Lévi, F., Fournier, C., Hecqet, B., Lemaigre, G., Roulon, A., Mathé, G. and Reinberg, A. (1990) *J. Pharmacol. Exp. Ther.* **255**, 672-679.
199. Paukovits, W.R., Guigon, M., Binder, K.A., Hergl, A., Laerum, O.D. and Schulte-Hermann, R. (1990) *Cancer Res.* **50**, 328-332.
200. Nienhuis, A.W. (1988) *N. Engl. J. Med.* **318**, 916-918.
201. Hrushesky, W.J.M. (1987) *J. Biol. Response Mod.* **6**, 587-598.
202. Lévi, F., Misset, J-L., Brienza, S., Adam, R., Metzger, G., Itzakhi, M., Caussanel, J-P., Kunstlinger, F., Lecouturier, S., Descorps-Declere, A., Jasmin, C., Bismuth, H. and Reinberg, A. (1992) *Cancer* **69**, 893-900.
203. Lévi, F., Zidani, R., Vannetzel, J-M., Perpoint, B., Focan, C., Faggiuolo, R., Chollet, P., Garufi, C., Itzakhi, M., Dogliotti, L., Iacobelli, S., Adam, R., Kunstlinger, F., Gastiaburu, J., Bismuth, H., Jasmin, C. and Misset, J-L. (1994) *J. Natl Cancer Inst.* **86**, 1608-1617.

Chapter 24
The mammalian Golgi apparatus during M-phase

Tom Misteli

Cold Spring Harbor Laboratory, 1 Bungtown Road, Cold Spring Harbor, NY 11724, USA

The Golgi apparatus in mammalian cells disassembles into several thousand vesicles as cells enter M-phase. Disassembly is dependent on the action of cdc2-kinase and at least two pathways contribute to the fragmentation: One involves the budding of COP-coated vesicles from Golgi cisternae with concomitant inhibition of fusion with their target membranes, the other is a less well characterised COP-independent pathway. During telophase, the Golgi fragments reassemble and fuse into a fully functional Golgi stack, using at least two distinct fusion pathways. The morphological changes of the Golgi apparatus during M-phase offer an ideal system to study how cellular organelles are generated and how their structure is maintained during interphase.

INTRODUCTION

A hallmark of eukaryotic cells is their high degree of intracellular compartmentalisation. The physical separation between functionally distinct domains inside the cell is achieved by sub-dividing the cellular space with membranes. Cellular organelles formed in this manner execute specialised functions with higher efficiency than if their contents were freely distributed in the cytoplasmic space. Our understanding of how separate organelles are formed, how their structure is maintained and changed, and how the appearance of a given organelle is linked to its function is very limited. One approach to address these questions is to disassemble the organelles into their smallest units and to then reassemble them into functional entities. Morphological and biochemical analysis of these steps should reveal mechanisms involved in the reorganisation and the maintenance of the organelle. Fortunately, the difficult experimental task of disassembling and reassembling organelles is performed regularly by most mammalian cells. Many organelles, such as the nuclear envelope (NE), the endoplasmic reticulum (ER), the Golgi complex, and the endocytic membrane system disassemble into vesicular structures as cells enter M-phase and reassemble into fully functional organelles as cells exit M-phase (1). Amongst the cellular organelles which undergo mitotic disassembly, the Golgi apparatus is the most extensively fragmented and is therefore an ideal model system to study the morphogenesis and the structure-function relationship of cellular organelles.

THE GOLGI APPARATUS AND VESICULAR TRAFFIC

The Golgi apparatus lies at the centre of the exocytic membrane transport pathway in mammalian cells. Secretory proteins and cellular membrane proteins enter the exocytic pathway by translocation across the ER membrane (2, 3). The nascent proteins are frequently modified by the addition of high-mannose oligosaccharides and a variety of quality control systems ensure that only properly folded molecules are transported to the next destination in the secretory pathway, the Golgi apparatus (4). The functional unit of the Golgi apparatus is the stack of typically 3-6 closely apposed, flattened cisternae 1-2µm in diameter and ~30nm in thickness (Fig. 1). Each stack is polarised in a cis-to-trans fashion with regards to the general direction of secretion with cargo entering the stack at its cis-side and leaving it at the trans-face (5). The cis-, medial-, and trans-compartments are distinct in their set of resident Golgi enzymes, many of which, are involved in the sequential post-translational modifications of proteins and lipids as these are transported from cisterna to cisterna (5).

The Golgi apparatus appears to be a single copy organelle in mammalian cells. Although several stacks can usually be observed in a thin-section through a cell (Fig. 1), scanning electron microscopy (6), high voltage electron microscopy (7) and conventional electron microscopy using semi-thin sections (8) reveal that the apparently distinct stacks are connected by tubular membrane extensions that link equivalent cisternae (arrow in Fig. 2). This gives the Golgi apparatus the appearance of a ribbon, which frequently bifurcates and rejoins, making it into a compact Golgi reticulum. Associated with the Golgi apparatus is a large number of vesicles of 50-100nm in diameter, some of which can be seen to bud from the rims of Golgi cisternae (9, 10). The vesicles bear a distinctive protein coat, the COP-coat (11, 12), and have been suggested to mediate transport between the ER and the Golgi apparatus, and also between cisternae within the Golgi stack (Fig. 2) (13). The COP coat is ~10nm thick, appears fuzzy and is morphologically and in its molecular

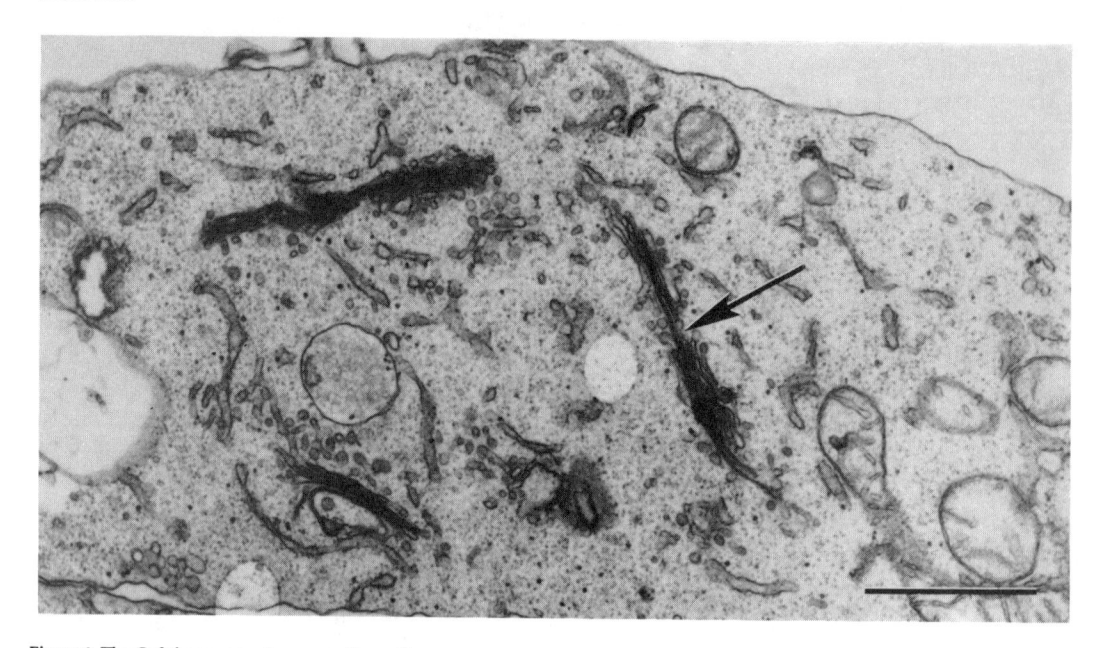

Figure 1. The Golgi apparatus in mammalian cells
A thin-section through Golgi stacks in CHO cells as seen by transmission electron microscopy. Associated with the stacks are small vesicles, most likely transport vesicles. The arrow indicates a possible tubular connection between equivalent cisternae in two stacks. Bar: 1μm

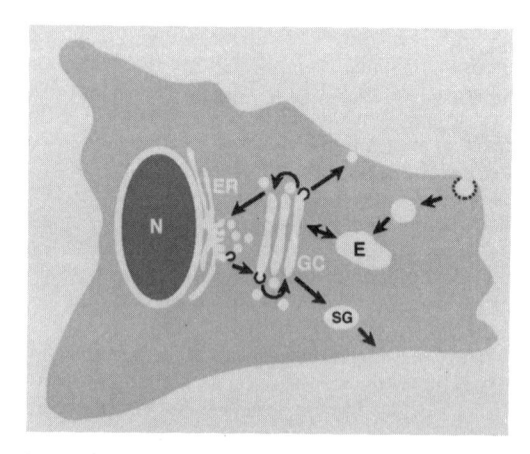

Figure 2. Membrane trafficking pathways during interphase
Secretory and membrane proteins are translocated into the endoplasmic reticulum and transported along the secretory pathway via vesicles to their destinations. Membrane is internalized from the plasma membrane by endocytosis and may reach the Golgi apparatus at its trans-side. Abbreviations: N, nucleus; ER, endoplasmic reticulum; GC, Golgi complex; SG, secretory granules; E, endosomal system.

composition distinct from the clathrin coat of vesicles that mediate endocytic traffic (Fig. 2). COP-coated vesicles budding from Golgi cisternae were first identified based on their morphological appearance (11). These vesicles are now known as COP I-coated vesicles, since it has become clear recently that there are several populations of COP-coated vesicles which are morphologically indistinguishable but are composed of different coat molecules (14). In yeast, two populations of COP coated vesicles, COP I and COP II, have been described (15). Both bud from the ER (16), but the two populations have distinct molecular compositions and possibly mediate separate transport events between the ER and the Golgi apparatus. In mammalian cells, COP I-coated vesicles have been implicated in ER-Golgi transport (17), in retrograde transport from the Golgi apparatus to the ER (18), and in anterograde as well as retrograde transport through the Golgi stack (13). The COP-II population of vesicles has been characterised in semi-intact mammalian cells and is suggested to be involved in ER-Golgi transport (Fig. 2) (19).

Biogenesis of a COP-coated vesicle requires first the association of the cytoplasmic ARF (ADP-ribosylation factor) GTPase in the GTP-bound form with the organelle membrane (20). This association then triggers the binding of COP I components. Assembly of the coat on the cisternal rim induces the deformation of the membrane and, ultimately, the fusion of the inner layers of the membrane bilayer results in breaking off of the coated vesicle (21). After budding of a COP-coated vesicle, GTP on ARF is hydrolysed, the coat shed and the components of the COP-coat are recycled (22). As the uncoated vesicle is docked to the target membrane a multi-protein fusion complex is formed. The complex is made up of an ATPase called NSF (N-ethyl-maleimide sensitive factor) (23), SNAPs (soluble NSF attachment protein) (24), and two trans-membrane receptors, termed SNAREs (SNAP-receptor) (25, 26), one present in the vesicle membrane (v-SNARE), the other in the target membrane (t-SNARE) (27).

Additionally, a member of the rab family of small GTP-binding proteins is involved in the formation of the fusion complex, possibly contributing to the specificity of the fusion event (28).

THE FUSION/FISSION EQUILIBRIUM AND MORPHOLOGY

Obviously, budding of vesicles from Golgi cisternae removes membrane from the stack. This loss of membrane is counterbalanced by vesicles fusing with the stack from the ER or the endocytic membrane compartment and also intra-Golgi vesicles (Fig. 2). Therefore, the Golgi apparatus is in a continuous equilibrium between exiting membrane and entering membrane and the morphological appearance of the organelle is a direct reflection of the functional activity of the stack. This becomes apparent when the regular sequence of events in the formation or consumption of COP-coated vesicles is perturbed. For example, slowing down of ER-Golgi transport at reduced temperature causes the Golgi apparatus to become smaller and vesicles to accumulate in the vicinity of the Golgi apparatus (29). Similarly, inhibition of formation of the fusion complex of docking vesicles by N-ethyl-maleimide completely vesiculates the Golgi apparatus (30) as does microinjection of a mutant rab1a isoform, which prevents fusion of ER-Golgi transport vesicles (31), or overexpression of mutant rab6, which is required for intra-Golgi transport (32). The Golgi apparatus is severely vesiculated in a conditional lethal mammalian cell line, which carries a defect in ε-COP, an essential component of the COPI-coat (33). On the other hand, interference with the formation of COP-coated vesicles, either by addition of the fungal metabolite BFA *in-vivo* (34) or low concentrations of ATP *in-vitro* (35), results in the formation of tubular networks, most likely due to random and uncontrolled fusion events. In yeast sec7 mutants, the Golgi apparatus has lost its typical stacked structure and is dramatically tubulated (36). This is not surprising since sec7 is required for COP-coated vesicle budding and absence of sec7 will shift the equilibrium between budding and fusing vesicles towards fusion.

All these observations suggest that in the absence of vesicle fusion, membranes vesiculate spontaneously, while in the absence of vesicle formation membranes fuse promiscuously into tubular networks.

THE MORPHOLOGICAL APPEARANCE OF THE GOLGI APPARATUS DURING M-PHASE

During M-phase the Golgi apparatus in all mammalian cell-types undergoes a dramatic morphological transformation. Upon entry into M-phase, the Golgi apparatus loses its stacked appearance and by the time cells enter anaphase the organelle is found in several thousand vesicles dispersed throughout the cytoplasm. During telophase, the reverse process takes place as the stacks are re-formed and the Golgi reticulum is re-established. A more detailed picture of the morphological changes which occur during this cycle of disassembly and re-assemble has emerged by using sensitive stereological techniques combined with immuno-cytochemical identification of the Golgi membranes by electron microscopy.

Disassembly of the Golgi reticulum most likely starts with the severing of the tubular connections between the single stacks. This process has not been convincingly demonstrated in mammalian cells, most likely due to the complex three-dimensional architecture of the Golgi reticulum. Indirect evidence comes from the observation that in immature *Xenopus laevis* oocytes, which are arrested in meiotic prophase, the Golgi apparatus exists as a number of small, dispersed and, importantly, functionally active stacks (37). This is in contrast to the juxta-nuclear reticulum found in somatic cells and suggest that the dispersed, stacked form of the Golgi apparatus is an early disassembly intermediate. In mammalian cells, the only indication of alteration during this phase of disassembly is the change in cellular localisation of the Golgi reticulum from a peri-centriolar to a more dispersed, perinuclear localisation (38, 39). In the earliest stages of pro-phase, coinciding with the onset of mitotic chromosome condensation, the stacks of the Golgi apparatus shorten and vesicles with the size and appearance of COP-coated vesicles accumulate at the periphery of the stacks (Fig. 3 A-B). As the chromatin condenses in the nucleus and the nuclear envelope fragments, the cisternae in the stacks become increasingly shorter and a larger number of vesicles accumulate in the periphery (Fig. 3 C-D) (40). By the time cells enter late pro-phase, stack have completely disappeared and instead 50-300 Golgi clusters can be seen (Fig. 3 E) (8, 41). The clusters are typically ~200nm -~1µm in diameter and consist of aggregates of small vesicles (~50nm in diameter), larger vesicles (~200nm) and tubular elements which are occasionally branched and interconnected (8, 40). The vesicular profiles are then lost from the clusters as suggested by the observation that cells with small clusters have a larger number of free vesicles and vice-versa (Fig. 3 F) (42). Indeed, before anaphase is reached, the clusters have completely dissolved and their membrane components randomly dispersed throughout the mitotic cytoplasm. For the period of time in which the genetic material of the cell is partitioned between the daughter cells, virtually no traces of the Golgi apparatus can be detected morphologically. At the onset of telophase, the Golgi apparatus reassembles by essentially the reverse of the disassembly process. The first indication of a re-forming Golgi apparatus

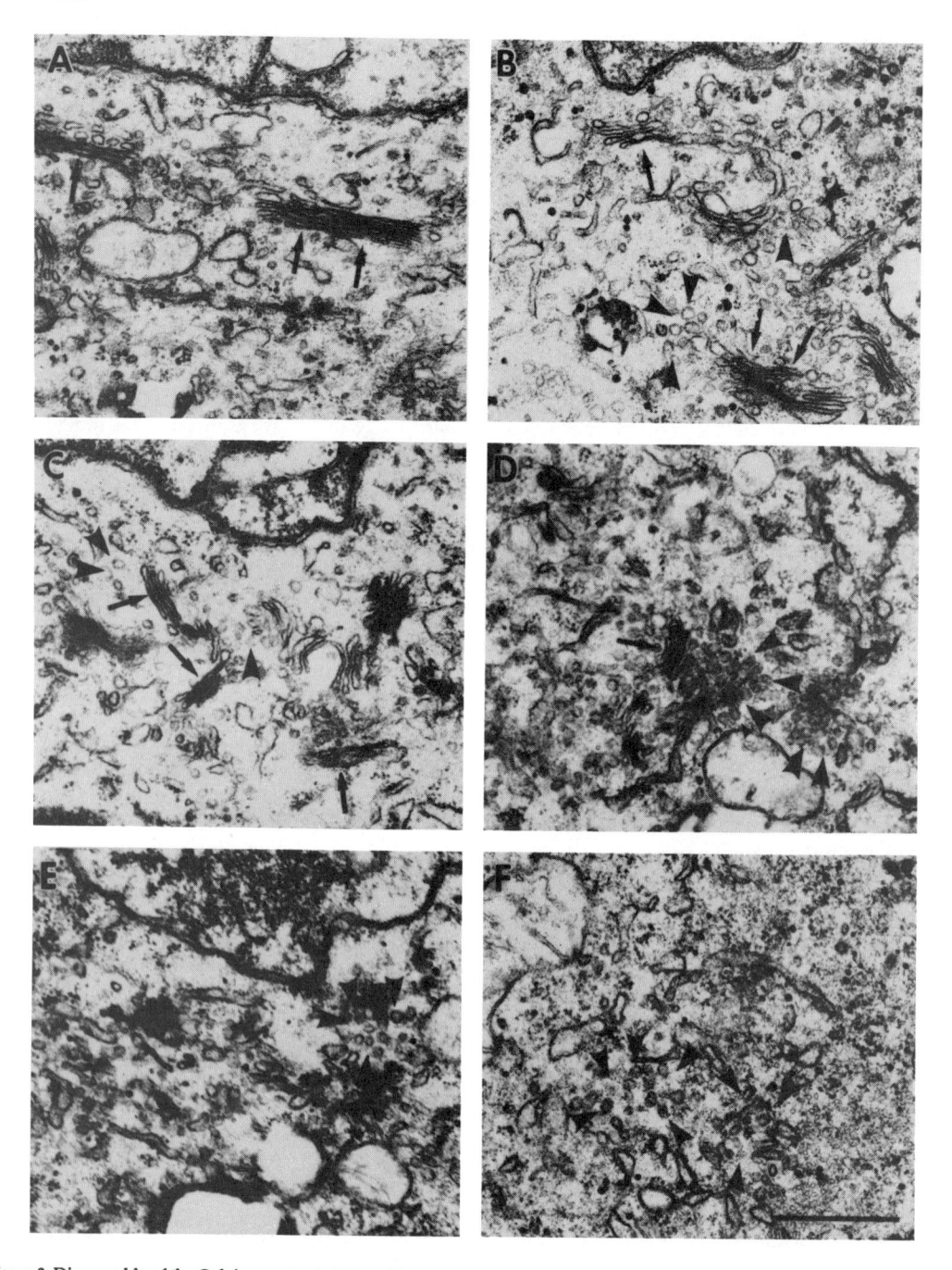

Figure 3. Disassembly of the Golgi apparatus in HeLa cells
Six morphological stages of Golgi disassembly observed by transmission electron microscopy in synchronised murine FT210 cells. The position of the cell in M-phase was determined by the degree of chromatin condensation and nuclear envelope breakdown (40). Cells are in (A) G2, (B) early pro-phase, (C, D) mid-prophase, (E) late prophase, (F) pro-metaphase. The Golgi stacks (arrows) become shorter in cross-section and Golgi vesicles (arrowheads) accumulate in the vicinity of the stack (A-D). Unstacking then takes place and fragmentation is completed. During the late stages (E-F) more vesicles are lost from the cluster than are produced by the fragmentation, decreasing the size of the cluster. Bar: 1µm.

is the accumulation of vesicles in clusters similar to the ones formed during disassembly (Fig. 4 A) (42, 43). Stacks are re-formed from these clusters by fusion amongst the fragments to ultimately form the Golgi reticulum (Fig. 4 B-D) (42, 43). Functionally, transport of glycolipids and proteins appears to resume concomitantly with formation of the stack (43, 44).

MOLECULAR MECHANISMS OF DISASSEMBLY

Given the dramatic and complex changes to Golgi apparatus architecture during mitosis it is both interesting and important to understand the molecular mechanisms underlying these processes. Disassembly of the Golgi apparatus in mammalian cells is a rapid process, estimated to occur within 20-40 minutes of cells entering M-phase (45). This and the extreme difficulty in obtaining sufficient numbers of cells in the earliest stages of M-phase make the more detailed morphological and biochemical analysis of disassembly and the molecular dissection of the process *in-vivo* extremely difficult. A cell-free system which mimics the disassembly process in the test tube has allowed a more detailed morphological and biochemical analysis of the disassembly mechanisms (46). In the *in-vitro* system, isolated rat liver Golgi stacks are incubated with cytosol from synchronised pro-metaphase cells. Upon incubation the intact Golgi stacks fragment in a time-, energy- and temperature-dependent manner into vesicles and tubules which are morphologically indistinguishable from Golgi fragments observed in mitotic cells. In such incubations membrane is lost from stacks at an estimated rate of 3-4 μm^2/min during the initial stages of fragmentation. As a consequence of loss of membrane from the cisternae, the stacked cisternae shorten in diameter from ~1μm to ~0.3μm within 20 minutes (Fig. 5, top left). Membrane is selectively lost from the periphery of the cisternae by budding of membrane fragments. The shortened cisternae remain stacked up to this point and most likely contain the majority of Golgi resident enzymes. Only then does unstacking occur very rapidly, showing it to be a late step in the break-up of the Golgi apparatus (Fig. 5, bottom left) (46). It is possible that unstacking is regulated independently from cisternal vesiculation, but no molecules have been identified to date which are bona-fide stacking components and could act as possible mediators of unstacking.

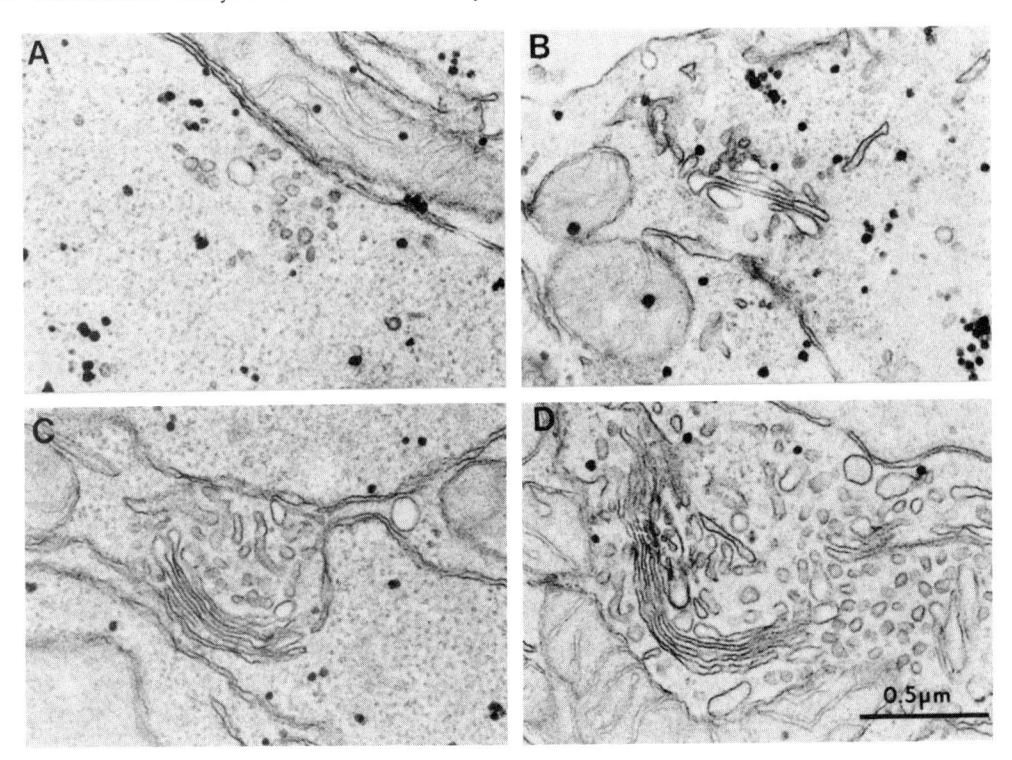

Figure 4. Re-assembly of the Golgi apparatus in HeLa cells
Small clusters of mitotic fragments are formed (A) which fuse with each other to form small stacks (B). As more vesicle fuse with the re-forming stack, the stack grows to its interphase size (C, D). Bar: 0.5µm. (Courtesy of Dr. Marc Pypaert, University of Louvain. Reproduced from The Journal of Cell Biology, 1993, 122, p.533 by copyright permission of The Rockefeller University Press)

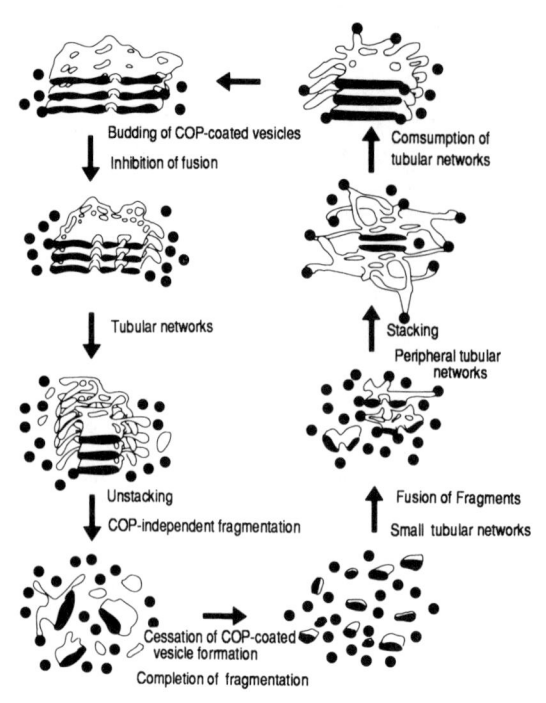

Figure 5. The cycle of fragmentation and re-assembly of the mammalian Golgi apparatus

The membrane which is lost from the stack can entirely be accounted for by the formation of two populations of vesicles: One population of almost perfectly spherical vesicles with an average diameter of ~50nm and a second, more heterogeneous population of more oblate vesicles with a diameter between 100-200nm (46, 47). The dominant population of small vesicles, which constitutes typically more than 60% of total fragmented membrane, consists in fact of the same vesicles which under interphase conditions bud from the rims of Golgi cisternae, the COP-coated vesicles. This becomes obvious when uncoating of COP-coated vesicles is inhibited by GTPγS, which results in a dramatic accumulation of COP-coated vesicles without altering the kinetics of the fragmentation. Indeed, the number of COP-coated vesicles which bud from the rims of Golgi cisternae is the same under interphase as under mitotic conditions, suggesting that budding of COP-coated vesicles is a default activity of Golgi cisternae (46). However, what is different during M-phase is that the COP-coated vesicles that form do not fuse with their target membrane. Under mitotic conditions, the fusion incompetent vesicles accumulate and contribute significantly to the loss of membrane from the Golgi stack. In support of a major role of COP-coated vesicles, fragmentation does not take place in mitotic cytosol depleted of components of the COP-coat (46, 47).

However, not the entire Golgi stack is consumed by COP-coated vesicles (47). A significant amount of membrane appears as larger vesicles, which are not generated by the same mechanism which forms COP-coated vesicles (Fig. 5 left). The larger vesicles appear to form somewhat later in the fragmentation process, at a time when the Golgi resident enzymes leave the core-stacks and are packaged into membrane fragments. It is, therefore, possible that the two populations of vesicles not only represent two distinct fragmentation pathways, but that they also represent two distinct domains found in Golgi cisternae. Scanning electron microscopy has revealed morphological differences between the periphery of cisternae which is typically highly fenestrated and often tubulated, and the inner region of the cisternae, which is more solid and has a smoother membrane appearance (48). Consistent with the observations made on disassembling Golgi membranes, COP-coated vesicles only form from the periphery of Golgi cisternae *in-vitro* (48).

The molecules involved in the second fragmentation pathway are completely unknown. It is also not entirely clear whether both pathways are essential for complete fragmentation or whether they are redundant, although isolated Golgi stacks can be consumed completely by the COP-independent pathway in the absence of the COP-dependent pathway (47). The fragmentation under those conditions is much slower and less efficient than when the COP-dependent pathway is also active. Interestingly, during COP-independent fragmentation large tubular networks are formed which undergo slow fragmentation into vesicles of 100-200nm, suggesting that they are intermediates in the COP-independent pathway. Similar, although smaller, tubular networks can be detected during Golgi apparatus fragmentation *in-vitro* (in the presence of coat proteins) (47) and *in-vivo* (40). One possible explanation for the origin of these networks is a process termed periplasmic fusion (21, 47). In order to bud a vesicle from the Golgi membrane the inner layers of the membrane bilayer must come in close proximity and fuse. This event is termed periplasmic fusion as opposed to the fusion of the cytoplasmic layers of the bilayer, which is termed cytoplasmic fusion and occurs as a vesicle fuses with its target membrane. During initial stages of fragmentation structural restrictions imposed by cytoskeletal elements or stacking matrices are released and the membranes of the Golgi cisternae have more freedom for dynamic movement. This increases the chance of random collisions between the inner layers of the Golgi membranes. Each such event results in the formation of a "hole", or fenestration, in the cisternae. With time an increasing number of periplasmic fusion events take place and the fenestrations eventually fuse with each other to give rise to tubular networks (Fig. 5 left). These tubular networks are then consumed in the later stages of fragmentation by the COP-independent and also partially by the COP-dependent pathway (47).

MOLECULAR MECHANISMS OF RE-ASSEMBLY

At the beginning of telophase, the cell is confronted with the tantalising task of rebuilding the structures which were lost during mitosis. The nuclear envelope re-forms, the microtubular system is reorganised and the membrane compartments of the exocytic pathway must be re-established. The Golgi apparatus poses a particularly difficult problem. Not only do the membrane fragments have to re-fuse, but they have to re-form the stack, and re-assemble in a manner that re-establishes the polarity of the stack. Finally, the Golgi reticulum needs to be placed in the peri-centriolar region of the cell.

As for the disassembly process, the development of cell-free systems to study the re-assembly process has given valuable insight into the responsible molecular mechanisms. Warren and colleagues have taken mitotic Golgi fragments produced by incubation with mitotic cytosol, re-isolated them by centrifugation and re-assembled them in the test-tube (49). Alternatively, Malhotra and colleagues have used a semi-intact cell system in which Golgi membranes are first fragmented by treatment of intact cells with the sponge metabolite ilimaquinone, and then re-assembled in permeabilised cells (50). In both systems, Golgi stacks with similar morphology as interphase Golgi stacks re-form in a time-, energy-, and temperature-dependent manner. The re-assembly process is essentially a reversal of the disassembly and involves two overlapping events: Fusion of membrane fragments and reorganisation of the fused fragments into an ordered stack (Fig. 5 right). In permeabilised cells Golgi fragments of 60-90nm in diameter associate with each other and fuse into larger vesicular structures of 200-300nm in diameter (50). It is not clear whether these larger vesicles then change their shape to give a small cisternae or whether additional fusion events between small and larger fragments take place to result in a small cisternae. In the *in-vitro* system, this step is not observed, but membrane fragments fuse directly to give small cisternae (49). Associated with the rims of this early cisternae are small, tubular networks, similar to the ones formed as intermediates during fragmentation. This is most likely a reflection of the high rate of fusion events between membrane fragments and fragments with the re-forming cisternae. While more membrane is being added to the peripheral tubular networks of the re-forming Golgi apparatus, stacking begins. This is possibly achieved by the lateral association of several re-forming single cisternae or by folding back of portions of the peripheral tubular networks. Additional fragments do not appear to fuse directly to the core of the re-forming Golgi cisternae, but rather to the peripheral tubular networks. From there membrane is redistributed to the cisternae, resulting in a significant increase in cisternal length at this stage of re-assembly (49). The tubular networks are flattened and joined into sheets of cisternae to give the final product of the re-assembly process, a stack of 3-6 flattened Golgi cisternae (Fig. 5 top right).

Biochemical analysis using these *in-vitro* systems indicates that two distinct, but similar, fusion mechanism participate in the re-assembly reaction (51, 52). As expected from the fact that the majority of mitotic fragments are COP-coated vesicles, the molecules involved in the fusion of COP-coated vesicles to their target membranes are required in the fusion leading to the re-assembly of the stack. Addition of combinations of characterised components of the fusion complex of COP-coated vesicles with their target membranes, including the NEM-sensitive factor (NSF) and SNAPs, to Golgi fragments promoted complete or partial re-formation of Golgi stacks *in-vitro* (51, 52). But just as the COP-coated vesicles are not the only product of Golgi disassembly, the NSF-mediated pathway of fusion is not the sole fusion mechanism. p97, like NSF, a member of the family of NEM-sensitive ATPases, is also required for the complete re-formation of Golgi stacks from fragments. The p97 ATPase was first functionally characterised in *S. cerevisiae* by means of its homologue cdc48. Cdc48/p97 is required for the fusion of the nuclear envelope in yeast after mating (53). The exact roles of the two distinct ATPase-mediated pathways in Golgi re-assembly is not clear at present. One possibility is that they act at temporally different points in the re-assembly pathway. Acharya et al. suggested that NSF is required for the first fusion step of single vesicles into small cisternae, while the p97 ATPase is involved in the later formation of the stack (51). Alternatively, the two ATPases could mediate the fusion of the two fragmentation products: NSF promotes the fusion of COP-vesicles, p97 the fusion of the larger, more heterogeneous fraction of membranes (52).

The ultimate step in the re-formation of the functional Golgi ribbon in telophase cells is virtually a white spot on the map. Once the single Golgi stacks have re-formed from mitotic fragments, the stacks will have to find each other and fuse to give rise to a single Golgi organelle. This is possibly facilitated by movement of stacks along micro-tubules towards the microtubule-organising centre during telophase using microtubule motor molecules (54).

INHIBITION OF MEMBRANE FUSION EVENTS CAUSES ORGANELLE FRAGMENTATION

From morphological and biochemical analysis of the Golgi apparatus dis-assembly and re-assembly it seems that by controlling the fusion reaction between vesicles and the Golgi apparatus, the morphology of the organelle can be controlled in a simple and elegant way. Fragmentation can be triggered by preventing fusion of budded vesicles with target membranes and re-assembly can be triggered by the reversion of this inhibition (55, 56). A similar mitotic inhibition behaviour has been observed for the fusion of secretory granules to the plasma membrane in mast cells (57) and the fusion of isolated endocytic vesicles with each other in-vitro (58). Inhibition of vesicle fusion might be the universal trigger to fragment all cellular organelles during mitosis. For example, COP-coated vesicles budding from the ER, not able to fuse with the Golgi apparatus will contribute to the fragmentation of the ER. The ER does indeed fragment during M-phase, although the mitotic ER fragments are generally larger in size than the majority of Golgi fragments (1). This could be a reflection of the much larger surface area and the sub-compartmentalisation of the ER. In the case of the ER, COP-coated vesicles only bud from specialised regions of the organelle, the transition elements. These areas might well be consumed by COP-coated vesicles at the onset of mitosis, while the vast majority of the ER membrane is consumed by a COP-independent pathway, resulting in larger fragments.

THE REGULATION OF INHIBITION OF FUSION

What regulates the inhibition of fusion and what are the immediate targets of inhibition? Not surprisingly, the mitotic master-regulator cdc2 kinase is key in the inhibition of membrane fusion during M-phase and also in the disassembly of the Golgi apparatus in-vitro (46, 59, 60). Addition of recombinant cyclin A to interphase HeLa cytosol mimicked the effect of mitotic cytosol and resulted in complete fragmentation of Golgi stacks in-vitro (46). More direct evidence came from the use of a temperature sensitive cell-line with a point mutation in the cdc2 gene. FT210 cells exhibit normal cdc2 kinase activity at their permissive temperature of 32°C, but at the non-permissive temperature of 39°C, cdc2 kinase is degraded within 4 hours and cells arrest at the G2/M boundary (61). Golgi stacks fragment efficiently in FT210 cytosol supplemented with recombinant cyclin A, but not in the same cytosol after cdc2 kinase has been destroyed by pre-incubation of the cytosol at 39°C (46). The effect could be reversed by re-addition of purified cdc2 kinase, but not CDK2 kinase, indicating that cdc2 kinase was involved in the initiation of mitotic Golgi disassembly in a cell-free system (46). Whether cdc2 kinase acts directly on Golgi targets or indirectly via other kinases is unclear. However, addition of okadaic acid, an inhibitor of members of the serine-threonine phosphatase families 1 and 2A, to interphase cytosol resulted in fragmentation of the stack without an increase in histone phosphorylation, an indicator of cdc2 kinase activity (T. Misteli and G. Warren, unpublished observation). The most simple model for this effect is that PP1/2A is part of a phosphatase/kinase cycle downstream of cdc2 kinase. The phosphatase could directly counteract the cdc2-dependent kinase responsible for phosphorylation of the mediator of inhibition of fusion or it could act as a positive or negative effector of a regulating kinase (see (62)). Alternatively, PP1/2A could act as an antagonist of a cdc2 kinase-independent kinase which acts on the same target as cdc2 kinase. Nothing is known about the downstream targets of cdc2 kinase which mediate Golgi disassembly or further cell-cycle specific regulatory factors. Of potential interest in this context is the recent finding that cyclin B2 is primarily associated with the Golgi apparatus, in contrast to B1 which is found associated with microtubules (63).

As cells enter M-phase, fusion of not only Golgi vesicles but virtually every membrane fusion step in the cell is inhibited (Fig. 6). Several of these fusion reactions, including endocytic fusion and fusion of Golgi derived vesicles, can be inhibited by activation of cdc2 kinase. It is therefore likely that a general factor or different, but similarly acting factors, involved in fusion reactions are the target for mitotic inhibition of fusion. The most obvious candidates are the members of the family of SNAP proteins, which are essential components of the fusion complex and contain cdc2-consensus phosphorylation sites. However, to date there is no evidence that these sites are specifically used at the onset of M-phase or at any other point during the cell cycle. Another attractive candidate group is the family of small GTP-binding proteins of the rab family. Rabs have been widely implicated in contributing to the specificity of vesicular transport as each vesicle-mediated transport step appears to require one specific rab protein (64). Indeed, some of the rab proteins, namely rab1a, rab1b, and rab4 involved in ER-Golgi transport, intra Golgi transport and endocytosis, respectively, are mitotically phosphorylated (65, 66). However, no direct evidence has linked the phosphorylation of rabs to structural changes in organelle morphology. In contrast, rab6, involved in intra-Golgi transport is not phosphorylated during mitosis, yet transient overexpression fragments the Golgi apparatus (32).

A new candidate has emerged very recently in the form of p115, a cytosolic protein needed for intra-Golgi transport. Cloning and sequencing have shown

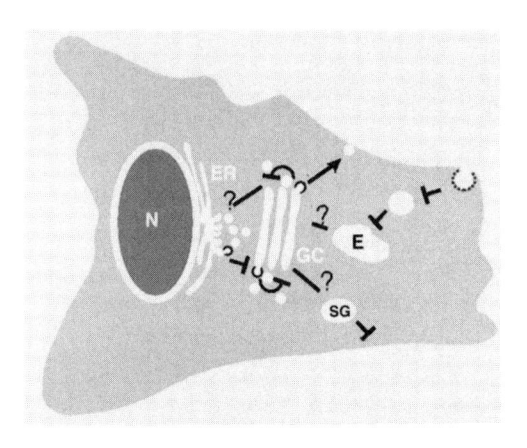

Figure 6. Membrane trafficking pathways during M-phase
A number of vesicular transport steps are inhibited during M-phase. Question marks indicate that no information about the mitotic activity of the indicated transport step is available. Abbreviations: N, nucleus; ER, endoplasmic reticulum; GC, Golgi complex; SG, secretory granules; E, endosomal system.

that this protein is identical to a factor involved in the docking of transcytotic vesicles with the plasma membrane and that it is homologous to a yeast protein, Uso1p, needed for transport between the endoplasmic reticulum and the Golgi apparatus (67, 68). These observations make it likely that p115 is a common factor in vesicle docking and as such is a candidate for mitotic inhibition. Levine and Warren have recently demonstrated that binding of p115 to mitotic Golgi membranes is significantly inhibited when compared to interphase membranes and that an, as yet, unidentified p115 receptor is reversibly inactivated. This inhibition appears to be cdc2 kinase dependent and to involve a downstream kinase/phosphatase cycle (T. Levine and G. Warren, unpublished observation). In support of a role for p115 in the mitotic fragmentation of the Golgi stack, addition of excess recombinant p115 to fragmentation incubations slowed down the fragmentation of stacks. One possibility is that the p115 receptor is inactivated during M-phase and will prevent vesicles from docking with their target membranes and, therefore, ultimately from fusing.

WHY DISASSEMBLE THE GOLGI APPARATUS?

Apart from the Golgi apparatus, other organelles such as the nuclear envelope, the endoplasmic reticulum and the endocytic membrane system fragment at the onset of mitosis. A possible reason for organelle fragmentation arises from the need of cells not only to partition chromosomes, but also all other essential cellular components equally between daughter cells. This might be achieved by a stochastic process in which two daughter cells of equal volume, containing equal amounts of every cellular component dispersed throughout the cytoplasmic space, are produced. In the case of multi-copy organelles, such as lysosomes and mitochondria, equal partitioning is ensured by their random distribution in the cell. However, for single-copy organelles, such as the nuclear envelope, the ER, the endosomal system and the Golgi apparatus, to be randomly distributed, a fragmentation and dispersal process must first take place. The accuracy of partitioning is then only dependent on the number of fragments generated. The number of fragments for ER and Golgi apparatus in mammalian cells have been estimated to be >10, 000 (42, 69). This results in a 99.99% theoretical chance of both daughter cells receiving $50 \pm 0.6\%$ of total membrane (1) and therefore would assure equal partitioning.

Such a stochastic method of equal partitioning appears to be highly efficient in mammalian cells, however, does not seem to be universal. In *S. pombe*, the Golgi apparatus is made up of one or several small stacks. During cell division, the stacks do not appear to fragment but are still equally distributed between the two daughter cells (70). In *S. cerevisiae*, the Golgi apparatus does not exist as a single copy organelle, but as a multi-copy organelle with a number of single cisternae randomly distributed throughout the cytoplasm (71). However, the number of cisternae is too low as that a purely stochastic process would be sufficient to ensure equal partitioning. Other mechanisms must contribute to the proper segregation of the membrane components which make up the yeast Golgi apparatus. The different strategies for partitioning might reflect the differences in the cell division machinery between higher and lower eukaryotes.

UNDERSTANDING THE INTERPHASE GOLGI APPARATUS

The study of the morphological changes of the Golgi apparatus during M-phase also gives us information as to how the Golgi apparatus functions and how it maintains its structure during interphase. The development of cell-free systems to study the molecular basis for the morphological changes of the Golgi stack has put us in the position of the watchmaker apprentice who has taken his timepiece apart and by way of putting it together will learn how its parts work. It has become clear from these recent results that the structure of the Golgi apparatus is directly linked to the balance of incoming and exiting membrane. It is now also clear that other mechanisms than the previously described COP-coated vesicle mediated pathway are involved in removing membrane from cisternae and possibly in transport through or from the Golgi apparatus in interphase cells. Using organelle fragments it will be possible to study the role of subpopulations of these vesicles in the context of mitotic dis-assembly and re-assembly, but also their function in interphase cells.

Similarly, study of the mitotic re-assembly has revealed a novel class of fusion molecules and it will be of importance in the future to identify their role in interphase cells.

SUMMARY

The mammalian Golgi apparatus undergoes a dramatic cycle of fragmentation and re-assembly within the short period of M-phase (summarised in Fig. 5). Fragmentation is brought about by at least two mechanisms: The continued budding of COP-coated vesicles from Golgi cisternae with concomitant inhibition of their fusion, and the shedding of membrane via an ill-characterised COP-independent pathway. Similarly, fusion of the fragments during re-assembly involves at least two different pathways, one involving the NSF-ATPase and one the cdc48/p97-ATPase pathway. The complexity of the morphological changes during this cycle are astonishing and we are only beginning to understand what they mean and how they are brought about on a molecular level. However, from the glimpse that we have taken so far it is once more most satisfying to realise that the cell uses its intrinsic tools, such as continued budding of COP-coated vesicles with concomitant inhibition of their fusion, to fulfil complex tasks very elegantly and efficiently.

ACKNOWLEDGEMENTS

I thank Marc Pypaert, Louvain, Belgium, for providing Fig. 4. Drs. Tim Levine and Graham Warren for communicating unpublished results and Drs. Graham Warren and Catherine Rabouille for critically reading of the manuscript.

REFERENCES

1. Warren, G. (1993) *Annu. Rev. Biochem. Sci.* 323-348
2. Blobel, G. and Dobberstein, B. (1975) *J. Cell Biol.* **67**, 835-851
3. Gilmore, R. (1994) *Cell* **75**, 589 - 592
4. Hammond, C. and Helenius, A. (1994) *J. Cell Biol.* **126**, 41-52
5. Roth, J. (1991) *J. El. Micr. Tech.* **17**, 121 - 131
6. Tanaka, K. and Fukudome, H. (1991) *J. El. Micr.Tech.* **17**, 15 - 23
7. Rambourg, A. and Clermont, Y. (1990) *Eur. J. Cell Biol.* **51**, 189-200
8. Lucocq, J. M., Pryde, J. G., Berger, E. G. and Warren, G. (1987) *J. Cell Biol.* **104**, 865-874
9. Jamieson, J. D. and Palade, G. E. (1967) *J. Cell Biol.* **34**, 577 - 596
10. Dalton, A. J. and Felix, M. D. (1954) *Am. J. Anat.* **94**, 171 - 207
11. Orci, L., Glick, B. S. and Rothman, J. E. (1986) *Cell* **46**, 171-84
12. Serafini, T., Stenbeck, G., Brecht, A., Lottspeich, F., Orci, L., Rothman, J. E. and Wieland, F. T. (1991) *Nature* **349**, 215-20
13. Rothman, J. E. (1994) *Nature* **372**, 55-63
14. Pelham, H. R. B. (1995) *Cell* **79**, 1125-1127
15. Barlowe, C., Orci, L., Yeung, T., Hosobuchi, M., Hamamoto, S., Salama, N., Rexach, M. F., Ravazzola, M., Amherdt, M. and Schekman, R. (1994) *Cell* **77**, 895-907
16. Bednarek, S. Y., Ravazzola, M., Hosobuchi, M., Amherdt, M., Perrelet, A., Schekman, R. and Orci, L. (1995) *Cell* **83**, 1183-1196
17. Hosobuchi, M., Kreis, T. E. and Scheckman, R. (1992) *Nature* **360**, 603 - 605
18. Letourneur, F., Gaynor, E. C., Hennecke, S., Demolliere, C., Duden, R., Emr, S. D., Riezman, H. and Cosson, P. (1994) *Cell* **79**, 1199-1207
19. Aridor, M., Bannykh, S. I., Rowe, T. and Balch, W. E. (1995) *J. Cell Biol.* **131**, 875-893
20. Palmer, D. J., Helms, J. B., Beckers, C. J. M., Orci, L. and Rothman, J. E. (1993) *J. Biol. Chem.* **268**, 12083-12089
21. Rothman, J. E. and Warren, G. (1994) *Curr. Biol.* **4**, 220-233
22. Tanigawa, G., Orci, L., Amherdt, M., Ravazzola, M., Helms, J. B. and Rothman, J. E. (1993) *J. Cell Biol.* **123**, 1365-1371
23. Block, M. R., Glick, B. S., Wilcox, C. A., Wieland, F. T. and Rothman, J. E. (1988) *PNAS* **85**, 7852 - 7856
24. Whiteheart, S. W., Brunner, M., Wilson, D. W., Wiedman, M. and Rothman, J. E. (1992) *J. Biol. Chem.* **267**, 12239 - 12243
25. Wilson, D. W., Whiteheart, S. W., Wiedmann, M., Brunner, M. and Rothman, J. E. (1992) *J Cell Biol* **117**, 531-8
26. Söllner, T., Whiteheart, S. W., Brunner, M., Erdjument-Bromage, H., Geromanos, S., Tempst, P. and Rothman, J. E. (1993) *Nature* **362**, 318-324
27. Söllner, T., Bennett, M. K., Whiteheart, S., Scheller, R. H. and Rothman, J. E. (1993) *Cell* **75**, 409-418
28. Søgaard, M., Tani, K., Ye, R. R., Geramanos, S., Tempst, P., Kirchhausen, T., Rothman, J. E. and Söllner, T. (1994) *Cell* **78**, 937-948
29. Saraste, J. and Kuismanen, E. (1984) *Cell* **38**, 535-549
30. Orci, L., Malhotra, V., Amherdt, M., Serafini, T. and Rothman, J. E. (1989) *Cell* **56**, 357-368
31. Pind, S. N., Nuoffer, C., McCaffrey, M. J., Plutner, H., Davidson, H. W., Farquhar, M. G. and Balch, W. E. (1994) *J. Cell Biol.* **125**, 239 -252
32. Martinez, O., Schmidt, A., Salamero, J., Hoflack, B., Roa, M. and Goud, B. (1994) *J. Cell Biol.* **127**, 1575-1584

33. Guo, Q., Vasile, E. and Krieger, M. (1994) *J. Cell Biol.* **125**, 1213-1224
34. Lippincott-Schwartz, J., Yuan, L. C., Bonifacino, J. S. and Klausner, R. D. (1989) *Cell* **56**, 801-813
35. Cluett, E. B., Wood, S. A., Banta, M. and Brown, W. J. (1993) *J. Cell Biol* **120**, 15-24
36. Rambourg, A., Clermont, Y. and Képès, F. (1993) *Anat. Rec.* **237**, 441-452
37. Colman, A., Jones, E. A. and Heasman, J. (1985) *J. Cell Biol.* **101**, 313-318
38. Cajal, S. R. (1914) *Trab. Lab. Invest. Biol.* **12**, 134-227
39. Burke, B., Griffiths, G., Reggio, H., Louvard, D. and Warren, G. (1982) *EMBO J* **1**, 1621-8
40. Misteli, T. and Warren, G. (1995) *J. Cell Sci.* **108**, 2715-2727
41. Lucocq, J. M. and Warren, G. (1987) *EMBO J.* **6**, 3239-3246
42. Lucocq, J. M., Berger, E. G. and Warren, G. (1989) *J. Cell Biol.* **109**, 463-474
43. Souter, E., Pypaert, M. and Warren, G. (1993) *J. Cell Biol.* **122**, 533-540
44. Collins, R. N. and Warren, G. (1992) *J. Biol. Chem.* **267**, 24906-24911
45. Zieve, G. W., Turnbull, D., Mullins, J. M. and McIntosh, J. R. (1980) *Exp.Cell Res* **126**, 397-405
46. Misteli, T. and Warren, G. (1994) *J. Cell Biol.* **125**, 269-282
47. Misteli, T. and Warren, G. (1995) *J. Cell Biol.* **130**, 1027-1039
48. Weidman, P. J., Melançon, P., Block, M. R. and Rothman, J. E. (1989) *J. Cell Biol.* **108**, 1589-1596
49. Rabouille, C., Misteli, T., Watson, R. and Warren, G. (1995) *J. Cell Biol.* **129**, 605-618
50. Acharya, U., McCaffery, J. M., Jacobs, R. and Malhotra, V. (1995) *J. Cell Biol.* **129**, 577-589
51. Acharya, U., Jacobs, R., Peters, J.-M., Watson, N., Farquhar, M. G. and Malhotra, V. (1995) *Cell* **82**, 895-904
52. Rabouille, C., Levine, T. P., Peters, J.-M. and Warren, G. (1995) *Cell* **82**, 905-914
53. Latterich, M., Fröhlich, K.-U. and Schekman, R. (1995) *Cell* **82**, 885-894
54. Corthésy-Theulaz, I., Pauloin, A. and Pfeffer, S. R. (1992) *J. Cell Biol.* **118**, 1333-1346
55. Warren, G. (1989) *Nature* **342**, 857-858
56. Warren, G. (1985) *Trends Biochem. Sci.* **10**, 439-443
57. Hesketh, T. R., Beaven, M. A., J, R., Burke, B. and Warren, G. (1984) *J. Cell Biol.* **98**, 2250-2254
58. Tuomikoski, T., Felix, M. A., Doree, M. and Gruenberg, J. (1989) *Nature* **342**, 942-945
59. Woodman, P. G., Adamczewski, J., Hunt, T. and Warren, G. (1993) *Mol. Biol. Cell.* **4**, 541-553
60. Mackay, D., Kieckbusch, R., Adamczewski, J. and Warren, G. (1993) *FEBS Lett.* **336**, 549-554
61. Th'ng, J. P., Wright, P. S., Hamaguchi, J., Lee, M. G., Norbury, C. J., Nurse, P. and Bradbury, E. M. (1990) *Cell* **63**, 313-24
62. Paulson, J. R., Ciesilski, W. A., Schram, B. R. and Mesner, P. W. (1994) *J. Cell Sci.* **107**, 267-273
63. Jackman, M., Firth, M. and Pines, J. (1995) *EMBO J.* **14**, 1646-1654
64. Zerial, M. and Stenmark, H. (1993) *Curr. Op. Cell Biol.* **5**, 613-620
65. Lapetina, E. G., Lacal, J. C., Reep, B. R. and Molina-y-Vedia, L. (1989) *PNAS* **86**, 3131-3134
66. Van der Sluijs, P., Hull, M., Webster, P., Male, P., Goud, B. and Mellman, I. (1992) *EMBO J.* **11**, 4379-4389
67. Sapperstein, S. K., Walter, D. M., Grosvenor, A. R., Heuser, J. E. and Waters, M. G. (1995) *Proc. Natl. Acad. Sci. USA* **92**, 522-527,
68. Barroso, M., Nelson, D. S. and Sztul, E. (1995) *Proc. Natl. Acad. Sci. USA* **92**, 527-531
69. Zeligs, J. D. and Wollman, S. H. (1979) *J. Ultrastruct. Res.* **66**, 53-77
70. Kanbe, T., Kobayashi, I. and Tanaka, K. (1989) *J. Cell Sci.* **94**, 647-656
71. Preuss, D., Mulholland, J., Franzusoff, A., Segev, N. and Botstein, D. (1992) *Mol. Biol. Cell* **3**, 789-803

Contributors

ARNAUD Lionel, (11, 107-114) Department of Molecular Biology, Sciences II, University of Geneva, 30 Quai Ernest-Ansermet, CH-1211 Geneva, Switzerland.

BARATTE Blandine, (13, 129-135) CNRS Station Biologique, BP 74, 29682 Roscoff cedex, France.

BERRY Lynne D., (10, 99-105) Howard Hughes Medical Institute, Department of Cell Biology, Vanderbilt University, Nashville, TN 37212, USA.

BLOW Julian, (8, 83-90) DNA Replication Control Laboratory, ICRF Clare Hall Laboratories, South Mimms, Herts EN6 3LD, United Kingdom.

BOJANOWSKI Krzysztof, (22, 229-239) Department of Surgery and Pathology, Children's Hospital and Harvard Medical School, 300 Longwood Ave., Boston, MA 02115, USA.

BRANDSEN Jeroen, (14, 137-145) Department of Chemistry and E.O. Lawrence Berkeley National Laboratory, University of California, Berkeley, CA 94720, USA.

BUCHKOVICH Karen J., (18, 187-195) University of Illinois at Chicago, Department of Pharmacology, 835 S. Wolcott Avenue, Chicago, IL 60612, USA.

CARIOU Sandrine, (4, 37-47) Unité de Recherches Hepatologiques, INSERM U49, Hopital Pontchaillou, 35033 Rennes, France.

CHONG James P. J., (8, 83-90) DNA Replication Control Laboratory, ICRF Clare Hall Laboratories, South Mimms, Herts EN6 3LD, United Kingdom.

CHUN Kristin T., (12, 115-127) Department of Biochemistry and Molecular Biology, and The Walther Oncology Center, 635 Barnhill Drive, Medical Sciences Building, Indiana University School of Medicine, Indianapolis, IN 46202-5122, USA.

CORLU Anne, (4, 37-47) Unité de Recherches Hepatologiques, INSERM U49, Hopital Pontchaillou, 35033 Rennes, France.

DAYTON Jennifer S., (21, 217-228) Department of Pharmacology, Duke University Medical Center, Durham, NC 27710, USA.

DIEDERICH Ludger, (15, 147-163) Institut de Biologie Structurale J. -P. Ebel, 38027 Grenoble Cedex 1, France.

EILERS Martin, (7, 73-82) Zentrum für Molekulare Biologie Heidelberg (ZMBH), Im Neuenheimer Feld 282, 69120 Heidelberg, Germany.

FAN Saijun, (16, 165-173) Laboratory of Molecular Pharmacology, Room 5C-25, Bldg 37, National Cancer Institute, National Institutes of Health, Bethesda, MD 20892, USA.

FILGUEIRA DE AZEVEDO Jr. Walter, (14, 137-145) Department of Chemistry and E.O. Lawrence Berkeley National Laboratory, University of California, Berkeley, CA 94720, USA.

FOTEDAR Arun, (15, 147-163) La Jolla Institute for Allergy and Immunology, La Jolla, CA 92037, USA.

FOTEDAR Rati, (15, 147-163) Institut de Biologie Structurale J. -P. Ebel, 38027 Grenoble cedex 1, France.

GLAISE Denise, (4, 37-47) Unité de Recherches Hepatologiques INSERM U49, Hopital Pontchaillou, 35033 Rennes, France.

GOEBL Mark G., (12, 115-127) Department of Biochemistry and Molecular Biology, and The Walther Oncology Center, 635 Barnhill Drive, Medical Sciences Building, Indiana University School of Medicine, Indianapolis, IN 46202-5122, USA.

GOLSTEYN Roy M., (11, 107-114) Institute Curie, Section de Recherche., 26, Rue d'Ulm, F-75231 Paris, France.

GOULD Kathleen L., (10, 99-105) Howard Hughes Medical Institute, Department of Cell Biology, Vanderbilt University, Nashville, TN 37212, USA.

GUGUEN-GUILLOUZO Christiane, (4, 37-47) Unité de Recherches Hepatologiques, INSERM U49, Hopital Pontchaillou, 35033 Rennes, France.

HOUGHTON Janet A., (17, 175-185) Department of Molecular Pharmacology, St. Jude Children's Research Hospital, 332 N. Lauderdale, Memphis, TN 38105-2794, USA.

HOUGHTON Peter J., (17, 175-185) Department of Molecular Pharmacology, St. Jude Children's Research Hospital, 332 N. Lauderdale, Memphis, TN 38105-2794, USA.

ILYIN Guenadi, (4, 37-47) Unité de Recherches Hepatologiques, INSERM U49, Hopital Pontchaillou, 35033 Rennes, France.

CONTRIBUTORS

JOHN Peter C. L., (6, 59-72) Plant Cell Biology Group, Research School of Biological Sciences, Australian National University, Canberra, ACT 2600, Australia, and Collaborative Research Centre for Plant Science, GPO Box 475 ACT 2601, Australia.

KIM Sung-Hou, (14, 137-145) Department of Chemistry and E.O. Lawrence Berkeley National Laboratory, University of California, Berkeley, CA 94720, USA.

L'ALLEMAIN Gilles, (5, 49-58) Centre de Biochimie-CNRS, Université de Nice, Parc Valrose, 06108 Nice, France.

LANE Heidi A., (11, 107-114) Swiss Institute for Experimental Cancer Research, 155 Chemin des Boveresses, CH-1066 Epalinges, Switzerland.

LARSEN Annette K., (22, 229-239) Department of Structural Biology and Pharmacology, Institut Gustave Roussy PR2, 94805 Villejuif cedex, France.

LAVOIE Josée N., (5, 49-58) Centre de Biochimie-CNRS, Université de Nice, Parc Valrose, 06108 Nice, France.

LECLERC Vincent, (19, 197-204) URA 671 CNRS, BP28, 06230 Villefranche-sur-mer, France.

LÉOPOLD Pierre, (19, 197-204) URA 671 CNRS, BP28, 06230 Villefranche-sur-mer, France.

LOYER Pascal, (4, 37-47) Unité de Recherches Hepatologiques, INSERM U49, Hopital Pontchaillou, 35033 Rennes, France.

MARTÍN-CASTELLANOS Cristina, (3, 29-35) Instituto de Microbiología Bioquímica, Departamento de Microbiología y Genética CSIC/Universidad de Salamanca, Edificio Departamental, Avda. del Campo Charro s/n, 37007 Salamanca, Spain.

MASUI Yoshio, (1, 1-13) Department of Zoology, University of Toronto, Toronto, Ontario M5S 3G5, Canada.

MATHIAS Neal, (12, 115-127) Department of Biochemistry and Molecular Biology, and The Walther Oncology Center, 635 Barnhill Drive, Medical Sciences Building, Indiana University School of Medicine, Indianapolis, IN 46202-5122, USA.

McKEON Frank, (9, 91-97) Department of Cell Biology, Harvard Medical School, 240 Longwood Avenue, Boston, MA 02115, USA.

MEANS Anthony R., (21, 217-228) Department of Pharmacology, Duke University Medical Center, Box 3813, Durham, NC 27710, USA.

MISTELI Tom, (24, 267-277) Cold Spring Harbor Laboratory, 1 Bungtown Road, Cold Spring Harbor, NY 11724, USA.

MORENO Sergio, (3, 29-35) Instituto de Microbiología Bioquímica, Departamento de Microbiología y Genética, CSIC/Universidad de Salamanca, Edificio Departamental, Avda. del Campo Charro s/n, 37007 Salamanca, Spain.

MÜLLER Daniel, (7, 73-82) Zentrum für Molekulare Biologie Heidelberg (ZMBH), Im Neuenheimer Feld 282, 69120 Heidelberg, Germany.

MUNDT Kirsten E., (11, 107-114) Swiss Institute for Experimental Cancer Research, 155 Chemin des Boveresses, CH-1066 Epalinges, Switzerland.

NANTHAKUMAR Nanda N., (21, 217-228) Department of Pharmacology, Duke University Medical Center, Durham, NC 27710, USA.

NIGG Erich A., (11, 107-114) Dept. of Molecular Biology, Sciences II, University of Geneva, 30, Quai Ernest-Ansermet, CH-1211 Geneva, Switzerland.

O'CONNOR Patrick M., (16, 165-173) Laboratory of Molecular Pharmacology, Room 5C-25, Bldg 37, National Cancer Institute, National Institutes of Health, Bethesda, MD 20892, USA.

POUYSSÉGUR Jacques, (5, 49-58) Centre de Biochimie-CNRS, Université de Nice, Parc Valrose, 06108 Nice, France.

REED Steven I., (2, 15-27) Department of Molecular Biology, MB-7, The Scripps Research Institute, 10666 North Torrey Pines Road, La Jolla, CA 92037, USA.

RIVARD Nathalie, (5, 49-58) Centre de Biochimie-CNRS, Université de Nice, Parc Valrose, 06108 Nice, France.

RUDOLPH Bettina, (7, 73-82) Zentrum für Molekulare Biologie Heidelberg (ZMBH), Im Neuenheimer Feld 282, 69120 Heidelberg, Germany.

SCHULZE-GAHMEN Ursula, (14, 137-145) Department of Chemistry and E.O. Lawrence Berkeley National Laboratory, University of California, Berkeley, CA 94720, USA.

SKLADANOWSKI Andrzej, (22, 229-239) Department of Structural Biology and Pharmacology, Institut Gustave Roussy PR2, 94805 Villejuif cedex, France. and Department of Pharmaceutical Technology and Biochemistry, Technical University of Gdansk, Narutowicza St 11, 80-952 Gdansk, Poland.

SMAALAND Rune, (23, 241-266) Department of Oncology, Haukeland Hospital, University of Bergen, N-5021 Bergen, Norway.

STEINER Philipp, (7, 73-82) Whitehead Center for Medical Research, 9 Cambridge Center, Cambridge. MA, USA.

TANG Damu, (20, 205-216) Department of Biochemistry, The Hong Kong University of Science and Technology, Clear Water Bay, Kowloon, Hong Kong.

TOURRET Jérôme, (9, 91-97) Department of Cell Biology, Harvard Medical School, 240 Longwood Avenue, Boston, MA 02115, USA. and: École Normale Supérieure, 45 rue d'Ulm, 75005 Paris, France.

VOGEL Lee, (13, 129-135) CNRS Station Biologique, BP 74, 29682 Roscoff cedex, France.

WANG Jerry H., (20, 205-216) Department of Biochemistry, The Hong Kong University of Science and Technology, Clear Water Bay, Kowloon, Hong Kong.

index

6-dimethylaminopurine 84; 125
6-DMAP 125
7q36 302

Alzheimer 212
Anticancer agents 168; 175-185
Antimetabolite 175-185
Apoptosis 43; 44; 78-80; 147-163; 167-170
Arabidopsis 60
Aspergillus 217-228
Ataxia Telangiectasia 156
ATP binding pocket 139-141; 144
Auxin 64

Bcl-2 148; 149
Biological oscillators 241
Bone marrow 249-255
Brain proline-directed kinase 206

Ca^{++}/calmodulin-dependent kinase 222-224
Caffeine 232
Caldesmon 211
Calmodulin 217-228
cAMP 245
Cancer chemotherapy 175-185; 241-266
Cancer therapy 165-173
Cdc10 29; 30
Cdc13 30; 100; 101; 190
Cdc17 188; 189
Cdc18 87
Cdc2 60; 65; 99-105
CDC20 124; 125
Cdc 25 102; 103
Cdc27 124
CDC34 121
CDC5 110
CDC6 87, 92; 93
Cdk2 crystals 137-145
Cdk2-CKShs1 co-crystal structure 131
Cdk4 17; 20; 21
Cdk5 205-216
Cdk8 197-204
Cdk inhibitor 32; 33
ced 149
Cell cycle checkpoints 103
Cell death 175-185
Cell proliferation 75-77
Chemical inhibition of CDK2 137-145
Chemotherapy 168; 169
Chlamydomonas 68; 69
Chromatin assembly 232; 233
Chromatin condensation 232; 233
Chromosome condensation 233
Chromosome segregation 233
Chronotherapy 258-261
Cig1 29-31
Cig2 29-31; 33

CIM3 123
Cip1 20; 167
Circadian control of the cell cycle 241-266
Circadian rhythm 241-266
Cks 129-135
CLN 16; 18; 19; 118-121
c-myc 73-82
CnA 220; 221
Cornea 246
CSF 6; 8
CTD-kinase 198-200
Cyclin C 197-204
Cyclin D 40
Cyclin D1 51-54
Cyclin E 76; 77
Cyclins in plants 63
Cyclosporin A 219
Cytokinin 64-66

dbf4 110
DNA damage checkpoint 165-173
DNA replication checkpoint 91-97
DNA topoisomerase II 229-239
Drosophila 109

Epidermis 246; 247
Euglena 245
Euplotes 189

FAR1 30
Fas 147; 148
FK506 219
Flavopiridol 140; 143
Fnk 109; 112
Folates 176-178

G1 cyclins 16; 118-121
G1 progression 29-35; 37-47
G1/S transition 15-27
G2 catenation checkpoint 231; 232
G2/M transition 218
GADD45 166; 167
GI-tract 248
Golgi apparatus 267-277

Hematopoietic system 255; 256
Hepatocytes 37-47
Hepatocyte growth factors 38; 39
Hepatocyte growth inhibitors 39

ICE 149; 150
Isopentenyadenine 140; 141; 143

Kip1 20; 21; 55; 56
Kip2 20; 21

INDEX

Licensing factor 83-90
Liver cell proliferation 246
liver regeneration 37

MAP kinases 49-58
Max 74
Mcm 84-86
MDM2 166
Metaphase-anaphase transition 233; 234
Mik1 1; 91-97; 102
MPF 1; 4 -10; 68-70
M-phase 269; 270
Munc-18 211
Myc 73-82; 154
Myosin 224
Myt1 95; 102

Nck5a 205-216
Nclk 205-216
Neurofilaments 211; 212
Nicotiana 68
Nim1/Cdr1 101; 102
NIMA 218
NIME 218
NIMX 218
NIN1 122; 123
Nuclear envelope 88

Okadaic acid 274
Olomoucine 140-142
Oocyte maturation 1; 3
Origin recognition complex 87; 88

p107 76; 78
$p13^{suc1}$ 129-135
$p27^{Kip1}$ 76; 77
p53 152-156; 165-171; 178-183
p67 211
Pentoxifylline 170; 171
Plant cell cycle 59-72
Plk1 111; 112
Plo1 110
Polo 107-114
pRb 23; 155; 156
Preprophase band 67; 68
PRG1 123
Programmed cell death 147-163
Proteasome 116; 117; 122

PSSALRE 206
Puc1 33

RAD9 190
Reaper 148
Restriction point 39; 40
RNA polymerase II 197-204
$rum1^+$ 29-35

Saccharomyces cerevisiae 18; 19; 110
Sak-a/b 109; 112
Schizosaccharomyces pombe 29-35; 110
Sec7 269
Sic1 22; 121
Snk 109; 112
Spindle formation 67
Srb10/Srb11 201; 202
START 15; 16; 18
Suc1 129-135

T14/Y15 Phosphorylation 99-105
Tau 212
Telomerase 187-195
Telomeres 187-195
Tetrahymena 187-195
TFIIH 201; 202
TGF-β 20; 39; 151; 152
Thymidylate synthase 175-178
Thymidylate synthase inhibitors 179-181
Thymineless death 178
Topo II inhibitors 235; 236
Topoisomerases II 229-239
Tradescantia 69
Tumorigenesis 73-74

UBC9 123; 124
UCN-01, 170; 171
Ubiquitin-activating enzyme (E1) 115; 116
Ubiquitin-dependent proteolysis 115-127

Vesicular traffic 267; 268

Wee1 91-97; 101

Xenopus 83-90